Microwave Radiometry and Remote Sensing of the Earth's Surface and Atmosphere

T0186550

Microwave Radiometry and Remote Sensing of the Earth's Surface and Atmosphere

Editors:

P. Pampaloni and S. Paloscia

CRC Press
Taylor & Francis Group
Boca Raton London New York

CRC Press is an imprint of the
Taylor & Francis Group, an **informa** business

Contents

Preface   xi
List of Reviewers   xiii

1. REMOTE SENSING OF THE EARTH'S SURFACE   1

1. 1 Ocean surface and wind speed

Sea surface parameters retrieval by passive microwave polarimetry
A. Kuzmin, M. Pospelov and Y. Trokhimovski   3

Comparison of the sea surface brightness temperature measured during
the Coastal Ocean Probing Experiment (COPE'95) from a blimp with
model calculations
Y. G. Trokhimovski, V. G. Irisov, E. R. Westwater, L. S. Fedor and V. E. Leuski   13

Determination of ocean surface wind speeds from the
TRMM Microwave Imager
L. N. Connor and P. S. Chang   21

Ocean winds measured by an imaging, polarimetric radiometer
B. Laursen, S. Søbjærg and N. Skou   29

Interference effects in freshwater and sea ice
K.-P. Johnsen and G. Heygster   39

A comparison of the errors associated with the retrieval of the latent heat
flux from individual SSM/I measurements
D. Bourras, L. Eymard and C. Thomas   47

1.2. Land surface

Airborne passive microwave measurements on agricultural fields
G. Macelloni, S. Paloscia, P. Pampaloni, R. Ruisi, C. Susini and J.P. Wigneron   59

A combined two-scale model for microwave emissivities of land surfaces
K. Paape, R. Bennartz and J. Fischer   71

Modelling of the electromagnetic response of geophysical media with
complicated geometries: analytical and numerical approaches
A. Sihvola, R. Sharma and J. Avelin   81

Faraday rotation and passive microwave remote sensing of soil moisture
from space
D. M. Le Vine and S. Abraham   89

On simultaneous optical and passive microwave remote sensing of soil,
water and vegetation
D. Pearson, E. J. Burke, R. J. Gurney and L. P. Simmonds   97

Application of physically-based models of land surface processes in the
interpretation of passive microwave radiometry
L. P. Simmonds and E. J. Burke   107

Detection of ice sheet on asphalt roads
G. Macelloni, R. Ruisi, P. Pampaloni and S. Paloscia   119

2. REMOTE SENSING OF ATMOSPHERE   127

2.1. Ground based observations

Resolution and accuracy of a multi-frequency scanning radiometer for
temperature profiling
E. R. Westwater, Y. Han and F. Solheim   129

Implications of an improved atmospheric absorption model on water
vapor retrievals
C. S. Ruf   137

Analysis of tip cal methods for ground-based microwave radiometric
sensing of water vapor and clouds
Y. Han and E. R. Westwater   145

Observations of integrated water vapor and cloud liquid water at the
SHEBA ice station
J. C. Liljegren   155

Measurement of plentiful water vapor and cloud liquid water around
precipitation area
G. Liang, H.-C. Lei, Y. Li, Z. Shen and C. Wei   165

Characteristics of atmospheric water on measurements of ground based
microwave radiometer and rain recorder
C. Wei, Y. Wu, P. Wang and Y.-J. Xuan   173

Using a micro-rain radar to assess the editing of ground-based microwave
radiometer data
L. Gradinarsky, G. Elgered and Y. Xue   183

Microwave ground based unattended system for cloud parameters monitoring
A. Koldaev, E. Kadygrov and A. Mironov   193

Statistical analysis of the tropospheric radio-path delay using radiosonde
and ground-based radiometric data
E. Fionda, F. Barbaliscia and P. G. Masullo 203

First results of tropospheric transmission measurements at 94 and 212 GHz
A. Lüdi, L. Martin, A. Magun, C. Mätzler, N. Kämpfer, P. Kaufmann,
J. E. R. Costa, G. C. Gimenez De Castro, M. Rovira and H. Levato 213

2.2. Satellite observations

Influence of cloud temperature on the retrieval of integrated cloud liquid
water content from space radiometry
A. Guissard 223

Analysis of cloud liquid water content characteristics from SSMI and
ship-borne radiometer
L. Eymard 235

Estimation of water vapor vertical distribution over the sea from Meteosat
and SSM/I Observations
F. Porcú, D. Capacci and F. Prodi 247

Effects of AMSU-A cross track asymmetry of brightness temperatures
on retrieval of atmospheric and surface parameters
F. Weng, R. R. Ferraro and N. C. Grody 255

Comparison of Advanced Microwave Sounding Unit observations with
global atmospheric model temperature and humidity profiles and their impact
on the accuracy of numerical weather prediction
C. Poulsen, S. English, A. Smith, R. Renshaw, P. Dibben and P. Rayer 263

Radiometric measurements in the 10/50 GHz band for application to radio
wave propagation
C. Capsoni, A. V. Bosisio and M. Mauri 271

3. REMOTE SENSING OF CLOUDS AND PRECIPITATIONS 281

3.1. Radiative transfer models

Effects of aspherical ice and liquid hydrometeors on microwave brightness
temperature
E. Moreau, C. Mallet and C. Klapisz 283

Effects of heterogeneous precipitating atmospheres on simulated brightness
temperatures
C. Mallet, C. Klapisz and N. Viltard 291

Determination of cloud structure from spaceborne microwave observations
I. G. Rubinstein 299

Modelling and measurements of Stokes vector microwave emission and
scattering for a precipitating atmosphere
*A. Hornbostel, A. Schroth, A. Sobachkin, B. Kutuza, A. Evtushenko and
G. Zagorin*   313

Polarimetric radiometry of rain events: theoretical prediction and
experimental results
*A. Camps, M. Vall-Llossera, N. Duffo, F. Torres, J. Barà, I. Corbella
and J. Capdevila*   325

3.2. Satellite and ground based observations

Application of AMSU for obtaining hydrological parameters
N. Grody, F. Weng and R. Ferraro   339

Meteorological applications of precipitation estimation from combined
SSM/I, TRMM and infrared geostationary satellite data
*F. J. Turk, G. D. Rohaly, J. Hawkins, E. A. Smith, F. S. Marzano,
A. Mugnai and V. Levizzani*   353

SOM network-based retrieval algorithms for determining precipitable
water and rainfall over oceans from the SSM/I measurements
H. Chen, J. Bian, P. Yang and D. Lu   365

Analysis of selected TRMM observations of heavy precipitation events
*A. Tassa, S. Di Michele, E. D'Acunzo, C. Accadia, S. Dietrich, A. Mugnai,
F. Marzano, G. Panegrossi and L. Roberti*   371

Multisensor analysis of Friuli flood event (October 5-7, 1998)
*S. Dietrich, R. Bechini, E. D'Acunzo, S. Di Michele, R. Fabbo, A. Mugnai,
S. Natali, F. Porcú, F. Prodi, L. Roberti and, A. Tassa*   379

SSM/I data analysis for retrieving cloud properties: comparisons with
ground-based measurements
*G. D'Auria, N. Pierdicca, P. Basili, S. Bonafoni, P. Ciotti and
F. S. Marzano*   387

Rainfall retrieval from ground-based multichannel microwave radiometers
F. S. Marzano, E. Fionda, P. Ciotti and A. Martellucci   397

4. NEW RADIOMETRIC INSTRUMENTS AND MISSIONS   407

4.1. Microwave sensors and calibration

An airborne submm radiometer for the observation of stratospheric trace
gases
*M. Von König, H. Bremer, V. Eyring, A. Goede, H. Hetzheim, Q. Kleipool,
H. Küllmann and K. Künzi*   409

EMCOR: A new radiometer for the measurement of minor constituents
in the frequency range of 201 to 210 GHz
D. Maier, N. Kämpfer, W. Amacher, M. Wüthrich, J. De La Noë,
P. Ricaud, P. Baron, G. Beaudin, C. Viguerie, J.-R. Pardo,
J.-D. Gallego, A. Barcia, J. Cernicharo, B. Ellison, R. Siddans,
D. Matheson, K. Künzi, U. Klein, B. Barry, J. Louhi, J. Mallat,
M. Güstafsson, A. Räisänen and A. Karpov   417

The radiometer for atmospheric measurements (RAM)
K. Lindner, B. Barry, U. Klein, J. Langer, B.-M. Sinnhuber,
I. Wohltmann and K. F. Künzi   427

Automatic self-calibration of ARM microwave radiometers
J. C. Liljegren   433

Analog correlator for HUT polarimetric radiometer
J. Lahtinen, O. Koistinen and M. Hallikainen   443

A method for calibrating the ground-based triple-channel microwave
radiometer
H.-B. Chen   453

Calibration methods in large interferometric radiometers devoted to Earth
observation
F. Torres, A. Camps, I. Corbella, J. Barà, N. Duffo and M. Vall-Llossera   459

4.2. Remote sensing missions

The Soil Moisture and Ocean Salinity Mission: an overview
Y. H. Kerr, P. Waldteufel, J.-P. Wigneron, J.-M. Martinuzzi, B. Lazard,
J. Goutoule, C. Tabard and A. Lannes   467

Field of view characteristics of a 2-D interferometric antenna, as illustrated
by the MIRAS/ SMOS L-band concept
P. Waldteufel, E. Anterrieu, J. M. Goutoule and Y. Kerr   477

Retrieval capabilities of L-band 2-D interferometric radiometry over land
surfaces (SMOS Mission)
J.-P. Wigneron, P. Waldteufel, A. Chanzy, J.-C. Calvet, O. Marloie,
J.-F. Hanocq and Y. Kerr   485

The role of microwave radiometry in the CLOUDS project
B. Bizzarri and P. Spera   493

Non-linear inversion of Odin sub-mm observations in the lower stratosphere
by neural networks
C. Jiménez, P. Eriksson and J. Askne   503

4.3. AMSR

Progress in AMSR snow algorithm development
A. Chang and T. Koike   515

Retrieval of soil moisture from AMSR data
E. Njoku, T. Koike, T. Jackson and S. Paloscia   525

AMSR/AMSR-E algorithm developments in NASDA—water vapor,
cloud liquid water, sea surface wind speed, and SST
A. Shibata and T. Hayasaka   535

Resampling of AMSR observations for spatial resolution matching
P. Ashcroft and F. J. Wentz   541

Author Index   551

PREFACE

This book contains a selection of refereed papers presented at the 6th Specialist Meeting on Microwave Radiometry and Remote Sensing of the Environment held in Florence, Italy on March 15-18, 1999, and sponsored by the IEEE Geoscience and Remote Sensing Society and by the Electromagnetics Academy. It has been eleven years since this Conference, the second of its kind organized in Florence by the Centre for Microwave Remote Sensing and the Research Institute on Electromagnetic Waves, was last held in this city: in the meantime it has been hosted in Boulder (CO) in 1992, Rome, Italy, in 1994 and Boston (MA) in 1996.

Since the first edition of the Specialist Meeting, held in Rome in 1983, passive microwave remote sensing has made considerable progress, and has achieved significant results in the study of the Earth's surface and atmosphere. Many years of observations with ground-based and satellite-borne sensors have made an important contribution to improving our knowledge of many geophysical processes of the Earth's environment and of global changes. The evolution in microwave radiometers aboard satellites has increased steadily over recent years. A series of excellent instruments, such as the Special Sensor Microwave Imager (SSM/I) and the four channel Microwave Sounding Unit (MSU), have operated for many years with very high stability and precision. They have made it possible to observe land and atmosphere features and to monitor even the smallest climatic changes. A multi-frequency imaging radiometer called TMI (TRMM Microwave Imager) was flown by NASA aboard the Tropical Rainfall Measuring Mission (TRMM) in November 1997. In May 1998, the National Oceanic and Atmospheric Administration (NOAA) began launching the next generation operational satellites (NOAA-K, L, M and N) that carry on board the Advanced Microwave Sounding Unit (AMSU), a multi-channel instrument designed primarily to improve the accuracy of temperature soundings beyond that of MSU.

At the same time, many investigations have been carried out both to improve the algorithms for the retrieval of geophysical parameters and to develop new technologies. Significant examples are the developments in radio-polarimetry and in aperture-synthesis radiometers. Indeed, the knowledge acquired on the polarisation state of the measured brightness temperature has added valuable information on the physics of the observed object, while interferometric techniques will make it possible to simplify operation from space by reducing antenna mass and volume.

The book is divided into four main sections: three of these are devoted to the observation of the Earth's surface and atmosphere, and the fourth, to future missions and new technologies. The first section deals with the study of sea and land surfaces, and reports recent advances in remote sensing of ocean wind, sea ice, soil moisture and vegetation biomass, including electromagnetic modelling and the assimilation of radiometric data in models of land surface processes. The following two sections are devoted to the measurement of atmospheric quantities which are of fundamental importance in climatology and meteorology, and, since they influence radio-wave propagation, they also impact on several other fields, including geodesy, navigational satellite and radioastronomy. The last section presents an overview of new technologies and plans for future missions. The main objective of the Soil Moisture and Ocean Salinity (SMOS) mission is to deliver two crucial variables of land and ocean, while the role of microwave radiometry in the CLOUDS project is to investigate cloud interior parameters. A most important mission for the future of microwave radiometry is the Advanced Microwave Scanning Radiometer (AMSR). It will be launched in November 2000 aboard NASDA's a.m. platform ADvanced Earth Observation Satellite-II (ADEOS-II) and in December 2000, aboard NASA's p.m. platform, Earth Observing System (EOS-PM). Compared with other sensors, such as SSM/I, AMSR will have two additional frequency channels and a considerably larger antenna. Besides providing geophysical parameters on the basis of standard algorithms, the instruments presents very attractive characteristics for research in the coming years

Paolo Pampaloni & Simonetta Paloscia
Firenze, September 1999

Acknowledgements

The papers included in this book have been selected after an accurate PEER review. This involved many renowned scientists, who completed this difficult task in an extremely short time and to whom we are very grateful.

We wish to express our gratitude to our colleagues of the Scientific Steering Committee: Giovanni d' Auria, *Università La Sapienza, Roma*, Martti Hallikainen, *Helsinki University of Technology*, Jin A. Kong, *MIT, Cambridge, MA*, Anatolij Shutko, *IRE, Moscow*, Domenico Solimini, *Università Tor Vergata, Roma*, Calvin T. Swift, *University of Massachusetts, Amherst, MA*, Ed R. Westwater, *University of Colorado, Boulder, CO*, for their encouragement and valuable support in organizing the conference.

We are also very grateful to Professor Francesco Adorno, President of the Accademia Toscana di Scienze e Lettere "La Colombaria", who generously hosted our meeting.

Lastly a special acknowledgment goes to Ms. Beatrice Picchi, for her skilful collaboration in the editing of this volume.

List of Reviewers

The editors wish to thank the following reviewers who have contributed their time and talent to the success of this book:

Ashcroft P.
Bauer P.
Bennartz R.
Bizzarri B.
Bourras D.
Buehler S.
Camps A.
Capacci D.
Capsoni C.
Cavalieri D. J.
Chang A.
Chang P. S.
Chen H.
Ciotti P.
D'Auria G.
Dietrich S.
Elgered G.
England A.
English S. J.
Eymard L.
Ferraro R.
Ferrazzoli P.
Fionda E.
Gasiewski A. J.
Grody N. C.
Guissard A.
Hallikainen M.
Hewison T.
Hornbostel A.
Johnsen K.-P.
Karam M. A.
Kilham D.
Klapisz C.
Kutuza B. G.

Le Vine D. M.
Liljegren J. C.
Macelloni G.
Mallet C.
Marzano F. S.
Moreau E.
Mugnai A.
Njoku E. G.
Paloscia S.
Pearson D.
Petty G. W.
Pierdicca N.
Porcu' F.
Pospelov M.
Prodi F.
Pulliainen J.
Rott H.
Rubinstein I. G.
Ruf C. S.
St. Germain K.
Schiavon G.
Schlüssel P.
Schmugge T.
Shibata A.
Sihvola A.
Skofronick Jackson G.
Skou N.
Smith D. F.
Swift C. T.
Turk F. J.
Waldteufel P.
Westwater E. R.
Wigneron J.-P.

1. Remote sensing of the Earth's surface

1.1 Ocean surface and wind speed

Microw. Radiomet. Remote Sens. Earth's Surf. Atmosphere, pp. 3–11
P. Pampaloni and S. Paloscia (Eds)
© VSP 2000

Sea surface parameters retrieval by passive microwave polarimetry

ALEXEY KUZMIN, MICHAEL POSPELOV, YURI TROKHIMOVSKI

Space Research Institute, Moscow 117810, Russia

Abstract - Thermal microwave emission of water surface is partially polarized and may be described with Stokes parameters. Wind-induced anisotropy of ripples results in noticeable azimuthal variations of Stokes parameters. The azimuthal dependence of Stokes parameters may be approximated by a series of harmonic functions. At nadir view angle, the first harmonic coefficient (upwind-downwind asymmetry) is equal to zero, and increases with the view angle. The second harmonic coefficient (upwind-crosswind asymmetry) is approximately constant at view angles of up to 20 degrees off nadir; its dependence on wind speed may be described by the quadratic law. These experimental results made it possible to propose an algorithm of wind vector retrieval from polarimetric measurements; examples of such retrieval from airborne polarimeter data are presented. The accuracy of wind speed and direction measurements was about 1 m/s and 10 degrees. Moreover, allowing for the remotely retrieved wind speed, the accuracy of the sea surface temperature (SST) retrieval from radiometric measurements may be improved by 20%.

1. INTRODUCTION

In passive remote sensing of the sea surface, only two basic polarizations, vertical and horizontal, have been traditionally used. This approach was applied in satellite radiometers, like SMMR and SSM/I, as well as in a great majority of airborne radiometers. However, recent research has demonstrated that full polarization measurements may improve significantly the potential of microwave radiometers.

This principal conclusion follows from the fact that sea surface geometry is azimuth-dependent, due to the wind effect. The azimuthal anisotropy of the surface gives rise to the same anisotropy of microwave brightness temperature. This dependence was first proposed by *Stogryn* [1], who relied upon the calculations of sea-surface microwave emissivity made in Kirchhoff approximation. It was originally assumed that owing to different wave slopes in along- and cross-wind directions, the azimuth dependence of microwave emission from wind-disturbed sea surface would arise. However, further investigation has demonstrated that Kirchhoff approximation is not adequate at microwave frequencies. More adequate is a composite model developed initially to explain microwave backscatter from a real sea surface. Being applied for sea surface emissivity, this model accounts for effects of both wind-induced gravity-capillary waves and large-scale waves. The evolution of the knowledge of ocean surface thermal microwave emission, based on both theoretical and experimental results, such as [2-4], led to the idea of polarimetric radiometry application for sea surface remote sensing [5].

During the last decade, a number of aircraft experiments aimed at sea-surface polarimetric remote sensing have been made. Measurements with airborne polarimeters were performed over a wide range of wavelengths (from 3 mm up to 8 cm) at various view angles, from grazing to nadir [6-9]. A comprehensive review of recent results in polarimetric measurements and modeling is presented in [10-11].

In this paper, the results obtained with airborne nadir-looking Ka-band polarimeter are presented. Surface wind speed and direction were retrieved from the measurements of the second and the third Stokes parameters. S-band radiometer data were used for sea surface temperature (SST) remote measurements. Remotely measured surface parameters were compared with true meteorological data.

2. PHYSICAL BACKGROUND

It is well known that water surface radiates partially polarized thermal electromagnetic emission. Such emission can be completely described by four real values, namely, the four Stokes parameters, which may also be treated as a four-component pseudo-vector. The Stokes vector may be represented either as a combination of electric field components (traditionally used in optics) or as a combination of brightness temperatures measured at several basic polarizations (more common for remote sensing):

$$\vec{S} = \begin{bmatrix} I \\ Q \\ U \\ V \end{bmatrix} = \begin{bmatrix} <E_v E_v^*> + <E_h E_h^*> \\ <E_v E_v^*> - <E_h E_h^*> \\ 2\,\mathrm{Re}<E_v E_h^*> \\ 2\,\mathrm{Im}<E_v E_h^*> \end{bmatrix} = c \begin{bmatrix} T_v + T_h \\ T_v - T_h \\ T_{+45} - T_{-45} \\ T_l - T_r \end{bmatrix} \quad (1)$$

where I, Q, U, V – components of the Stokes vector; E_v, E_h – orthogonal components of electric field; $<>$ - means statistical averaging; $*$ - means complex conjugate; T_v, T_h – brightness temperatures measured at linear orthogonal polarizations; T_{+45}, T_{-45} – the same in the basis rotated through 45°; T_l, T_r – brightness temperatures measured at left and right circular polarizations; c - coefficient.

There are several different physical mechanisms responsible for the formation of partially polarized emission. First, for the flat-water surface, a regular difference T_v-T_h exists because the Fresnel reflection coefficients for vertical and horizontal polarization are different; this difference is azimuth isotropic and depends on elevation view angle. Second, long surface waves have various mean slopes depending on wind direction; this results in azimuthal dependence of the Stokes parameters. Third, short gravity-capillary waves contribute to the partially polarized emission [12]; this emission is also azimuth dependent.

The Stokes parameters dependence on azimuth angle φ may be approximated by a series of harmonic functions ($\varphi=0$ corresponds to upwind direction):

$$I = I_0 + I_1 \cos\varphi + I_2 \cos 2\varphi; \quad Q = Q_0 + Q_1 \cos\varphi + Q_2 \cos 2\varphi;$$
$$U = U_1 \sin\varphi + U_2 \sin 2\varphi; \quad V = V_1 \sin\varphi + V_2 \sin 2\varphi \quad (2)$$

All the coefficients in (2) are functions of frequency band, view angle, and parameters of surface waves depending, in turn, on near-surface wind. The latter dependence allows using microwave polarimetry for surface wind vector retrieval. The problem of surface wind vector remote measurements is of great importance for both basic studies and practical needs. Those needs call for further studies of passive polarimetry application for surface wind monitoring.

In our experimental studies, we used airborne microwave polarimeters installed at nadir view angle. The choice of nadir angle results from the following reasons.

The relative contribution of surface waves of various scales to the brightness contrast depends on the elevation angle of observation. This statement is supported by model calculations of microwave emission from waved sea surface.

An example of such modeling is presented in Fig.1. The brightness contrast at Ka-band (frequency 37 GHz) vertical and horizontal polarization is plotted *versus* view angle. By the term "brightness contrast", we mean the difference between the waved and flat sea surface brightness temperatures. The contributions of long gravity waves and short gravity-capillary waves to the contrast are calculated separately. For the modeling, we used the surface wave spectrum proposed in [13] for the wind speed of 7 m/s. Long wave contrast was calculated using the Kirchhoff approximation. For short waves, the small perturbation method (SPM) was applied. In this plot, the contrasts are averaged over all azimuth angles, so the azimuthal dependence of brightness temperature is not taken into account. Foam influence and atmospheric attenuation are neglected as well. It is obvious from the Fig.1 that the brightness contrasts at near-nadir angles are caused mainly by the ripples. At grazing angles, the influence of long gravity waves and swell is dominant. At medium angles, contributions from both mechanisms are comparable.

Since gravity-capillary waves have very short lifetime, they are closely related to instantaneous airflow over sea surface. That is why near-nadir view angles are preferable for wind measurements.

Another argument in favor of near-nadir sounding is that such observations are less sensitive to the atmospheric attenuation of microwaves. The greater is the view angle, the greater is the additional polarization contrast due to atmosphere emission reflected from the surface. This may exceed wind contrasts by an order of magnitude.

On the other hand, nadir sounding has an obvious drawback: it is impossible to discriminate between upwind and downwind directions since at nadir all the first harmonic coefficients in (2) are equal to zero. This 180° ambiguity may be removed using any additional information (air pressure maps, ship or buoy data, etc.).

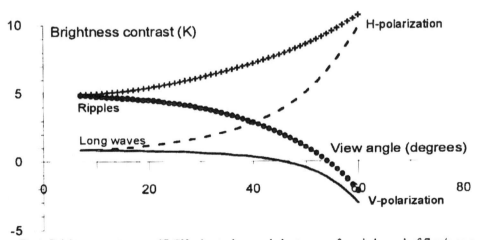

Fig.1. Brightness contrasts at 37 GHz due to long and short waves for wind speed of 7 m/s as a function of view angle.

3. ALGORITHMS OF WIND VECTOR AND SST RETRIEVAL

In the following sections, the experimental results obtained with airborne radiometers and polarimeters are presented. The microwave instruments were installed aboard TU-134 research aircraft at nadir view angle.

For wind speed and direction retrieval, we used Ka-band polarimeter data. As it was pointed out, at nadir $Q_0=I_1=Q_1=U_1=V_1=0$. From (2) it follows that the angle between wind direction and basic linear polarization marked v (that coincides with the aircraft axis) may be defined as:

$$\varphi = 1/2\,\mathrm{ArcTan}(U/Q) \tag{3}$$

To remove the 180° ambiguity, remotely measured wind direction was compared with *in situ* measurements once in every flight.

Wind speed W may be defined as a function of the polarized component in partially polarized emission. At nadir view angle, the thermal emission of flat sea surface is unpolarized, and when the wind blows, a polarized component of thermal emission arises. Our previous experiments have demonstrated that for low and moderate winds (up to approximately 12 m/s) the intensity of the polarized component may be approximated as a quadratic function of the wind speed. Consequently,

$$W = A(U^2 + Q^2)^{1/4} \tag{4}$$

here $A=7.24$ is an empirical coefficient; the circular polarized component V is neglected.

For SST reconstruction, S-band radiometer data were used because at S-band the sea-surface brightness dependence on the physical temperature is strong enough. The measurement of absolute brightness with reasonable accuracy is rather a difficult problem. To increase accuracy, we performed relative measurements using *in situ* data as a reference point. Another problem is that the brightness temperature at S-band also depends on wind speed. To decrease this dependence, the data were corrected using wind speed W retrieved from polarimeter data. Accordingly, the SST t_s determined from radiometer data is:

$$t_s = \left(\frac{\Delta T_{Br} - \gamma W}{\beta}\right) + t_{ref}, \tag{5}$$

here ΔT_{Br} is brightness temperature variation with respect to the reference point; t_{ref} is reference temperature defined from actual surface data once during every flight; $\beta=0.4$ and $\gamma=0.38$ K/(m/s) are experimentally determined scale factors.

4. SUMMARY OF THE EXPERIMENTAL RESULTS

The described algorithms were tested during the joint US/Russia experiment JUSREX conducted in the Northern Atlantic in 1992. During the experiment, TU-134 performed ten flights over the test area approximately 60 nmi south of Long Island, NY. Nadir-looking microwave radiometers were used for SST and wind vector remote measurements. The aircraft was also equipped with optical camera and Ku-band side-looking radar (SLAR) that took two swaths of approximately 13-km width from the left and right aircraft board, each at both VV and HH polarization. *In situ* data were collected aboard "Akademik Ioffe" research vessel.

Some experimental results of JUSREX were reported previously [14-15]. In this paper, we give an example of sea surface parameters retrieval. The results of a comparison between remotely measured and true surface parameters are presented. Special attention is paid to the reasons that may cause the deviations between remote and contact measurements.

Fig.2 shows flight data of July 24. SST and wind fields were reconstructed from four parallel passes. The passes, the "Akademik Ioffe" position, and the contour of radar swath are marked off as well. An increase of wind speed from the east to the west is visible in both radar and radiometer images. The research vessel reported air temperature to be about 16.5°C. In the western part of the radar image, the convection cells and the wind strips are seen, that is the evidence of unstable atmospheric stratification in that region. The transition from neutral to unstable atmospheric stratification exhibited in this radar image coincides with the quantitative data on the SST map.

A comparison of remotely measured and *in situ* data was made when the distance between the aircraft and "Ak. Ioffe" was less than 15 km. The total number of points compared during the experiment was 22. The agreement of data is quite satisfactory, except for several wind direction values under weak wind conditions.

The rms deviation between remotely measured and true surface parameters was: for wind direction 31°, for wind speed 2.5 m/s, for SST 2.1°C.

5. ERROR ANALYSIS

When discussing the discrepancy between remote and contact measurements, it is necessary to keep in mind the following consideration. Structures of these two data sets are quite different, due to different principles of sampling. *In situ* data are obtained at a fixed point and undergone time averaging (over 10 min. in this case), whereas remote sensing data are spatially averaged (along approximately 6-km pass). This difference is particularly important for wind vector data, which vary significantly in both space and time.

The SST data were related to true surface data once every flight. Therefore, only 16 points are independent. For those points, the rms deviation is 2.1°C without wind correction and 1.7°C with wind correction (5) based on polarimetric data. Fig.3. shows remotely retrieved SST compared with *in situ* data. Although the improvement in the algorithm is evident, the deviation from *in situ* data remains significant. A probable reason for this deviation is the fact that S-band radiometer was installed in the pressurized volume of the aircraft. The losses in the glass window at S-band were over 2 dB, so any uncontrollable variations of fuselage temperature would disturb the radiometer output.

Wind direction can be determined from polarimeter data with reasonable accuracy. The rms deviation for all 22 points under consideration is 31°. Such serious deviation stems from two reasons. First, at low wind speed, wind direction is known to be extremely variable in time and space. Different principles of averaging (temporal for shipborne anemometer and spatial for airborne polarimeter) may significantly affect the measured wind direction. The second reason is evident from formula (3). At low wind speed, the polarized component of sea-surface emission becomes relatively small. For small values of the second and the third Stokes parameters, the equation (3) leads to a large uncertainty in angle determination. This is illustrated by Fig. 4, where the value $\delta_\varphi = |\varphi - \varphi_0|$ is plotted *versus* the linear polarized component $\Delta T_a = (Q^2 + U^2)^{1/2}$ (here φ - remotely retrieved wind direction; φ_0 - true surface data). It is obvious that, as the anisotropy approaches the sensitivity threshold 0.15 K, the discrepancy rises sharply. Conversely, the greater the wind speed, the better the accuracy is. If we assess the accuracy of wind direction retrieval for the case $\Delta T_a \geq 0.35$ K, that corresponds to $V \geq 6.7$ m/s under neutral stability, the standard deviation will be 11°.

A. Kuzmin et al.

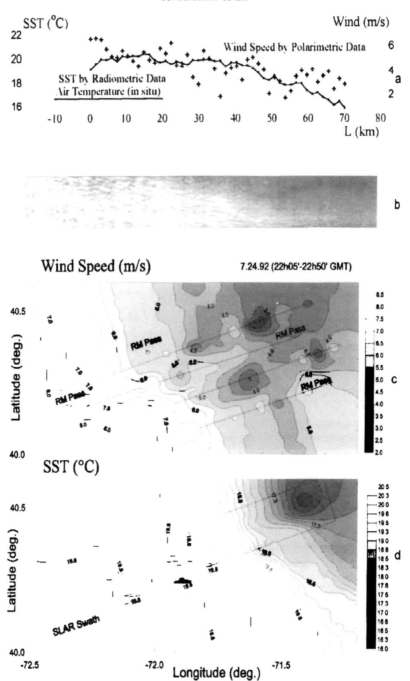

Fig.2. Flight data of July 24,1992: (**a**) wind speed (crosses) and SST (dots) profiles along pass 1; (**b**) SLAR image, VV polarization; (**c**) wind-speed field reconstructed from passes 1-4 (top to bottom); (**d**) SST field reconstructed from passes 1-4; SLAR image contour and "Ak. Ioffe" position are marked off.

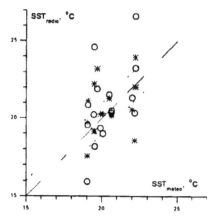

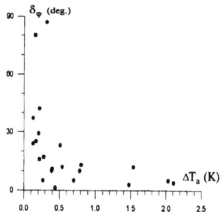

Fig.3. Comparison of remotely retrieved (radio) and *in situ* (meteo) SST. Dashed lines indicate ±1°C range. Circles/asterisks mean without/ with wind speed correction.

Fig.4. Wind direction error *versus* amplitude of the polarized component.

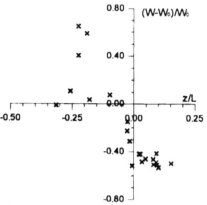

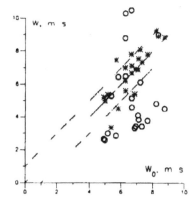

Fig.5. Wind speed relative error dependence on the atmosphere stability.

Fig.6. Comparison of remotely retrieved (W) and *in situ* (W₀) wind speed. Dashed lines indicate ±1m/s range. Circles/ asterisks mean without/with stability correction.

The accuracy of wind speed measurements is strongly dependent on atmospheric stratification. Fig.5 illustrates this. The value of relative error $(W-W_0)/W_0$ (here W_0 is true surface wind speed) is plotted *versus* stability parameter z/L. Negative parameter z/L corresponds to unstable boundary layer stratification, positive – to stable one. Polarimeter-derived wind speed is obviously overestimated if atmospheric stratification is unstable, otherwise it is underestimated. The dependence of sea surface polarized emission on atmospheric stability has been specially analyzed in [16] where a procedure for wind speed algorithm correction for atmospheric stability has been suggested. According to this algorithm, wind speed W is:

$$W = A \cdot \Delta T_a^{\,B} \cdot \exp(C \cdot z / L) \qquad (6)$$

where coefficients (A=9.83; B=0.37; C=0.89) were derived from experimental data obtained during JUSREX experiment; stability parameter z/L was calculated from *in situ* measurements of wind speed and sea and water temperatures.

While the standard deviation of remotely retrieved wind speed from *in situ* measurements for the 22 points mentioned was 2.5 m/s, after this correction the deviation has been decreased to 0.9 m/s. This improvement is also illustrated by Fig.6 where *in situ* data observations and their retrievals are compared.

6. CONCLUSION

Airborne microwave radiometers and polarimeters make it possible to retrieve the SST and surface wind vector fields. The comparison of remotely measured surface parameters with *in situ* data shows reasonable agreement. Round mean square deviation amounted to: 11° for wind direction (for wind speed over 6.7 m/s); 0.9 m/s for wind speed (with stability correction); 1.7 K for SST (with wind correction).

ACKNOWLEDGMENT
The authors wish to thank two anonymous referees for their constructive comments and M.Pospelova and T.Bocharova for editing the manuscript. The work was supported by INTAS grant 97-10569 and RFBR grant 97-02-17275.

REFERENCES
1. A.Stogryn. The apparent temperature of the sea at microwave frequencies. *IEEE Trans. Ant. Prop.*, **AP-15**, (2), 278-286 (1967).
2. E.A.Bespalova, V.M.Veselov, A.A.Glotov, Y.A.Militskiy, V.G.Mirovskiy, I.V.Pokrovskaya, A.E.Popov, M.D.Raev, E.A.Sharkov, and V.S.Etkin. Sea-ripple anisotropy estimates from variations in polarized thermal emission of the sea. *Oceanology*, **21**, 213-215 (1981) [Transl. from Russian: *Doklady Akad. Nauk SSSR*, **246**, (6), 1482-1485 (1979)].
3. V.G.Irisov, Y.G.Trokhimovski, and V.S.Etkin. Radiothermal spectroscopy of the ocean surface. *Sov. Phys. Docl.*, **32**, (11), 914-915 (1987) [Transl. from Russian: *Doklady Akad. Nauk SSSR*, **297**, 587-589 (1987)].
4. V.G.Irisov, A.V.Kuzmin, M.N.Pospelov, Y.G.Trokhimovski, and V.S.Etkin. The dependence of sea brightness temperature on surface wind direction and speed. Theory and experiment. *Proc. Int. Geosci. Remote Sensing Symp. (IGARSS'91)*, Espoo, Finland, 1297-1300 (1991).
5. M.S.Dzura, V.S.Etkin, A.S.Khrupin, M.N.Pospelov, and M.D.Raev. Radiometers-polarimeters: Principles of design and applications for sea surface microwave emission polarimetry. *Proc. Int. Geosci. Remote Sensing Symp. (IGARSS'92)*, Houston, USA, 1432-1434 (1992).
6. Y.G.Trokhimovski, G.A.Bolotnikova, V.S.Etkin, S.I.Grechko, and A.V.Kuzmin. The dependence of S-band sea surface brightness temperature on wind vector at normal incidence. *IEEE Trans. Geosci. Remote Sens.*, **33**, (4), 1085-1088 (1995).
7. S.H.Yueh, W.J.Wilson, F.K.Li, S.V.Nghiem, and W.B.Ricketts. Polarimetric measurements of sea surface brightness temperature using an aircraft K-band radiometer. *IEEE Trans. Geosci. Remote Sensing*, **33**, (1), 85-92 (1995).
8. D.B.Kunkee and A.J.Gasiewski. Airborne passive polarimetric measurements of sea

surface anisotrophy at 92 GHz. *Proc. Int. Geosci. Remote Sensing Symp. (IGARSS'94)*, 2413-2415 (1994).

9. V.G.Irisov, A.V.Kuzmin, Y.G.Trokhimovski, and V.S.Etkin. Azimuthal dependence of sea surface microwave emission at grazing observation angles. *Sov. J. Remote Sensing*, **8**, (6), 1043-1056 (1991) [Transl. from Russian: Issledonania Zemli iz Kosmosa, (6), 99-103 (1990)].

10. Y.G.Trokhimovski and V.G.Irisov. Wind speed and direction measurements using microwave polarimetric radiometers. *NOAA Tech. Mem. ELP ETL-250* (1995).

11. P.Coppo, J.T.Johnson, L.Guerriero, J.A.Kong, G.Macelloni, F.Marzano, P.Pampaloni, N.Pierdicca, D.Solimini, C.Susini, G.Tofani, and Y.Zhang. Polarimetry for passive remote sensing. *ESA Rep. 1146/95/NL/NB* (1996).

12. Y.A.Kravtsov, E.A.Mirovskaya, A.E.Popov, I.A.Troitskiy, and V.S.Etkin. Critical effects in the thermal radiation of a periodically uneven water surface. *Academia Nauk SSSR, Izvestia, Atmospheric and Oceanic Physics*, **14**, (7), 522-526 (1978) [Transl. from Russian: *Physica Atmosphery & Okeana*, **14**, (7), 733-738 (1978)].

13. J.R.Apel. An improved model of the ocean surface wave vector spectrum and its effects on radar backscatter. *J. Geophys. Res.*, **99**, 16269-16291 (1994).

14. V.S.Etkin, A.V.Kuzmin, M.N.Pospelov, A.I.Smirnov, and V.V.Yakovlev. The determination of sea surface wind and temperature with airborne radiometric data (Joint US/Russia Internal Waves Remote Sensing Experiment). *Proc. Int. Geosci. Remote Sensing Symp. (IGARSS'93)*, Tokyo, Japan, 1622-1624 (1993).

15. V.S.Etkin, A.V.Kuzmin, M.N.Pospelov, A.I.Smirnov, and Y.G.Trokhimovski. Investigation of sea surface temperature and wind fields in Joint US/Russia Internal Waves Remote Sensing Experiment. *Proc. Int. Geosci. Remote Sensing Symp. (IGARSS'94)*, Pasadena, CA, 750-752 (1994).

16. M.N.Pospelov. Surface wind speed retrieval using passive microwave polarimetry: The dependence on atmospheric stability. *IEEE Trans. Geosci. Remote Sensing*, **34**, (5), 1166-1171 (1996).

Microw. Radiomet. Remote Sens. Earth's Surf. Atmosphere, pp. 13–20
P. Pampaloni and S. Paloscia (Eds)

Comparison of the sea surface brightness temperature measured during the Coastal Ocean Probing Experiment (COPE'95) from a blimp with model calculations

YURI G. TROKHIMOVSKI,[1] VLADIMIR G. IRISOV,[2] ED R. WESTWATER,[2]
LEN S. FEDOR,[2] and VLADIMIR E. LEUSKI[2]

[1]*NOAA/ERL Environmental Technology Laboratory, Boulder, Colorado (visiting scientist)*
[2]*Cooperative Institute for Research in Environmental Sciences, Environmental Technology Laboratory Boulder, Colorado*

Abstract — Experimental data collected by a microwave dual-frequency radiometer at 23.87 / 31.65 GHz and a polarimeter at 37.0 GHz installed on an airship blimp during the Coastal Ocean Probing Experiment 1995 are analyzed. Observations of the downwelling atmospheric radiation as a function of angle were used to achieve a high accuracy of absolute calibration. The dependencies of the brightness temperature of the sea surface upon nadir and azimuthal angles are considered. The comparison of the experimental data with theoretical calculations based on the several models of sea wave spectra is made.

1. INTRODUCTION

The microwave brightness temperature contrast of the sea surface, defined as the difference between the brightness temperature observed at some sea state and the brightness temperature of the smooth water surface with the same physical parameters (temperature and salinity), depends in general on surface roughness and foam generated by wind flow. If the wind speed is not too high, the main contribution to the brightness temperature contrast comes from surface waves of different scales. In general microwave radiometric measurements carried out at different angles, wavelengths and polarization give an opportunity to retrieve parameters of sea surface such as the slope distribution of long waves and wave spectrum in the gravity-capillary interval [1,2]. These parameters are anisotropic relative to the wind direction and as consequence microwave polarimetric radiometers are being considered as comprehensive technique for wind vector retrieval from space. The development of this technique demands quantitative experimental data collected under well-known environmental condition.

New radiometric studies supported by quantitative *in situ* measurements of the air and sea surface parameters were undertaken during the Coastal Ocean Probing Experiment (COPE) in September-October of 1995. Two radiometers were mounted on a blimp, previously used for radar and boundary layer research [3] (Fig.1). In this paper we summarise the most important results of the radiometric measurements associated with the study of wind waves and compare experimental data with prediction of a composite model.

Figure 1. Blimp used during COPE'95 for microwave scatterometric and radiometric measurements.

2. COASTAL OCEAN PROBING EXPERIMENT

COPE was performed off the coast of the northern Oregon to investigate remote sensing signatures of the ocean surface and air-sea interaction in a coastal environment [4]. The goal of blimp operation during COPE, with UW/APL coherent X-band radar and NOAA/ETL microwave radiometers, was to measure microwave emission and scattering from ocean waves. The blimp was operated from the former Naval air base in Tillamook. The main part of measurements was made in the region of the Floating Instrument Platform (*FLIP*) to minimize the separation distance between *in situ* and microwave remote sensing data. Over the ocean, blimp altitudes ranged from 70 to 400m, and were known to at least 10 m.

3. INSTRUMENTATION

Dual-frequency radiometer. A dual-frequency radiometer at 23.87 and 31.65 GHz was developed by NOAA/ETL. The two frequencies share a common antenna (offset parabola), and have equal half-power beam widths of 3.6°. Each channel has a double-side band receiver with a total band width of 1 GHz resulting in a measured 1 s rms sensitivity of 0.064 K. The radiometer switches between antenna, cold, and hot reference loads, with the time for one switch state of 20 msec.

37-GHz polarimetric radiometer. This instrument was designed in cooperation between NOAA/ETL and Lebedev Physical Institute, Moscow, Russia, in 1995. Its corrugated horn antenna has 6° beam width at 3-dB level. A ferrite rod installed into a circular waveguide is used as polarization shifter. A current through the control coil of the polarization shifter is switched periodically to ensure the reception of -45°, 0°, and +45° polarized signals. The brightness temperature at the horizontal polarization discussed below was calculated from these values. The polarimeter sensitivity referenced to 1 s time constant is 0.02 K.

Both radiometers were mounted on two specially constructed platforms on the left side of the blimp cabin. The elevation angle scans were limited by obstruction from the blimp and were conducted from about 15° from the nadir to about 45° in elevation above the horizon. An additional video camera and two infrared radiometers were also installed on the platforms.

The scanning of radiometers at elevation angles above the horizon provides the best calibration of the microwave radiometers in field measurements. During COPE the absolute values of brightness temperature were determined by this "tip cal" techniques [5] using the output signals from the atmosphere measured at different angles and the temperature of the reference load of the radiometer. The same measurements give the absorption in the atmosphere above the blimp level.

4. MODELING

To calculate the sea surface brightness temperature in presence of long and short waves we applied a composite model. The contribution of long waves was calculated using the Kirchoff method with accounting for single re-reflection and self-shadowing. Brightness temperature contrasts from small gravity capillary waves were calculated using small perturbation technique [6]. The separation between short and long waves was made at wavenumber k_L taken from the condition $k_L = 0.05 k_0$, where k_0 is the wavenumber of the electromagnetic radiation.

A critical issue for modeling is the approximation of the curvature spectrum of sea waves. There are more than ten different models developed last twenty years, and these models can differ up to one order in magnitude of the predicted spectral density in the gravity-capillary region. We select for calculations only four models developed by *Donelan and Pierson* [7], *Apel* [8], *Romeiser et al.* [9], and *Elfouhaily et al.* [10]. In the remainder of this paper, we will label these models as D, A, R, and E. The omnidirectional spectrum $B(k)$ is determined in such a way that:

$$B(k,\varphi) \cdot kdkd\varphi = B(k) \cdot W(\varphi) \cdot kdkd\varphi , \qquad \int_{-\pi}^{\pi} W(\varphi)d\varphi = 1,$$

where $B(k,\varphi)$ is the two-dimensional symmetrical curvature spectrum, $B(k)$ is the dimensionless omnidirectional curvature spectrum, and $W(\varphi)$ is the spreading function.

The spreading functions proposed by different authors are different and are given in original publications. In our modeling we account both for contrast associated with a modulation of the microwave radiation by the sea surface roughness and a modulation of the reflected sky radiation. We neglect the contribution from foam and spray, since our data were collected under low and moderate wind speed conditions and our visual observations and video records from the blimp indicate that white caps were small.

5. RESULTS

Brightness temperature dependence on nadir viewing angle. To analyze the dependence on elevation angle the data taken on September 19, 20, 22 and 25 were used. In Fig. 2 mean values of the observed brightness temperature during September 22 are

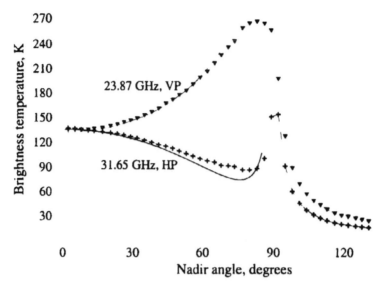

Figure 2. Averaged brightness temperature measured by dual-frequency radiometer 23/31 as a function of nadir view angle. September 22, 1995. Lines - calculation for smooth water surface. Wind speed 3.3 m/s, $u_* = 17$ cm/s, bulk water temperature 17.3 °C, skin water temperature 16.3 °C, air temperature 16.3 °C.

plotted as a function of nadir angle for 23.87 GHz (vertical polarization) and 31.65 GHz (horizontal polarization). Data are averaged over an angular interval of three degrees. In the same figure we plot the angular dependence calculated for smooth water surface based on dielectric permitivity models proposed in [11].

It was concluded that the retrieval of the brightness temperature at horizontal polarization from the -45°, 0°, and +45° polarized brightness temperatures could result in a relatively large error. The retrieved brightness temperature at horizontal polarization 37 GHz is appropriate for the study of azimuth dependence, but we don't include these data in the analysis of brightness temperature contrast of surface waves.

The brightness temperature contrasts were averaged for stable (September 20, 22) and unstable (September 19, 25) ocean-atmosphere temperature difference. We consider these contrasts as mean values averaged over azimuthal angle because the majority of data included in the averaging procedure were obtained with azimuthal scanning or different blimp orientation relative to the wind direction.

Comparison with modeling results is given in Fig. 3. Qualitatively, all spectra predict similar angular dependence for the horizontal polarization, but a quite different dependence for the vertical one. Spectra D and R result in a positive maximum of the brightness temperature contrast at angles of 50-55° from the nadir. The COPE experimental data, similar to results [12,13], demonstrate that the brightness temperature contrast does not have a maximum at the nadir angle of 50-55° and is definitely smaller then at near-nadir.

The large values at 70-80° as predicted by modeling with D, A, and R spectra both for vertical and horizontal polarization are connected first of all with large value of mean squared slope. We agree with the conclusion made in [10], that D, A, and R models

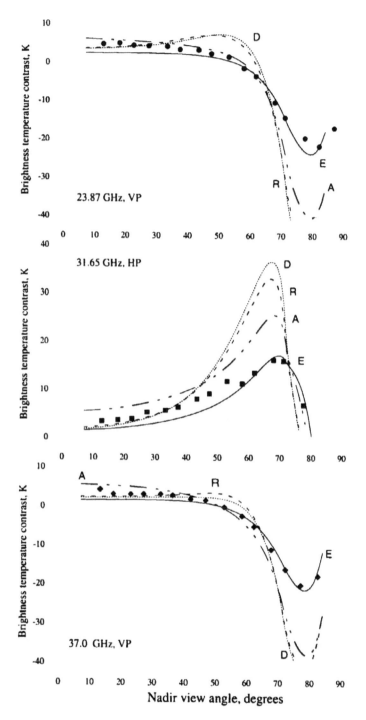

Figure 3. Brightness temperature contrast under unstable conditions (mean for data taken Sep. 19 and Sep. 25). Modeling results for wind speed 7.3 m/s, $u_* = 24.0$ cm/s are shown by lines.

overestimate the mean squared slope for surface waves, because they don't satisfy the maximum wave slope criteria by *Plant* [14]. Our radiometric experimental data confirm this statement.

In general, we established that the spectrum [10] explains better such COPE results as the brightness temperature of the wavy sea surface averaged over azimuthal angle, but it seems that spectrum density in the vicinity of phase velocity minimum, which provides the dominant contribution to brightness contrast at small angles of incidence, is somewhere in-between E and A, at least during the unstable conditions of COPE.

Azimuthal dependence of the brightness temperature. The brightness temperature of the sea surface exhibits azimuthal variations because of the surface waves which propagate mostly in downwind direction. Such anisotropy of the surface roughness results in an azimuthal dependence of the microwave radiation. Usually the brightness temperature at vertical and horizontal polarizations (and the first and the second Stokes parameters respectively) is approximated by two azimuthal harmonics:

$$T_b = A_0 + A_1 \cos(\varphi) + A_2 \cos(2\varphi),$$

where φ is the angle between observation plane and near-surface wind direction. The first harmonic is related with up-down wind difference of the brightness temperature, and the second harmonic – with up-cross wind difference. The third and the fourth Stokes parameters, which are the differences of the brightness temperatures at +45°, -45° linear polarizations and right and left circular polarizations respectively, are approximated by "sin" functions. For a nadir viewing angle the first harmonic vanishes and the second harmonic describes the dependence of the brightness temperature upon the angle between the polarization plane and near-surface wind.

In this paper we do not consider the first harmonic dependence because it cannot be described by a power spectrum of the sea surface – one needs to introduce the asymmetrical features such as wave breaks, ripple modulation by long waves, etc. In contrast, the second harmonic is closely related with the spreading function of the spectrum and we can compare the observed values of the second harmonic of the brightness temperature with predictions given by various spectrum models.

The result of data processing and comparison with modeling is presented in Fig.4 for 37 GHz vertical polarization (a), horizontal polarization (b), and the third Stokes parameter (c). The lines correspond to the theoretical calculations based on D (short dashed line), A (dashed-dotted line), R (dashed line), and E (solid line) spectrum models.

The agreement with R and E models is rather good. The E model seems to agree better with the radiometric data than R although the accuracy of our measurements does not allow us to make an ultimate conclusion.

6. CONCLUSION

The comprehensive *in situ* and remote sensing observations of the ocean during COPE give important information about the sea surface waves. The comparison of the view angular dependencies of the brightness temperature with the model calculations permits us to choose the spectrum model which is in the best agreement with the experimental radiometric observations.

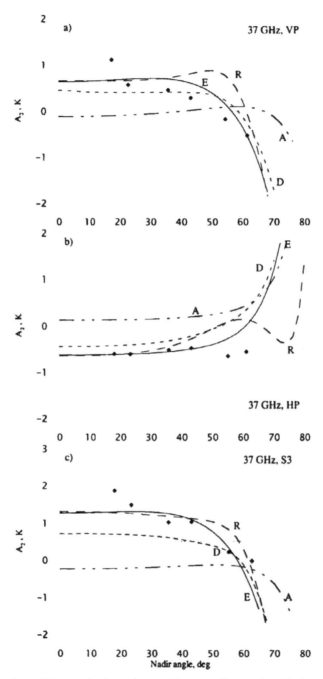

Figure 4. Comparison of the angular dependence of the second harmonic with theoretical calculations. Wind speed 10 m/s. See text for explanation of model lines.

It appears that spectrum [10] better explains the observed brightness temperature of a wavy sea surface averaged over azimuthal angle, but the agreement is not absolutely perfect.

The analysis of the azimuthal dependencies allowed us to compare the amplitude of the second harmonic with the prediction given by different spectrum models. Again we found that the azimuthal dependence based on the E spectrum model is in the best agreement with our data, although our accuracy does not permit us to distinctly discriminate between this model and the model proposed by *Romeiser et al.* [1997]. Additional measurements of the azimuthal dependencies at vertical polarization and viewing angle 45-50° are necessary to determine which directional spectrum is closer to reality.

Finally we would like to emphasize the "sensitivity" of the radiometric data to the model spectra of sea waves. The combination of elevation angular and azimuthal measurements allows us to estimate models for omnidirectional curvature wavenumber spectrum and spreading function of the gravity-capillary waves. We believe that the polarimetric *K*-, *Ku*- and *Ka*-measurements of the sea surface brightness temperature at various angles would be very appropriate for retrieval of the mean squared slope of long waves and the spatial wave number spectrum of gravity-capillary waves.

ACKNOWLEDGMENTS

This work was supported by Advanced Sensor Application Program, Department of Defense and partially by grant INTAS 97-10569. We are very grateful to UW/APL team and personally Drs. W. Plant and W. Keller for their help during the experiment and data processing.

REFERENCES

1. V. G. Irisov, Yu. G. Trokhimovskii and V. S. Etkin. Radiothermal Spectroscopy of the Ocean Surface. *Sov. Phys. Dokl.*, **32**, 914-915, (1987).
2. Yu., G. Trokhimovski. The model for microwave thermal emission of sea surface with waves. *Earth observation and Remote Sensing*, (1), 39-49, (1997).
3. T. V. Blanc,, W. J. Plant and W. C. Keller. The Naval Research Laboratory's Air-Sea Interaction Blimp Experiment. *Bull. Am. Met. Soc.*, **70**, 353-365, (1989).
4. R. A. Kropfli and S. F. Clifford. The Coastal Ocean Probing Experiment: Future studies of air-sea interactions with remote and in-situ sensors. *Proc., IGARSS'96*, 1739-1741, (1996).
5. Y. Han, J. B. Snider, E. R. Westwater, S. H. Melfi and R. A. Ferrare. Observations of water vaper by ground-based microwave radiometers and Raman lidar. *J. Geophys. Res.*, **99**, 18695-18702, (1994).
6. V. G. Irisov. Small-slope expansion for thermal and reflected radiation from a rough surface. *Waves Rand. Media*, (7), 1-10, (1997).
7. M. A. Donelan and W. J. P. Pierson. Radar scattering and equilibrium ranges in wind-generated waves with application to scatterometry. *J. Geophys. Res.*, **92**, 4971-5029, (1987).
8. J. R. Apel. An improved model of the ocean surface wave vector spectrum and its effects on radar backscatter. *J. Geophys. Res.*, **99**, 16269-16291, (1994).
9. R. Romeiser, W. Alpers and V. Wismann. An improved composite surface model for the radar backscattering cross section of the ocean surface 1. Theory of the model and optimization/validation by scatterometer data. *J. Geophys. Res.*, **102**, 25237-25250, (1997).
10. T. Elfouhaily, B. Chapron, K. Katsaros and D. Vandemark. A unified directional spectrum for long and short wind-driven waves. *J. Geophys. Res.*, **102**, 15781-15796, (1997).
11. L. A. Klein and C. T. Swift. An improved model for the dielectric constant of sea water at microwave frequencies. *IEEE Trans. Antennas Propag.*, **25**, 104-111, (1977).
12. J. P. Hollinger. Passive microwave measurements of sea surface roughness. *Trans. Geosci. Electron.*, **9**, 165-169, (1971).
13. C. T. Swift. Microwave radiometer measurements of the Cape Cod Canal. *Radio Sci.*, **9**(7), 641-653, (1974).
14. W. J. Plant. A relationship between wind stress and wave slope. *J. Geophys. Res.*, **87**, 1961-1967, (1982).

Microw. Radiomet. Remote Sens. Earth's Surf. Atmosphere, pp. 21–27
P. Pampaloni and S. Paloscia (Eds)
© VSP 2000

Determination of ocean surface wind speeds from the TRMM Microwave Imager

LAURENCE N. CONNOR[1] and PAUL S. CHANG[2]

[1]*UCAR Visiting Scientist Program, Camp Springs, MD 20746, USA*
[2]*NOAA/NESDIS/ORA, Camp Springs, MD 20746, USA*

Abstract - An analysis of brightness temperature data from the TRMM Microwave Imager (TMI) is presented with regard to the retrieval of ocean surface wind speeds using standard regression techniques with *in situ* meteorological buoy measurements. Comparisons to similar satellite radiometer data from the Special Sensor Microwave/Imager (SSM/I) are also presented to help quantify atmospheric contributions to the surface wind retrievals. Particular emphasis is placed upon the use of the 10.7 GHz channels aboard the TMI in overcoming the contamination in the ocean surface brightness temperature measurements caused by precipitation and water vapor in the propagation path. The resulting wind retrieval improvements permit a relaxation in the rain flag definitions used to determine precipitation interference cutoff criteria, allowing accurate wind speed retrievals over a wider range of precipitation conditions.

1. INTRODUCTION

The successful launch of the Tropical Rainfall Measuring Mission (TRMM) observatory and subsequent data collection provide an interesting opportunity for investigating and improving satellite based microwave radiometer determinations of ocean surface wind speeds. It is well established that radiometric measurements of the ocean surface brightness temperature may be used to determine wind speeds near that surface [1]. Using brightness temperature measurements from SSM/I and colocated ocean buoy wind measurements, Goodberlet et al. [2] developed a D-matrix based SSM/I wind speed retrieval algorithm capable of 2 *ms*[-1] accuracy under rain free conditions. Much of the work presented follows the strategy used with SSM/I in the development of the D-matrix approach which assumes a linear relationship between the measured brightness temperatures and a desired environmental parameter; namely, ocean surface winds. The implementation of such an approach has proven successful with SSM/I measured brightness temperatures, but is significantly hampered by the presence of precipitation and significant water vapor. The additional 10.7 GHz channels on the TRMM Microwave Imager (TMI) provide a more transparent window to the ocean surface, while still maintaining sensitivity to the wind modulated ocean surface brightness temperatures.

Several months of TMI brightness temperature measurements are examined with regard to the retrieval of ocean surface wind speeds and the accuracy and operational improvements offered by including the 10.7 GHz channels. Colocated SSM/I measurements of total precipitable water (TPW) and liquid water path (LWP) are used to identify the physical mechanisms associated with D-matrix rain flag designations while colocated wind speed measure-

Table 1. D-matrix rain flag definitions [2]. The accuracy column indicates the wind speed accuracy expected from SSM/I D-matrix wind retrievals for data associated with the corresponding rain flag.

Rain Flag	Criteria				Accuracy (m/s)
0		$T_B(37V) - T_B(37H)$	>	50 K	< 2
		And $T_B(19H)$	$\leq$	165 K	
1	37 K $\leq$	$T_B(37V) - T_B(37H)$	<	50 K	2 - 5
		or $T_B(19H)$	$\geq$	165 K	
2	30 K $\leq$	$T_B(37V) - T_B(37H)$	<	37 K	5 - 10
3		$T_B(37V) - T_B(37H)$	<	30 K	> 10

ments from offshore buoys are employed in regression analyses with TMI brightness temperatures. The increased precipitation transparency provided by the 10.7 GHz channels suggests a relaxation in the rain flag definition is possible, providing an operational enhancement in the form of greater acceptable wind speed estimate coverage.

2. DATA SOURCES

The TMI is a nine-channel microwave radiometer with horizontal and vertical polarization channels at 10.7 GHz, 19.4 GHz, 37.0 GHz and 85.5 GHz and a vertical polarization channel at 21.3 GHz [3]. The channels are based on a similar configuration found on the SSM/I, but with the additional 10.7 GHz channels and an offset in the 21.3 GHz water vapor channel to avoid saturation in TRMM's tropical orbit. The TMI's antenna beam pattern, conical scanning geometry, and integration times result in a cross-track effective field of view (EFOV) ranging from 4.6 km to 9.1 km and a down-track EFOV ranging from 63.2 km to 7.2 km. TRMM has a 350 km circular orbit with a 35° inclination angle, providing excellent coverage of the tropical latitudes (38° S to 38° N) and completing a 24-hour day in 16 orbits.

Comparative SSM/I precipitation measurements are derived from brightness temperature pixels meeting a TMI pixel matchup criteria of surface level separation distance less than 5 km and time separation less than 1 minute. Ground truth measurements of the ocean surface winds were obtained from NOAA National Data Buoy Center offshore buoys. Match up constraints for TMI brightness temperatures and buoy wind speeds required a surface level separation of less than 25 km and a temporal separation of less than 30 minutes. To avoid terrestrial brightness temperature contamination, only matchups greater than 100 km from any land mass were considered. All buoy wind speed measurements were adjusted to a height level of 19.5 meters.

3. SSM/I PRECIPITATION COMPARISONS

Following the strategy employed by Goodberlet et al. [2], all TMI brightness temperature matchup samples were separated into one of four rain flag categories defined in Table 1. The accuracy specified in the right column is the wind speed accuracy expected from the SSM/I

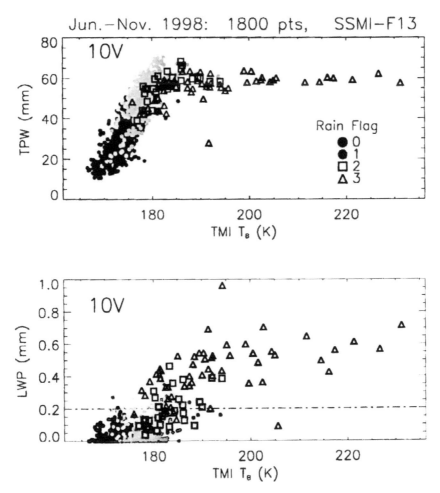

Figure 1. Top panel shows SSM/I derived total precipitable water (TPW) plotted as a function of TMI 10V brightness temperature and separated according to rain flag. Bottom panel shows similar plot for SSM/I derived liquid water path (LWP).

D-matrix retrievals for data associated with the corresponding rain flag. Clearly retrievals from the rain flag 0 category offer excellent wind speed estimates, while increasing rain flag number indicates a strong increase in wind speed signature contamination.

Using colocated measurements of TMI brightness temperatures and SSM/I derived precipitation parameters, it is possible to gain some insight into the physical phenomena governing the rain flag classification and the usefulness of the TMI 10.7 GHz channels. The top panel of Figure 1 shows SSM/I derived TPW measurements [4] plotted as a function of the TMI 10.7 GHz vertical polarization (10V) brightness temperature for different rain flags. The bottom panel follows a similar format, plotting SSM/I determined LWP [5] against the TMI 10V brightness temperature. Examination of these precipitation parameters in terms of rain flag designation reveals that points associated with rain flags 0 and 1 tend to fall in a well structured band of increasing TPW with increasing brightness temperature, rain flag 0 points found mostly in the lower TPW region and rain flag 1 points in that of higher TPW. Rain flag 2 and 3 points show little dependency of TPW on brightness temperature, while these points do seem to possess a strong liquid water content, as seen in the lower panel, the majority meeting the rainfall criteria of having LWP greater than 0.2 mm.

4. REGRESSION ANALYSIS

Using 4000 TMI matchups with ocean buoy wind measurements collected from an 11 month data set, two multiple linear regression analyses were carried out, fitting buoy wind speeds onto TMI brightness temperatures. One analysis included the same four brightness temperature channels used for the SSM/I D-matrix algorithm, namely 19V, 21V, 37V, and 37H (note that the 21V channel is slightly offset from the 22V channel found on SSM/I). The second analysis employed these same four frequencies, but also included the 10V and 10H channels. Regression coefficients were calculated for both sets using only those points that met the rain flag 0 or 1 criteria. Figure 2 shows the scatter plot results of these fits using rain flag 0 data withheld from the original 11 month data set. The bottom panel scatter plot associated with the six component (plus constant term) D-matrix, that includes the 10V and 10H channels, clearly shows a fit improvement over the four channel scatter plot with a 0.23 ms^{-1} reduction in residual RMS value.

A more quantitative measure of performance may be determined by examining residual parameters as a function of the D-matrix wind estimate. In particular, the points shown in Figure 2 were sorted by their D-matrix wind value and separated into 2 ms^{-1} bins. The mean (bias) and standard deviation of the residual values (D-matrix wind speed minus buoy wind speed) were then calculated for each bin, effectively producing error and bias estimates as a function of wind estimate. The broken and dashed lines in Figure 3 show the results of this operation, here labeled as "Old Flags". The top set of lines, with data points indicated by squares, show the standard deviation calculations. The bottom set, with the triangle points, show the bias calculations. The six component D-matrix demonstrates a clear and consistent improvement in the standard deviation of the residual over the four component D-matrix, ranging from 0.1 to 0.3 ms^{-1}. The bias behaves roughly the same for both representations.

5. OPERATIONAL IMPROVEMENTS

Since the atmosphere is more transparent to the 10.7 GHz channels, it is desirable to use the information provided by these channels to increase wind retrieval coverage over areas of significant precipitation. If the old D-matrix wind residual errors associated with the four SSM/I brightness temperatures is considered acceptable, then a simple approach is to relax the rain flag 0 conditions for the regression analysis until the six component D-matrix standard deviation performance is raised to the level of the four component D-matrix performance. This proved feasible by establishing new rain flag 0 criteria, namely: TB(37H) - TB(37H) > 42 K and TB(19H) < 200 K (relative to the old thresholds in Table 1: 50 K → 42 K and 165 K → 200 K). For the withheld matchups used for performance assessment, this resulted in a conversion of 85% of the rain flag 1 points over to rain flag 0 points. The regression performance resulting from this new set of rain flags and their associated six component D-matrix is represented by the solid lines in Figure 3 relative to the old four and six component D-matrix performances. The new D-matrix performs very similarly to the old six component D-matrix for wind estimates less than or equal to 7 ms^{-1}, but matches the performance of the old four component D-matrix for estimates greater than 7 ms^{-1}. Table 2

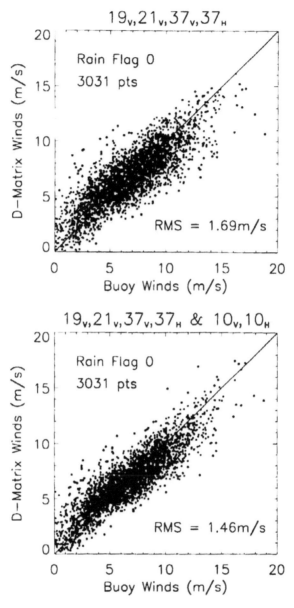

Figure 2. The top panel shows a scatter plot of rain flag 0 TMI/buoy matchups plotting buoy measured wind speed against D-matrix wind speeds determined from the four component algorithm. The bottom panel shows a similar scatter plot using the six component algorithm that includes the 10.7 GHz channels.

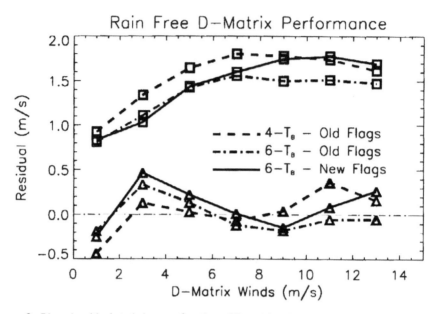

Figure 3. Binned residual statistics as a function of D-matrix wind speed estimate. Upper set of lines square data points) indicates standard deviation of residual. Lower set of lines (triangle data points) shows bias of residual. Dashed lines correspond to four component D-matrix performance with old flags. Broken and solid black lines correspond to six component D-matrix performance with old and new rain flags, respectively.

Table 2. New D-matrix coefficients

Channel	Coefficient	Value
Conts.	C_0	113.6780
10V	C_1	0.2578
10H	C_2	0.2325
19V	C_3	-0.0026
21V	C_4	-0.0226
37V	C_5	-1.2330
37H	C_6	0.6195

shows the coefficients for the new six component D-matrix to retrieve wind speeds at the 19.5 meter level using the formula:

$$wind\ speed[ms^{-1}] = C_0 + C_1 T_B(10V) + C_2 T_B(10H) + C_3 T_B(19V)$$
$$+ C_4 T_B(21V) + C_5 T_B(37V) + C_6 T_B(37H)$$

6. CONCLUSIONS

It is clear that including the 10.7 GHz channels in the TMI regression analysis of wind speed retrieval significantly reduces the estimate error. Interpreting this improvement as a result of the 10.7 GHz signal's decreased susceptibility to atmospheric interference, it is possible to expand the criteria for acceptable "rain free" data samples. Thus, an operational improvement in the form of increased wind retrieval coverage through higher levels of atmospheric contamination is realized. Matchups for higher wind speeds are required to extend this analysis beyond 15 ms^{-1}.

REFERENCES

1. F. T. Ulaby, R. K. Moore, and A. K. Fung. *Microwave Remote Sensing: Active and Passive*, Vol. 3, Artech House, Norwood, MA, (1986).

2. M. A. Goodberlet, C. T. Swift, and J. C. Wilkerson. Remote sensing of ocean surface winds with the special sensor microwave/imager. *J. Geophys. Res.*, **94**(C10), 14547-14555, (1989).

3. C. Kummerow, W. Barnes, T. Kozu, J. Shiue, and J. Simpson. The tropical rainfall measuring mission (TRMM) sensor package. *J. Atmos. Oceanic Technol.*, **15**, 809-817, (1998)

4. R. R. Ferraro, F. Weng, N. Grody, and A Basist. An eight-year (1987-1994) time series of rainfall, clouds, water vapor, snow cover, and sea ice derived from SSM/I measurements. *Bull. Amer. Meteor. Soc.*, **77**(5), 891-905, (1996)

5. F. Weng and N. Grody. Retrieval of cloud liquid water using the special sensor microwave/imager (SSM/I). *J. Geophys. Res.*, **99**(D12), 25535-25551, (1994)

Microw. Radiomet. Remote Sens. Earth's Surf. Atmosphere, pp. 29–37
P. Pampaloni and S. Paloscia (Eds)
© VSP 2000

Ocean winds measured by an imaging, polarimetric radiometer

BRIAN LAURSEN, STEN SØBJÆRG, and NIELS SKOU

Technical University of Denmark, DK 2800, Lyngby, Denmark.

Abstract– An airborne experiment, with the aim of measuring wind direction over the ocean using an imaging polarimetric radiometer, is described. A polarimetric radiometer system measuring all four Stokes brightness parameters has been designed and built. It is based on the correlation type of radiometer, and imaging is achieved using a 1 m aperture conically scanning antenna. The polarimetric azimuthal signature of the ocean is known from modeling and circle flight experiments. Combining the signature with the measured brightness data enables the wind direction to be determined on a pixel by pixel basis in the radiometer imagery.

1. INTRODUCTION

For some years it has been known that ocean wind direction can be assessed with polarimetric radiometer systems measuring the full set of Stokes parameters. Airborne experiments and model work have been carried out, and such activities are still ongoing. Most experimental work has been carried out employing a staring radiometer pointing towards the sea surface while the aircraft makes full 360° turns. Thus the radiometric signatures as a function of azimuthal observation angle relative to the wind direction were investigated [1 – 4].

However, finding the complete 360° polarimetric signature using a staring radiometer with a long integration time, hence good radiometric sensitivity, is a necessary and important task. It is a quite different task to determine the wind direction pixel for pixel based on data from a future spaceborne imager with it's inherent short integration time, hence poorer sensitivity. Furthermore, the full 360° response will not be available, and the retrieval must rely on experimentally or modeled signatures.

Airborne experiments, designed to simulate the space instrument situation, are highly warranted, and is the subject of the present paper. A very successful intermediate step was reported in 1998 [5]. Here the objective was also to retrieve wind directions from airborne imaging radiometer data, but in this case the full 360° signatures were available due to a unique scan configuration. The experiment to be discussed in the following is rather realistic as the imager only has a limited view of the scene and the wind directions are retrieved pixel for pixel only using the associated azimuthal observation angle.

2. POLARIMETRIC SIGNATURES OF THE OCEAN

Generally, the radiation from an object is partly polarized meaning that the vertical brightness temperature T_V is different from the horizontal T_H. A well known example is the sea surface. To deal with partial polarization it is convenient to use the Stokes parameters. The (brightness) Stokes vector is:

$$
\overline{T}_B = \begin{pmatrix} I \\ Q \\ U \\ V \end{pmatrix} = \begin{pmatrix} T_V + T_H \\ T_V - T_H \\ T_{45^\circ} - T_{-45^\circ} \\ T_l - T_r \end{pmatrix} = \frac{\lambda^2}{k \cdot z} \begin{pmatrix} \langle E_V^2 \rangle + \langle E_H^2 \rangle \\ \langle E_V^2 \rangle - \langle E_H^2 \rangle \\ 2 \ \mathrm{Re} \ \langle E_V E_H^* \rangle \\ 2 \ \mathrm{Im} \ \langle E_V E_H^* \rangle \end{pmatrix}
$$

where z is the impedance of the medium in which the wave propagates, λ is the wavelength and k is Boltzmanns constant. T_{45° and T_{-45° represent orthogonal measurements skewed 45° with respect to normal and T_l and T_r refer to left-hand and right-hand circular polarized quantities. I represents the total power, and Q the difference of the vertical and horizontal power components. The first and second Stokes parameters are measured using vertically and horizontally polarized radiometer channels, followed by addition or subtraction of the measured brightness temperatures.

The third Stokes parameter can be measured with a 2-channel radiometer connected to an orthogonally polarized antenna skewed 45°, and subtracting the measured brightness temperatures, or it can be found as the real part of the cross correlation of the vertical and horizontal electrical fields. The fourth Stokes parameter can be measured with the 2-channel radiometer connected to a left-hand / right-hand polarized antenna system, or it can be found as the imaginary part of the cross correlation of the vertical and horizontal electrical fields.

It is seen that all Stokes parameters are immediately measured by a 2-channel correlation radiometer (employing a complex correlator) connected to a traditional horizontally and vertically polarized antenna system. This is how the radiometer system, to be described later, operate.

The brightness temperature of the ocean depends on the wind speed [6]. At incidence angles around 50°, and at Ku and Ka bands, the sensitivity to wind speed is around 0.5 K per m/sec wind at vertical polarization, and 1 – 1.5 K per m/sec at horizontal polarization. This means that a radiometer measurement with an accuracy of 1 K enables determination of the wind speed to better than 1 m/sec (excluding other error sources). But the measurement of the Stokes parameters place more stringent requirements to the accuracy of the radiometers. Typical variations in the second, third, and fourth Stokes parameters due to wind direction are shown in Figure 1.

The x-axis is the azimuth angle ψ_o normalized to the wind direction so that $\psi_o = 0$ is the upwind direction. The incidence angle is 50°. The curves are based on results from previous experiments [1 – 4]. The second Stokes parameter, Q, has a typical variation of 3 K around a mean value of 63 K. This mean value is dependent on frequency, wind speed, atmospheric attenuation, and other geophysical parameters. The curve has a first and second harmonic cosine shape with maximum in the upwind direction (local maximum downwind) and minimum in crosswind. U has the same typical peak to peak variation but around zero mean. It has a first and second harmonic sine shape with steep zero crossing in the upwind direction and maximum and minimum close to crosswind. The fourth Stokes parameter signal is always somewhat smaller than Q and U, but with a very clean second harmonic sine shape.

The small signals require the radiometer sensitivity and absolute accuracy to be fractions of a Kelvin, which is challenging.

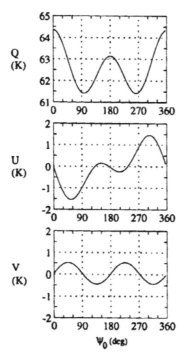

Figure 1. Typical variation in ocean Stokes parameters with azimuth angle ψ_o.

3. OCEAN WIND RETRIEVAL

A polarimetric radiometer system for ocean wind vector measurements will typically use several frequencies (here Ku and Ka band), and the 3 Stokes parameters Q, U, and V. Some signals will be strongly dependent, for example U at Ku and Ka band, while others will be orthogonal like Q and U at the same frequency. The wind direction retrieval is theoretically completely free of direction ambiguities due to this orthogonality. But noise, be it radiometric or from other sources, will give uncertainties in the retrieval, especially in the downwind directions where the Q and U signal variations are smallest. The resulting errors in wind direction do not just resemble Gaussian noise but include possible ambiguities, i.e. directions way off like for example 180°.

In order to carry out the wind retrieval from a given set of polarimetric radiometer measurements, the azimuthal signature of the sea has to be established corresponding to the actual conditions of the measurements: frequency, incidence angle, atmospheric conditions, sea temperature, wind speed. This is called the model function, and it will typically look like Figure 1. It can be determined by modeling, or by carrying out a proper 360° circle flight on the day of the measurement. The first is what is needed in a future operational system, while the latter is a viable option in the experimental phase. The wind retrieval used here is based on a non-linear weighted least-squares minimization of the following error expression:

$$e(\psi_0) = \sum_{j=1}^{n} \sum_{i=1}^{4} \frac{(\overline{T}_{i,j}(\psi_0) - T_{i,j})^2}{\Delta T_{i,j}^2}$$

where n is the number of frequency bands used (here 2), $\overline{T}_{i,j}(\psi_0)$ is the model function and $T_{i,j}$ is the measurement at the j'th frequency concerning the i'th Stokes parameter. The weighting functions $\Delta T_{i,j}$ does not only include radiometric noise, but also noise from uncertainties in antenna pointing (incidence angle, polarization mixing). Thus, in the retrieval, preference is given to less noisy channels.

It was found of advantage to make a slight modification to the error expression in order to reflect the fact that some Stokes parameters have larger peak to peak response ($\overline{T}_{i,jpp}$) than others, that is, put extra weight to channels with large response:

$$e(\psi_0) = \sum_{j=1}^{n} \sum_{i=1}^{4} \frac{(\overline{T}_{i,j}(\psi_0) - T_{i,j})^2}{\Delta T_{i,j}^2} \cdot (\overline{T}_{i,jpp})^2$$

I is very dependent on cloud conditions (atmospheric attenuation), Q is quite dependent on cloud conditions, while U and V are only marginally dependent on clouds. This is not surprising bearing in mind that I is the sum of T_V and T_H, both being dependent on atmospheric conditions. Q is the difference between T_V and T_H but changes due to varying atmospheric conditions do not quite cancel since T_V is much larger than T_H (typically 63 K, see Figure 1). In contrast to this, U and V can be interpreted as differences between channels having almost the same signal strength (see the Stokes vector definition and Figure 1). Thus, for example, $T_{45°}$ increases with atmospheric attenuation but $T_{-45°}$ will increase with the same amount, and U will to first order be unaffected. Therefore, in cloudy conditions with unusually severe attenuation (actually monitored using the I channel) the weighting function for Q ($\Delta T_{2,j}$) was enhanced proportionally to put less weight on Q in the retrieval (rely more on U and V). In practice this was done by comparing the measured I values with the expected I values, calculated using the actual wind speed, sea temperature, salinity, but standard atmosphere without clouds. For differences 0 - 20 K, $\Delta T_{2,j}$ was enhanced proportionally resulting eventually in a weight on Q as low as 5% compared with U and V.

In short, the retrieval procedure is:
- establish the model function $\overline{T}_{i,j}(\psi_0)$
- establish the peak to peak response $\overline{T}_{i,jpp}$
- establish the weighting function $\Delta T_{i,j}$
- compare the measurements $T_{i,j}$ with the model function (properly weighted) and find the wind angle ψ_u that results in the smallest difference $e(\psi_u)$.

4. EMIRAD POLARIMETRIC RADIOMETER SYSTEM

The airborne, imaging, polarimetric EMIRAD system employs Ku and Ka band polarimetric radiometers of the correlation type. They are of identical design.

The correlation radiometer uses two receivers that are connected to the vertical and the horizontal outputs of a dual polarized feed horn, see Figure 2. The receivers are total power radiometers, and fast switches (latching circulators) are included for frequent calibration: two calibration loads are observed by the radiometers whenever the scanning antenna reverses it's scan direction at the swath edges.

The correlation technique requires phase coherence between the two super-heterodyne receivers in order to measure the complex correlation between input signals. This is achieved by using one common local oscillator.

The analog detectors are tunnel diode detectors followed by integrators having 8 msec integration time.

The digital correlator employs three level (2 bit) A to D conversion at the input and proper multiplications. The sampling rate is 1540 MHz and an 8 msec averaging is carried out on the correlator outputs.

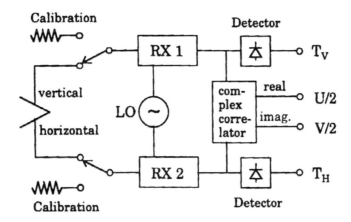

Figure 2. EMIRAD correlation radiometer

The radiometric sensitivity for the vertical (or horizontal) channel is 0.35 K. Thus the sensitivity for the first and second Stokes parameters is $\sqrt{2}\cdot0.35 = 0.5$ K. The sensitivity of the correlation measurements, i. e. the third and fourth Stokes parameters is 0.57 K.

The imaging antenna is based on an offset 1m aperture parabolic reflector scanning around a vertical axis, and illuminated by microwave horns pointing upwards along this axis. The beamwidth of the antenna is 2° at Ku band. The cross polarization due to the offset parabolic reflector is below 25 dB. The maximum scan angle is +/- 25°, and the incidence angle on the ground is constantly 50° for this conically scanned system. Due to the fixed feed horn and the scanning reflector, significant polarization mixing takes place when scanning away from straight aft. This is corrected in the data analysis, following the considerations in [7].

The antenna and the receivers are mounted on a cargo pallet, which is positioned on the loading ramp of a C-130 aircraft. The ramp is closed during take-off, landing, and transit, but lowered to its horizontal position when measurements are to be carried out. The antenna thus has an unobstructed view of the scene below and aft of the aircraft. The ground speed of the aircraft is 70 m/sec during operation with the ramp open. The flight altitude is 2000 m, and the +/-25° scan then results in a swath of 2000 m. For the 2000 m altitude the footprint on ground is roughly 100 m at Ku band, and half of that at Ka band.

The polarimetric signatures of the wind driven sea only show small azimuthal variations in the range of a few Kelvin. At the same time they are quite dependent on incidence angle and pointing geometry which may result in polarization rotation. Thus aircraft attitude must be carefully monitored, and the radiometer system includes an inertial navigation unit mounted directly on the antenna frame. The antenna attitude is thus measured to within 1/10 of a degree. Using this information data is corrected for unwanted attitude variations (typically less than 1°).

5. OCEAN FLIGHT EXPERIMENT

The polarimetric radiometer data to be discussed in this section were acquired on October 21, 1998 over a target area centered at 55° 40' N, 4° 46' E in the middle of the North Sea. The test site was chosen between two offshore oil platforms from where in-situ wind data was acquired. One platform was approximately 8 km to the north of the target center, the other some 8 km to the south. Thus wind shadow effects in the test area, stemming from the oil platforms, are avoided under predominantly westerly winds. At the time of data collection the winds in the area were 17 m/sec with a direction of 216°. The winds are normalized to standard meteorological reference, and they are 5 min averages disregarding gusts. The wind speed and direction were quite stable for many hours before and during the experiment. Cloud conditions were: overcast with heavy clouds and even showers.

Figure 3 shows details about the flight pattern. The left-hand part of the Figure summarize the imaging parameters, while the right-hand part shows the actual flight pattern centered at the target center coordinates. 8 passes are carried out, 4 legs each flown back and forth. Each pass is approximately 8 km. The 8 passes correspond to headings: 1) 270°, 2) 90°, 3) 315°, 4) 135°, 5) 00°, 6) 180°, 7) 45°, 8) 225°. Flying the passes two by two back and forth simulates a possible fore-and-aft look situation for a future spaceborne instrument. The time delay between two observations of the target center is around 5 min. fitting well with the time between the fore and the aft look for a typical space instrument. During the passes, aircraft roll and pitch are kept to a minimum, generally below a few tenth of a degree.

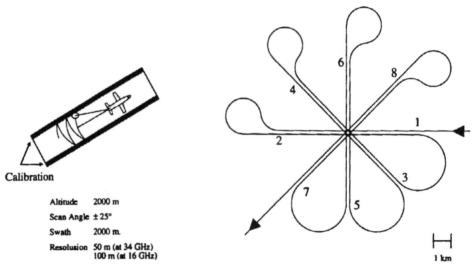

Calibration

Altitude	2000 m
Scan Angle	± 25°
Swath	2000 m.
Resolution	50 m (at 34 GHz)
	100 m (at 16 GHz)

1 km

Figure 3. Flight pattern.

In addition to the primary flight pattern described above, a circle flight was carried out to check the instrument, and to confirm the model function. Such measurements are carried out with the radiometer staring at the sea surface without scanning, while the aircraft makes full 360° turns around the target center. Actually, in the present case these measurements are used to establish the mean value of the second Stokes parameter model function as this is dependent on atmospheric conditions not being measured otherwise.

Figure 4 shows an example of the radiometric data acquired and the retrieved wind directions.

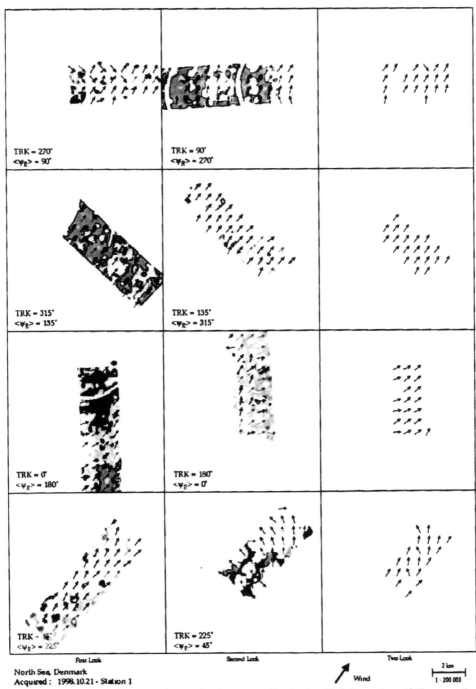

TRK = 270°
<ψ_R> = 90°

TRK = 90°
<ψ_R> = 270°

TRK = 315°
<ψ_R> = 135°

TRK = 135°
<ψ_R> = 315°

TRK = 0°
<ψ_R> = 180°

TRK = 180°
<ψ_R> = 0°

TRK = 45°
<ψ_R> = 225°

TRK = 225°
<ψ_R> = 45°

First Look   Second Look   Two Look

North Sea, Denmark
Acquired : 1998.10.21 - Station 1   Wind   2 km   1 · 200 000

DCRS, Dept. of Electromagnetic Systems, Technical University of Denmark

Figure 4. Radiometric I data and retrieved winds. TRK is aircraft track angle, while <ψ_R> is average azimuth look angle (TRK-180° due to aft looking)

The 8 panels in the two left-hand columns correspond to the 8 passes over the target area. Recalling that the flight consisted of 4 legs each flown back and forth, it is seen that the left-hand column can be regarded as four examples of a fore look and the mid column is the corresponding aft looks (180° apart from the first). The 16 GHz I data is shown as the background. All panels have north pointing towards the top of the page, and with the target center at the panel center. The recorded swaths are shown properly oriented and located in the appropriate panels. The average in-situ wind direction is shown at the bottom of the Figure. The wind directions, as retrieved from the recorded radiometric data, are shown as arrows on top of the 16 GHz I data. Before the retrieval, the data is averaged to 600 x 600 m ground resolution, in order to improve radiometric sensitivity and to avoid measurements on individual waves, i. e. obtain average surface conditions.

The third (right-hand) column is an attempt to combine the wind retrievals from the simulated fore and aft looks into a two-look situation. A rather simple, automatic selection algorithm has been implemented: if the difference between directions of the two single look retrievals is smaller than 60° , the average value has been calculated and plotted (the value 60° is a rather arbitrary choice, found to work well during the data processing). Differences larger than 60° indicate ambiguity problems, and a selection of one direction instead of the other must be performed. As already stated the upwind condition is favorable compared with the downwind condition due to larger polarimetric signatures in that direction. Hence, if one look has an estimated wind vector in the upwind sector, and the other has one in the downwind sector, the upwind is considered more reliable and hence selected. Note that all these operations are incorporated in the data processing software and are performed automatically, i.e. without human intervention.

In general the retrieved wind directions compare well with the actual wind direction with relatively few ambiguities. The worst errors and ambiguities are seen in the lower row (45° / 225° passes) where the second look is seriously hampered by a massive, rainy cloud that shows up as white in the underlying I image. An interesting observation can be made in the top row (270° / 90° passes): the first look has many 90° ambiguities due to relatively heavy clouds and downwind look geometry, while the second look has no ambiguities, despite even heavier clouds, due to upwind look geometry.

Table 1 summarizes the wind retrievals.

	all wind vectors		ambiguity removed	
passes	wind dir.	stdev.	wind dir.	stdev.
270/90	210°	26°	207°	8°
315/135	218°	4°	218°	4°
0/180	240	12°	240°	12°
45/225	198	22°	204°	14°

Table 1. Wind retrieval statistics.

The combined wind directions, as illustrated in the right-hand column of Figure 4, have been averaged and the standard deviation calculated. Recall that the actual wind direction was 216°. The wind direction is found quite well, but the standard deviation is very sensitive to ambiguities. Looking at the wind vectors in Figure 4, it is quite obvious that even a very simple "geophysical filter" can remove the ambiguities in most cases: it is not meaningful with one wind vector pointing 90° (or even 180°) away from the others amidst a wind field. Manually removing such obvious ambiguities in the upper and the lower row seriously improve the statistics as seen in the second half of Table 1.

This paper deals with data from one experiment. Another experiment was carried out later the same day with practically the same weather conditions and the result was almost identical. However, other flight campaigns with different conditions have also been performed. The results are being evaluated at the time of writing.

6. CONCLUSIONS

Based on previous experience, an imaging, polarimetric radiometer system has been designed and built. Of important design issues are radiometric stability, in order to measure the relatively small Stokes parameters, and antenna attitude monitoring, in order to untangle polarization mixing and incidence angle variations. With this instrument an experiment was carried out over the North Sea on a day with good winds (17 m/sec), but rather poor (but realistic!) weather conditions for radiometer measurements: heavy clouds and rain showers. Despite the weather, wind direction was found with good fidelity and few ambiguities in most cases, especially using a combination of 2 looks 180° apart. Removing obvious ambiguities, further improves the results, and the wind direction is on average found to be within 12° of the true value, with a standard deviation of 10°.

From the data it is clear that also one-look retrieval is possible, with more ambiguities to handle, however. This is of importance when designing a future spaceborne polarimetric system: the inclusion of both a fore and an aft look is no problem for the imaging radiometer itself, but it certainly imposes constraints on the spacecraft design to ensure unobscured looks fore and aft. Further analysis is needed in order to assess more accurately the benefits of having 2 looks for the retrieval.

REFERENCES

1. S. H. Yueh, W. J. Wilson, S. V. Nghiem, F. K. Li and W. B. Ricketts. Polarimetric Measurements of Sea Surface Brightness Temperatures using an Aircraft K-Band Radiometer. *IEEE Trans. on Geoscience and Remote Sensing*, Vol 33, No. 1, January 1995.
2. S. H. Yueh, W. J. Wilson, S. V. Nghiem, F. K. Li and W. B. Ricketts. Polarimetric Brightness Temperatures of Sea Surfaces Measured with Aircraft K- and Ka-Band Radiometers. *IEEE Trans. on Geoscience and Remote Sensing*, Vol 35, No. 5, September 1997.
3. N. Skou and B. Lauersen. Measurement of ocean wind vector by an airborne, imaging polarimetric radiometer. *Radio Science*, Vol. 33, No. 3, May-June 1998.
4. S. H. Yueh, W. J. Wilson, S. J. Dinardo, and F. K. Li. Polarimetric Microwave Brightness Signatures of Ocean Wind Directions. *IEEE Trans. on Geoscience and Remote Sensing*, Vol 37, No. 2, March 1999.
5. J. R. Piepmeier, A. J. Gasiewski, M. Klein, V. Boehm, and R. C. Lum. Ocean Surface Wind Direction Measurement by Scanning Polarimetric Microwave Radiometry. *Proceedings IGARSS'98*, p 2307.
6. Y. Sasaki, I. Asanuma, K. Muneyama, G. Naito, and T. Suzuki. The dependence of sea-surface microwave emission on wind speed, frequency, incidence angle, and polarization over the frequency range from 1 to 40 GHz. *IEEE Trans. on Geoscience and Remote Sensing*, Vol. GE-25, No. 2, March 1987.
7. A. J. Gasiewski and D. B. Kunkee. Calibration and application of polarization-correlating radiometers. *IEEE Trans. Microwave Theory Tech.*, 41(5), 767-773, 1993.

Microw. Radiomet. Remote Sens. Earth's Surf. Atmosphere, pp. 39–46
P. Pampaloni and S. Paloscia (Eds)
© VSP 2000

Interference effects in freshwater and sea ice

K.-P. JOHNSEN and G. HEYGSTER

Institute of Environmental Physics, University of Bremen, Germany

Abstract— In order to better understand the radiometric signal of sea ice and lake ice between 1 and 100 GHz we combine two different model approaches - the Strong Fluctuation Theory (SFT) and a Radiative Transfer Theory - and present the Combined radiative transfer Strong Fluctuation Theory (CSFT). Due to the coherent Fresnel reflection coefficients, the SFT shows oscillations of the brightness temperature *e.g.* with frequency and ice thickness. These oscillations were observed over freshwater ice in a tank experiment. They are reduced in sea ice due to the large variations of the ice thickness within the footprint of groundbased and spaceborne radiometers. The CSFT calculates the non scattering contribution to the emissivity with a radiative transfer theory, shows no oscillations, and is in good agreement with observed sea ice emissivities up to 40 GHz. Comparisons with radiometric *in-situ* measurements taken within the Arctic Ocean and from a tank experiment allow to obtain further knowledge about the parameters (snow and ice thickness, liquid water content within the snow) which govern the microwave signal of ice.

1. INTRODUCTION

Although there is a wide literature on *in-situ* observations of sea ice properties [1] there are not yet distinct descriptions of different sea ice types in terms of those microphysical parameters which govern their emissivity properties. However the crucial parameters have been identified in a recent study [2] [3], namely snow water content, snow density and grain size, and air bubble diameter and salinity.

We present radiometric *in-situ* measurements taken during the campaign ARK-XII/1 of the german R.V.Polarstern in the Kara and Laptev Sea from 12th of July to 23rd of September 1996, and thin lake ice measurements taken in a tank experiment at Lake Ladoga (Russia). We explain the measurements with an incoherent radiative transfer theory as well as with the coherent Strong Fluctuation Theory (SFT).

2. IN-SITU MEASUREMENTS

During the cruise ARK-XII/1 of R.V.Polarstern, a record of general ice conditions was taken based on observations from the ship's bridge [4]. The field program on the ice included the taking of ice cores and thickness measurements as well as measurements of snow depth, temperature, density and microwave brightness temperatures [5]. During the entire expedition, summer firstyear ice, which has survived the previous summer, was found. It was covered with snow either remaining from the previous winter, or new snow accumulating during the second half of the expedition. The microstructure of the snow (grain size, density), the distinct boundary between snow and ice as well as the low electric conductivity of the melted snow indicate this to be aged snow, rather than decomposed surface ice as commonly observed on summer Arctic sea ice [6] [7].

Table 1. Mean snow and ice characteristics observed on horizontal profiles from 31 stations during the cruise of R.V.Polarstern.

Region	Ice thickness [m]	Snow thickness [m]
Kara Sea	1.64 ± 80	0.08 ± 0.10
Nansin Basin	2.37 ± 1.14	0.07 ± 0.09
Transpolar Drift, west	2.50 ± 0.95	0.21 ± 0.16
Transpolar Drift, east	3.02 ± 1.00	0.26 ± 0.15
Transpolar Drift, south	2.13 ± 0.66	0.14 ± 0.12
Laptev Sea	1.29 ± 0.75	0.10 ± 0.06
Snow density	410 ± 70 kg/m^3	
Snow grain size	1.80 ± 0.45 mm	
Electric conductivity of snow	18 ± 21 μS/cm	

The mean snow characteristics taken at 31 stations during the cruise are given in Table 1. Generally, the snow showed fairly coarse grains; based on an analysis of 39 samples taken at these 31 stations using an image processing system. The mean major and minor axes of snow grains were found to be 2.3 and 1.1 mm, respectively. New snow also observed during the campaign from station 232 (19th of August) onward had smaller grain size of around 0.5 mm. Furthermore temperatures at the interfaces of ice/snow and snow/air were recorded.

Surface based passive microwave sensors measure the brightness temperatures which depend on surface emissivity, physical temperature and the reflected sky radiation. To obtain a better understanding of the corresponding microwave signatures with respect to the snow thickness we have performed groundbased passive microwave measurements with Dicke radiometers operating at 11, 21 and 35 GHz along transects every 53 cm on level parts of ice floes (Figure 1). The incidence angle was set to 50° to match the SSM/I onboard the DMSP F13 spacecraft observing angle of 51.7°. Radiometric measurements together with snow thickness were taken along profiles of different lengths with 19 to 100 points. With the tipping curve method the brightness temperature of the cosmic background is derived. It is used as the cold load: A plane-parallel atmosphere is assumed. The sky brightness temperatures T_{Sky} were determined under incidence angles of $\theta = 0°$ and 60° and extrapolated with $T_{sky} = T_a(1 - \exp(-\tau_0 \sec \theta))$ for $\tau_0 = 0$. T_a is the effective brightness temperature of the atmosphere. More details are given in [8].

In Figure 2a a radiometric measurement at 37 GHz taken over an open air tank filled with Lake Ladoga water is shown. The size of the tank was 110×300×45 cm^3. Before the onset of crystallization the water surface was cooled down to 0°C and was seeded by small snow crystals. During the experiment, air temperature dropped from −5.5°C to −7.5°C and then increased to −5.0°C. The freezing ice was observed with a passive microwave radiometer at 37 GHz in both polarizations. The radiometer with a footprint of 58 cm was mounted 1.5 m above the ice surface. It was calibrated periodically using a blackbody at ambient temperature and with the sky as cold load using also the tipping curve method. The incidence angle was also set to 50°. The ice grew up to a thickness of about 1.7 cm. The brightness temperatures of the thin ice show strong oscillations with time.

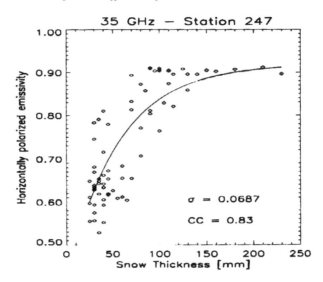

Figure 1. Radiometric measurements taken at 35 GHz at station 247 of ARK-XII/1. The line shows the model of Ulaby and Stiles. σ is the standard deviation and CC the correlation coefficient between the experimental and theoretical data. The physical temperature at the snow/air boundary was −1.0°C and at the ice/snow boundary −2.6°C.

3. MICROWAVE EMISSIVITY MODELS OF LAKE AND SEA ICE

Microwave sea ice signature models (Winebrenner et al., 1992) may be grouped depending on how they treat scattering, either they model backscattering or volume scattering, the latter often means to solve the radiative transfer equation. But in sea ice the assumption of a non-dense medium where the scatterers are in the far field of each other is not true; Therefore several modifications have been proposed which use the analytic wave theory based on Maxwells equations. They can be grouped into discrete scatterer approaches (e.g. [10]) and the random medium approaches (e.g. [11]).

Here we use the radiative transfer theory of Ulaby and Stiles [12] as well as random medium theories (SFT and CSFT) to explain the radiometric measurements.

The model originally proposed by Ulaby and Stiles (1980) for snow over ground assumes that the brightness temperature measured by the radiometer consists of two components, one due to the emission by the ice and the underlying water T_B^{ice}, and the other due to emission by the snow layer T_B^{snow}:

$$T_B = T_B^{ice} + T_B^{snow} \tag{1}$$

Ulaby and Stiles assumed plane-parallel snow and ice layers, neglected multiple scattering at the boundaries and derived for the emissivity

$$\varepsilon = \frac{T_B}{T_0} = A + B\exp(-vd_s\alpha_e \sec\theta) \tag{2}$$

with T_0 as the physical temperature of the sea ice and

$$A = Y_{as}\frac{\alpha_a}{\alpha_e} \tag{3}$$

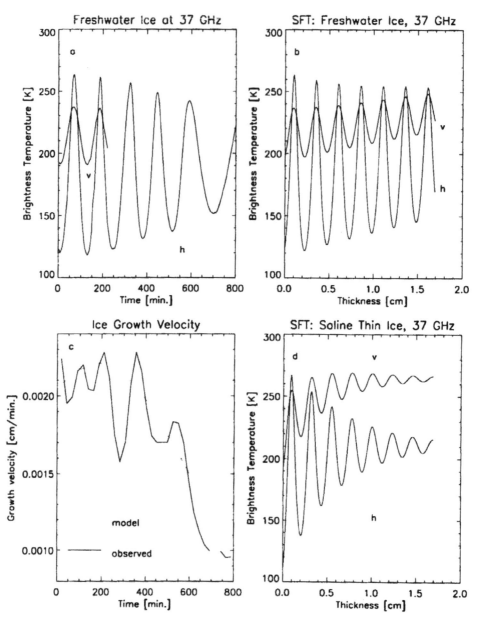

Figure 2. Passive microwave measurements at 37 GHz of freshwater ice grown from Ladoga Lake water (a) and the modelling with the SFT (b). The maximum ice thickness was 1.7 cm. (c): Ice growth velocity as calculated from the microwave measurements of the freshwater ice and with a thermodynamical model of Maykut [9]. (d): Horizontally polarized brightness temperature of thin saline ice as calculated with the SFT. All parameter are identical to the thin lake ice except of the salinity, which is 5 ppt. v means the vertically polarized brightness temperature, h the horizontally polarized.

$$B = Y_{as}Y_{si} - Y_{as}\frac{\alpha_a}{\alpha_e} \tag{4}$$

$$\alpha_e = \alpha_a + \alpha_s, \tag{5}$$

here Y_{as} and Y_{si} are the transmission coefficients at the interfaces from air to snow and from snow to ice, respectively, α_a is the absorption coefficient, α_s the scattering coefficient and α_e the extinction coefficient of the snow layer in units of cm^{-1}. The increase of the emissivity with increasing snow thickness at station 247 can be explained by absorption due to a high free water content (approximately 5 ppt) within the snow.

3.1. Strong Fluctuation Theory (SFT)

The SFT solves Maxwell's equations and accounts for the magnitude and the phase of the electromagnetic signal reflected within the layered structure of the ice and snow. With Kirchhoff's law the emissivity ε_a at the polarization a ($a = h$ or $a = v$) can be written as [13]

$$\varepsilon_a = 1 - |R_a|^2 - \frac{1}{4\pi}\int (\gamma_{ah}(\vec{k}_0,\vec{k}) + \gamma_{av}(\vec{k}_0,\vec{k}))\sin\theta d\theta d\phi. \tag{6}$$

The SFT determines the bistatic scattering coefficients γ_{ah} and γ_{av} for a layered model. Each layer is described by the parameters temperature, thickness, density, salinity of the ice and water layers, the diameter of the air bubbles, ice and snow grain size, the liquid water content of the snow, the angle of the brine pockets with respect to the vertical and the ratio of the length and diameter of the brine pockets [14] [11].

The SFT shows for thin ice an oscillatory behaviour of the brightness temperature with the ice thickness (Fig. 2b and [9]). The amplitude, frequency and the phase of these oscillations depend on the microwave frequency, and on the dielectric constant of the ice. This allows to determine the ice thickness (modulo the wavelength of the oscillations) and the ice growth velocity (Fig. 2c and [9]). The dielectric constant was derived from the lake ice parameters and the SFT as to be $\varepsilon = 3.05 + i0.011$. Thus the penetration depth is about 21 cm. Similar oscillations were observed over sea ice with a salinity between 0.7 and 1.0 ppt and a snow cover up to 15 cm at 610 MHz [15]. Here, due to the lower frequency the oscillations showed larger thickness periods. Hallikainen [16] has shown that at 37 GHz and an ice thickness of about 25 cm scattering has also to be taken into account.

The decreasing growth velocity can be explained thermodynamically with a decreasing heat transport through the increasing ice cover [17]. For saline ice, the amplitude of the oscillations decreases with thickness due to an increase in dielectric loss (Fig. 2d).

3.2. Combined Radiative Transfer-Strong Fluctuation Theory (CSFT)

If coherence effects are averaged out by the variations of the horizontal ice structure within the footprint of the radiometer, an incoherent model, e.g. a radiative transfer technique would be more appropriate than a coherent one. The inclusion of scattering is necessary because (1) it can contribute much to the emissivity of snow for frequencies above approximately 30 GHz and (2) the density of the uppermost layer of multiyear ice may be so small [1] that its scattering contributes considerably to the total emissivity according to equation (6).

In order to model the emissivities we combine the SFT, which calculates the scattering part in the near field with a radiative transfer model: The third term on the right hand side of equation (6) (integral over the bistatic scattering coefficients) is calculated with the fluctuating part of the electric field of the SFT [14]. The oscillations of the SFT are caused by the Fresnel reflection coefficient R_a, the second term in equation (6). It is calculated in the SFT [2] with a Riccatti differential equation from homogeneous layers with a mean dielectric constant.

Brekhovskikh [18] has shown that R_a is the same as that from a layer model *without* scattering which uses a coherent radiative transfer approach. In the theory of radiative transfer it was shown [19] that the incoherent approach allows quite similar results like the coherent approach but without the oscillations. Here, we include scattering into the model as formulated by Burke et al. [20]. They calculated the polarized brightness temperature $T_{B,a}$ of a layered medium as

$$T_{B,a}(\theta_0) = \sum_{i=1}^{N} PT_i(1 - \exp(-\gamma_i(\theta_0)\Delta z_i)) \times (1 + R_{a,i+1}(\theta_0)\exp(-\gamma_i(\theta_0)\Delta z_i)) \qquad (7)$$

with

$$P = \prod_{j=1}^{i}(1 - R_{a,j}(\theta_0))\exp-(\sum_{j=2}^{i}\gamma_{j-1}(\theta_0)\Delta z_{j-1}).$$

In these equations N means the number of layers and θ_0 the incidence angle. Δz_i is the thickness of i-th layer of sea ice. The absorption coefficients γ_i follow from the Poynting-Theorem [20] as $\gamma_i = 2\omega k'_{zi}/c$ with k'_{zi} the imaginary part of the z-component of the wave number in the i-th layer, c the speed of light and ω the angular frequency. From equation (7) we calculate the emissivity for the i-th isothermal layer as

$$\varepsilon_{a,RAD} = \frac{T_{B,a}}{T} \qquad (8)$$

with $T = T_i$ as one constant physical temperature for all layers $i = 1,..,N$. In general, the emissivity of a body of non-uniform temperature is not defined. We take T of the layer with the greatest contribution to the emanated radiation. Note that in case of a large temperature gradient and the radiation coming from a broad range of depths this temperature is not necessarily the one of the uppermost layer and the temperature of the appropriate layer must be determined carefully. In order to estimate the resulting error from a temperature gradient of 20 K within the snow, the parameter set of station 247 (Figure 3 left) was used without free water content within the snow. The brightness temperature difference between the cases with and without temperature gradient was less than ±2 K between 1 and 100 GHz.

Because scattering of a dense medium is not implemented into the radiative transfer model, the reflectivity follows from the emissivity according to

$$| R_{a,RAD} |^2 = 1 - \varepsilon_{a,RAD}. \qquad (9)$$

The equation does not contain the term for the transmissivity because it is zero. Under the sea ice layers the model assumes a half space of sea water. Now, the idea of the CSFT is to replace the reflectivity $| R_a |^2$ in equation (6) by $| R_{a,RAD} |^2$ using (8) and (9):

$$\varepsilon_{a,CSFT} = \frac{T_{B,a}}{T} - \frac{1}{4\pi} \int (\gamma_{ah}(\vec{k}_0, \vec{k}) + \gamma_{av}(\vec{k}_0, \vec{k})) \sin\theta d\theta d\phi. \qquad (10)$$

The scattering term provides the dense medium contribution to the emissivity.

Figure 3 left compares the mean emissivities according to the SFT and the CSFT, respectively, for the summer firstyear ice [22] of Figure 1 and Figure 3 right for dark nilas without snow. For details about layers and their microphysical description see [2]. Above approximately 40 GHz the theories show similar values for the emissivities. For the lower SSM/I-frequencies (19, 22 GHz) the combined approach reproduces with good accuracy the mean value of the oscillatory emissivity of the SFT. The small oscillations of the CSFT emissivities result from the scattering contribution.

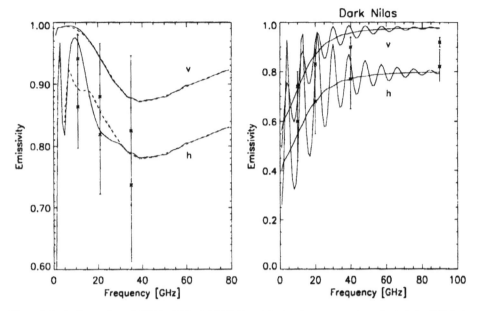

Figure 3. Left: Comparison of SFT (solid line), CSFT (dotted line) and measurements made at station 247 during ARK 12/1 with R.V.Polarstern in the Laptev Sea. h: Horizontally polarized emissivity, v: vertically polarized emissivity. The dashed lines are the mean values of the emissivities over 15 different snow thicknesses with constant change of thickness between 93 and 107 mm, calculated with the SFT. The mean snow thickness was 10 cm and the ice thickness 1.1 m. **Right:** Comparison of the emissivities according to the SFT (oscillating), the CSFT (smooth) and radiometric measurements of dark nilas from Eppler [21]. A salinity of 10 ppt, a thickness of 1 cm, and an air bubble diameter of 1.2 mm were assumed for the dark nilas.

4. CONCLUSIONS

Radiometric *in-situ* observations of sea ice together with sea ice parameters in the Arctic and Antarctic are quite sparse. Nevertheless these combined measurements are necessary to derive physical models and to understand spaceborne radiometric data. Especially the snow thickness, liquid water content within the snow, the snow grain size and the ice density and salinity should be measured because of their strong influence on the microwave signal [2].

The statistical variations of the sea ice parameters in the Arctic are so large that an incoherent approach to describe spaceborne passive microwave data is necessary. Using the precise description of the scattering part within the strong fluctuation theory we presented a combined radiative transfer strong fluctuation theory. The coherence effect was analysed: It was shown that especially for thin ice this effect is important and can be used to derive the ice thickness under laboratory conditions. Salinity reduces the oscillations of the brightness temperature of growing thin ice.

REFERENCES

1. W. B. Tucker, D. K. Perovich, A. J. Gow, W. F. Weeks, and M. R. Drinkwater. Physical Properties of Sea Ice Relevant to Remote Sensing. In F. D. Carsey, editor, *Microwave Remote Sensing of Sea Ice*, Geophysical Monograph 68, chapter 4, pages 9–28. American Geophysical Union, 1992.

2. R. Fuhrhop, G.Heygster, K.-P.Johnsen, P.Schlüssel, M.Schrader, and C.Simmer. Study of Passive Remote Sensing of the Atmosphere and Surface Ice. *Final report for ESA Contract No.11198/94/NL/CN, Berichte aus dem Institut für Meereskunde, Kiel, Germany*, 297, 1997.

3. R. Fuhrhop, T. C. Grenfell, G. Heygster, K.-P. Johnsen, P. Schluessel, M. Schrader, and C. Simmer. A combined radiative transfer model for sea ice, open ocean, and atmosphere. *Radio Science*, 33(2):303–316, March–April 1998.

4. M. Lensu, C. Haas, F. Cottier, C. Friedrich, J. Weissenberger, K.Abrahamsson, A. Ekdahl, A. Darovskikh, and K.-P. Johnsen. Arctic '96: Polarstern Ice Station Report. M-214, 1996.

5. E. Augstein and Cruise Participants. In *Die Expedition Arctic'96 des F.S. Polarstern, ARK XII mit der Arctic Climate System Study (ACSYS)*. Berichte zur Polarforschung, volume 234, Alfred-Wegener-Institut für Polar- und Meeresforschung, Bremerhaven, 1997.

6. T. C. Grenfell. Surface-based passive microwave studies of multiyear sea ice. *J. Geophys. Res.*, 97(C3):3485–3501, 1992.

7. H. Eicken, M. Lensu, M. Leppaeranta, W.B. Tucker, A. J. Gow, and O. Salmela. Thickness, structure and properties of level summer multiyear ice in the Eurasian sector of the Arctic ocean. *J. Geophys. Res.*, 100(C11):22697–22710, 1995.

8. K.-P. Johnsen. *Radiometric Measurements in the Arctic Ocean - Comparison between Theory and Experiment*. PhD thesis, University of Bremen, Germany, 1998.

9. A. Darovskikh, K.-P. Johnsen, V. Fedotov, K. Tyshko, H. Eicken, and G. Heygster. Growth velocity of freshwater ice and air bubble sizes linked to microwave radiometer measurements. In *The Proceedings of the 14th International Symposium on Ice*. Editor: Hung Tao Shen, Vol.1, No.2, pages 391-395, Potsdam, New York, USA, 1998.

10. T. Rother and K. Schmidt. The discrete Mie-formalism for plane-wave scattering on axisymmetric particles. *Journal of Electromagnetic Waves and Applications*, 10:273–297, 1996.

11. A. Stogryn. An Analysis of the Tensor Dielectric Constant of Sea Ice at Microwave Frequencies. *IEEE Transactions on Geoscience and Remote Sensing*, GE-25:147–158, 1987.

12. F. T. Ulaby and W. H. Stiles. The active and passive microwave response to snow parameters - 2. water equivalent of dry snow. *J. Geophys. Res.*, 85(C2):1045–1049, 1980.

13. D. P. Winebrenner et al. Microwave sea ice signature modeling. In F. D. Carsey, editor, *Microwave Remote Sensing of Sea Ice*, Geophysical Monograph 68, chapter 4, pages 47–71. American Geophysical Union, 1992.

14. A. Stogryn. A study of the microwave brightness temperature of snow from the point of strong fluctuation theory. *IEEE Transactions on Geoscience and Remote Sensing*, GE-24(2):220–231, 1986.

15. M. Hallikainen. A new low-salinity sea-ice model for UHF radiometry. *International Journal of Remote Sensing*, 4(3):655–681, 1983.

16. M. Hallikainen. The brightness temperature of sea ice and fresh-water ice in the frequency range 500 MHz to 37 GHz. *Digest of the 1982 International Geoscience and Remote Sensing Symposium (IGARSS'82)*, Munich, 1-4 June, year = 1982, volume = II, Paper TA-8/2.

17. G. A. Maykut. Energy Exchange Over Young Sea Ice in the Central Arctic. *J. Geophys. Res.*, 83(C7):3646–3658, 1978.

18. L. M. Brekhovskikh. *Waves in layered media*. Moskow, 1973.

19. F. T. Ulaby, R. K. Moore, and A. K. Fung. *Microwave remote sensing Artech House, Norwood, MA02062*, 1:84ff., 1981.

20. W. J. Burke, T. Schmugge, and J. F. Paris. Comparison of 2.8- and 21-cm Microwave Radiometer Observations Over Soils with Emission Model Calculations. *J. Geophys. Res.*, 84(C1):287–294, 1979.

21. D.T. Eppler et al. Passive Microwave Signatures of Sea Ice. In F. D. Carsey, editor, *Microwave Remote Sensing of Sea Ice*, Geophysical Monograph 68, chapter 4, pages 47–71. American Geophysical Union, 1992.

22. World Meteorological Organization. WMO Sea-Ice Nomenclature. Technical Report WMO/OMM/BMO – No. 259 Supplement No. 5, World Meteorological Organization, Geneva, April 1989.

Microw. Radiomet. Remote Sens. Earth's Surf. Atmosphere, pp. 47–56
P. Pampaloni and S. Paloscia (Eds)
© VSP 2000

A comparison of the errors associated with the retrieval of the latent heat flux from individual SSM/I measurements

DENIS BOURRAS, LAURENCE EYMARD and CECILE THOMAS

CETP/UVSQ, 10-12 avenue de l'Europe 78140 Vélizy Villacoublay, France

Abstract- we compare three latent heat flux estimation methods: the Liu Niiler [1984] method and two statistical algorithms respectively based on a linear regression and on a neural network approach. To apply the different methods, we use a global dataset that groups ECMWF analyses and SSM/I microwave brightness temperatures. The root mean square deviation between satellite and surface data is about 50 W/m². It is slightly less important for the neural network than for the two other methods. In order to analyze the error, the methods are applied to particular subsets of the global dataset on which the estimation error is different, and to data of the SEMAPHORE, TOGA-COARE and CATCH/FASTEX experiments.

1. INTRODUCTION

The turbulent latent heat flux (L_E) is useful for numerous studies in meteorology. For instance, it is one of the major terms of the surface energy budget at the ocean surface. To calculate L_E, bulk type parametrizations are often used [1]:

$$L_E = \rho \; C_E \; Lv \; U \; (Q_S - Q_A) = \rho \; C_E \; Lv \; U.\Delta Q \qquad (1)$$

In (1), ρ is the density of air, C_E is the Dalton number (which depends on wind speed and surface layer stability) and U is the scalar horizontal wind at height z_A; Q_S and Q_A stand for the specific humidity at the surface (simple function of the SST) and at a height of z_A meters.

In a global study, Weare [2] finds that the flux determination error is 30 W/m² in standard deviation, i.e. 20% for a 150 W/m² typical mean flux value, at a global scale. Only sparse flux measurements are available at a global scale, since they are performed inboard Research Vessels (R/V hereafter). In order to get a larger spatial and temporal coverage, we try to use spaceborne sensors to estimate the flux. In this comparison, we use data of the SSM/I microwave radiometer since it is sensitive to the flux parameters Q_A [3] and U [4]. The SSM/I channels used are 19, 22 and 37GHz in vertical and horizontal polarization (excepted for T22V, in vertical polarization only). We assume that Q_S is a known quantity because it can be derived from sea surface temperature measured by infrared spaceborne sensors [5]. We do not consider the effects of the stability of the surface boundary layer: C_E is taken equal to $1.2e^{-3}$. Thus, the two unknowns of our problem are U and Q_A.

We compare three flux estimation methods: in the first one (the Liu and Niiler method [3], or LN hereafter), we obtain the flux by applying a bulk parametrization to SSM/I derived U and Q_A values. The second method is a statistical algorithm based on a multilinear regression, whose inputs are TB combinations and a SST information ([6], [7] and [8]). The third one is a neural network approach (NN in the following), with the same inputs as in the algorithm above.

In the next two sections, we describe the methods and the datasets: a global dataset and three local ones corresponding to the SEMAPHORE [9], TOGA-COARE [10] and CATCH/FASTEX [11] experiments.
In the last two parts, we compare the methods on every dataset. To analyze the estimation error on the global dataset, the methods are applied to subsets of GI78, obtained by selecting particular ranges of U and ΔQ.

2.   DESCRIPTION OF THE METHODS

LN method: U and Q_A are estimated separately from the SSM/I measurements (TB); then the formula (1) is applied, to obtain L_E. The Goodberlet [4] algorithm provides wind estimates, whereas Alishouse's [12] gives water vapor contents W (which are related to Q_A values, see [3]).

Algorithm (multilinear regression): the inputs of the regression are Q_S and TB combinations ([6] and [7]). The flux estimates provided by this method should be at least as accurate as the fluxes derived from the LN method, because both use the sensitivity to U and Q_A. Nevertheless, the algorithm we proposed in [8] also takes account of the direct sensitivity of the H channels to the $U.Q_A$ product. Indeed, we use a simple model of the H channels as a function of U and QA, by linearizing the radiative transfer equation:

$$TB_H = \alpha\, Q_A + \beta\, U + \gamma\, U.Q_A \qquad (2)$$

$\qquad\qquad\qquad (I)\qquad (II)\quad (\alpha, \beta \text{ and } \gamma \text{ are functions of the frequency})$

The algorithm inputs are Q_S and TB combinations sensitive to U, Q_A and $U.Q_A$. Each of these input parameters (fig. 1) is used to estimate either $U.Q_S$ or $U.Q_A$ (the difference between these two quantities is proportional to L_E, according to equation 1). On one hand, the $U.Q_A$ and $U.Q_S$ products are linearized; i.e. they are approximated by a linear combination of U and Q_A (branches (A) and (B) in fig. 1). On the other hand, $U.Q_A$ is obtained by using a linear combination of H channels and TB combinations sensitive to U

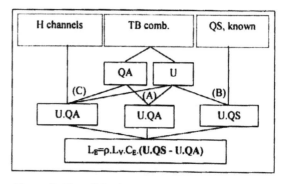

Figure 1. Use of the input parameters (dotted) in the flux algorithm. The approximations "linear combination / product" are used to estimate U.QA and U.QS.

and Q_A (branch C in fig. 1). In this case, U and Q_A are provided to compensate terms (I) and (II) in equation (2).

Some of the TB combinations of fig. 1 are TB polarization ratios (equation 3) at 19 and 37GHz for the wind sensitivity. For the others, [8] show that the 22V channels and the TB differences 19V-22V and 22V-37V depend almost linearly on Q_A, if lower than 19 $g.kg^{-1}$.

$$p(v) = \frac{280 - TB_V(v)}{280 - TB_H(v)} \qquad (v \text{ is for frequency}) \quad (3)$$

Finally, the multilinear algorithm has the following form:

$$L_E = \alpha_1.T_{19H} + \alpha_2.T_{37H} + \alpha_3.T_{22V} + \alpha_4.(T_{19V} - T_{22V}) \qquad (4)$$
$$+ \alpha_5.(T_{22V} - T_{37V}) + \alpha_6.p(19) + \alpha_7.p(37) + \alpha_8.Q_S$$

Neural network approach: to our knowledge, [13] were the first to use a NN approach to retrieve LE. In their study, two input parameters (W and SST) feed the NN that provides estimations for T_A (the temperatures at z_A meters) and Q_A. Then a bulk method is applied to T_A, Q_A and to SSM/I derived U values, to get the flux estimations.

In this study, the NN approach is considered as an enhanced multilinear regression method which can take into account some of the non linearities of the « input parameters/flux » relationship. We use the same input parameters as in the regression method, in order to compare the two approaches later.

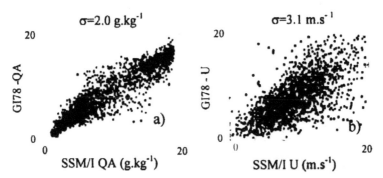

Figure 2. Comparison between the SSM/I derived data and the corresponding surface data in GI78, for the wind speed (right) and the humidity at 10m (left).

3. DATASETS

We use SSM/I measurements and two types of surface data: either GCM output fields (for the global dataset), or measurements performed inboard R/Vs (local datasets).

The global dataset: GI78 (Global Instantaneous dataset) groups 12 ECMWF global analyses distributed over 1997 and 98, at the resolution of 1.125 square degrees, and the corresponding SSM/I orbits. The latter are projected on a grid at the ECMWF resolution. Then, the values are averaged on each cell where there is enough data (at least ten values).

The maximum time difference between SSM/I and ECMWF data is 45 minutes. Equation (1) is used to compute the GI78 surface fluxes used in this study. We do not use the fluxes provided by the model since they are integrated over 6 hours and they are computed with a stability dependent CE.

In order to check -at least partially- the validity of this dataset, we use the algorithms of [4], [3] and [12] to compute U and QA estimates, that we compare (fig. 2) to the surface GI78 data; the algorithms are reported in (5), (6) and (7). The Root mean square Deviation between the Satellite derived quantities and the corresponding Surface data - hereafter referred as RDSS- are larger (3.1 m.s^{-1}and 2 g.kg^{-1}) than in [4] and [12] (2 m.s^{-1} for U and 1.1 g.kg^{-1} for Q_A). On figure 2, we notice that the SSM/I overestimates significantly the scalar wind (the bias equals 2 m.s^{-1}).

$$U(\text{m.s}^{-1})=1.0969\ T_{19V}-0.4555\ T_{22V}-1.760\ T_{37V} \tag{5}$$
$$+0.7860\ T_{37H}+147.90$$

$$W(\text{kg.m}^{-2})=-0.148596\ T_{19V}-1.829125\ T_{22V}-0.36954\ T_{37V} \tag{6}$$
$$-0.006193\ (T_{22V})^2+232.89393$$

$$Q_A(\text{g.kg}^{-1})=3.818724\ W+1.897219e^{-1}\ (W)^2 \tag{7}$$
$$+1.891893e^{-1}\ (W)^3-7.549036\ e^{-2}(W)^4+6.088244\ e^{-3}(W)^5$$

The local datasets: the ship measurements are averaged over two hours. In this way, they can be compared to instantaneous but spatially averaged satellite data (the SSM/I footprint is 40 km for the low frequency channels) In fact, this equivalence between time and space averages is exact only for a constant synoptic wind speed value of 5.5 m.s^{-1} (and if we assume that the ship data are homogeneous horizontally within a distance of 20 km).

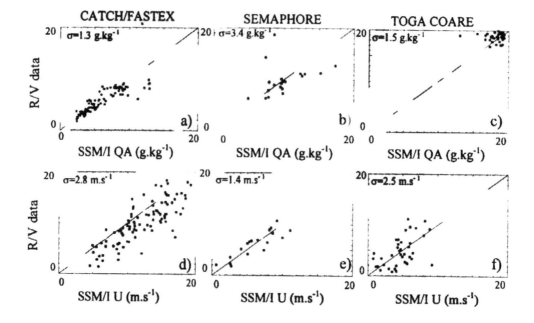

Figure 3. Comparison between the scalar wind estimated by SSM/I and the R/V reference win values (bottom). At the top, the same comparison in terms of specific humidity at ten meters.

- The CATCH/FASTEX field experiment was performed in Jan.-Feb. 1997 in North Atlantic (Lat/Lon 35-46W, 45-52N). During CATCH, there was many depressions, i.e. the meteorological conditions were very variable (see [11]). We use the surface data from the R/V « Le Suroît ». The mean U and Q_A values are 11.1 m.s^{-1} and 5.6 g.kg^{-1}. The maximum difference in time and space between the SSM/I orbits and the R/V data is ± 45 min. in time, and ± 0.5° in lat./lon. When comparing the ship U and Q_A data to the values derived from the algorithms of [4] and [12], the scatter is large (fig.3), especially for U (we notice a positive 2 m.s^{-1} bias in fig 3c).

- During the SEMAPHORE [9] experiment (Oct to Nov. 1993), the R/V « Le Suroît » was near the Azores islands (Lat/Lon 20-28W, 30-38N), in North Atlantic. The weather was mainly anticyclonic, and there was systematically a strong inversion at the top of the boundary layer. The U and Q_A mean values -calculated on the overall dataset- are close to those of GI78 (7 m.s^{-1} and 9 g.kg^{-1}). For Q_A, the RDSS is important   (3.4 g.kg^{-1}), due to the small number of points (the scatter is particularly large to the high values, as shown on fig. 3a).

- We use the surface data of the R/V « The Vickers » sailing in the equatorial pacific during TOGA-COARE [10] (92-begining of 1993, Lat/Lon 120-220W, 20S-20N). The climate was tropical, with convection and a very humid atmosphere at anytime. On this dataset, the mean U and Q_A values are 4.6 m.s^{-1} and 18.8 g.kg^{-1}. As shown on fig. 3c, the Q_A estimation error is wide (1.5 g.kg^{-1}) because large amounts of cloud liquid water are present in the atmosphere that debases the Q_A:W relationship. Moreover, the Q_A:W relationship saturates at high Q_A values. Indeed, when the humidity is saturated (i.e. Q_A is almost constant) W can vary according to the maximum altitude where humidity can be found.

These three datasets provide very different meteorological conditions, which is of interest to evaluate the methods. On one hand, the U and QA mean values on CATCH and TOGA COARE are very far from the annual mean values. On the other hand, SEMAPHORE is in a subtropical zone where the flux is a priori difficult to retrieve because of the subsidence, and of the strong temperature inversion associated with the humidity abrupt decrease, at the top of the boundary layer.

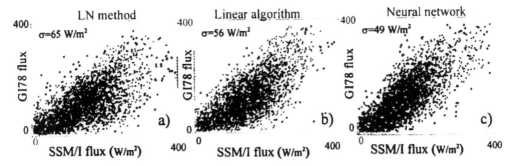

Figure 4. Comparison between different flux estimation methods on the GI78 global dataset : On the left panel, the LN method is used with algorithms whose coefficients are adjusted on the GI78 dataset.

4. COMPARISON ON THE GLOBAL DATASET

In this section, we compare the flux derived from satellite data to reference surface fluxes derived from ECMWF analyzes. We cannot obtain the exact flux retrieval error, but only a RDSS (see in section three) : actually, this deviation consists of –at least- four components: the error in the satellite data, the error due to the retrieval method, the error in the surface reference data and the error linked to the different sampling of data.

The global dataset is used as a training dataset for the three flux estimation methods. As there is a lack of representation of the high values in GI78, we replicate them in the dataset, when building algorithms and training neural networks on GI78.

LN method: we adjust the coefficients of the U algorithm to GI78, since it must be a training dataset for all the methods to be compared. In this purpose, we build a U retrieval algorithm on GI78, with the same input parameters as in [4]. The same way, we build a QA retrieval algorithm with the following parameters which are sensitive to QA (as shown in [8]). We preferred tó build a new QA algorithm than adjusting the coefficients of the [12] W retrieval algorithm and the Q_A:W relationship of [3], since it gives better results. Therefore, equation (1) is used to calculate the LN flux values.

The RDSS is 65 W/m² with respect to the GI78 flux (fig. 4a). The reason for such scatter is that the RDSS for Q_A and U are increased when performing the U.ΔQ product in equation (1); it is the major problem of the LN method. Note that, if we take the original coefficients of [4] and [12], the error is 72 W/m².

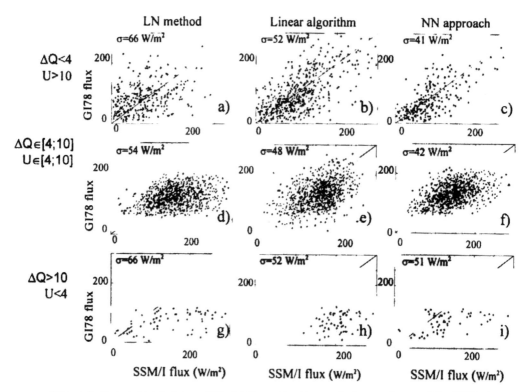

Figure 5. Comparison between different flux estimation methods on subsets of the GI78 global dataset (the selection criteria are U and ΔQ): on the left figure, the LN is used with algorithms whose coefficients are adjusted on the GI78 dataset.

Regression algorithm: when applying this method to GI78, we obtain a flux RDSS of 56 W/m^2 (fig. 4b). This discrepancy is a bit lower than with the LN method, especially to the strong flux values. On a dataset with no replication of the strong flux values, [8] show that the algorithm underestimates the strong fluxes because of the linearization of the U.Q$_A$ and U.Q$_S$ products, and that the RDSS is small at the intermediate flux range. In Fig. 4b, the behavior is not the same: we observe a large dispersion since we use an important replication of the high flux values in GI78, in order to constrain the high flux to fit the GI78 flux for the three methods. Therefore, the non-linearity of the regression algorithm is converted into a large scatter in the intermediate flux range, whereas the high flux values are estimated correctly.

Neural network inversion: we use a backpropagation network where each neuron contains a log-sigmoïd transfer function. We use three layers only, in order that the weights of the NN are less dependent of the training dataset. There are five neurons on the central layer. On fig. 4c, we show that the deviation is lower than with the other methods (49 W/m^2).

5. ERROR ANALYSIS AND VALIDATION ON LOCAL DATASETS

Principle: in this section, the flux error is analyzed as a function of U and ΔQ ranges (these are two flux parameters of equation 1). Each (U, ΔQ) range is named a class in the following. The flux RDSS is different according to the class selected. Actually, the partial derivative of LE with respect to U (resp. to ΔQ) may be neglected in some cases. Besides, the TB may be more sensitive to U or Q$_A$.

Let us differentiate equation (1), we obtain: $dL_E = \rho\ C_E\ Lv\ (\Delta Q.dU + U.d\Delta Q)$. If $d\Delta Q = dU$, then $\partial L_E / \partial U$ (resp. $\partial L_E / \partial \Delta Q$) is negligible when $U > \Delta Q$ (resp. $U < \Delta Q$); i.e. the sensitivity of L$_E$ to ΔQ is greater if ΔQ is low.

The sensitivity of the TB to U and Q$_A$ is different according to the studied class. For instance, the U RDSS observed on GI78 is more important in strong wind conditions (but not negligible at low wind speeds, see on fig. 2b). The RDSS is smaller at low Q$_A$ values since there are few amounts of columnar water vapor in this case; as most of the water vapor is located in the low atmosphere, the Q$_A$:W relationship is reliable. On the contrary, the Q$_A$:W relationship saturates at high Q$_A$ values (see section 3). At the intermediate Q$_A$ range, the scatter is very large because all types of humidity profile are mixed.

Comparison of the methods on subsets of GI78: we extract three subsets from GI78, that correspond to: two extreme classes (high U, low ΔQ) and (low U, high ΔQ), and the class the most represented in GI78: (intermediate U, intermediate ΔQ).

- (U>10 m.s^{-1}, ΔQ<4 g.kg^{-1}): as shown on fig. 5a, the scatter is large with the LN method. The NN provide a small RDSS (41 W/m^2). In fact, the microwaves are very sensitive to the flux for this class: $\partial L_E / \partial U$ is negligible and the SSM/I is very sensitive to the low ΔQ values (i.e. low Q$_A$ values). Even so, the large wind RDSS (especially in this strong wind case) debases the flux estimation. In this particular case, we might retrieve L$_E$ values only by using Q$_A$ and a satellite derived mean U value (to get the order of magnitude of the wind speed).

- (U$\in$[4;10], ΔQ$\in$[4;10]): the NN is the most accurate (fig. 5e): the RDSS is 42 W/m^2. In [8], we show that the algorithm should be very efficient on this class, but this

D. Bourras et al.

behavior is not marked on figure 5e because of the replication of the high flux (see above). The LN method is not accurate, since the Q_A and U RDSS are important.

- ($\Delta Q > 10, U < 4$) : whatever the method we use, the correlation is low, as expected in the last section. The reasons are the saturation of the Q_A:W relationship and the large sensitivity of L_E to the wind (whose RDSS is high).

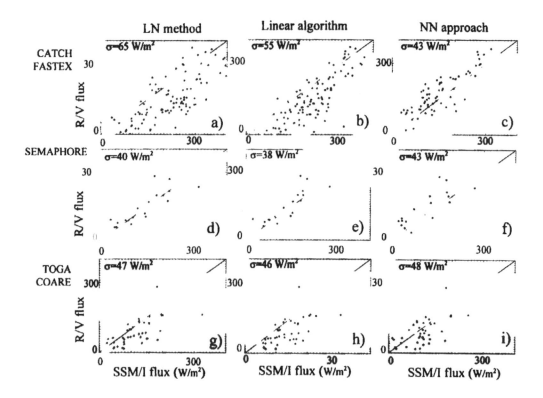

Figure 6. Comparison between different flux estimation methods on the local datasets; on the left figure, the LN is used with algorithms whose coefficients are adjusted on the GI78 dataset.

Comparison on the local datasets: the three local datasets are nearly representative of the classes studied above (i.e. the mean values of parameters such as U are close). However, we cannot find exactly the same results on both types of datasets since the time and space variability in our local datasets is smaller than in the corresponding subsets of GI78.

- CATCH/FASTEX: the RDSS is very low when applying the NN (figs. 6a), whereas we observe a flux overestimation for the LN method and the regression algorithm, because of the overestimation of U retrievals (see on fig. 3). This comparison shows that only the NN takes into account correctly the high sensitivity of the microwaves to the flux on this class, as suggested above (this method provides a greater weight to the sensitivity to QA).

- SEMAPHORE: The low scatter found for the three methods is explained, since the humidity profile is not very variable on this dataset. The algorithm gives the best results on this dataset: 38 W/m².

- TOGA COARE: the U and Q_A algorithms whose coefficients fit the GI78 dataset saturate on TOGA data. Indeed, the ECMWF analyses underestimates the QA values in tropical zones, as shown on fig. 2a. In spite of the large bias due to the Q_A underestimation in the ECMWF model, the RDSS is quite close for the three methods.

On figure 7, we show the scatterplots obtained if use the original algorithms (5),(6) and (7) (i.e. not adjusted on GI78) to apply the LN method. As shown on figure 7a, the RDSS is lower than in figure 6a, because the Goodberlet wind provide better estimates than the wind algorithm whose coefficients are adjusted on GI78 (note that the bias of figure 3d has no strong impact here in part because of the distribution of the cluster of points). On the other hand, there is no more bias in fig. 7c (unlike in fig. 6g), which supports that the GI78 QA fields are biased at high humidities (see above).

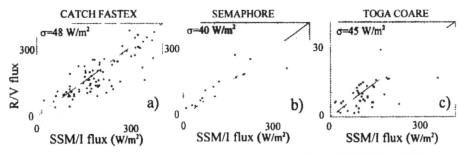

Figure 7. Application of the LN method with the algorithms of [3], [4] and [12].

6. CONCLUSION

We have compared three methods to infer latent heat flux measurements from SSM/I measurements. The three methods have been adjusted on a training dataset which groups ECMWF analysis and SSM/I brightness temperatures. The methods have been applied to particular ranges of wind and vertical humidity gradient of the global dataset, and to the datasets of three different experiments.

We show that the instantaneous flux retrieval error (or RDSS) is about 50 W/m², for a global and three local datasets. The error is 38 W/m² at the least, which is consistent with the error found by [14]. The algorithm underestimates the strong fluxes because of the linearization of the radiative transfer equation. We show that the neural network approach is of potential interest to retrieve the latent heat flux. For the most part of our datasets, this approach provides a slightly lower RDSS than the LN method does.

As we analyze the flux estimation error according to particular ranges of U and ΔQ, we show that the behavior of the methods is consistent whatever the type of dataset: subset of the global dataset or independent local dataset. In particular, we show that the neural network approach provides a low RDSS both on the "(strong U, low ΔQ) class" of the global dataset, and on the CATCH dataset.

Despite the fact that the neural network approach seem to be a good choice to retrieve the latent heat flux in several cases, the impact of the inversion technique choice is of lower importance as compared to geophysical noise and problems in validation strategy. Indeed, the gain in standard deviation of error is never greater than 10 W/m² with the neural network approach on local datasets.

Nevertheless, the comparison on the local datasets suggests that the latent heat flux can be retrieved correctly on particular climate zones. Therefore, it is of interest to study the spatial variations of the latent heat flux from space, see -for instance- the relationship between the flux and the horizontal gradients of SST at mesoscale. A method such as the neural network approach could help to take into account the spatial variations of the flux in the classical vertical inversion methods.

ACKNOWLEDGMENTS

This study was carried out at CETP/UVSQ under contract with DGA/DRET. The satellite data were provided by NOAA, the GCM output fields by the ECMWF and the TOGA/COARE data by COAPS (Florida State University). The authors are grateful to S. Thiria from LODYC/Jussieu University and to the reviewers for their helpful comments about this study.

REFERENCES

1. Businger, J. A., J. C. Wyngaard, and Y. Izumi, Flux profile relationships in the atmospheric surface layer, *J. Atmos. Sci.*, 28, 181-189, 1971.
2. Weare, B. C., Uncertainties in estimates of surface heat fluxes derived from marine reports over the tropical and subtropical oceans, Tellus, 41A, 357-370, 1989.
3. Liu, W. T., and P. P. Niiler, Determination of monthly mean humidity in the atmospheric surface layer over oceans from satellite data, J. Phys. Oceanogr., 14, 1451-1457, 1984.
4. Goodberlet, M. A., C. T. Swift, J. C. Wilkerson, Ocean surface wind speed from the Special Sensor Microwave/Imager (SSM/I), IEEE Trans. Geosci. Remote Sens., 28, 5, 823-828, 1990.
5. Njoku, E. G., Satellite remote sensing of sea surface temperature, G. C. Geernaert and W. J. Plant, 2, Kluwer Academic Publishers. 311-338, 1990.
6. Liu, W. T., Remote sensing of surface turbulent heat flux, Surface Waves and Fluxes, G. C. Geernaert and W. J. Plant, 2, Kluwer Academic Publishers, 293-309, 1990.
7. Bourras, D., and L. Eymard, Direct retrieval of the surface latent heat flux from satellite data; comparison to Liu et al [1984] method on data of the SEMAPHORE experiment, 9th conference on Satellite Meteorology and Oceanography, AMS, France, Vol. 1, 278-282, 1998.
8. Bourras, D., and L. Eymard, Physical insights in methods for retrieving the instantaneous latent heat flux over oceans from SSM/I measurements, submitted to the JGR Ocean, march 1999.
9. Eymard, L., et al., Study of the air-sea interactions at the mesoscale: The SEMAPHORE experiment, Ann. Geophysicae, 14, 986-1015, 1996.
10. Schulz, J., J. Meywerk, S. Ewald, and P. Schlüessel, Evaluation of satellite derived latent heat fluxes, J. Climate, 10, 2782-2795, 1997.
11. Eymard, L., et al., Surface fluxes in the North Atlantic current during the CATCH/FASTEX experiment, submitted to the QJRMS, 1998.
12. Alishouse, J. C., S. Snyder, J. Vongsathorn, and R. R. Ferraro, Determination of oceanic total precipitable water from the SSM/I, IEEE Trans. Geosci. Remote Sens., 28, 811-822, 1990.
13. Gautier, C., C. Jones, and P. Peterson, A new satellite method to compute monthly ocean air temperature, specific humidity, and latent heat flux, 9th AMS conference on satellite meteorology and Oceanography, AMS, Paris, Vol. 1, 118-120, 1998.
14. Esbensen, S. K., D. B. Chelton, D. Vickers, and J. Sun, An analysis of errors in Special Sensor Microwave Imager evaporation estimates over the global oceans, J. Geophys. Res., 98, C4, 7081-7101, 1993.

1. Remote sensing of the Earth's surface

1.2 Land surface

Microw. Radiomet. Remote Sens. Earth's Surf. Atmosphere, pp. 59–69
P. Pampaloni and S. Paloscia (Eds)

Airborne passive microwave measurements on agricultural fields

G.MACELLONI[1], S.PALOSCIA[1], P. PAMPALONI[1], R. RUISI[1], C. SUSINI[1]
and J.P. WIGNERON[2]

IROE-CNR, Firenze,Italy

Abstract - This paper presents an overview of the results obtained in May 1997 on the agricultural site "Les Alpilles " (France), which is a test-site for the EC ReSeDA project. The main objective of this project was the monitoring of soil and vegetation processes using multisensor and multitemporal observations. The radiometric measurements were carried out in the framework of the STAAARTE program, using the French aircraft ARAT equipped with two channels (6.8 and 10 GHz) of the IROE (Instrument for Radio Observation of the Earth). Two different flights, at incidence angles equal to 20 and 40 degrees, were carried out on May 1st and 25th to observe the development of agricultural crops on several test fields. Crop and soil properties were measured on the ground by the group of INRA -Bioclimatology (Avignon). The obtained results confirm a good sensitivity of microwave emission to crop types and biophysical characteristics. A classification algorithm , developed on the basis of data collected in a small training area, made possible the separation of six vegetation classes of the whole area. The surface soil moisture was estimated using the normalized brightness temperature measured at C-band and low incidence angle, whereas the plant water content and the leaf area index of wheat were retrieved from the polarization index measured at X-band and high incidence angle, using a semi-empirical model validated with ground-based measurements carried out in past years.

1. INTRODUCTION

The sensitivity of microwave emission to soil moisture and vegetation biomass has been proved in several experimental and theoretical studies [1-6]. For a vegetation-covered soil, the sensitivity to soil moisture is higher at the lower frequencies (around 1-3 GHz), whereas, as the frequency increases, the contribution from vegetation to the radiation is significant and the emission can be related to vegetation type and plant water content.

A remote sensing airborne campaign with microwave radiometers was carried out in May 1997 to observe different growth stages and soil moisture conditions of several fields on the agricultural test area of Les Alpilles. The airborne campaign was organized within the framework of the STAAARTE (Scientific Training and Access in Aircraft for Atmospheric research Throughout Europe) Project. This gave us the opportunity of free flying IROE sensors onboard the French aircraft ARAT, and provided all the necessary technical assistance. The campaign was also a part of the EC Project ReSeDA (Assimilation of multi-sensor & multitemporal Remote Sensing Data to monitor vegetation and soil functioning). The main objective of the ReSeDA project was to monitor soil and vegetation processes using multi-sensor and multitemporal observations. The role of IROE was to evaluate the biophysical characteristics of different agricultural crops (i.e. moisture, biomass) by using microwave sensors.

This paper is a report on the results obtained from the airborne measurements.

2. THE MEASUREMENTS

The IROE sensor consisted of two microwave channels at 6.8 and 10 GHz in H and V polarizations, a thermal infrared radiometer (8-14 μm) for estimating surface temperature, and a TV camera used for ground reference. The microwave instruments were self-calibrating radiometers with an internal calibrator based on two loads at different temperatures (250 ± 0.2 K and 370 ± 0.2 K). The main characteristics of the sensors are summarized in Table I. Radiometric digital and analog signals were acquired by a PC with a dedicated program for data processing and storage on disk. The instruments were installed on ARAT in "side looking" configuration, and operated at two incidence angles: 20° and 40° (Fig.1).

The test site was a flat area mainly cultivated with corn, wheat, alfalfa and sunflower; a few bare soils were also present at the time of the flights. Crop and soil properties were investigated with an intensive ground measurement campaign, carried out by the INRA - Avignon group.

Table I- Characteristics of the microwave radiometers

frequency/ wavelength	6.8 GHz	10 GHz	8-14 μm
Polarization	Horizontal and Vertical	Horizontal and Vertical	
IFOV (degs)	16	16	2
Sensitivity	0.3 K	0.3 K	0.1 K
Accuracy	1 K	1 K	0.5 K

Two flights were carried out, on May 1st and 25th, at suitable altitude (150 m) and speed (150 knots) for obtaining reasonable spatial (<100m) and temporal resolutions. Six flight lines were chosen by INRA, to make possible the comparison of radiometric data with the data collected by other active sensors for the ReSeDA Project. These lines guaranteed complete coverage of the 12 'calibration' and 'validation' fields, where intensive ground data collection was made during the entire plant growth cycle. For each flight line, two passes at two different incidence angles (θ=20° and 40°) were carried out, in order to point out different emission characteristics of the surfaces observed. Each flight line was followed using the navigation system of the aircraft, while the GPS system recorded the

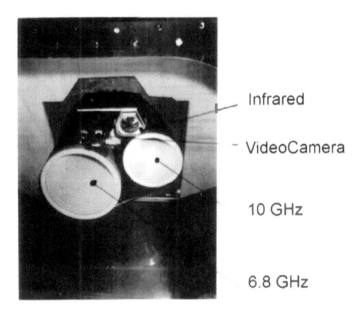

Figure 1. The IROE sensors installed on the ARAT

route. The first flight was disturbed by a strong wind, and some data were lost due to a failure in the data acquisition system. The second flight gave excellent results. Data processing was performed according to the following steps:
- calibration control of the brightness temperature data,
- retrieval of flight lines from GPS data,
- localization of antenna footprint by correcting data for the incidence angle and the systematic GPS error and identification of the brightness temperature of each field.

3. EXPERIMENTAL RESULTS

3.1 Crop discrimination
As expected from previous results obtained on the same crops in a different development stage [7], the normalized   brightness temperature Tn, obtained by the ratio between microwave and infrared brightness temperatures. was found to be very sensitive to the type of surface cover at both frequencies, and could be used to separate fields and to discriminate crop types.

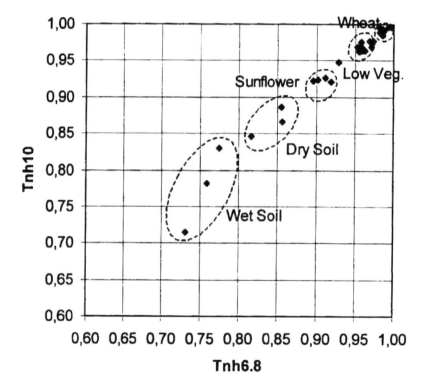

Figure 2. The normalized temperature at 10 GHz as a function of normalized temperature at 6.8 GHz ($\theta = 40°$, H pol.)

By representing Tn data collected at two frequencies, polarizations and incidence angles in bi-dimensional diagrams, we can isolate clusters for the purpose of establishing useful criteria for separating crops. For example, we see in Fig. 2 that there is a significant correlation between C- and X- band emission, at $\theta = 40°$ incidence angle and H polarization, and that there is a correspondence between crop type and Tn range. This means that even a single frequency/polarization system could allow the discrimination of a few surface categories. However, a more detailed study has shown that the use of the full set of data appreciably improves the discrimination accuracy. Using a set of two-frequency, two-incidence angle ($\theta = 20°$ and 40°), H polarized normalized temperatures, collected on a few test fields, a simple algorithm for separating six vegetation classes (dry bare soils, wet bare soils, low vegetated fields, sunflower, well developed wheat and alfalfa) has been implemented based on the measurements on training area containing 10 fields. A block diagram of the algorithm is represented in Fig. 3.The map of correctly classified fields (80%) over the whole area containing more than 50 fields is represented in Fig. 4.

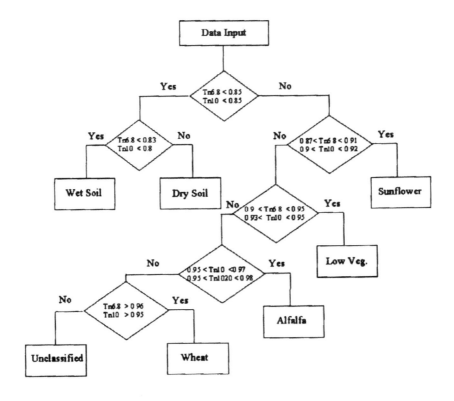

Figure 3. Block diagram of the classification algorithm

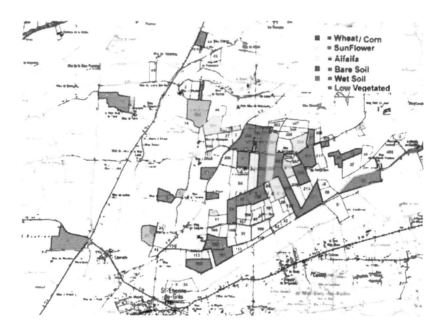

■ = Wheat/ Corn
= SunFlower
= Aifaifa
■ = Bare Soil
■ = Wet Soil
= Low Vegetated

Figure 4. Crop classification map

3.2 Retrieval of vegetation biomass

Previous investigations had pointed out that the sensitivity of Tb to vegetation biomass depends on crop type and observation parameters (frequency, polarization, incidence angle) [2], so that a simple observation at a single frequency single polarization can be of small relevance. A more significant parameter for detecting vegetation biomass was found to be the polarization index PI = (Tbv-Tbh) /½ (Tbv+Tbh), which has a high value on moist bare soil and decreases as the vegetation grows. Experimental results conducted on various crop types and theoretical analyses suggested relating PI to the leaf area index (LAI) of crop according to the following equation [8]:

$$PI(LAI) = PI(0,\mu) \ e^{(-LAI / y\mu\sqrt{\lambda})} \tag{1}$$

where: $PI(0,\mu)$ is the PI of bare soil, $\mu = Cos \ \theta$ is the observation direction, y is a crop factor [8], and λ is the wavelength. This equation can be easily inverted to obtain LAI from PI.

$$LAI = y \ \mu \ \sqrt{\lambda} \ \ln [\ PI(0,\mu) \ /PI(LAI)] \tag{2}$$

A comparison of the model with experimental data collected at 10 GHz on wheat fields is given in Fig. 5, while Fig. 6 represents the relationship between measured and retrieved LAI.

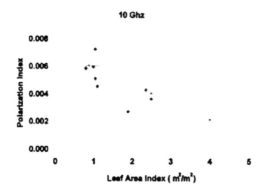

Figure 5. The polarization index as a function of LAI

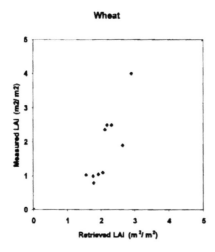

Figure 6. Measured versus retrieved LAI

3.3 Retrieval of soil moisture from measurements at low incidence angle

At the flight dates, the agricultural area was rather dry and the surface soil moisture was very low (<10%) except for the irrigated fields. These two levels were very well identified by microwave radiometers, especially at 6.8 Ghz and $\theta = 20°$. Since ground data of irrigated fields, collected simultaneously with the radiometric measurements, were not available, we derived the SMC of these wet bare fields from Tb at 6.8 GHz, using the following empirical model [6].

$$SMC \ (\%) = 167 \cdot (1 - \ Tn \ ) \qquad\qquad (3)$$

where Tn is the normalized brightness temperature measured at 6.8 GHz and $\theta = 20°$.

The retrieved soil moisture values of irrigated fields were close to 40%. The validity of this approach is confirmed by the diagram of Fig. 7, which compares the measured normalized temperature at 10 GHz as a function of SMC retrieved from Eq.3.with the following empirical model obtained from other data [9].

$$Tn \ (10 \ GHz) = \ 0.95 - 0.0041 \cdot SMC \ (\%) \qquad\qquad (4)$$

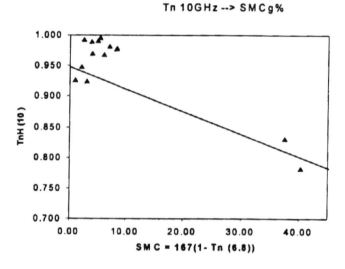

Tn 10GHz --> SMCg%

Figure 7. Normalized temperature at 10 Ghz as a function of SMC estimated from Tn at 6.8 Ghz through Eq 3 . The continuous line represent a model obtained in [9].

A SMC map of the investigated fields, obtained from measurements at 6.8 GHz by means of equation (3) is given in Fig. 8.

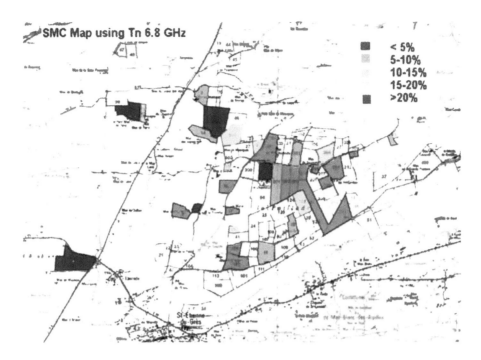

Figure 8. Soil moisture map of the investigated fields obtained from measurements at 6.8 GHz by means of equation 3 in the text.

3.4 Retrieval of soil moisture from measurements at high incidence angle

When measurements at low incidence angle are not available, such as in the case of satellite observations, the effect of vegetation is important and strongly reduces the sensitivity of Tn to SMC. In this case an alternative approach is based on the use of polarization indexes measured at 6.8 and 10 GHz. Experimental investigations have shown that, for measurements at 6.8 GHz and $\theta = 40°$, the polarization index can be related linearly to SMC. Since vegetation cover affects the sensitivity of microwave emission to SMC, the retrieval of the latter parameters in vegetated fields can be improved by evaluating the biomass through the polarization index at 10 GHz.

The relationship: between PI at 6.8 GHz and SMC can be written as follows:

$$PI_{(6.8)} = N \cdot SMC \qquad (5)$$

where N depends on $PI_{(10)}$ according to the following relationship:

$$N = 0.092 \, Ln \, PI_{(10)} \qquad (6)$$

A comparison of SMC data estimated from Equation 5 and 6 with SMC retrieved from equation 3 and assumed as reference value is shown in Figure 9.

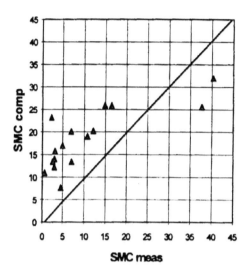

Figure 9. SMC pomtuted from Polarizion Index es through Eq. 5 and 6 as function of SMC retrieved from Equation 3

CONCLUSIONS

The airborne campaign made it possible to collect passive microwave remote sensing data on agricultural fields on two dates. These data enabled us to separate six classes of surface cover over the entire area. The sensitivity to soil moisture and vegetation biomass, already established in previous experiments, has been confirmed. The suggested retrieval algorithms seem able to give significant results. However, further investigations on a wider scale are needed in order to evaluate the operational capability of the method.

Acknowledgments
The airborne campaign was supported by the EC within the framework of the STAAARTE Project. We wish to express our gratitude to Mr. Guy Penazzi ARAT Facilitator, and the staffs of INSU and IGN (France) for their capable assistance.

The ground measurements on the test site were carried out by INRA - Bioclimatology (Avignon-France) within the framework of the EC ReSeDA Project. We sincerely thank the coordinator Frederic Baret and his colleagues.

REFERENCES

1. F. T. Ulaby, M. Razani and C. Dobson, "Effects of vegetation cover on the microwave radiometric sensitivity to soil moisture" *IEEE Trans. Geosci. Remote Sensing*, vol 21, pp. 51-61, 1983.

2. P. Pampaloni and S. Paloscia, "Experimental relationships between microwave emission and vegetation features," *Int. J. Remote Sensing*, Vol. 6, pp. 315-323, 1985.

3. R. Huppi, E. Stotzer, and E. Schanda, "Calibrated microwave signature measurements of soil and wheat, in Proc. 3rd Intern. *Coll. on Spectral Signatures of Objects in Remote Sensing*, Les Arcs, France. ESA SP-247, pp. 351-355, 1985.

4. S. Paloscia and P. Pampaloni, "Microwave vegetation indexes for detecting biomass and water conditions of agricultural crops," *Remote Sens. Environ.*, **40**, pp. 15-26, 1992.

5. J. P. Wigneron, J. C. Calvet, Y. Kerr, A. Chanzy, "Microwave emission of Vegetation: Sensitivity to Leaf characteristics," *IEEE Trans. Geosci. Remote Sensing*, vol 31, pp. 716-726, 1993.

6. T.J. Jackson, T. Schmugge and P. O'Neill, "Passive Microwave Remote Sensing of Soil Moisture from an Aircraft Platform", *Remote Sens. Environ.* 14: 135-151,1984

7. Ferrazzoli, P., L. Guerriero, S. Paloscia and P. Pampaloni, 1995, "Modeling X and Ka band emission from leafy vegetation," Journal of Electromagnetic Waves and Applications, vol 9, N 3, pp 393-406.

8. S. Paloscia and P. Pampaloni "Microwave polarization index for monitoring vegetation growth" *IEEE Trans. Geosci. Remote Sensing*, GE-26, 5, pp.617-621,1988

9. Pampaloni P., S. Paloscia, L. Chiarantini. 1985 b. "Contribution of passive microwave remote sensing in soil moisture and evapotranspiration measurements". Proceedings of Conference on Parameterization of Land Surface Characteristics, Roma, 2-6 December 1985. ESA SP-248, p. 327-331.

Microw. Radiomet. Remote Sens. Earth's Surf. Atmosphere, pp. 71–80
P. Pampaloni and S. Paloscia (Eds)
© VSP 2000

A combined two-scale model for microwave emissivities of land surfaces

KATJA PAAPE, RALF BENNARTZ and JÜRGEN FISCHER

Institute for Space Sciences, Free University Berlin, Fabeckstr. 69, 14195 Berlin, Germany

Abstract – This paper presents a model to simulate the emissivity behaviour of land surfaces by considering the effects of surface roughness at two different scales, which are related to the detector wavelength. Besides other parameters surface roughness is important to determine surface emissivity. Surface roughness in the scale of the detecting wavelength is responsible for the surface reflectivity. Roughness heights at scales larger than the detector wavelength cause changes in the hemispherical distribution of emitted and reflected energy and an energy exchange between both polarizations. Thus, dependent on the wavelength under consideration, surface roughness is split into a small-scale and a large-scale part, cutting off between the two scales at the double value of the detector wavelength. Based on the small-scale roughness the emissivity can be calculated as a function of zenith angle. From the large-scale roughness facets are calculated. Based on those facets a facet model is used to account for the changes in the zenith dependence of the emissivity, introducing a dependence on azimuth and an exchange between the two polarisations. First modelling results show a better reproduction of observed emissivities, especially for higher zenith angles. The results further indicate that the emissivity and the brightness temperature of significantly structured surfaces show a strong dependence on the azimuth angle. Available data sets and own experiments allow a model verification for a broad frequency range from 2 to 157 GHz.

1. INTRODUCTION

The emissivity of land surfaces is a key parameter for the interpretation of satellite microwave data over land. Especially for the derivation of atmospheric parameters it is important to consider the influence of the underlying surface on the measured satellite signal. To be able to evaluate this effect it is necessary to use models for the simulation of emissivity characteristics of different land surface types. Especially sensors suitable for retrieval of atmospheric parameters like precipitation provide spectral bands within a broad range of frequency up to 160 GHz. So models should be designed to cover the range from 1 to 160 GHz, that they can be used to simulate emissivities at frequencies provided by most satellite sensors.

Many models published so far, especially those based on a semi-empirical or empirical parameterisation of surface roughness effects, show significant differences compared to ground based measured emissivities [1,2]. A common problem is an inaccurate representation of the emissivity dependence on the zenith angle as well as on frequency.

In case of vegetation covered surfaces, the structure and the water content of the vegetation dominate the emissivity characteristics. The influence of vegetation compared to the influence of features of the underlying soil depends on the vegetation cover fraction as well as on frequency and observation angle [3]. Considering bare soils, surface emissivity at lower frequencies mainly depends on soil moisture content and is dominated by surface roughness characteristics at higher frequencies [4]. An uncertainty in the modeling of emissivity is the way of treating the effect of surface roughness. Obviously the impact of surface roughness on the reflection characteristics of a surface depends on the detector frequency, because mainly roughness heights in the range of the wavelength corresponding to the detector frequency directly determine reflectivity and emissivity of the surface. Roughness heights at large scales cause a change in the angular dependence of the emissivity and further an exchange between the energy amounts of horizontal and vertical polarisation.

2. THE TWO-SCALE MODEL

The model presented here accounts for two scales of surface roughness which both determine the surface emissivity. To calculate the parts contributing to each scale, the surface roughness profile or relief is split up into a small scale and a large scale part using the Fast Fourier Transformation (FFT). In case of profiles a one-dimensional FFT, in case of reliefs a two-dimensional FFT is applied. From the resulting power spectrum spectrum all values lower (larger) than a certain cut-off value are removed to obtain the large (small) scale part of the roughness. After the inverse FFT two resulting reliefs or profiles are obtained that contain the large scale and the small scale roughness respectively. As the definition of the scales depends on the observer frequency, the wavelength corresponding to the detector frequency determines the cut-off value. In the results shown here the double value of the considered wavelength has been chosen as cut-off value. This value has been found empirically by comparing model results based on the small-scale roughness (calculated for different cut-off values) to ground-based measurements.

From the large-scale roughness facets are calculated, where the horizontal size of each facet again is defined by the cut-off value. The facets are further characterised by a slope angle and an azimuth angle.

To summarise, the large-scale roughness is used to calculate facets. Then the emissivity is calculated for the half-space of each facet, the orientation of the half space is defined by the orientation (slope and azimuth) of the facet. As input parameter for the emissivity model the small-scale roughness is used. The combination of the emissivity model with the facet model is schematically shown in Figure 1.

Surface emissivity ε for each facet may be defined by surface reflectivity Γ assuming thermal equilibrium:

$$\varepsilon_{h,v}(\theta) = 1 - \Gamma_{h,v}(\theta) \tag{1}$$

$$\Gamma_h(\theta) = \frac{1}{4\pi \cos(\theta)} \int_0^{2\pi} \int_0^{\pi/2} [\sigma_{hh}(\theta,\theta_S,\phi_S) + \sigma_{vh}(\theta,\theta_S,\phi_S)] \sin(\theta_S) d\theta_S d\phi_S \tag{2}$$

In the above equation surface reflectivity Γ is defined for horizontal polarisation (indicated by index h), but can be defined for vertical polarisation (v) respectively by exchanging the indices for the bistatic scattering coefficients σ. This definition assumes that all energy which is not scattered from the surface is absorbed and emitted. The bistatic scattering coefficients depend on the incident zenith angle θ, the scattering zenith angle θ_S and the scattering azimuth ϕ_S. Two published emissivity models are used to calculate the reflectivity for each facet. One is a physical optics model which is known as the scalar approximation of the Kirchhoff model [5]. This model calculates the reflectivity corresponding to Equation (2). The second one is a model, which accounts for the effect of surface roughness using a semi-empirical method developed by Choudhury et al. [6] and Wang and Choudhury [7]. This method calculates the reflectivity of a rough surface from the Fresnel reflectivity using an empirical relationship between the reflectivity of a smooth and a rough surface. The model of Dobson et al. [8] has been used to calculate the soil dielectric properties, based on the volumetric soil moisture as input parameter. By definition, the emissivity $\varepsilon(\theta)$ is a function of the local zenith angle θ of each facet.

In a next step the outgoing brightness temperature $T_B{}^L(\theta)$ is calculated for each facet according to the formulation for $T_B{}^L$ given in Figure 1. The outgoing surface brightness temperature consists of an emission and a reflection term. It is calculated in the local co-ordinate system of each facet, indicated by an *L*. If a facet is now seen from an external observer located in a global co-ordinate system, the local brightness temperature field of the facet has to be transformed into the global reference co-ordinate system. This transformation is caried out according to the slope and the azimuth of the facet. The azimuth of a facet is defined here as the direction into which the surface slope is oriented. The orientation of the facet causes a change in the angular distribution of the facet's brightness temperature field and introduces a dependence on azimuth, when the facet is seen in global co-ordinates. In addition an exchange between horizontally and vertically polarised temperatures occurs because the local polarisation vectors h^L and v^L change to global vectors h^G and v^G, when the reference co-ordinate system changes to global co-ordinates. This transformation of the local brightness temperature field to a global co-ordinate system is performed using the facet model as described in Figure 1. The facet model includes the change in the contribution to each polarisation.

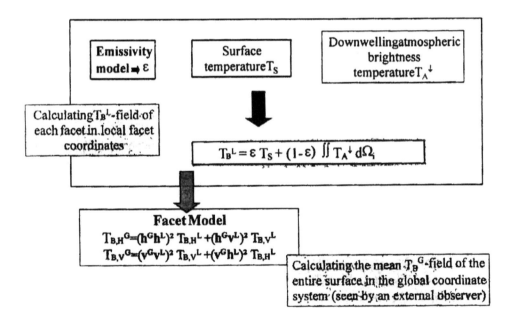

Figure 1. Schematic representation of the combination between the emissivity model and the facet model.

Applying the described method, the local brightness temperature field of each facet is transformed into a global reference co-ordinate system. To obtain the average global brightness temperature field of the entire surface, the global temperature fields of all facets are averaged.

3. MODEL RESULTS FOR DIFFERENTLY STRUCTURED SURFACES

Using the two-scale model the brightness temperature fields of different surfaces have been simulated at 23 GHz. For these simulations the semi-empirical model of Choudhury [6,7] is applied to calculate the emissivity of the facets of each surface. The following characteristics are unchanged for all surfaces:

- surface temperature   300 K
- soil moisture   20%
- sand fraction   30%
- loam fraction   20%

Four different surface reliefs have been used as model input. Three reliefs represent ideal surfaces and one represents a natural rough surface. The resulting brightness temperature fields are displayed as polar plots to visualize the dependence on both zenith and azimuth angle. In these polar plots the black dashed lines correspond to zenith angles of 30° (inner circle), 60° (middle circle) and 90° (outer circle).

Figure 2 shows that the brightness temperature field of the smooth surface still depends on the zenith angle only. As the surface has no slope, the local temperature field $T_B^L(\theta)$ equals the global temperatures $T_B^G(\theta,\phi)$ for all azimuth angles. Looking at the rough surface in Figure 3, the brightness temperatures do not show significant dependence on the azimuth angle, too, because the surface has been created assuming a normal distribution of the roughness heights. Accordingly, the slopes and azimuths of all facets are normally distributed. Compared to the smooth surface the dependence of the rough surface temperature field on zenith angle decisively changed for both vertical and horizontal polarisation. The difference which can be observed is due to the effect of the included facet model.

Further the brightness temperature field of a smooth inclined surface is simulated and displayed in Figure 4. The surface is characterised by a constant slope of 19° and an azimuth of 180°. At a viewing azimuth angle of 180° the temperature pattern for both polarisations corresponds to the one of the smooth surface, but is shifted by 20° into the direction of the surface azimuth. Looking at the surface from more sideward directions at azimuth viewing angles up to 90° and 270°, especially the pattern of vertical polarisation differs from the one of the surface without inclination. This difference is due to the effect of polarisation mixing between horizontal and vertical polarisation for viewing azimuth angles representing sideward viewing directions. This effect of polarisation mixing leads to an increase in brightness temperatures for horizontal polarisation and a decrease for vertical polarisation compared to the flat surface. The region around 0° azimuth and zenith angles higher than 71° is below the surface.

In Figure 5 the model results for a smooth row structured surface are shown. This surface consists of three rows within one metre length, the height of each row is about 20 cm. The ridges of the rows are oriented parallel to the axis between 90°-270° azimuth. The resulting brightness temperature field for horizontal polarisation shows a pattern similar to the one of the flat surface (without rows) for viewing directions parallel to the row ridges, but with increased temperature values especially at higher zenith angles. Considering vertical polarisation lower brightness temperatures compared to the flat surface can be observed for

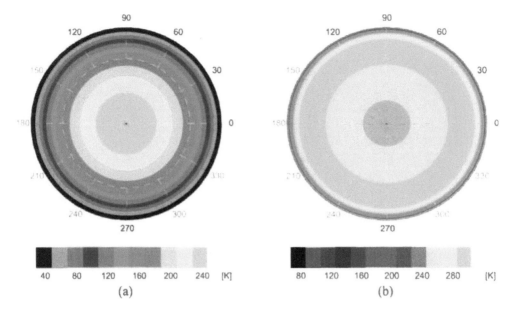

Figure 2. Horizontally (a) and vertically (b) polarised brightness temperature field of a smooth surface.

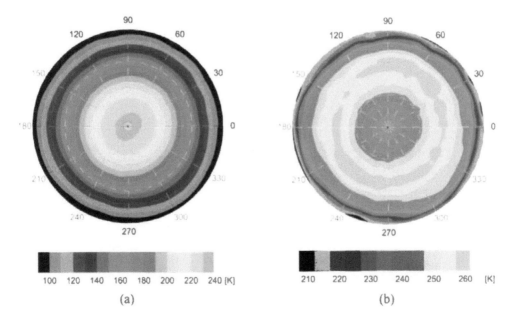

Figure 3. Horizontally (a) and vertically (b) polarised brightness temperature field of a rough surface with roughness heights up to 8cm.

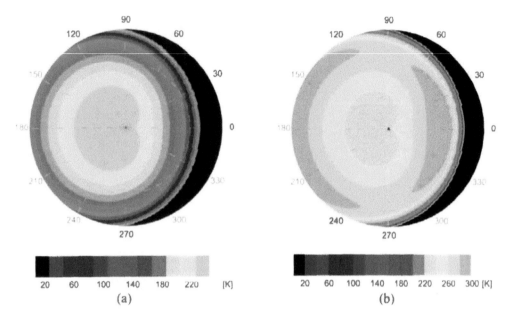

Figure 4. Horizontally (a) and vertically (b) polarised brightness temperature field of a smooth surface with a slope of 19° and an azimuth of 180°.

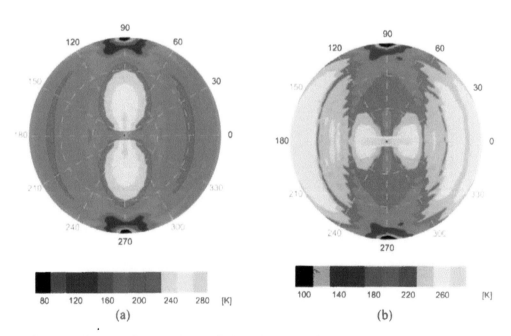

Figure 5. Horizontally (a) and vertically (b) polarised brightness temperature field of a smooth surface with row structure (ridges of the rows are oriented along the 90°-270° axis).

the viewing direction parallel to the ridges. In directions perpendicular to the ridges the patterns differ significantly from the ones of the flat surface for both polarisations. It can be observed that especially at higher zenith angles the horizontally polarised temperatures strongly increase and the vertically polarised temperatured strongly decrease. For this perpendicular viewing direction temperature differences of about 100 K occur compared to the flat surface, mainly caused by the change of the angular dependence when changing from local to global co-ordinates. This makes clear that on the one hand a significant transfer from horizontally to vertically polarised temperatures took place for azimuthal viewing directions parallel to the rows (around 90° and 270°) due to rotation of the polarisation plane. On the other hand changes in the temperature pattern for azimuthal viewing directions perpendicular to the rows (close to 180°) can be explained by the change in the angular dependence of the temperature field when both slope and azimuth of the facets are considered.

Significant modifications in the emission pattern of a surface due to a row structure have also been found by [9].

4. FIRST VALIDATION RESULTS WITH GROUND BASED AND IN-SITU MEASUREMENTS

For a first validation of the combined two-scale model results have been compared to ground based radiometer measurements from the comprehensive RASAM data set [11]. As the RASAM catalogue archives all relevant parameters besides the radiometer data, the model calculations were performed using the specific characteristics (soil moisture, texture, surface temperature, roughness profiles) of the considered field. The available roughness profiles were used to create the surface relief.

Figure 6 gives an example for comparison between ground based measured brightness temperatures at 11 GHz from the RASAM-data catalogue and model results for 11 GHz. Part (a) of Figure 6 shows ground based measured brightness temperatures taken from the RASAM data set [11]. The results of the combined two-scale model are shown in part (b) and (c) as triangles and stars.

The results presented in (b) are calculated using the semi-empirical emissivity model [6,7] for emissivity calculation. Both, the dashed and the solid curves do not include the facet model, they are computed applying the emissivity models only. The dashed lines show the result of the emissivity models only, excluding the facet model, using the small-scale roughness (standard deviation of only the roughness heights in the scale of the double wavelength) as input parameter. The solid lines give the results of the emissivity model (again without the facet model) using the classical rms-height (standard deviation of the complete relief) as input parameter instead of the small-scale roughness. Hence, the difference between the dashed and solid curves represents the variation of brightness temperature when two different roughness parameters are used as input for the emissivity models, in one case the classical rms-height (solid line) and in the other the small-scale roughness (dashed line).

The results presented in (c) are based on the physical optics (PO) model [5] for emissivity calculation. As in part (b) the different linestyles represent the results of the PO model (without facet model) using different roughness parameters as described for part (b).

The results demonstrate that the complete facet model resembles the RASAM observations quite well. The use of only the rms-height as input parameter to the emissivity models without the facets leads to emissivities close to $\varepsilon=1.0$, resulting in significantly too large brightness temperatures as can be seen from the solid lines in parts (b) and (c) of Figure 6. Using the small-scale roughness as input parameter to the emissivity models gives more reasonable brightness temperature values, but does not well represent the dependence on zenith angle when the effect of the facets is not considered (as displayed in dashed lines). Looking at the results for the combined emissivity and facet model the effect of the facet model is demonstrated quite well by a significant improvement in the dependence on zenith angle. The combined model using the Choudhury-model for emissivity calculation shows good results compared to the RASAM measurements. However, the modelled temperatures are about 17 K lower than the measured ones at a zenith angle of 0°. This difference between modelled and measured data increases with increasing zenith angles and reaches a maximum value of 25 K for both polarisations at a zenith angle of 50°. Then the difference decreases again to a value of about 15 K at 70° zenith angle. The PO-model in general gives lower emissivities than the

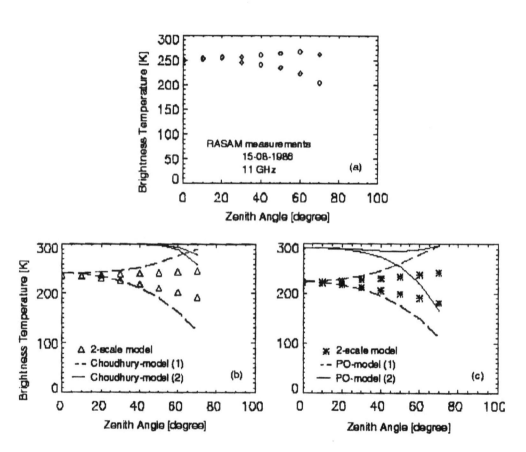

Figure 6. Model results compared to ground based measured data; (a) measurements from the RASAM data set. (b) results of the two-scale model using the semi-empirical emissivity model [6,7]. (c) results of the two-scale model using the physical optics (PO) model [5] for emissivity calculation. The dashed curves indicated by (1) represent the results of the emissivity model only (excluding the facet model) based on the small-scale roughness as input, (2) represents the emissivity based on the classical roughness parameter rms-height.

semi-empirical Choudhury-model. So the results of the combined model show larger differences compared to the RASAM-data when the PO-model is used for emissivity calculation. The differences between model results and measurements are about 30 K for a zenith angle of 0° and decrease to about 25 K at 70°. Comparing both versions of the combined model it can be stated that the difference is highest at a zenith angle of 0° and decreases with increasing zenith angle. For an angle of 70° both model versions give identical results for vertical polarisation, but still show a difference of 10 K for horizontal polarisation.

5. CONCLUSIONS

The presented results of the combined two-scale model show that considering the effect of large-scale roughness leads to a more realistic representation of the emissivity of rough surfaces. Especially for surfaces with a certain orientation as in case of a constantly inclined surface or a row structured surface with slopes in two dominant directions, the introduced dependence on azimuth angle is significant. This is important when considering emissivities at local scales (e.g. single fields) as well as when emissivities at large scales (topography) are investigated [10].

In further investigations, we will focus on the validation of the model at different frequencies. For validation different data sets are available covering a range of frequency from 2-157 GHz. One important data set is the RASAM-data catalogue (2-12 GHz) [11], which also includes roughness profiles. First results of comparison between measurements from the RASAM data set and modelled brightness temperatures show a quite good correspondance. Considering the use of the two-scale model in the interpretation of satellite data especially the much better description of emissivitiy dependence on zenith angle is important, because common microwave satellite sensors like the SSM/I or AMSU provide viewing angles up to 60°. To summarise, the combined two-scale model leads to an improvement in the representation of the emissivity dependency on zenith angle as well as to a satisfying correspondance between modelled and measured brightness temperatures.

In addition the VALEM (VAlidation of Land surface Emissivity Models) experiment has been carried out in autumn 1998 by the Free University Berlin in co-operation with the UK MetOffice. This experiment provides airborne microwave radiometer measurements (23-157 GHz) together with ground truth data (including roughness reliefs) for different test fields.

Acknowledgements
The authors wish to thank Prof. C. Mätzler (University of Bern) and Dr. U. Wegmüller (Gamma Remote Sensing) for generously providing the RASAM data set.

REFERENCES

1. S. S. Saatchi, E.G. Njoku and U. Wegmüller. "Synergism of active and passive microwave data for estimating bare soil surface moisture", in: *Passive microwave remote sensing of land-atmosphere interactions*, B.J. Choudhury et al. (eds.), VSP, Utrecht, The Netherlands (1995)
2. U. Wegmüller, C. Mätzler and E.G. Njoku. "Canopy opacity models", in: *Passive microwave remote sensing of land-atmosphere interactions*, B.J. Choudhury et al. (eds.), VSP, Utrecht, The Netherlands (1995)
3. S. Paloscia, "Microwave emission from vegetation", in: *Passive microwave remote sensing of land-atmosphere interactions*, B.J. Choudhury et al. (eds.), VSP, Utrecht, The Netherlands (1995)

4.   F.T. Ulaby, R.K. Moore and A.K. Fung. *Microwave remote Sensing, Active and Passive, Volume II, Radar remote sensing and surface scattering and emission theory*, Addison-Wesley (1982).

5.   F.T. Ulaby, R.K. Moore and A.K. Fung. *Microwave remote Sensing, Active and Passive, Volume III, From Theory to Applications*, Artech House, Dedham, MA (1986).

6.   B.J. Choudhury, T.J. Schmugge, A. Chang and R.W. Newton. "Effect of surface roughness on the microwave emission from soils", *J. Geophys. Res.*, **84**, 5699-5706 (1979).

7.   J.R. Wang and B.J. Choudhury. "Remote sensing of soil moisture content over bare field at 1.4 GHz frequency", *J. Geophys. Res.*, **86**, 5277-5282 (1981).

8.   M.C. Dobson, F.T. Ulaby, M.T. Hallikainen and M.A. El-Rayes. "Microwave dielectric behaviour of wet soil. Part II: Dielectric mixing models", *IEEE Trans. Geosci. Remote Sensing*, **23**, 35-46 (1985).

9.   J.R Wang, R.W. Newton and J.W. Rouse. "Passive microwave remote sensing of soil moisture: the effect of tilled row structure", *IEEE Trans. Geosci. Remote Sensing*, **GE-18 (4)**, 296-302 (1980).

10. C. Mätzler and A. Standley. "Relief effects for passive microwave remote sensing", submitted to *Int. J. Remote Sensing* (1998)

11. U. Wegmüller, C. Mätzler, R. Hüppi and E. Schanda. "Active and passive microwave signature catalogue on bare soil (2-12 GHz)", *IEEE Trans. Geosci. Remote Sensing*, **32**, 698-702 (1994).

Microw. Radiomet. Remote Sens. Earth's Surf. Atmosphere, pp. 81–88
P. Pampaloni and S. Paloscia (Eds)

Modelling of the electromagnetic response of geophysical media with complicated geometries: analytical and numerical approaches

ARI SIHVOLA, REENA SHARMA and JUHA AVELIN

Electromagnetics Laboratory, Helsinki University of Technology,
P.O. Box 3000, 02015 HUT, Finland.
Tel: +358-9-4512261; Fax: +358-9-4512267; E-mail: ari.sihvola@hut.fi

Abstract—The aim in this paper is to estimate the effective dielectric constant of a composite material made up of small inclusions embedded in a homogeneous matrix. Basic classical mixing rules are discussed as well as the bounds between which the effective permittivity has to stay. But for special geometries of the microstructure of the mixture, few results are available. The present contribution presents findings regarding mixtures with complicated geometries for which only numerical efforts can bring results.

1. INTRODUCTION

Microwave remote sensing of the environment is based on the interaction of electromagnetic radiation with the dielectrically random structure of natural objects, like snow, ice, vegetation canopies, and various earth layers. Because of limitations of calculational capacity, in the analysis of the radiometric signals, the whole randomness and small-scale variations in matter has to be averaged in some reasonable sense. The homogenization of the spatial dielectric variations—electromagnetic mixing formulas—is a very old topic in electromagnetics, and in general in mathematical physics. Mixing rules are recipes to calculate the effective, or macroscopic, permittivity of a heterogeneous medium, so that in a large scale the medium could be replaced by a sample of homogeneous material.

A heterogeneous medium made up of spherical inclusions in a host medium can be treated as an effective homogeneous medium with its constitutive parameters being characterized by effective permittivity and effective permeability operators (and, if necessary, also more general operators accounting for magnetoelectric coupling). Of course, an effective model disregards very much of the exact description of the material. Hence, the effective parameters can be meaningfully used in the case when the resolution power of the electromagnetic fields is smaller than the microstructure of the heterogeneous medium. In simple terms, this means that the wavelength of the incident field has to be large compared to the inclusion size and correlation length of the medium. The effective permittivity of a mixture is therefore a low-frequency concept.

In the following we assume that the heterogeneous medium to be studied is non-magnetic; its relative permeability is the same as that of the free space ($\mu_r = 1$). We also discuss only isotropic dielectric materials, and show how the effective permittivity ϵ_{eff} of such mixtures can be calculated as a function of the parameters of the host and the inclusion materials and their fractional volume. The media are considered to consist of two discrete components: the environment (host) with permittivity ϵ_e, and the inclusions (guest) which occupy a volume fraction f in the mixture that have permittivity ϵ_i.

2. CLASSICAL MIXING RULES

Perhaps the most common mixing rule is the *Maxwell Garnett* formula [1] which assumes that the inclusions are spheres, randomly located in the environment:

$$\epsilon_{\text{eff}} = \epsilon_e + 3f\epsilon_e \frac{\epsilon_i - \epsilon_e}{\epsilon_i + 2\epsilon_e - f(\epsilon_i - \epsilon_e)} \tag{1}$$

This formula is in wide use in very diverse fields of application. The beauty of the Maxwell Garnett formula is in its simple appearance. It satisfies the limiting processes for vanishing inclusion phase

$$f \to 0 \qquad \Rightarrow \qquad \epsilon_{\text{eff}} \to \epsilon_e \tag{2}$$

and vanishing background

$$f \to 1 \qquad \Rightarrow \qquad \epsilon_{\text{eff}} \to \epsilon_i \tag{3}$$

Another form of this formula is the so-called Rayleigh mixing rule:

$$\frac{\epsilon_{\text{eff}} - \epsilon_e}{\epsilon_{\text{eff}} + 2\epsilon_e} = f \frac{\epsilon_i - \epsilon_e}{\epsilon_i + 2\epsilon_e} \tag{4}$$

The perturbation expansion of the Maxwell Garnett rule gives the mixing equation for dilute mixtures ($f \ll 1$):

$$\epsilon_{\text{eff}} = \epsilon_e + 3f\epsilon_e \frac{\epsilon_i - \epsilon_e}{\epsilon_i + 2\epsilon_e} \tag{5}$$

Figure 1 shows the prediction of the Maxwell Garnett formula for different values of the dielectric contrast ϵ_i/ϵ_e. Shown is the susceptibility ratio

$$\frac{\epsilon_{\text{eff}} - \epsilon_e}{\epsilon_i - \epsilon_e}$$

which vanishes for $f = 0$ and is unity for $f = 1$, independently of the inclusion-to-background contrast. The figure shows the fact that the effective permittivity function becomes a very nonlinear function of the volume fraction for large dielectric contrasts.

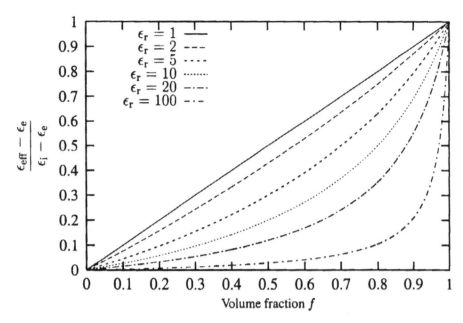

Figure 1. The susceptibility ratio $(\epsilon_{\text{eff}} - \epsilon_e)/(\epsilon_i - \epsilon_e)$ for the Maxwell Garnett prediction of the effective permittivity of a mixture with spherical inclusions of permittivity ϵ_i in a background medium of permittivity ϵ_e.

In the remote sensing community many other mixing rules have a respected place in addition to the Maxwell Garnett rule. In fact, many of these rules can be shown to be contained in a family of mixing rules which can be given with the unified formula [2]

$$\frac{\epsilon_{\text{eff}} - \epsilon_e}{\epsilon_{\text{eff}} + 2\epsilon_e + \nu(\epsilon_{\text{eff}} - \epsilon_e)} = f \frac{\epsilon_i - \epsilon_e}{\epsilon_i + 2\epsilon_e + \nu(\epsilon_{\text{eff}} - \epsilon_e)}. \tag{6}$$

This formula includes a dimensionless parameter ν. For different choices of ν, various mixing rules are recovered: for the simplest case, $\nu = 0$ gives the Maxwell Garnett rule.

But if we choose $\nu = 2$, the formula gives the so-called Polder–van Santen formula [3],

$$(1 - f)\frac{\varepsilon_e - \epsilon_{\text{eff}}}{\varepsilon_e + 2\epsilon_{\text{eff}}} + f \frac{\varepsilon_i - \epsilon_{\text{eff}}}{\varepsilon_i + 2\epsilon_{\text{eff}}} = 0 \tag{7}$$

In theoretical electromagnetics research, this formula is called as *Bruggeman formula* [4], and also the name *Böttcher formula* [5] can be found in the literature.

Furthermore, the case $\nu = 3$ gives us the Coherent potential mixing formula [6]:

$$\epsilon_{\text{eff}} = \varepsilon_e + f(\varepsilon_i - \varepsilon_e)\frac{3\epsilon_{\text{eff}}}{3\epsilon_{\text{eff}} + (1 - f)(\varepsilon_i - \varepsilon_e)} \tag{8}$$

The effective permittivity given by the above three mixing rules obtained by putting $\nu = 1$, 2, and 3 are shown in Figure 2 for a dielectric contrast of 10. It is worth noting that for dilute mixtures ($f \ll 1$), all three mixing rules, Maxwell Garnett, Polder–van Santen, and Coherent potential, predict the same results. Up to the first order in f, the formulas are the same, given by Equation (5).

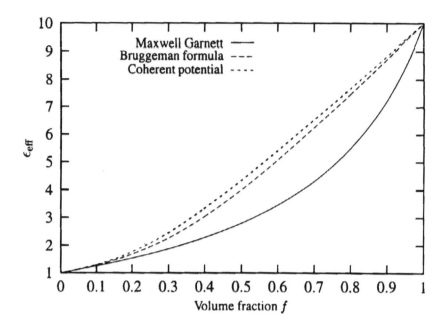

Figure 2. The effective permittivity variation with volume fraction for Maxwell Garnett, Bruggeman and Coherent potential mixing rule. The dielectric contrast is $\epsilon_i/\epsilon_e = 10$.

3. BOUNDS AND LIMITS

If it is difficult to predict exactly what is the effective permittivity for a given mixture, we still are able to say something about the bounds that the permittivity cannot theoretically exceed. For lossless materials (the permittivities are real-valued), the obvious limits are given by the permittivities of the two components themselves:

$$\min\{\epsilon_e, \epsilon_i\} \le \epsilon_{\text{eff}} \le \max\{\epsilon_e, \epsilon_i\} \tag{9}$$

But these limits can be tightened into the so-called *Wiener* bounds:

$$\epsilon_{\text{eff,max}} = f\epsilon_i + (1 - f)\epsilon_e \tag{10}$$

and

$$\epsilon_{\text{eff,min}} = \frac{\epsilon_i\epsilon_e}{f\epsilon_e + (1 - f)\epsilon_i} \tag{11}$$

In circuit theory, these two cases correspond to capacitors (conductances) that are connected in parallel or series in a circuit, respectively. In the first case, the effective permittivity is the volume average of the component permittivities, and in the latter case the inverses of the permittivities are averaged by volume basis.

But such mixtures are not statistically isotropic. For isotropic mixtures, still stricter bounds are given by the *Hashin–Shtrikman* conditions [7]. For $\epsilon_i > \epsilon_e$,

$$\epsilon_{\text{eff,max}} = \epsilon_i + \frac{1 - f}{\dfrac{1}{\epsilon_e - \epsilon_i} + \dfrac{f}{3\epsilon_i}} \tag{12}$$

and

$$\epsilon_{\text{eff},\min} = \epsilon_e + \cfrac{f}{\cfrac{1}{\epsilon_i - \epsilon_e} + \cfrac{1-f}{3\epsilon_e}} \tag{13}$$

In fact, these limits correspond exactly to the Maxwell Garnett mixing rule for the mixture and its complement (meaning that the guest and host phases are interchanged). This emphasizes the extremum character of the spherical Maxwell Garnett rule.

4. NUMERICAL APPROACHES

The models discussed in Section 2 assumed that the inclusions in the mixture were spherical. The assumption of spherical (or ellipsoidal) inclusions has been very appealing in the classical mixing theories because the response of a single particle is simple. For spheres and ellipsoids the internal field generated in a uniform external field is also uniform. Therefore the polarisability of such an inclusion is a simple formula depending on the permittivity and the volume; for a sphere the polarisability (the ratio between the dipole moment and the external field) is

$$\alpha = 3\epsilon_e V \frac{\epsilon_i - \epsilon_e}{\epsilon_i + 2\epsilon_e} \tag{14}$$

where V is the volume of the sphere.

But if the inclusion is not of such a shape, its polarisability has to be solved numerically. For example, the polarisability of another canonical shape, the cube, has not been solved in satisfactory manner, in our opinion.

One approach to this problem that we have taken is to treat the inclusion within a fully dynamical electromagnetic field, and then go to the low-frequency limit. It is well known that a sphere can be replaced by an ideal dipole (under quasistatic approximation) and its polarisability is related to its Radar cross section (RCS), σ_{rcs} by the following equation:

$$\frac{\alpha}{\epsilon_e V} = \frac{\sqrt{4\pi\sigma_{rcs}}}{k_o^2 V} \tag{15}$$

where k_o is the free space wavenumber, V is the volume and σ_{rcs} is the RCS of an electrically small dielectric sphere. We have used the same definition as in equation (15) to calculate the polarisability of the cube by first calculating its RCS. Since cube is a three-dimensional object with sharp corners, the Volume Integral Method (VIM) is chosen to calculate its RCS, even though it is computationally more intensive and time consuming compared to the surface integral equation. In this method the volume of the scatterer is discretized using cubical cells [8]. To circumvent the singularity of the Green's function the volume is divided into two parts, one without the singularity and another volume V_ϵ inside an infinitesimal sphere, in the neighbourhood of the singular point. The integration now over the entire volume can be done properly. The method followed in the code for handling of singularity of the Green's function is the same as that described in [9]. A cube of dimensions much smaller than the operating wavelength is discretized and the RCS is computed. This value of the RCS of the cube is then used to calculate its polarisability using equation (15) with V now being the volume of the cube.

Another way with which we have tried to evaluate the polarisability of a cube has been through a surface integral equation for the static potential $\phi(\mathbf{r})$ (see, for example, [10]):

$$\phi_i(\mathbf{r}) = \frac{\tau+1}{2}\phi(\mathbf{r}) + \frac{\tau-1}{4\pi}\int_S \phi(\mathbf{r}')\frac{\partial}{\partial n'}\left(\frac{1}{|\mathbf{r}-\mathbf{r}'|}\right) dS' \qquad (16)$$

where $\tau = \epsilon_i/\epsilon_e$ and the integration is over the surface of the cube. Here $\phi_i(\mathbf{r}) = -E_e z$ is the known incident potential for the uniform external field $\mathbf{E}_e = E_e\mathbf{u}_z$, where $\mathbf{u}_z$ is the unit vector in the field direction. Once the potential function $\phi(\mathbf{r})$ is known at the surface, the dipole moment of the cube comes from

$$\mathbf{p} = \alpha\mathbf{E}_e = -(\tau-1)\epsilon_e\int_S \phi(\mathbf{r})\mathbf{u}_n\, dS \qquad (17)$$

which gives us the desired polarisability α.

The polarisability was evaluated for different dielectric constants of the cube and the results were compared with that given by the surface integral equation. Table 1 gives the comparison of values of $\alpha/(V\epsilon_e)$ obtained using the surface integral method and the volume integral method. The percentage deviations are also shown in the table. From the table it can be seen that the results for the polarisability of a cube are quite accurate as they almost give the same values by using two different approaches.

τ	$\frac{\alpha}{V\epsilon_e}$ (16)	$\frac{\alpha}{V\epsilon_e}$ from VIM	Percentage Error
2.0000	0.7602	0.7610	0.1046
2.5000	1.0236	1.0245	0.0871
5.0000	1.8295	1.8298	0.0186
7.5000	2.2501	2.2503	0.0088
10.0000	2.5110	2.5123	0.0528
12.5000	2.6891	2.6929	0.1420
15.0000	2.8186	2.8268	0.2919
17.5000	2.9170	2.9307	0.4682
20.0000	2.9944	3.0064	0.4010

Table 1: Comparison in the values of $\alpha/(V\epsilon_e)$ using the surface and volume integral formulations

From the table we can see that a sphere is indeed a minimum shape regarding the polarisability. Given two inclusions of the same material, one a sphere and the other a cube, the polarisability of the sphere

$$\frac{\alpha}{\epsilon_e V} = 3\frac{\epsilon_i - \epsilon_e}{\epsilon_i + 2\epsilon_e} \qquad (18)$$

is always smaller, and so also the dipole moment induced in it in a given electric field (the maximum of normalized polarisability is 3 for a sphere, as $\epsilon_i/\epsilon_e \to \infty$).

If and when one knows the polarisability of certain inclusions, one is able to proceed in determining the effective permittivity of a mixture where these inclusions are embedded in a background medium. For example, the Maxwell Garnett mixing formula (1), written in terms of the polarisability rather than the inclusion permittivity, reads

$$\epsilon_{\text{eff}} = \epsilon_e + \frac{n\alpha}{1 - n\alpha/(3\epsilon_e)} \tag{19}$$

This form of the Maxwell Garnett rule is obviously more suitable to mixtures where the inclusions have arbitrary shapes. For a mixture where cubic dielectric inclusions of permittivity $\epsilon_i = 20\epsilon_e$ are located in environment with permittivity ϵ_e, we can see from Table 1 that the normalized polarisability is $\alpha/(V\epsilon_e) \approx 3$, whereas for a sphere the corresponding figure is 2.59.

To compare mixtures, we can take as an example a mixture where the inclusion phase occupies 5% of the total volume. Then the Maxwell Garnett prediction for the effective permittivity is $\epsilon_{\text{eff}} \approx 1.158\epsilon_e$ if the inclusions are cubes, and $\epsilon_{\text{eff}} \approx 1.135\epsilon_e$ for the case of spherical inclusions.

In the dilute-inclusion limit, the dependence of the effective permittivity on the polarisability is simple:

$$\epsilon_{\text{eff}} \approx \epsilon_e + n\alpha \tag{20}$$

where n is the number density of the inclusions ($1/m^3$) in the mixture.

One might raise the question how the orientation of the cubical inclusions affect the permittivity of the mixture they form the guest phase. Does the mixture become anisotropic? It is true that the symmetry of a cube is less than the perfect isotropic symmetry of a sphere. And also the analysis above is valid if the cubes are randomly oriented in the mixture. But more can be said. If we limit ourselves to the dyadic (second-order) description of the polarisability response, the eigenvalues of the cubical geometry are the same in the three orthogonal directions, and the isotropy is retained. Therefore, the orientation distribution does not break the isotropy, and the effective permittivity remains the same as that for randomly oriented cubes. Of course, this is strictly valid in statics: in higher-order description, a small anisotropy for an ordered cubic lattice can be observed.

5. DISCUSSION

One of the problems in the modelling of heterogeneous materials (regardless of whether the response to be modelled is dielectric, magnetic, magnetoelectric, thermal, or elastic, etc.) is the requirement of suppressing many degrees of freedom in the description of the material. To solve that problem, one may use brute assumptions and approximations in the calculation of the response of a single scatterer and also in taking into account the interaction between the neighboring scatterers. One example of such approaches leads to the Maxwell Garnett formula with the advantage of easy usage but with a limited range of applicability. Widely in the literature, it has been pointed out that the validity regime of Maxwell Garnett formula is the dilute mixture limit and low contrast between the permittivities of the inclusions.

One may also try to solve more rigorously one of the subproblems in the modelling project. One example of such an approach has been given Section 4 of this paper where the interaction problem was of the same degree of rigour as the Maxwell Garnett theory but the response of the individual scatterer was given fuller attention. In principle, using the procedures mentioned in Section 4, the polarisability of a single scatterer can be solved with any desired accuracy. Then the mixture problem can be attacked, but the degree of rigour is lost very easily; this happens for example if the inclusions in the mixture are not identical in shape.

Finally, a "third way" of solving the mixing problem is gaining popularity presently and certainly in the future. This is the full numerical approach. One does not separate inclusions as scatterers that have their own distinct polarisability properties but rather, keeps them as integral parts of the complex sample itself. By discretising the whole mixture, the full electrostatic or even electromagnetic problem can be solved with an accuracy that depends on computing capabilities and cleverness in transforming the sample geometry into a form that the numerical method can accept. One preliminary example is the evaluation of the effective permittivity of a two-dimensional mixture with the electromagnetic FDTD (Finite Difference Time Domain) approach [11]. This study has shown that in fact, a dynamical procedure can—with reasonable amount of computing resources—produce results for the quasistatic mixing problem that are difficult reach with static-based numerical methods.

REFERENCES

1. J. C. Maxwell Garnett. Colours in metal glasses and metal films. *Trans. of the Royal Society,* **Vol. CCIII**, 385-420, London (1904).
2. A. Sihvola. Self-consistency aspects of dielectric mixing theories. *IEEE Trans. , Trans. Geosci. Remote Sensing.*, **GE-27**, (4), 403-415 (1989).
3. D. Polder and J. H. van Santen. The effective permeability of mixtures of solids. *Physica,* Vol. **XII**, (5), 257-271, (1946).
4. D. A. G. Bruggeman. Berechnung verschiedener physikalischer Konstanten von heterogenen Substanzen, I. Dielektrizitäts konstanten und Leitfähigkeiten der Mischkörper aus isotropen Substanzen. *Annalen der Physik,* 5. Folge, Band 24, 636-664, (1935).
5. C. J. F. Böttcher. *Theory of electric polarization,* Elsevier Amsterdam, (1952).
6. L. Tsang, J.A. Kong and R.T. Shin, *Theory of microwave remote sensing*, Wiley Interscience, New York, (1985), p. 475.
7. Z. Hashin and S. Shtrikman. A variational approach to the theory of the effective magnetic permeability of multiphase materials. *Journal of Applied Physics,* **33**, (10), 3125-3131 (1962).
8. J. J. H. Wang. Generalized Moment Methods in Electromagnetics: Formulation and computer solution of integral equations. *John Wiley and Sons,* New York (1991).
9. D. E. Livesay and Kun–Mu Chen. Electromagnetic fields induced inside arbitrarily shaped biological bodies. *IEEE Transactions on Microwave Theory and Techniques* **MTT–22**, (12), 1273–1280, (1974).
10. J. Van Bladel. Electromagnetic Fields. *Hemisphere Publishing Corporation,* New York (1985).
11. O. Pekonen, K. Kärkkäinen, A. Sihvola and K. Nikoskinen. Numerical testing of dielectric mixing rules by FDTD method. *J. Electromagnetic Waves and Applic,* **13**, (1), 67-87 (1999).

Microw. Radiomet. Remote Sens. Earth's Surf. Atmosphere, pp. 89–96
P. Pampaloni and S. Paloscia (Eds)
© VSP 2000

Faraday rotation and passive microwave remote sensing of soil moisture from space

DAVID M. LE VINE[1] and SAJI ABRAHAM[2]

[1] *Code 975, Goddard Space Flight Center, Greenbelt, Maryland   20771*
[2] *Raytheon ITSS, Lanham, Maryland , 20706*

ABSTRACT: Rotation of the polarization vector of microwave radiation as it propagates from the earth surface through the ionosphere (Faraday rotation) can be a source of error in passive microwave remote sensing from space. The error can be significant at L-band (1.4 GHz), the frequency window likely to be used in future sensors in space to measure soil moisture. This paper presents calculations of the potential error using the International Reference Ionosphere (IRI-95) to compute the electron density and the International Geomagnetic Reference Field (IGRF) to compute the magnetic field. Examples are given assuming a sensor in an orbit representative of missions proposed to measure soil moisture (6 am sun-synchronous orbit at 675 km). Errors are presented in terms of change in brightness temperature using a flat surface to model the emissivity of the land. At 6am and modest incidence angles (30 degrees or less), the error in brightness temperature is less than 1K; however, for other local times and scan geometry, the errors can be significant.

1. INTRODUCTION

Recent developments in the technology of passive microwave remote sensing from space have led to increased interest in remote sensing of soil moisture from space [6,10]. These measurements are best made at the low frequency end of the microwave spectrum (L-band, 1.4 GHz) and as a result large antennas are required in space to achieve adequate resolution. However, advances in technology such as the development of aperture synthesis for earth remote sensing [7,14] and progress in the deployment of light weight mesh antennas and inflatable structures [11], have increased the possibility of flying an affordable sensor in space. In fact, proposals have recently been made to NASA and ESA for a sensor at 1.4 GHz to map soil moisture [8,9].

At 1.4 GHz, Faraday rotation is not negligible and is a potential source of error in the retrieval algorithm. Errors occur because radiation that is horizontally or vertically polarized at the surface changes polarization as the signal propagates through the ionosphere as shown in Eqn. 1. In these expressions, $T_h(\Delta\Omega)$ and $T_v(\Delta\Omega)$ are the brightness

$$T_h(\Delta\Omega) \quad = \quad [e_h \cos^2(\Delta\Omega) + e_v \sin^2(\Delta\Omega)] \, T_o \qquad\qquad 1a$$

$$T_v(\Delta\Omega) \quad = \quad [e_v \cos^2(\Delta\Omega) + e_h \sin^2(\Delta\Omega)] \, T_o \qquad\qquad 1b$$

temperatures from horizontally and vertically polarized antennas on the spacecraft; $\Delta\Omega$ is angle of Faraday rotation; and e_v and e_h are the emissivity of the surface. T_o is the physical temperature of the surface.

Estimating the effect of Faraday rotation is complex because $\Delta\Omega$ depends on local time, season, location (latitude and longitude), the solar cycle, altitude and even the scan mode of the sensor. The objective of this paper is to assess the problem in the context of future missions to measure soil moisture from space. For this purpose the orbit of a recently proposed mission called, HYDROSTAR [8], will be used to compute examples. The orbit is driven by scientific requirements for the measurement and the orbit selected for HYDROSTAR (sun-synchronous with a 6am/6pm equatorial crossing and an altitude of 675 km) is similar to the orbit of other proposed missions.

2. FARADAY ROTATION

At L-band the frequency of observation is much greater than both the plasma frequency (~10 MHz) and the gyrofrequency (~ 1.4 MHz) in the ionosphere and the expression for the rotation angle, $\Delta\Omega$, has the simplified form [15]:

$$\Delta\Omega \quad = \quad (\pi/cv^2) \int v_p^2(s) \, v_b(s) \, \cos(\Theta_b(s)) \, ds \qquad\qquad 2$$

where $\Delta\Omega$ is in radians; $v_p \approx 9\sqrt{N_e}$ is the electron plasma frequency; $v_b = q_e B/(2\pi M_e)$ is the gyro frequency of electrons in the earth magnetic field, B; and Θ_b is the angle between the direction of propagation and the earth magnetic field:

$$\cos(\Theta_b) \quad = \quad \cos(\Theta) \sin(I) \, - \, \sin(\Theta) \cos(I) \cos(\varphi - D) \qquad\qquad 3$$

The angles, Θ and φ, are polar coordinates of a line from the spacecraft to the observation point on the surface. In this coordinate system, the z-axis is normal to and pointing into the surface and the x-axis is pointing to geographic north. The angles I and D specify the direction of the magnetic field, B. The dip angle, I, is the angle the magnetic field makes with the local horizontal (plane perpendicular to the z-axis). The declination angle, D, is the angle the plane containing B and the z-axis makes with geographic north (x-axis).

The objective is to compute the rotation angle, $\Delta\Omega$, for a sensor in the HYDROSTAR orbit and then to substitute $\Delta\Omega$ into Eqn. 1 to estimate the effect of Faraday rotation on the observed brightness temperature. To accomplish this, the International Reference Ionosphere (IRI-95 [2,3,13]) will be used to model the electron density profile and the International Geomagnetic Reference Field (IGRF [1,5]) will be used for the magnetic field. To simplify the many calculations needed to map $\Delta\Omega$ globally, Eqn.2 has been rewritten by making the change of variables ds = sec(Θ)dz where z is the normal to the surface at the sub-satellite point and replacing B by its vector value at an altitude of 400 km. One obtains:

$$\Delta\Omega \quad = \quad (K/v^2) \int B(z) \cos(\Theta_b(z)) \sec(\Theta) \, N_e(z) \, dz \qquad\qquad 4a$$

$$= \quad 6950 \, B(400) \cos(\Theta_b) \sec(\Theta) \, VTEC \qquad\qquad 4b$$

where VTEC = $\int N_e(z)dz$ is the vertical total electron content at the sub-satellite point. This approximation makes it easier to identify the factors important in determining $\Delta\Omega$ and simplifies the computations because VTEC is provided directly by IRI-95. (B is in telsa, VTEC is in total electron content units, 10^{16} electrons/m^2, and $\Delta\Omega$ is now in degrees.)

It is clear from Eqn. 4 that the rotation angle depends on the total electron content, VTEC, and also the orientation of the sensor with respect to the magnetic field (i.e. Θ_b). Each of these can be quite variable. For example, VTEC varies with location (latitude and longitude), season, local time and solar activity. Figure 1 illustrates VTEC as function of latitude and local time. The data are for a very active sun, June, 1989, a solar maximum with sunspot number, Rz = 158. (Rz is a 12-month, running mean centered on the month of interest. It is the parameter for solar activity used in the IRI-95. The mean sunspot number for the month of June, 1989 was 196.) Figure 1 illustrates several characteristics of VTEC. First, notice that the total electron content is largest at noon when the solar radiation is most direct and smaller at night. Of particular importance for the remote sensing of soil moisture is the behavior at 6 am and 6 pm, the times most often proposed for satellite over pass. At 6 am the total electron content is near minimum. Also, notice the slight bias toward lower values in the southern hemisphere. This occurs because the data are for June, summer in the northern hemisphere. However, this pattern is not always true and winter values can be larger than summer values [4,12].

Figure 2 illustrates the effect of sensor orientation with respect to the magnetic field on Faraday rotation. Figure 2 is a plot of rotation angle, $\Delta\Omega$, for a conical scanning radiometer located at (35N, 75E). The calculations assume the HYDROSTAR orbit (sunsynchronous, 675 km altitude) and are for solar maximum conditions (June, 1989). The arrow indicates the satellite heading. It is assumed that the radiometer scans conically about the sub-satellite point and the circles indicate incidence angles of 10,20,30,40 and 50 degrees, respectively. The values at the X's along the circles are the values of $\Delta\Omega$ in degrees for a sensor looking in that direction. Notice that $\Delta\Omega$ increases with incidence angle and also notice the variation as the scan changes in azimuth. This is due to the change in Θ_b as the angle, φ, changes around the scan path (Equation 3). At a conical angle of 50 degrees, $\Delta\Omega$ changes from a minimum of less than 1 degree to a maximum of more than 11 degrees.

3. EFFECT ON BRIGHTNESS TEMPERATURE

The effect of Faraday rotation on brightness temperature is given by Eqn. 1. In the absence of Faraday rotation, $T_h(0) = e_h T_o$, and the surface parameter (soil moisture) is obtained by solving for e_h and inverting the emissivity for the desired parameter. Following the same procedure in the presence of Faraday rotation can result in an error because the received power is a function of both e_v and e_h which are different functions of the soil moisture. For purposes of deciding when the error might be significant, it can be represented as an equivalent error in brightness temperature ΔT by taking the difference $\Delta T(\Delta\Omega) = T(\Delta\Omega) - T(0)$:

$$\Delta T_h(\Delta\Omega) = [e_v - e_h] \sin^2(\Delta\Omega) T_o \qquad 5$$

To compute ΔT_h it will be assumed that the surface is flat and without roughness or vegetation canopy. In this case, the Fresnel reflection coefficients of the surface can be used to estimate the emissivity:

$$e_{h,v} = 1 - |R_{h,v}(\Theta)|^2 \qquad 6a$$

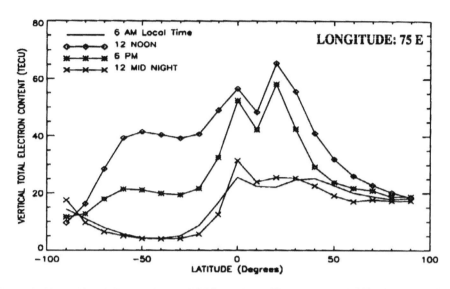

Figure 1: Vertical Total Electron Content, VTEC, for June, 1989 at longitude 75 E. The data are for a period of solar maximum (sunspot number Rz= 158).

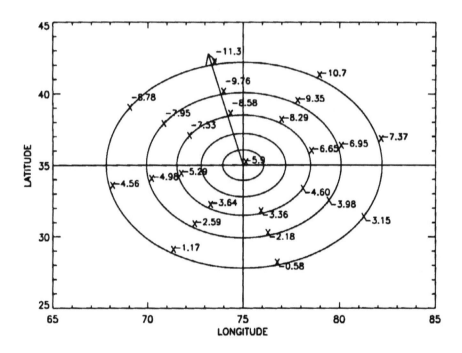

Figure 2: Faraday rotation for a conical scan geometry at incidence angles of 0, 10, 20, 30, 40 and 50 degrees. The numbers are Faraday rotation at the X's. Data are for (35 N, 75 E) and 6 AM. The vertical total electron content (675 km) is 27 TECU.

where the reflection coefficients, $R_{h,v}(\Theta)$ are:

$$R_v(\Theta) \;=\; \frac{\varepsilon_r \cos(\Theta) - \{\varepsilon_r - \sin^2(\Theta)\}^{1/2}}{\varepsilon_r \cos(\Theta) + \{\varepsilon_r - \sin^2(\Theta)\}^{1/2}} \qquad \text{6b}$$

$$R_h(\Theta) \;=\; \frac{\cos(\Theta) - \{\varepsilon_r - \sin^2(\Theta)\}^{1/2}}{\cos(\Theta) + \{\varepsilon_r - \sin^2(\Theta)\}^{1/2}} \qquad \text{6c}$$

where ε_r is the relative dielectric constant of the medium. The reflection coefficients, $R_{h,v}(\varepsilon)$ are related to soil moisture through ε_r using the relationship between dielectric constant and soil moisture proposed for L-band by Wang (1980):

$$\varepsilon_{real} \;=\; 3.1 + 17.36\,W + 63.12\,W^2 \qquad \text{7a}$$

$$\varepsilon_{imag} \;=\; 0.031 + 4.65W + 20.42\,W^2 \qquad \text{7b}$$

where ε_{real} and ε_{imag} are the real and imaginary parts of ε_r, respectively and W is the volumetric water content. In the examples presented below the values, W = 0.25 and T_o = 30C, are used.

Figure 3 illustrates the effect of Faraday rotation on brightness temperature as a function of solar activity using the HYDROSTAR orbit at a fixed location over land (35N, 75E). The values of VTEC are for the month of June. The local time is 6 am and the data are for the right side of a cross track scan (φ = 90 degrees), which is the worst case. The top panel shows vertical total electron content, VTEC, as a function of solar activity (6 am and June) at this location. The middle panel shows the corresponding Faraday rotation, $\Delta\Omega$, for several incidence angles between 0-50 degrees. The bottom panel shows the error in brightness temperature caused by this rotation in a horizontally polarized receiving antenna (Eqn. 5). Notice that the error increases with incidence angle and solar activity as one would expect and that, except in the extreme case of an incidence angle of 50 degrees, the error is less than 1K even at maximum solar activity. In part, this is because 6 am is a time of minimum VTEC. The maximum error, ΔT_h, would be much larger if the observations were made close to noon.

Figure 4 shows the global distribution of Faraday rotation (left) and brightness temperature error (right) for the ascending branch (6am) of the HYDROSTAR orbit. The brightness temperature error and $\Delta\Omega$, is shown for several values of incidence angle on the right side of the cross-track scan pattern during solar maximum (June, 1989). Brightness temperature errors are over land with soil moisture of 0.25 and temperature of 30C. For incidence angles less than 40°, the errors in brightness temperature are less than 1K. Notice that, to a first approximation, there is a north-south bias reflecting that the computation is for summer in the northern hemisphere and an east-west symmetry reflecting that the computation is for constant (6 am) local time. However, there are significant deviations from this simple approximation. For example, notice the peak in Faraday rotation and error in brightness temperature over northern Asia. These patterns reflect the dependence of Faraday rotation on geometry (Θ_b) and the complex variations of VTEC with latitude and longitude.

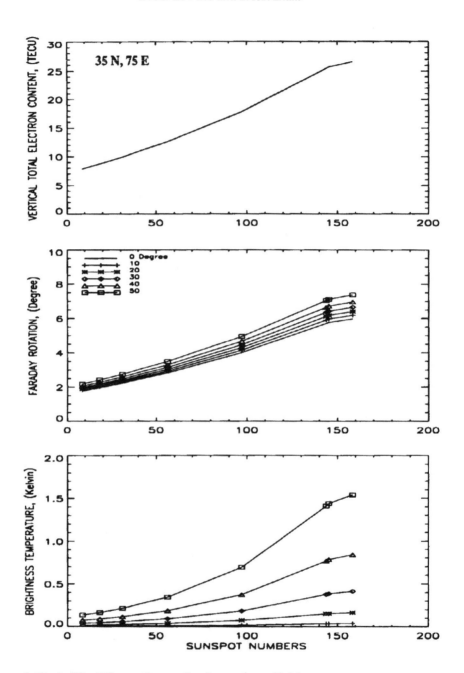

Figure 3: Vertical Total Electron Content, Faraday rotation and brightness temperature error as a function of solar activity (sunspot number) for (35 N, 75 E) at 6 AM and right side of a cross track scan. Top: VTEC at 675 km for June. Middle: Magnitude of Faraday rotation in degrees for incidence angles of 0–50 degrees in 10 degree steps. Bottom: Corresponding error in brightness temperature over land.

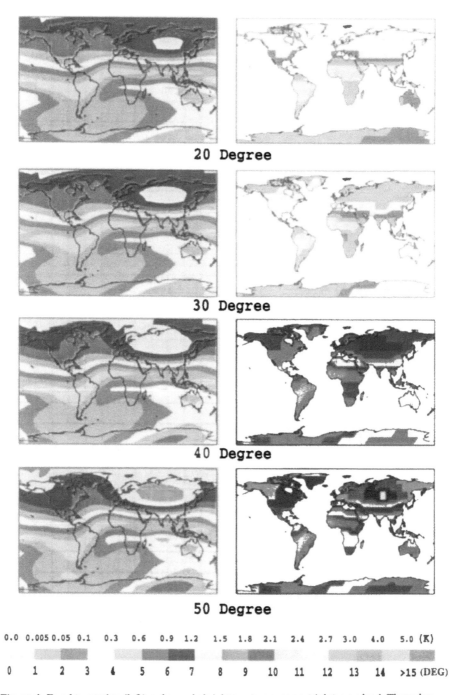

Figure 4: Faraday rotation (left) and error in brightness temperature (right) over land. These data are for June, 1989 (solar maximum, Rz= 158) and for a cross track scan at incidence angles of 20–50 degree. The surface has water content of 0.25 by volume and a physical temperature of 30 C.

4. CONCLUSIONS

The effect of Faraday rotation on the measurement of soil moisture and sea surface salinity from space depends very much on the parameters of the mission. Of importance are the time of observation (i.e. local time), the altitude of the sensor and even the scan geometry. Scan geometry is important because it determines both Θ_b and difference between e_v and e_h. In the case of the HYDROSTAR configuration, the observations are at a time of minimum electron content (close to 6am) and the incidence angles are relatively close to nadir and perpendicular to the local magnetic field. Hence, the Faraday rotation and associated errors are relatively small. Even in cases of maximum solar activity, the Faraday rotation is less than 10 degrees at incidence angles as large as 30-40 degrees and the resultant error in brightness temperature is less than 1K. Ignoring such an error in the measurement of soil moisture is probably acceptable, because the sensitivity to moisture is large (e.g. 2-4 K/percent moisture) and the dynamic range is many 10's of Kelvin. However, in other applications such as the measurement of sea surface salinity, where both the sensitivity and dynamic range are smaller, a correction probably will be needed.

5. ACKNOWLEDGEMENT

The authors wish to acknowledge the assistance of Dieter Bilitza in helping them get started running the IRI-95 and IGRF models.

6. REFERENCES

1. Barton, C.E., "International Geomagnetic Reference Field: The seventh generation", *J. Geomag. Geoelectr.*, 49, 123-148, (1997).
2. Bilitza, D., K. Rawer, L. Bossy and T. Gulyaeva, "International Reference Ionosphere – past, present, future", *Adv. Space Res.*, 13 (#3), 3-23, (1993).
3. Bilitza, D., "International Reference Ionosphere – status 1995/1996", *Adv. Space Res.*, 20(9), 1751-1754, (1997).
4. Huang, Y. and K. Cheng, "Solar cycle validation of the total electron content around equatorial anomaly crest region in east Asia", *J. Atmos. Terrestrial Phys.*, 57 (12), 1503-1511, (1995).
5. Langel, R.A.,"International Geomagnetic Reference Field, 1991 Revision," *J. Geomag. Geoelectr.*, 43, 1007-1012, (1991).
6. Le Vine, D.M., T.T. Wilheit, R.E. Murphy, and C.T. Swift, "A multifrequency microwave radiometer of the future", *IEEE Trans. Geosci. Remote Sens.*, 27, 193-199, (1989).
7. Le Vine, D.M., A.J. Griffis, C.T. Swift and T.J. Jackson, "ESTAR: A synthetic aperture microwave radiometer for remote sensing applications", *Proc. IEEE*, 82(12), 1787-1801, (1994).
8. Le Vine, D.M., "Synthetic aperature radiometer systems, IEEE MTT-S International Microwave Symposium", Anaheim, CA, June, (1999).
9. Martin-Neira, M., and J.M. Goutoule, "MIRAS-A two-dimensional aperture/synthesis radiometer for soil-moisture and ocean-salinity observations", ESA Bull.-European Space Agency, 92, 95-104, (1997).
10. Njoku, E.G. and D. Entekhabi, "Passive microwave remote sensing of soil moisture", *J. Hydrology*, 184, 85-99, (1996).
11. Njoku, E.G., Y. Rahmat-Samii, J.Sercel, W.J. Wilson and M. Moghaddam, "Evaluation of an inflatable antenna concept for microwave sensing of soil moisture and ocean salinity", *IEEE Trans. Geosci. and Remote Sensing*, 37 63-78, Part 1, (1999).
12. Rama Rao, P.V.S., P. Sri Ram, P.T. Jayachandran and S.S.V.V.D. Prasad, "Seasonal variation in ionospheric electron content and irregularities over Waltair, - A comparison with SLIM model", *Adv. Space Res.*, 18(6), 259-262, (1996).
13. Rawer, K. D. and P.A. Bradley, "IRI 1997 Symposium: New developments in ionospheric modeling and prediction", *Adv. Space Res.*, 22(6), 723-724, (1998).
14. Swift, C.T., D.M. Le Vine and C.S. Ruf, "Aperture synthesis concepts in microwave remote sensing of the earth", *IEEE Trans. Microwave Theory and Tech.*, 39(12), (1991).
15. Thompson, A.R., J.M. Moran and G.W. Swenson, Interferometry and synthesis in radio astronomy, New York: Wiley, (1986).
16. Wang, J.R., "The dielectric properties of soil water mixtures at microwave frequencies", *Radio Science*, 15, 970-985,(1980)

Microw. Radiomet. Remote Sens. Earth's Surf. Atmosphere, pp. 97–106
P. Pampaloni and S. Paloscia (Eds)
© VSP 2000

On simultaneous optical and passive microwave remote sensing of soil, water and vegetation

D. PEARSON[1], E.J. BURKE[2], R.J. GURNEY[1], L.P. SIMMONDS[3]

1. *Environmental Systems Science Centre, University of Reading, United Kingdom.*
   Email: dwcp@mail.nerc-essc.ac.uk
2. *Department of Hydrology and Water Resources, University of Arizona, USA.*
   Email: eleanor@hwr.arizona.edu
3. *Department of Soil Science, University of Reading, United Kingdom.*
   Email: asssimmo@reading.ac.uk

Abstract — Passive microwave remote sensing of near-surface soil moisture content is an established technique. New analyses undertaken by this group show that existing methods have demonstrated only a fraction of the potential for estimation of surface and sub-surface quantities. Time series of measured brightness temperature in the field can be used as targets in the optimisation of a water/energy transport model to generate estimates of soil hydraulic properties, with standard errors similar to those in traditional destructive methods of fieldwork. This method has worked even in the presence of vegetation — in this case, transport of substantive quantities is modelled by a SVAT, and that of microwave radiation by a simple radiative transfer model that requires an independent estimate of the state of the vegetation. Here we investigate biases that occur in the estimates of hydraulic parameters and fluxes, induced by uncertainty in measurements of the vegetation.

1. INTRODUCTION

It has long been recognised that the low emissivity of water at certain frequencies allows non-destructive measurement of the water content of porous media such as soil. Indeed, passive microwave remote sensing of near-surface soil moisture content is an established technique [1]. However, new analyses undertaken by this group show that existing methods have demonstrated only a fraction of the potential for estimation of surface and sub-surface quantities. The possibilities for land-based radiometry at L-band are correspondingly enhanced, and a better understanding of how airborne and spaceborne radiometric data may be interpreted is obtained.

An experimental method that is widely adopted is to establish an empirical relationship between the emissivity of the soil (usually expressed as a normalised brightness temperature) and an average near-surface water content (e.g. in a layer 5 cm deep) [2]. Ideally, one would obtain a clear functional relationship; but usually there is considerable scatter, even when measurements are taken over smooth, bare soil. This is often due to spatial variation of soil moisture, especially when point samples are measured and compared with those from a radiometer with a large field of view. However, we have shown that scatter in the relationship between near-surface water content and brightness temperature (at L-band) is due not only to spatial variations, but also largely due to diurnal hysteresis in the gradient of water content near the surface *at a single point*. The hysteresis loops only become apparent when the soil moisture and microwave emission are modelled in detail with close temporal sampling — sparse temporal sampling leads to apparent scatter like that in the experimental results [2].

Research is also being undertaken on the estimation of soil hydraulic properties by passive microwave remote sensing — the topic of the present paper.

2. ESTIMATION OF BARE SOIL HYDRAULIC PROPERTIES

Consider a land-based L-band radiometer that is taking measurements of the emission from bare soil. These measurements are recorded as a time series of brightness temperatures T_b. At any one time, the observed brightness temperature depends on many factors: the amount and distribution of water in the top regions of the soil, $\theta(z)$; the distribution of temperature, $T(z)$; the brightness temperature of the sky, T_{sky}; the reflectance of the soil, ρ_{soil}; the transmittance of the atmosphere, t_{air}; roughness and composition of the soil; the characteristics of the radiometer; and more. Let us assume that all these quantities are known except for $\theta(z)$ and $T(z)$. If one has a mathematical model that predicts these quantities and the resulting brightness temperature $\hat{T}_b$, one can then adjust the model to try to make $\hat{T}_b$ as close to T_b as possible. Here and throughout, all symbols with a hat (◦) denote modelled or estimated values of the corresponding hat-free quantities. T_b is referred to as the *target function* or *target time-series*. Specifically, let us assume that the quantities to be estimated are the saturated hydraulic conductivity k_{sat}; the air-entry potential ψ_e; the bulk density ρ_b; and the water-release characteristic b. T_b is measured in the field, and $\hat{T}_b(\hat{k}_{sat}, \hat{\psi}_e, \hat{\rho}_b, \hat{b})$ is modelled by the MICROSWEAT model (discussed in detail below). We then form a cost function:

$$C = \sum_i [T_b(t_i) - \hat{T}_b(t_i; \hat{k}_{sat}, \hat{\psi}_e, \hat{\rho}_b, \hat{b})]^2, \qquad \text{1.}$$

where the summation is over all or some of the time series. Minimisation of C with respect to the variables $(\hat{k}_{sat}, \hat{\psi}_e, \hat{\rho}_b, \hat{b})$ yields the values of these variables that best reproduce the measured brightness temperatures. These are then taken to be estimates of the true values in the field. This process is summarised in Fig. 1, where the box marked "Simulated process" should be ignored for now.

This procedure has been shown to generate good estimates of the soil hydraulic properties in several climatic conditions, for several soil types [3].

Another way to look at this problem is to note that the relation:

$$T_b = T_b(t; k_{sat}, \psi_e, \rho_b, b; other\ quantities) \qquad \text{2.}$$

may formally be inverted to obtain:

$$\begin{cases} \hat{k}_{sat} = \hat{k}_{sat}(T_b[t]; other\ quantities) \\ \hat{\psi}_e = \hat{\psi}_e(T_b[t]; other\ quantities) \\ \hat{\rho}_b = \hat{\rho}_b(T_b[t]; other\ quantities) \\ \hat{b} = \hat{b}(T_b[t]; other\ quantities). \end{cases} \qquad \text{3.}$$

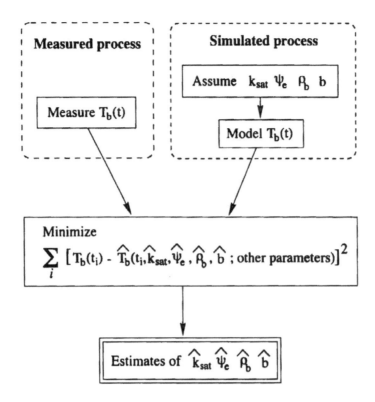

Figure 1. Flow diagram for estimating soil hydraulic properties. Starting in the box marked "Measured process", we see how a cost function is defined and minimised to form best estimates of the parameters in the field. Starting in the box marked "Simulated process", we see how field measurements can be simulated, and the *assumed* hydraulic parameters can be estimated.

Ideally, one would analyse the existence, uniqueness and stability of this inverse problem. However, that would be intractably difficult, and instead we investigate it intuitively and experimentally.

3. ESTIMATION OF VEGETATED SOIL HYDRAULIC PROPERTIES

If vegetation is present, the problem is made more complex by two sets of phenomena, which are represented by *"other parameters"* in Fig. 1. Firstly, the vegetation affects the water and energy budgets of the soil, changing $\theta(z)$ and $T(z)$. Secondly, the vegetation interferes with the microwave emission from the soil: it absorbs some of that radiation, reflects some, transmits some, and emits radiation itself. Nevertheless, if ground-based measurements of the vegetation's physical state are measured, these effects can be modelled (see the description of MICROSWEAT below), and estimation of the soil's hydraulic properties can be done [3]. However, in airborne or spaceborne remote sensing, detailed understanding of the vegetation will not be available.

Optical remote sensing offers the requisite estimation of above-ground vegetation properties. Various techniques are now available, or will become available in the near future. For example, simple spectral measurements (e.g. [4]) and directional spectral measurements (e.g. [5]) may be manipulated to estimate canopy characteristics. Limitations are being eroded with the introduction of new instruments and analytical techniques. Here we present a modelling study that investigates how errors in estimation of the properties of vegetation can bias estimation of soil hydraulic properties. A topic that we omit from the present paper, and which is a major problem in remote sensing, is the estimation of below-ground vegetation processes.

3.1. The models

Transport of heat and water is simulated by the SWEAT model, an overview of which is presented here. Daamen [6] provides further details. SWEAT is a one-dimensional model which simulates the movement of water and heat in the soil profile above a nominated depth (e.g. 2 m), up to a reference height in the air above the vegetation (e.g. at a height of 5 m). In the soil, fluxes of liquid water, water vapour and sensible heat are simulated using Richards' equation, thermally enhanced Fickian diffusion of vapour and the Philip and de Vries [7] approach for modelling heat flow. That is, the following coupled partial differential equations are solved:

$$Q_l = -k_l(\psi)[\nabla \psi + g],$$ 4.

$$Q_v = -D_v(\theta) e^*(T) \nabla h_r(\psi, T) - D_v(\theta) h_r(\psi, T) \eta(\theta) \frac{de^*}{dT} \nabla T,$$ 5.

$$Q_h = -k_T(\theta) \nabla T + \lambda Q_v,$$ 6.

where Q_l is the flux of liquid water, Q_v is the water vapour flux, Q_h is the heat flux, k_l is the hydraulic conductivity, ψ is the soil matric potential (negative suction), g is the acceleration due to gravity, D_v is the water vapour diffusivity in soil, θ is the volumetric water content, e^* is the saturated water vapour density, T is the temperature, h_r is the relative humidity, η is a semi-empirical enhancement factor, k_T is the thermal conductivity and λ is the latent heat of vaporisation of water. The coupled partial differential equations are solved by a fully implicit backward-difference scheme in time, approximately centred in the spatial grid. This numerical scheme can also be considered as a resistance-capacitance network. Spatial centering is approximate because the grid's spacing increases with depth. This enables accurate calculation of transport of water and heat close to the soil's surface, where changes are rapid because of fluctuations in forcing at the boundary. Non-linearity of the equations leads to non-linearity in the discretized equations, so a Newton-Raphson method is used to solve the implicit set of equations at each time step.

Non-linearity is apparent in the assumed dependence of the hydraulic conductivity and volumetric water content on the soil matric potential:

$$k(\psi) = k_{sat} \left(\frac{\psi_e}{\psi} \right)^{2+3/b}, \quad \theta(\psi) = \theta_{sat} \left(\frac{\psi_e}{\psi} \right)^{1/b},$$ 7.

where k_{sat} is the conductivity of saturated soil, ψ_e is the matric potential at which air begins to enter the soil, θ_{sat} is the saturated volumetric water content, and b is the water-release characteristic. Thus the hydraulic conductivity is a function of the volumetric water content.

In the canopy air space, fluxes of water vapour and sensible heat from the vegetation and soil are described using a two-source method developed from that of Shuttleworth and Wallace [8]. The vegetation is assumed to have a stomatal resistance r_{st} in series with a boundary layer resistance r_{lbl}. The latter terminates in the air space within the canopy, which is connected by a further resistance r_{scan} to the surface of the soil and by another resistance r_a to the atmosphere at the reference height. Thus the vegetation and the soil can interact within the canopy's air space — this is an important consideration in the sparse crops for which SWEAT was designed, where evaporation from the leaves can be increased by heating due to hot air rising from the soil between the plants, or reduced because of humidification of the within-canopy air space caused by evaporation from the soil's surface. Methods of calculating the resistances are taken from published literature. The movement of water from the bulk soil to the site of evaporation within leaves is modelled using an Ohm's law analogue of water flow in the soil-plant system, incorporating radial root and soil resistances (both of which depend on root-length density) in series with the hydraulic resistance in the shoot. Stomatal resistance and soil water are related by way of leaf water potential, there being a threshold value of the potential below which stomatal resistance increases steeply.

Solar and thermal radiative fluxes are simulated by a simple but accurate radiative transfer model [9]. The main inputs to this part of the model are the leaf area index A_L, the optical single-scattering albedo ϖ_0, and the extinction coefficient k_{ext}. These control the disposition of the available energy between the vegetation and soil, and so play in important rôle in the evolution of T_b.

The SWEAT model is forced by meteorological data at the reference height (temperature, incident solar radiation flux, relative humidity, wind speed and rainfall) and by an assumption of unit hydraulic gradient (i.e. gravity-driven flow with homogeneous matric potential) at the nominated depth in the soil that is greater than the depths of interest. The system of equations is closed by conservation of water and energy.

SWEAT is coupled with a microwave emission model to form MICROSWEAT [2]. MICROSWEAT calculates the dielectric constant of the soil at each depth of interest, using a transition-moisture-content semi-empirical method [10]. Brightness temperature of the soil is then calculated by a simple 1-dimensional (stratified) radiative transfer model [11], modified to take account of surface roughness by a semi-empirical model [12]. The effects of vegetation are modelled by a two-parameter radiative transfer model containing the microwave optical depth and the microwave single-scattering albedo [13]. This combined transport and radiation model is similar to that of Liou and England [14], augmented by the inclusion of vegetation.

3.2. Simulations

This work is purely a modelling study. However, it was thought necessary to use simulated data that are similar to those that could actually be encountered in a real experiment. We also wanted to examine a different soil-water régime from those that have been studied hitherto, where water has been abundant. For this reason, data from the HAPEX-Sahel experiment [15] were used to generate a simulated time-series of brightness temperatures corresponding to DOY 230–255 in 1992, at +13° latitude in Niger, West Africa. The soil profile was initiated using measurements of $T(z)$ and $\theta(z)$ available from the HAPEX-Sahel Information System [16]. Weather and solar radiation data were also available at HSIS, although a number of missing values had to be simulated by a cubic spline interpolant. Data on root distribution for a fallow savannah area were taken from the literature [17]: the root-density distributions were approximated by a smooth curve of the following form:

$$\text{root length density} \propto e^{-5z}, \qquad\qquad 8.$$

where z is the depth in metres, and the constant of proportionality was chosen to give a total of $1.2 \times 10^4 \, \text{m/m}^2$ roots. The leaf area index was assumed to be $A_L = 1.0$, with an optical extinction coefficient $k_{ext} = 0.5$ and optical single-scattering albedo $\varpi_0 = 0.18$.

The soil texture was assumed to be 88% sand, 3% silt and 9% clay [17]. The hydraulic properties were derived from the texture by pedo-transfer functions [18], hence: $k_{sat} = 0.049\,\mathrm{kgs\,m^{-3}}$, $\psi_e = -0.53\,\mathrm{J\,kg^{-1}}$, $\rho_b = 1.65\,\mathrm{Mg\,m^{-3}}$, and $b = 4.24$.

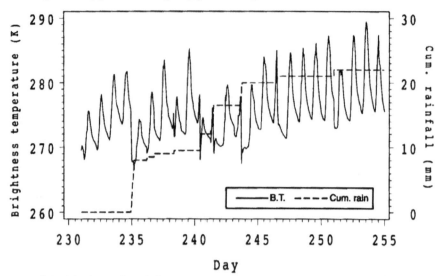

Figure 2. Modelled horizontally-polarised brightness temperature and cumulative rainfall from DOY 231.

The microwave optical depth and single-scattering albedos were assumed to take arbitrary but plausible values of 1.0 and 0.1, respectively. The evolution of the horizontally-polarized brightness temperature T_b^h (modelled by MICROSWEAT) is shown in Fig. 2, where the cumulative rainfall from DOY 231 is also plotted. This period represents the end of the long dry season in the Sahel, so the soil is initially very dry, the rainfall is sparse and evaporative demand is high. Thus each rainfall depresses T_b^h much less than has been observed in more temperate climates [2,3]. This time series of T_b is treated as if it were a dataset measured in the field; it is represented by the box labelled "Simulated process" in Fig. 1.

3.3. Estimation with errors

Small errors of ±5% in the extinction coefficient k_{ext} were introduced into MICROSWEAT. Then the cost function C was minimised by the downhill simplex method [19] as implemented in the NAG numerical library routine E04CCF, Mark 18. This algorithm is slow but robust. With small introduced errors such as these, remarkably good reproductions of the target time series could be achieved. Fig. 3 shows the residual $\hat{T}_b(t) - T_b(t)$ after C was minimised with k_{ext} perturbed to a value of $0.475\mathrm{m^{-1}}$. The resulting estimates of the soil hydraulic properties are shown in Table 1. Thus, assuming linearity for small biases in the vegetation parameters (units as in Table 1):

$$\frac{\partial \hat{k}_{sat}}{\partial k_{ext}} \approx 0.808, \quad \frac{\partial \hat{\psi}_e}{\partial k_{ext}} \approx 3.60, \quad \frac{\partial \hat{\rho}_b}{\partial k_{ext}} \approx -1.20, \quad \frac{\partial \hat{b}}{\partial k_{ext}} \approx 1.60. \qquad 9.$$

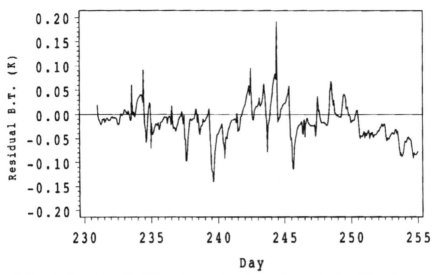

Figure 3. Error in time series of brightness temperature when an incorrect value of k_{ext} is assumed, but compensating errors are applied to soil properties. Observe that the error is very much smaller than the brightness temperature.

k_{ext}	$\hat{k}_{sat}$	$\hat{\psi}_e$	$\hat{\rho}_b$	$\hat{b}$
	(kg s m^{-3})	(J kg^{-1})	(Mg m^{-3})	
0.475	0.0323	-0.627	1.68	4.21
0.500	0.0488	-0.528	1.65	4.25
0.525	0.0726	-0.447	1.62	4.29

Table 1. Estimates of soil hydraulic properties when k_{ext} is perturbed from its assumed value of 0.500.

It was then assumed that k_{ext} was estimated less accurately, at $0.550\,\text{m}^{-1}$, i.e. a 10% error was added. This is an arbitrary value, but plausible in terms of estimation by optical remote sensing. Assuming linearity as a working approximation, the hydraulic parameters were then adjusted in accordance with equation 9, and SWEAT was run. Several diagnostic variables were compared with the "baseline" run, i.e. the run with no errors introduced. Fig. 4 shows that there was a considerable difference in the behaviour of the total water content of the soil column down to a depth of 2 m.

Furthermore, there was a difference between the behaviours of heat fluxes into the atmosphere in the two runs. Fig. 5 shows the baseline simulation of the latent heat flux λE, and the bias in the estimated values $\lambda \hat{E}$. The bias in the estimate of the sensible heat flux $\hat{H}$ behaves in the opposite manner, i.e. its bias is positive when the bias in $\lambda \hat{E}$ is negative, and vice versa. Thus, a bias is introduced into the estimate of the evaporative fraction.

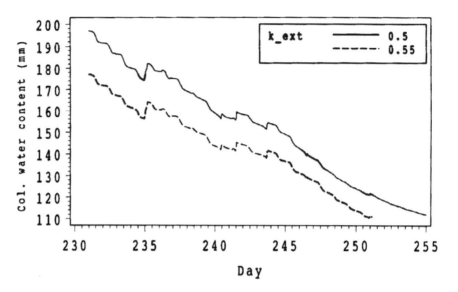

Figure 4. Comparison of column water content at the baseline value $k_{ext} = 0.5$ and with an introduced error of $k_{ext} = 0.55$.

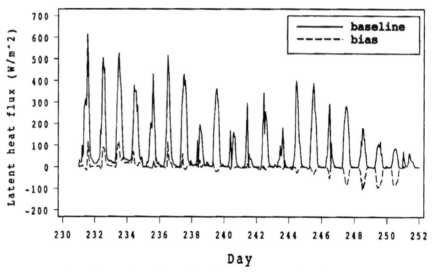

Figure 5. Latent heat flux at baseline value $k_{ext} = 0.5$; and the bias that occurs in this, induced by an introduced error of $k_{ext} = 0.55$.

4. DISCUSSION

We have demonstrated how the MICROSWEAT model can be inverted to obtain estimates of soil hydraulic properties, through least-squares fitting to a target function of measured or simulated brightness temperatures. We have shown that a 10% error in estimation of k_{ext} can lead to biases in the estimated water content of the soil and in the estimated heat fluxes. At first sight, this appears to render the technique problematical in application, because k_{ext} is dependent on the architecture of the vegetation canopy, which is difficult to quantify through remotely sensed data. However, this result may be peculiar to the extremely arid conditions assumed in this study, where T_B has a small dynamic response to wetting. Previous attempts at estimating soil hydraulic properties from passive L-band radiometry have been successful [3] in conditions of much greater dynamic response, although a detailed synthesis of optical and microwave sensing has not yet been undertaken in that context. It also remains to be seen how errors in other quantities interact and how significant the induced biases are. To address these problems, a wider range of simulations will now be undertaken. We will also obtain relevant experimental data using a passive L-band radiometer that is under procurement at the time of writing.

A plausible objection to the work presented here is that it is purely model-based, with no experimental data to confirm or refute it. However, all the components of the MICROSWEAT model have been shown to be good at reproducing experimental results [2, 3]. The water and energy transport module (SWEAT) is quite comprehensive and flexible. However, it may be that use of a more physically based microwave transfer model for vegetation (e.g. [20]) may be more robust in the inversion procedure discussed in this paper. Future work will investigate this question, and comparisons with results from the new UK-based L-band radiometer will be made.

ACKNOWLEDGEMENTS

The authors thank two anonymous referees for helpful comments that have enabled us to improve this paper. This work was funded by UK NERC grant number GR3/11551.

REFERENCES

1. T. Schmugge, P.E. O'Neill, J.R. Wang, "Passive microwave soil moisture research," *IEEE Trans. Geosci. Remote Sens.*, **GE-24**, 12–22 (1986).
2. L.P. Simmonds, E.J. Burke, "Application of a coupled microwave, energy and water transfer model to relate passive microwave emission from bare soils to near-surface water content and evaporation," *Hydrology and Earth System Sciences*, **3**, 31–38 (1999).
3. E.J. Burke, R.J. Gurney, L.P. Simmonds, P.E. O'Neill, "Using a modeling approach to predict soil hydraulic properties from passive microwave measurements," *IEEE Trans. Geosci. Remote Sens.*, **36**, 454–462 (1998).
4. E. Ridao, J.R. Conde, M.I. Minguez, "Estimating fAPAR from nine vegetation indices for irrigated and nonirrigated faba bean and semileafless pea canopies," *Remote Sens. Environment*, **66**, 87–100 (1998).
5. D.W. Deering, T.F. Eck, T. Grier, "Shinnery oak bidirectional reflectance properties and canopy model inversion," *IEEE Trans. Geosci. Remote Sens.*, **30**, 339–348 (1992).
6. C.C. Daamen, "Two source model of surface fluxes for millet fields in Niger," *Agric. For. Meteorol.*, **83**, 205–230 (1997).
7. J.R. Philip, D.A. deVries, "Moisture movement in porous materials under temperature gradients," *Trans. Am. Geophys. Union*, **38**, 222–232 (1957).
8. W.J. Shuttleworth, J.S. Wallace, "Evaporation from sparse crops — an energy combination theory," *Quart. J. Roy. Meteorol. Soc.*, **111**, 839–855 (1985).
9. D. Pearson, C.C. Daamen, R.J. Gurney, L.P. Simmonds, "Combined modelling of shortwave and thermal radiation for one-dimensional SVATs," *Hydrology and Earth System Sciences*, **3**, 15-30 (1999).

10. J.R. Wang, T.J. Schmugge, "An empirical model for the complex dielectric permittivity of soils as a function of water content," *IEEE Trans. Geosci. Remote Sens.*, **GE-18**, 288-295 (1980).

11. T.T. Wilheit, "Radiative transfer in a plane stratified dielectric," *IEEE Trans. Geosci. Electronics*, **16**, 138–143 (1978).

12. J.R. Wang, B.J. Choudhury, "Remote sensing of soil moisture content over bare field at 1.4 GHz frequency," *J. Geophys. Res.*, **86**, 5277–5282 (1981).

13. F.T. Ulaby, M.A. El Rayes, "Microwave dielectric spectrum of vegetation 2: dual dispersion model," *IEEE Trans. Geosci. Remote Sens.*, **GE-25**, 550–557 (1987).

14. Y-A Liou, A.W. England, "A land surface process/radiobrightness model with coupled heat and moisture transport in soil," *IEEE Trans. Geosci. Remote Sens.*, **36**, 273–286 (1998).

15. J.P. Goutorbe, T. Lebel, A.J. Dolman, J.H.C. Gash, P. Kabat, Y.H. Kerr, B. Monteney, S.D. Prince, J.N.M. Stricker, A. Tinga, J.S. Wallace, "An overview of HAPEX-Sahel: a study in climate and desertification," *J. Hydrol.*, **189**, 4–17, (1997).

16. http://www.orstom.fr/hapex/

17. S.R. Gaze, J. Brouwer, L.P. Simmonds, J. Bromley, "Dry season water use patterns under *Guiera senegalensis* L. shrubs in a tropical savanna," *J. Arid Environments*, **40**, 53–67 (1998).

18. Cosby, G.M. Hornberger, R.B. Clapp, T.R. Ginn, "A statistical exploration of the relationship of soil moisture characteristics to the physical properties of soils," *Water Resour. Res.*, **20**, 682–690 (1984).

19. J.A. Nelder, R. Mead, "A simplex method for function minimization," *Comput. J.*, **7**, 308–313 (1965).

20. J-P. Wigneron, J-C. Calvet, A. Chanzy, O. Grosjean, L. Laguerre, "A composite discrete-continuous approach to model the microwave emission of vegetation," *IEEE Trans. Geosci. Remote Sens.*, **33**, 201–210 (1995).

Microw. Radiomet. Remote Sens. Earth's Surf. Atmosphere, pp. 107–117
P. Pampaloni and S. Paloscia (Eds)
© VSP 2000

Applications of physically-based models of land surface processes in the interpretation of passive microwave radiometry

L.P. SIMMONDS[1] AND E.J. BURKE[2]

1. *Department of Soil Science, University of Reading, Whiteknights, Reading, RG6 6DW, UK. Email: asssimmo@reading.ac.uk*
2. *Department of Hydrology and Water Resources, University of Arizona, USA. Email: eleanor@hwr.arizona.edu*

Abstract- MICRO-SWEAT is a model for simulating the emission of microwave radiation from the land surface, based on knowledge of soil and vegetation properties and the prevailing meteorological conditions. Much of the work with MICRO-SWEAT to date has been carried out in conjunction with ground-based radiometry, with a typical footprint of around 1 m². This paper describes some preliminary findings from a study in collaboration with the Southern Great Plains '97 experiment in which MICRO-SWEAT was used to examine the impact of within-pixel variability in microwave emission on the interpretation of pixel-average brightness temperatures.

1. INTRODUCTION

Close coupling of remote sensing with physically-based modelling of land surface processes is an exciting area of development in the search for ways of making more effective use of remotely sensed information. This paper gives an overview of one such modelling approach. An established model of simultaneous heat and water flow through the soil-vegetation-atmosphere system (SWEAT) is coupled with equations to describe microwave transfer through a stratified medium to produce a model (MICRO-SWEAT) that predicts the time course of microwave emission from a vegetated land surface. Most of the work to date using MICRO-SWEAT has been done in conjunction with ground-based passive microwave radiometry (L and C band) with a footprint of the order of a few m². This paper concentrates on the application of physically-based modelling of microwave emission at larger scales, to investigate the impact of within-pixel heterogeneity on the interpretation of passive microwave radiometry.

2. MICRO-SWEAT

In brief, MICRO-SWEAT involves a sequential coupling of two simulation models. The first is a model of simultaneous heat and water in the soil-vegetation-atmosphere system

(SWEAT – [1]). The soil component includes consideration of Darcian water flow through the soil matrix, isothermal and thermally driven vapour flow, and the conduction of heat through the soil. When vegetation is present, transpiration and root water uptake are modelled assuming a simple electrical resistance analogue of soil/plant hydraulics. Water flow through the soil/plant system is coupled with the atmosphere via a stomatal response to leaf water potential. The link between subsurface processes and the atmosphere is made using the principles of the Shuttleworth and Wallace [2] approach to modelling the latent and sensible heat fluxes from two interacting evaporation surfaces. This approach requires the partitioning of net radiation between the canopy and soil surfaces, which is one of the novel aspects of the latest version of SWEAT, as described by Pearson et al. [3].

The outputs of SWEAT that are directly relevant to the modelling of microwave emission from the land surface are the vertical distributions of temperature and soil water content. The microwave component of MICRO-SWEAT makes use of the Wilheit [4] model of radiative transfer in stratified media to predict the microwave intensity emergent at the soil surface, and a simple two-parameter model (optical depth and single-scattering albedo) of the effects of the vegetation canopy on microwave transfer. The Wilheit model requires knowledge of the vertical distribution of temperature and dielectric constant. One of the outputs of the Wilheit model is the effective average physical temperature of the soil that is contributing to the emission from the soil surface. In MICRO-SWEAT, the dielectric constant of the soil is derived from the simulated soil water content using the Wang and Schmugge [5] semi-empirical mixing model, which is based on the proportions and dielectric properties of the soil constituents. Soil water is partitioned between bound water (with dielectric properties of ice) and free water according to the soil clay content. This is of particular relevance to the later discussion of the effect of soil type on apparent emissivity.

A number of recent papers have presented examples of the verification of MICRO-SWEAT in which predicted time courses of temperature have been compared with measurements using truck-mounted radiometry. These have included a range of bare soils and vegetated soils at both C (5.5 GHz) and L (1.4 GHz) band and at different look angles [6,7,8,9]. In most cases, the predicted time courses have done an excellent job of capturing both the diurnal variation in brightness temperature as well as the day-to-day changes as the soil surface wets and dries. Typically, the root mean square of the error in brightness temperature has been substantially less than 5K. Applications of MICRO-SWEAT at the 1 m^2 scale have included the development of physically-based algorithms for the retrieval of near surface water content [6,7]; the retrieval of soil hydraulic properties from detailed time courses of brightness temperature [7,8,9]; and the estimation of direct evaporation from the land surface [6].

3. SOUTHERN GREAT PLAINS '97 (SGP97)

Within-pixel heterogeneity in microwave emission is not an issue when interpreting ground-based radiometry with a footprint of the order of 1 m^2. However, proposed satellite missions are likely to have a resolution of the order of 20 km or more, and so will be subject to large within-pixel variability in the properties of both the soil and the vegetation, in conjunction with which will be variation in soil water status. The main objective of this paper is to present some preliminary results from a recent study which involves the application of physically-based models of microwave emission to investigate the impact of within-pixel variability on the retrieval of useful hydrological parameters. This study was part of the "Southern Great Plains '97" experiment (SGP97) run by USDA/NASA. SGP97 involved daily mapping of surface soil moisture for a month (18 June - 17 July, 1997) over 10 000 km^2 of Oklahoma using the Electronically Scanned Thinned Array Radiometer (ESTAR), an *L*-band passive microwave radiometer with a nominal resolution of 800 m. A description of the SGP97 experiment can be found at *http://daac.gsfc.nasa.gov/CAMPAIGN_DOCS/SGP97/sgp97.html.*

A site was selected for a detailed study of the impact on microwave emission of within-pixel variability in soil, vegetation and hydrology. This site (about 2.5 km^2) was within the USDA-ARS Grasslands Research Center at El Reno, Oklahoma, and consisted of a wheat field (ER15) separated from an area of rangeland (ER1-3) by a row of trees. In this paper, attention is concentrated on the large (~ 1600 m by 800 m) irregularly shaped wheat field. The senescing wheat was harvested midway through the experiment leaving short (<10 cm) stubble and a straw residue. There were large boggy patches in several parts of this field that remained wet throughout the experiment. Measurements within ER15 that are referred to in this paper included:

- Daily 0-5 cm gravimetric water content samples from a variable grid (14 locations) with a minimum separation of 100 m. A similar grid was located in ER1-3.

- Daily 0-6 cm theta probe (Delta-T Devices Ltd) soil moisture measurements from locations within each field based on a 7 by 7 grid, with each point separated by 100 m from each other (also in ER1-3).

- Soil particle size distribution, hydraulic conductivity @ 6 cm suction, water release characteristic and bulk density samples at each of the soil moisture sampling sites.

- The height of stubble and amount of trash after harvest at each soil moisture sampling location within ER15.

4. MODELLING FRAMEWORK

The original intention was to run MICRO-SWEAT in a distributed way in order to predict the time course of microwave emission from each sub-pixel unit. These predictions could then be verified by comparing the predicted emission aggregated over the pixel with that measured using ESTAR. This would provide the basis for exploring the extent to which the ESTAR signal is dominated by relatively small areas of the pixel.

However, there are drawbacks with this approach. First, uncertainties in the measurements of input parameters will lead to uncertainties in the predictions of spatial variability in microwave emission that are difficult to quantify, particularly because of non-linearities between model inputs and outputs. Second, the approach is limited to sites where detailed soil hydraulic properties are known. Third, the site selected for the fieldwork was chosen partly because of its complex hydrology, which involved significant lateral flow processes that are not incorporated into MICRO-SWEAT. A more satisfactory alternative was to use MICRO-SWEAT as an experimental tool to help identify a simpler,

L. Simmonds et al.

functional model of microwave emission that is driven by direct measurements of near-surface soil water content. An additional benefit is that a simple functional model is more appropriate than MICRO-SWEAT for exploring upscaling issues, and in the development of algorithms for retrieving soil water content information from microwave radiometry over heterogeneous land surfaces.

The brightness temperature measured using a radiometer (T_B) can be expressed conveniently as the product of the effective physical temperature of the emitter (T_{eff}) and the apparent emissivity (e_{app}). Hence a simple functional model for T_B requires estimation of T_{eff} and e_{app}. It is well known that the emissivity of soil decreases from over 0.9 when dry to under 0.6 when wet, reflecting the effect of water on the soil dielectric constant. The soil that is contributing to the emission from the soil surface will be at a range of temperatures and water contents, and will presumably be influenced by the shapes of the water content and temperature distributions. Hence we might expect a range of apparent emissivity associated with a given near-surface average water content, depending on whether there are steep temperature/water content gradients (around midday) or shallow gradients (at night). However, a remarkable conclusion from experimentation with MICRO-SWEAT was that, provided T_{eff} is well known (see later), there is a virtually unique relationship between e_{app} and the average near-surface water content [7]). Figure 1 shows examples for the average water content in the upper 2 cm of the profile (θ_{0-2}).

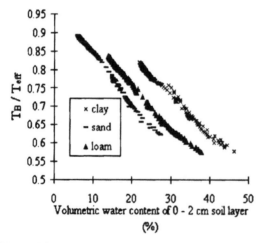

Figure 1. The relationship between apparent emissivity and near surface water content for contrasting soil textures

Part of the reason for this apparent stability is that near-surface gradients in water content tend to be steepest at times when there are also steep temperature gradients (e.g. around midday). There are compensatory 'trade offs' in the interactions between water content and temperature gradients and the relative weightings of different soil depths to the emission from the soil surface. Occasions when this relationship broke down were when there were rapid perturbations of near-surface water content (such as during rainfall) which caused the upper few centimetres of the profile to become, in effect, hydraulically decoupled from the soil below

(see [7]).

The simulated relationships between θ_{0-2} and e_{app} (obtained by extracting half-hourly values from MICRO-SWEAT simulations) were strongly soil dependent, with increasing clay content tending to increase the apparent emissivity associated with a given value of θ_{0-2} (Figure 1). One hypothesis was that this was attributable mainly to differences in the soil dielectric constant caused by differences in bound water/free water partitioning. Simulations were carried out for soils that differed in their hydraulic and thermal properties, but assuming a standardised dielectric constant/water content relationship. The θ_{0-2}/e_{app} relationships obtained from these various simulations coalesced (Figure 2) to give a virtually unique relationship that can be used as the basis of a functional microwave emission model [7]. The Wang and Schmugge [5] semi-empirical mixing model for predicting dielectric constant from soil composition was applied to, in effect, adjust the θ_{0-2}/e_{app} relationship for the 'standard' soil to give the corresponding relationship for the soil under consideration. This correction requires knowledge of the soil clay content that is used to predict the bound water/ free water partitioning.

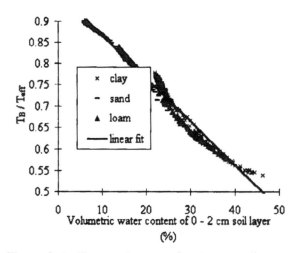

An important characteristic of the effect of soil texture on the θ_{0-2}/e_{app} relationship (Figure 1) is that the relationships are very close to being both linear and parallel. This has important consequences

Figure 2. As Figure 1, but assuming that the soils have an identical dielectric constant/water content relationship.

for upscaling microwave emission models, and for the interpretation of remote sensing of soil moisture over heterogeneous pixels. In particular, it can be shown easily that the microwave emission from a bare soil pixel (in terms of e_{app}) containing a wide range of soil texture and water content will be the same as for an equivalent uniform pixel that has the same areal average clay content and water content. The corollary is that basing a soil moisture content retrieval algorithm on the average clay content for the pixel will produce an estimate of water content that represents the true average for the pixel, irrespective of soil heterogeneity.

Accurate estimation of the effective soil physical temperature is an important element of any functional model to predict brightness temperature, or algorithm to retrieve soil moisture content from passive microwave remote sensing. Experiments with MICRO-SWEAT have shown that at 1.4 GHz, the effective weighted-average physical temperature of the soil is equivalent to the physical temperature at depths ranging from about 8 cm (wet soil) to about 18 cm (dry soil) [6,9]. An effective compromise is to assume that the temperature at 11 cm depth is representative of T_{eff} (e.g. Figure 3). The most promising of the various approaches we have attempted for developing a functional model for T_{eff} is to estimate the temperature at 11 cm depth from propagation of a sinusoidal temperature wave driven by the diurnal mean and amplitude of air temperature, assuming a relationship

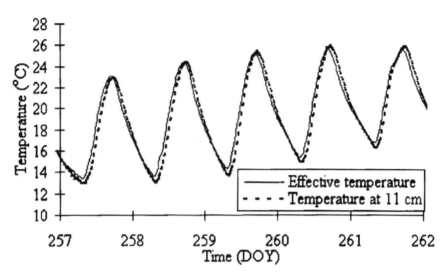

Figure 3. The time course of temperature at 11 cm depth, compared with the effective soil temperature, during a five-day drydown period following irrigation.

between soil thermal diffusivity and water content. Though crude and with obvious limitations (e.g. the inappropriateness of the sinusoidal function during the passage of air fronts, and the discrepancy between air and soil surface temperature), this simple approach seems not to introduce serious error into estimates of near-surface soil moisture content. This is in part because the temperature at 11 cm depth is well damped, and sensitivity analyses reveal that the sensitivity of soil water content retrievals to the estimate of T_{eff} is relatively small, being typically about 0.02% by volume water content per K.

The paragraphs above have outlined briefly the elements of the functional model developed to predict L band brightness temperature from knowledge of the near-surface soil water content, soil clay content, and daily maximum and minimum air temperature. To verify this model, data were obtained from a number of independent sources where there were high resolution measurements of brightness temperature (i.e. using radiometers mounted on a truck boom), soil clay content and near-surface water content. Figure 4 shows the comparison between the measured near-surface water content and that predicted by inverting our simple, functional brightness temperature model, which predicts soil moisture content to within 1.5 % by volume. This is equivalent to an uncertainty of about 5.5 K in brightness temperature, if the model is used to predict T_B from measured soil water content. Given the likely within-pixel variability in T_b, this provides a sufficiently accurate tool to study the effects of spatial variation in soil water and clay content on microwave emission from a bare soil.

Earlier work [9] has implied that the effects of vegetation on microwave emission can be accounted for adequately using a simple attenuation depth model to represent the effects of a vegetation canopy on the transfer of L band radiation. It was shown that that provided the optical depth of the canopy is well known, then useful information about the properties

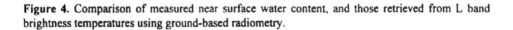

Figure 4. Comparison of measured near surface water content, and those retrieved from L band brightness temperatures using ground-based radiometry.

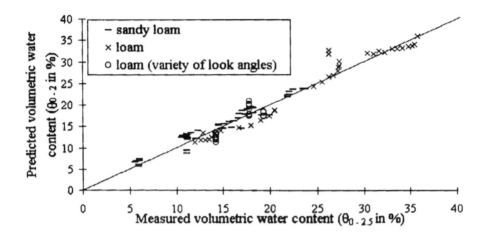

of underlying soil can be retrieved from L band radiometry even when the leaf area index is approaching 3. In a similar manner to Figure 1, Figure 5 shows the relationship for one soil type between the apparent emissivity and near-surface soil water content that was obtained by plotting the appropriate values from each half-hour timestep of a simulated 10-day dry down with a variety of canopy sizes. By contrast with Figure 1, the relationships between e_{app} and $\theta_{0.2}$ are non-linear, and the slopes are strongly affected by the size of the canopy. This has serious implications for the impact of within-pixel heterogeneity on the retrieval of soil moisture from passive microwave radiometry over vegetated surfaces.

The non-linear and non-parallel nature of the curves in Figure 5 imply that microwave emission from a land surface with heterogeneous vegetation cover could be substantially different to that predicted from an equivalent pixel with a uniform optical depth set to the same average value as the mixed pixel. The magnitude of this effect can be illustrated by the following example. Consider the fairly extreme case of a pixel which is 50% bare soil and 50% covered by a vegetation canopy with an optical depth of 0.6. For the purposes of illustration (though unrealistic), consider that the near surface water content is uniform across the pixel, and equal to 30% by volume. In this case, the apparent emissivities from the vegetated and bare soil areas of the pixel are expected to be 0.89 and 0.65 respectively (c.f. figure 5), giving a pixel-average apparent emissivity of 0.77. However, if a water content retrieval using figure 5 is attempted on the basis of pixel-averages for the apparent emissivity (0.77) and optical depth (0.3), then it would be deduced that the mean near surface water content was around 45% by volume (i.e. 1.5 times the 'true' value of 30% by volume). Repeating this exercise for different water contents reveals that the relative discrepancy between the 'true' pixel-average water content, and the value deduced from

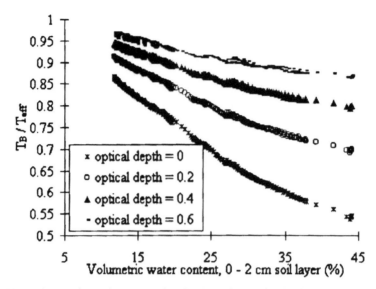

Figure 5. The effect of canopy optical depth on the relationship between near-surface water content and apparent emissivity.

the average optical depth is greatest when the soil water contents are high, because of the greater divergence between the optical depth curves in figure 5 at high soil water content. Of course, in reality the situation will be rather more complex than this, because there will be the strong confounding influence of interactions between canopy size and near surface. However, this simplistic calculation reveals the tendency for soil moisture content retrievals to become biased by sub-pixel heterogeneity in canopy optical depth.

5. SGP97 RESULTS

The soil properties within the wheat field were highly variable, with the clay content ranging from 5% to 33%, with consequent large variability in water retention and conductivity characteristics and in soil water content. Variogram analyses of the particle size distribution, water retention function parameters and the hydraulic conductivity data showed that in all cases the range of variation was of the order of 150 m, which was consistent with the soil water content data discussed later.

There was substantial within-field variability in near-surface soil water content on all dates. For example, on one date, the near surface water contents within the wheat field ranged from less than 0.15 to more than 0.40. Similarly, there was substantial variability between individual measurement locations in the magnitude of the changes in water content over each 24-hour period. Attempts to correlate either absolute water contents (or day-to-day changes in water content) with particle size distribution had limited success. The one occasion where there was significant correlation between near-surface water content and particle size distribution was within a day of rainfall (Figure 6), where the water content close to the surface was least affected by factors influencing lateral subsurface flows or antecedent evaporation. The presence of permanently boggy patches within the wheat field (presumably a consequence of topography and lateral flow processes) contributed to the lack of correlation between near-surface water content and soil texture. Also, there was large variability in trash cover at the scale of the water content measurement, with consequent variability in evaporation from the soil surface. Trash cover measurements made using 1 m^2 quadrats varied from 0 to 500 g fresh matter m^{-2}, caused by greater accumulation of residues between the wheel tracks of the combine harvester.

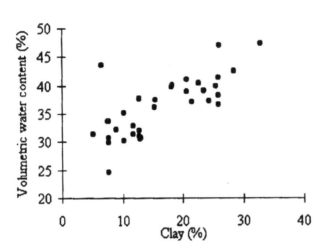

Figure 6. The relationship between near-surface water content and soil clay content on the day following rain.

Omnidirectional variogram analyses of near-surface water contents (0-5 cm depth) at a number of fields within the El Reno study area revealed that in all cases, the range of variation was between 150 and 200 m, in accord with the range of variation in soil hydraulic properties. The magnitude of the variance depended strongly on the time of sampling, with the variance just after rainfall tending to be much less (about half) that recorded after several days of continuous drying. Nugget variances were generally about 10% of the sill variance. Cross variograms revealed that spatial patterns in soil clay content and in near-surface soil water content were not related, except on the day following rainfall, for the reasons discussed above.

The effect of within field variation in soil texture on microwave emission from the harvested wheat field was examined using the simple functional model outlined above to predict T_B from knowledge of measured spatial variability in soil water and clay content. Figure 7 shows an example for one day of the deviation in e_{app} from the field average, grouping the data from the 40 sub-pixel locations into 10 classes of clay content. It is evident that there was no systematic effect of clay content on the contribution to the pixel-average microwave emission. Though areas with high clay content might be expected to have been wetter (and therefore have relatively low emissivity), this effect would be offset in part by there being a higher proportion of water that is tightly bound. However, the principal reason for the lack of evident systematic effect of clay content (or, indeed, any of the other particle size fractions) is the lack of strong correlation between soil texture and water content.

Figure 7. The deviations from the field-average emissivity for each 10 percent of the field area, ranked by clay content. The clay percentages shown are the averages for each class.

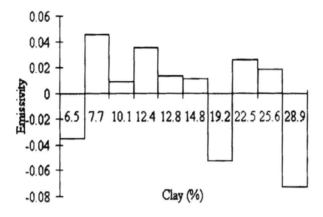

6. CONCLUSIONS

A major achievement of the project was to demonstrate how complex, process-based models of transfer processes can be applied to develop dedicated functional models for studying upscaling issues. In the case of bare soils it was shown that microwave brightness temperatures measured over heterogeneous areas are likely to be similar to equivalent uniform fields with the same average particle size distribution and near-surface water content. By contrast, heterogeneity in vegetation cover is likely to significantly bias estimates of near surface water content derived from microwave brightness temperatures using pixel-average values for canopy properties. Perhaps the principal conclusion to draw from this analysis is to emphasise the well-known need to account adequately for the effect of vegetation canopies on microwave emission when attempting to retrieve soil moisture information from passive microwave remote sensing.

The authors gratefully acknowledge the funding for this work from the UK Natural Environment Research Council (Grant GR3/11203).

REFERENCES

1. C.C.Daamen, L.P. Simmonds, "Measurement of evaporation from bare soil and its estimation using surface resistance," *Water Resources Research*, **32**, 1393-1402 (1996).
2. W.J. Shuttleworth, J.S. Wallace, "Evaporation from sparse crops — an energy combination theory," *Quart. J. Roy. Meteorol. Soc.*, **111**, 839–855 (1985).
3. D. Pearson, C.C. Daamen, R.J. Gurney, L.P. Simmonds, "Combined modelling of shortwave and thermal radiation for one-dimensional SVATs," *Hydrology and Earth System Sciences*, **3**, 15-30 (1999).
4. T.T. Wilheit, "Radiative transfer in a plane stratified dielectric," *IEEE Trans. Geosci. Electronics*, **16**, 138–143 (1978).
5. J.R. Wang, T.J. Schmugge, "An empirical model for the complex dielectric permittivity of soils as a function of water content," *IEEE Trans. Geosci. Remote Sens.*, **GE-18**, 288-295 (1980).

6. L.P. Simmonds, E.J. Burke, "Application of a coupled microwave, energy and water transfer model to relate passive microwave emission from bare soils to near-surface water content and evaporation," *Hydrology and Earth System Sciences*, **3**, 31–38 (1999).
7. L.P. Simmonds, E. J. Burke, "Estimating near-surface soil water content from passive microwave remote sensing - an application of MICRO-SWEAT," *Hydrol. Sci. J.*, **43** (4), 521-534, (1998).
8. E.J. Burke, R.J. Gurney, L.P. Simmonds, P.E. O'Neill, "Using a modeling approach to predict soil hydraulic properties from passive microwave measurements," *IEEE Trans. Geosci. Remote Sens.*, **36**, 454–462 (1998).
9. E.J. Burke, "Using a modelling approach to predict soil hydraulic parameters from passive microwave measurements for both bare and cropped soils." PhD Thesis, University of Reading, UK (1997)

Microw. Radiomet. Remote Sens. Earth's Surf. Atmosphere, pp. 119–125
P. Pampaloni and S. Paloscia (Eds)
© VSP 2000

Detection of Ice sheet on Asphalt Roads

G. MACELLONI, R. RUISI, P. PAMPALONI AND S. PALOSCIA

CNR- IROE, via Panciatichi 64, I 50127 Firenze, Italy

Abstract - An experimental study aimed at detecting different road conditions by means of millimeter wave radiometry has been conducted in an open-air laboratory using a sample of asphalt as target and two dual polarised radiometers at 10 and 37 Ghz. It was found that, when a thin (3 mm) ice crust is formed on the dry iced material, by spraying water, there is an increase in the brightness temperature of more than 15 Kelvin at 37 GHz horizontal polarisation. When, the sample becomes wet after de-icing, the brightness temperature steeply decreases by about 60 Kelvin at both frequencies and polarisations.

1. INTRODUCTION

Timely detection of road ice can be of vital importance for the road traffic safety. Classical approaches to the specific problem include the use of detectors embedded in the concrete or located nearby. These approaches may be expensive and/or unsatisfactory, since local road conditions may be different from the average status. On the other hand, only a few studies have been performed using remote sensors [1]

In this paper, we report an experimental study aimed at detecting different road conditions by means of microwave radiometry. The experiment was conducted in an open-air laboratory using a sample of asphalt as target and two dual polarised radiometers at 10 and 37 GHz. Experimental data have been interpreted by means of a simple radiative transfer model.

2. THE EXPERIMENT

The experimental setup was consisted of two dual polarized radiometers at 10 and 37 Ghz. The instruments were self-calibrating systems with an internal calibrator based on two loads at different temperatures (250 $\pm$ 0.2 K and 370 $\pm$ 0.2 K). Calibration checks in the 30 K - 300 K range were carried out during the experiments by means of an external blackbody and a noise source added to the sky emission. Sky emission was periodically measured by means of a reflecting plate placed above the target and subtracted from the total emission. The measurement accuracy (repeatability) achieved was better than 1.0 K, with an integration time of 1 sec. The beamwidth of the corrugated conical horns was 20° at -3 dB and 56° at -20 dB for both frequencies and both polarizations.

The target was a sample of road asphalt (40 X 60 X 12 cm) made of with the same material

120   G. *Macelloni* et al.

used in the construction of motor roads. The equipment geometry was arranged to meet the conditions of far-field operation at the observation angle θ = 40° (Fig.1).

Figure 1 The Experimental Equipment

To eliminate the background effects, the tile was placed over a reflecting plate whose dimensions were larger than the antenna side-lobes. Since the target was smaller than the antenna footprints (HPBW), its true brightness temperature Tbt was computed from the relationship :

$$Tbtot = v \ Tbt + (1-v) \ Tbb$$

where Tbtot and Tbb are, respectively, the measured and the background brightness temperatures and v is the beam filling factor. The latter was computed by measuring the brightness temperature of an eccosorb panel, with the same dimensions as the target, placed on the reflecting plate. The value of v obtained was 0.92 for the 37 GHz radiometer and of 0.96 for the 10 GHz.

Before measurement the target was iced in a freezer; the surface was then sprayed with water, in order to obtain a thin crust of ice. In both cases, to smooth the effect of temperature gradient between the asphalt surface and air, the target was placed in box of polystyrene: this material is a good thermal insulator, and is almost transparent to microwaves.

The brightness temperature and the polarisation index PI = 2*(Tbv-Tbh)/(Tbv+Tbh) of the radiation emitted from the target with dry and wet surfaces were measured as a function of surface temperature between -20 and +5 degrees centigrade.

Since the first measurements had shown that, below 0°C, the transmissivity of the target was appreciable, especially at 10 GHz, we placed it upon an eccosorb panel of the same dimensions. The resulting configuration was that of a slab (asphalt) upon an absorbing infinite half-space. The surface temperature of the asphalt was monitored with an infrared sensor, while additional temperature measurements were performed by means of thermoresistances (Pt 100) placed in the asphalt 3 cm below the surface and in the absorbing panel.

EXPERIMENTAL RESULTS

It was noted that, when the temperature of the dry asphalt increased from –20 °C to +5 °C, the brightness temperature Tb at 37 GHz slight increased (at both horizontal H and vertical V polarization components), while Tb at 10 GHz remained almost constant (Fig. 2).

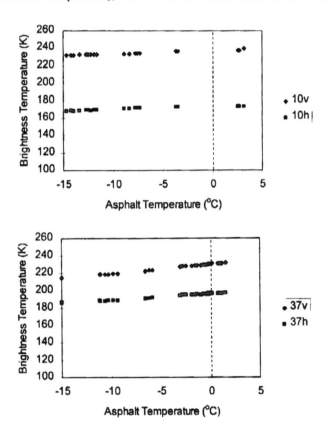

Figure 2 Brightness Temperature of asphalt at 10 GHz and 37 Ghz as a function of thermometric surface temperature

When a layer of 3 mm of ice was created upon the surface by spraying water on the target at -20° C, the brightness temperatures at 10 Ghz (V and H pol) and at 37 GHz (V pol) showed values and trends similar to those of the dry asphalt for target temperatures below 0° C, whereas the H component of brightness temperature at 37 GHz was about 15 K higher (Fig. 3).

When, after de-icing, the sample became wet, the brightness temperature steeply decreased by about 60K at both frequencies and polarisations. At the same time, the polarisation index increased from 0.10 to 0.25 at 37 Ghz. The values of Tb of the target in various conditions are summarized in Table I, which also includes the cases of dry and wet asphalt at about 20°C of surface temperature.

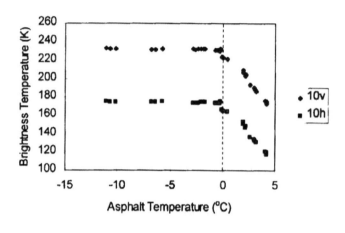

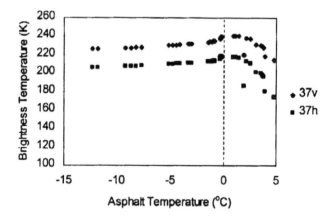

Figure 3 Brightness Temperature of the asphalt with a ice layer of 3 mm at 10 GHz and 37 Ghz as a function of the thermometric surface

Table I

Target	Tv10 (K)	Th10 (K)	Tv37 (K)	Th37 (K)	Tir (K)
Dry Asphalt	236.31	172.70	223.30	191.94	268
Asphalt + Ice	232.06	174.90	229.76	209.88	268
Dry Asphalt	247.09	181.15	247.18	213.35	295
Asphalt + Water	208.57	143.63	183.72	138.16	291

These results seem very promising for the realisation of a simple system (one frequency, single polarisation) able to provide a hazard warning to motorists and encourage the carrying out of further research on motor roads.

DATA ANALYSIS

In order to interpret the experimental results, we used a simple radiative transfer model in which ice and asphalt are characterized by single scattering albedo and extinction coefficient. The equation that describes the brightness temperature T_b of the asphalt tile covered by an ice layer and placed on the absorbing panel (Eccosorb) is the following:

$$T_b = \frac{1 - \Gamma_1}{1 - \Gamma_1 \Gamma_2 / L_2^2} \left\{ \left(1 + \frac{\Gamma_2}{L_2} \right) \left(1 - \frac{1}{L_2} \right) (1 - a_2) T_2 + \right.$$

$$\frac{1 - \Gamma_2}{L_2} (1 - \Gamma_1) \left[\left(1 - \frac{1}{L_3} \right) (1 - a_3) T_3 + \frac{1}{L_3} T_4 \right] \right\}$$

$$+ \Gamma_1 T_{sky}$$

where:

Γ_1 = Reflectivity at the air-ice interface;
Γ_2 = Reflectivity at the ice-asphalt interface ;
T_2 = Temperature of the ice layer ;
T_3 = Temperature of the asphalt layer ;
T_4 = Temperature of the absorbing layer (eccosorb);
T_{sky} = Brightness Temperature of sky;
a_2 = single scattering albedo of ice;
a_3 = single scattering albedo of asphalt
$L_2 = e^{k_{e2} d \, Sec \, \theta_2}$ = loss of ice
k_{e2} = extinction coefficient of the ice layer, d = thickness of the ice layer , θ_2 = incidence angle.
$L_3 = e^{k_{e3} D \, Sec \, \theta_3}$ = loss of asphalt

k_{e3} = extinction coefficient of the asphalt layer, D = thickness of the asphalt layer , θ_3 = incidence angle

The values of the single scattering albedo and extinction coefficient of asphalt have been retrieved through equation 1, by measuring the brightness temperature of the sample placed first on the reflecting plate and then on the absorbing panel. The ice values have been taken from previous works [2]

We note that at 10 GHz the model well fit the data (Fig. 4), but the ice layer is almost transparent and the result is similar to that obtained on the asphalt without ice. At 37 GHz the model represents the behavior of data with an overestimation of about 10 K for the vertical polarization and an underestimation of about 5 K for the horizontal polarization (Fig. 4).

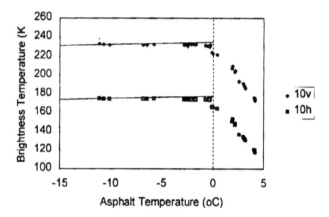

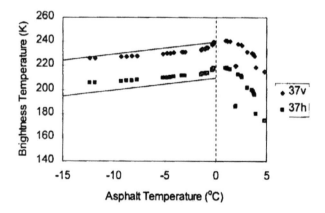

Figure 4 Brightness Temperature of the asphalt with ice as a function of the thermometric surface temperature (model = continuous line and data).

CONCLUSIONS

The measurements carried out on the target and the simple radiative transfer model have pointed out that a single polarization microwave radiometer at 37 Ghz is able to identify surface wetness of asphalt and to detect a crust ice of 3 mm. These results are very promising for the realization of a system able to provide a hazard warning to motorists and encourage the carrying out of further research on motor roads.

REFERENCES

[1] K. Sarabandi, E. S. Li, A. Nashashibi, *Modeling and Measurements of Scattering from Road Surfaces at Millimeter-Wave Frequencies,* IEEE Trans. On Antenna and Propagation, vol. 45,no. 11, pp. 1679-1688, November 1997.

[2] F. T. Ulaby, R. Moore, A. K. Fung , *Microwave Remote Sensing-Vol III.* Norwood, MA: Atrech House, 1986

2. Remote sensing of atmosphere

2.1 Ground based observations

Microw. Radiomet. Remote Sens. Earth's Surf. Atmosphere, pp. 129–135
P. Pampaloni and S. Paloscia (Eds)
© VSP 2000

Resolution and accuracy of a multi-frequency scanning radiometer for temperature profiling

ED R. WESTWATER [1], YONG HAN[1], and FRED SOLHEIM[2]

[1]CIRES, University of Colorado/NOAA Environmental Technology Laboratory, Boulder, CO 80303, USA.
[2]Radiometrics Corporation, Boulder, CO 80303, USA.

Abstract - A new multi-frequency scanning radiometer has been constructed by the Radiometrics Corporation. The accuracy of the system is evaluated for a climatology whose temperature profiles are difficult to recover by radiometric retrievals - Barrow, Alaska, USA. It is found that rms retrieval accuracies of less than 1 K are possible up to about 3 km. In addition, the vertical resolution of the system is evaluated using two variations of the Backus-Gilbert technique. The first uses only information from the radiometric weighting functions, while the second also includes a priori statistical information on temperature profiles.

1. INTRODUCTION

In experiments conducted on a FLoating Instrument Platform [1,2], at two Water Vapor Intensive Operating Periods at the Atmospheric Radiation Measurement (ARM) Program's Southern Great Plains Central Facility, and at the Boulder Atmospheric Observatory [3], an angular-scanning single-frequency radiometer accurately measured low altitude temperature profiles. ARM has purchased from the ATTEX Corporation, Moscow, Russia, and is currently operating a similar instrument on the North Slope of Alaska/Adjacent Arctic Ocean (NSA/AAO) Cloud And Radiation Testbed (CART)site. This instrument derives temperature profiles up to about 300 m with a rms accuracy of about 1.0 K. To extend the accuracy of such instruments to higher altitudes, Radiometrics Corporation, Boulder, Colorado, USA, is developing a multi-frequency scanning radiometer that operates in the frequency region of the 60 GHz oxygen absorption band. This paper describes the instrument as well as a theoretical evaluation of the technique.

2. DESCRIPTION OF INSTRUMENT

The Radiometrics ground-based TP/WVP-3000 portable water vapor and temperature profiling radiometer measures well-calibrated brightness temperatures from which are derived profiles of temperature, water vapor, and limited resolution profiles of cloud liquid water from the surface to 10 km. Descriptions of the system are given in [4,5] and only a short summary of instrument characteristics is given here. Novel characteristics of the system include a very stable local oscillator, an economical way to generate multiple frequencies, and the multi-frequency and scanning capability. The radiometer system consists of two separate subsystems in the same cabinet that share the same antenna and antenna pointing system. A highly stable synthesizer acts as receiver local oscillator and allows tuning to a large number of frequencies within the receiver bandwidth. The water vapor profiling (WVP) subsystem receives thermal emission at five selected frequencies between 22 and 30 GHz. The temperature profiling (TP) subsystem uses sky observations at seven selected frequencies between 51 and 59 GHz. Surface meteorological sensors measure air temperature, barometric pressure, and relative humidity. To improve measurement of water vapor and cloud liquid water density profiles, cloud base altitude information is obtained with an infrared thermometer. In this paper, we only evaluate the temperature sensing characteristics of the instrument. The salient characteristics of the temperature channels are shown in Table 1.

Table 1. Characteristics of Radiometrics TP-3000 Angular-Scanning Radiometer	
Frequencies (GHz) for water vapor and liquid sensing	22.235, 23.035, 23.835, 26.235, 30.00
Frequencies (GHz) for Temperature Sensing	51.25, 52.85, 53.85, 54.94, 56.60, 57.29, 58.80
Absolute Accuracy (K)	0.5
Sensitivity (K)	0.25
FWHP beamwidth (deg)	2.2 -2.4
Gain (dB)	36-37
Sidelobes (dB)	<-26

3. TEMPERATURE WEIGHTING FUNCTIONS

For the frequencies considered here, the main radiators in the atmosphere are water vapor (H_2O), oxygen (O_2), and cloud liquid droplets. Because the O_2 concentration in the troposphere is relatively stable and therefore, can be inferred from air pressure and temperature, the brightness temperature T_B is a function of the vertical profiles of air pressure p, temperature T, water vapor density ρ, and cloud water content ρ_L at specified frequency v and elevation angle θ, i.e., $T_B = T_{B,v,\theta}(p, T, \rho, \rho_L)$. The sensitivity of the brightness temperature T_b to any of the four parameters, represented by x, may be evaluated by the so-called weighting function $W_x(z)$, which is the response (change) of the brightness temperature to a unit positive perturbation of the profile in a 1-km-thick layer at a height z. The temperature weighting function W_T is given in [6]

$$W_T(z)=(1/\sin(\theta)) \, \alpha(z)e^{-\tau(0,z)/\sin(\theta)}$$

$$+(1/\sin(\theta)e^{-\tau(0,z)}\frac{\partial\alpha(z)}{\partial T}[T(z)-T_{bb}e^{-\tau(0,\infty)/\sin(\theta)} - \int_z^\infty T\alpha e^{-\tau(z,z')/\sin(\theta)} dz'] \quad (1)$$

In (1), α is the absorption coefficient, $\tau(z, z')$ is the optical depth between the heights z and z', T is the temperature, T_{bb} is the cosmic big bang brightness temperature (K), and θ is the elevation angle. The first term of the above equation is due to the temperature increase of the radiation source function in the layer at z, and the remaining three terms are due to the change of the absorption coefficient caused by the increase of the layer temperature. The first of the three terms is the increase (decrease) of radiation in the layer at z due to the increase (decrease) of the absorption coefficient; the other two terms count for the decrease (increase) of radiation, coming from higher layers and the space, caused by the increase (decrease) of the absorption at layer z. Examples of temperature weighting functions as a function of elevation angle are shown in Figure 1 for the most transparent and most opaque channel of Table 1. All temperature weighting functions for ground-based tropospheric remote sensing decrease rapidly

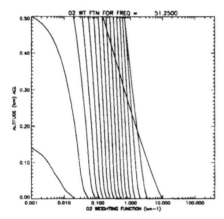 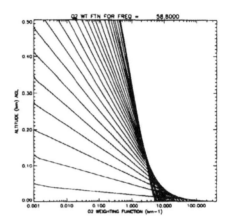

Figure 1. Temperature weighting functions as a function of elevation angle. (51.25 GHz) - the most transparent channel of Table 1 and (58.8 GHz) - the most opaque channel.

with height, but with different rates and strengths. It is the differentiation among the decreasing rates that makes possible temperature profile measurements with good vertical resolution below about 300 m. With the exception of the most transparent channels, the response to water vapor and clouds is negligible. Numerical evaluation of these weighting functions, as well as accuracy and resolution analysis that is presented later, is based on the millimeter wave propagation model of Liebe [7].

4. SINGULAR VALUE DECOMPOSITION OF WEIGHTING FUNCTIONS

As might be expected from the Figure 1, there is a high degree of redundancy between the various frequency and angular weighting functions. To characterize this redundancy, and also to compress the amount of useful information in the measurements, we performed a singular value decomposition (SVD) [8] of the weighting functions. We reduced the weighting functions to a matrix A based on an equally spaced quadrature of m = 301 points between 0 and 3 km. For the SVD, we write

$$A = U \Lambda V^T \tag{1}$$

where A is an n x m matrix, U is an n x m matrix whose columns contain the left singular vectors of A, Λ is an m x m diagonal matrix whose elements are the singular values of A, and V is and m x m matrix whose columns contain the right singular vectors of A. In an ill posed problem, in which many of the singular values of Λ are below the noise level, it is desirable to use only the vectors whose singular values are significant. For a selection of 50 angles and 7 frequencies; i. e., 350 measurements, it was determined that all of the information could be compressed to an accuracy of 1 part in 10^6 by 9 singular functions. The number of singular functions that are used depends on the minimum value below which the singular values are considered insignificant. This minimum value may be determined by the computational precision, as we did here.

5. ACCURACY ANALYSIS

The accuracy of a ground-based system is a function of the instrument characteristics, as well as the climatological regime in which the instrument operates. The primary instrumental characteristics are the absolute accuracy in measuring the brightness temperature and the beamwidth. The primary climatological factors include the strengths, heights, and frequency of occurrence of inversions, especially elevated inversions. We have performed an error analysis of the multi-frequency system for four diverse climatologies of the USA: Barrow, Alaska; Fairbanks, Alaska; the CART site in central Oklahoma, and Denver, Colorado. For brevity, we will only show results for the location with the highest variability: Barrow, Alaska. Our simulations showed that the accuracies could vary by almost a factor of two between the various climatologies.

About 5000 radiosondes over about 5 year period from each of the four sites are collected as statistical ensembles. Calculations of brightness temperatures at the frequencies shown in Table 1, as well as 50 elevation angles evenly distributed between 5 and 90 degrees were carried out. The lower limit of 5 degrees was chosen to be about 2 beamwidths, so that ground effects were small. Gaussian noise was added to the calculated brightness temperatures with the assumption that there are no correlations between the measurements. Surface air temperature Ts measurements were also included in the measurement vector. Figure 2 shows the expected accuracy of deriving temperature profiles from the seven channel radiometer, for an assumed noise levels of 0.1 and 0.5 K. For comparison, the accuracy derived from predicting the temperature by T_s is also shown.

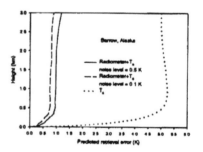

Figure 2. Estimated temperature retrieval accuracy as a function of altitude. Assumed noise level = 0.1 and 0.5 K rms. T_s = surface temperature.

6. RESOLUTION ANALYSIS

The characterization of vertical resolution in radiometric atmospheric profiling has been rather difficult to define. Definitions of resolution, such as the Rayleigh criterion that is applied in optics, or the width of a radar pulse volume in radar meteorology, are not readily applicable to the overlapping weighting functions of radiometry. One definition, as was applied by Backus and Gilbert [9] to remote sounding of the solid earth, is applicable, and, as extended by Rodgers

[10] to include **a priori** statistics, lends itself rather well to radiometry. The basic idea is, for each height at which a retrieval is desired, to construct a linear combination of weighting functions, that approximates as closely as possible, some function that has ideal resolution characteristics; e. g., a Dirac delta function, or a gaussian with small standard deviation centered about the point in question. We follow Backus and Gilbert and chose a function called the spread as a measure of resolution. A tradeoff between spread and error is a key element of their theory.

6.1 Minimum spread of the seven channel scanning radiometer

We evaluated minimum values of the spread for 11 altitudes between 0 and 3 km. First, we projected the weighting functions for 7 frequencies and 50 angles onto the first nine singular functions, and evaluated the spread for these projections. The results are shown in Table 2. We note that the resolution, without the input of **a priori** data, is roughly equal to the height in question.

Table 2. Minimum Spread of 7-Channel Angular-Scanning Radiometer	
height (km)	Spread (km)
0.025	0.020
0.050	0.046
0.100	0.097
0.150	0.151
0.250	0.232
0.500	0.469
0.750	0.789
1.000	0.953
1.500	1.651
2.000	1.770
2.500	3.544

6.2 Backus-Gilbert method using a priori statistics

As discussed in [10], information other than radiation measurements can be incorporated into the spread-error analysis. For example, we frequently have **a priori** climatological statistics, a forecast from a numerical model, or some other source of information. To use such information, the error characteristics of the source of information are needed. Here, we have introduced the covariance matrix of temperature fluctuations, conditioned on the surface temperature measurement. We evaluated the spread vs. error of the measurement system consisting of the seven-channel scanning radiometer (assumed noise level of 0.5 K) and the climatological mean, conditioned on T_S, determined from the **a priori** statistics of Barrow, Alaska. Again, the 350 weighting functions were projected onto the first nine singular functions. The tradeoff curves are shown in Figure 3. We note that, as expected from the weighting functions shown in Figure 1, the vertical resolution (spread) becomes poorer with height, but that good resolution and accuracy are achievable to about 300 m. As another figure of merit, we show in Figure 4 the spread at which the error standard deviation becomes 1 K rms.

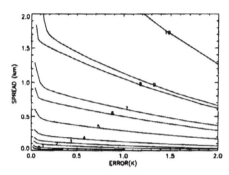

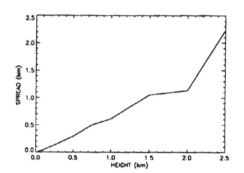

Figure 3. Spread vs. temperature error standard deviation as for several heights. The height labels from 0 to 10 indicate the heights: 0.025, 0.05, 0.1, 0.15, 0.25, 0.5, 0.75,1.0,1.5,2.0,2.5 km, and the assumed radiometric resolution is 0.5 K rms.

Figure 4. Spread vs. height for a temperature error standard deviation of 1.0 K rms.

7. CONCLUDING REMARKS

We evaluated the expected temperature profile retrieval accuracy during clear conditions, and for a severe arctic environment, of a scanning seven-frequency microwave radiometer, that has been developed for ARM by the Radiometrics Corporation. The results show that rms retrieval accuracies of better than 1 K rms are achievable up to 3 km with this system, although the vertical resolution degrades rapidly above 500 m. Although the results were obtained under the assumptions of clear conditions, the complete TP/WVP-3000 radiometer has water vapor and cloud channels, that should allow roughly the same temperature-profiling accuracy to be obtained during non-precipitating conditions. Previous experience obtained with a 6-channel zenith-viewing Radiometric Profiler showed that the effects of moisture could be taken into account with a dual-frequency water vapor radiometer [6]. The multi-frequency WVP should only improve on this accuracy.

Finally, we note that the Radiometric Corporation's TP/WVP-3000 was used at the NSA/AAO in Barrow, Alaska, during March 1999, along with two single-frequency scanning O_2-band radiometers. The data taken during this experiment should allow an excellent evaluation of the respective instruments during extreme arctic conditions.

REFERENCES

1. Y. G. Trokhimovski, E. R. Westwater, Y. Han, and V. Ye. Leuskiy, The results of air and sea surface temperature measurements using a 60 GHz microwave rotating radiometer, *IEEE Trans. Geosci. Remote Sensing*, **36**, 3-15, (1998).
2. E. R. Westwater, Y. Han, V. G. Irisov, V. Ye. Leuskiy, Yu. G. Trokhimovski, C. W. Fairall, and A. Jessup, Sea-air and boundary layer temperatures measured by a scanning 5-mm wavelength (60 GHz) radiometer: Recent results. *Radio Sc.*, **33**, 291-302, (1998).
3. E. R. Westwater, E. R., Y. Han, V. G. Erosive, V. Leuskiy, E. N. Kadygrov, and S. A. Viazankin, Remote sensing of boundary-layer temperature profiles by a scanning 5-mm microwave radiometer and RASS: A comparison experiment, *J. Atmos. Oceanic Technol.*,**16**, 805-818, (1999).
4. F. Solheim, J. Godwin, E. Westwater, Y. Han, S. Keihm, K. Marsh, and R. Ware, Radio metric profiling of temperature, water vapor, and liquid water using various inversion methods, *Radio Science*, **33**, 393-404, (1998).
5. F. Solheim, J. Godwin, and R. Ware, Passive ground-based remote sensing of atmospheric temperature, water vapor, and cloud liquid profiles by a frequency synthesized microwave radiometer, *Meteorologische Zeitschrift*, **7**, 370-376, (1998).
6. E. R. Westwater, Ground-based Microwave Remote Sensing of Meteorological Variables, Chapter 4 in *Atmospheric Remote Sensing by Microwave Radiometry*, J. Wiley & Sons, Inc., Michael A. Janssen, ed., 145-213, (1993).
7. H. J. Liebe, MPM: An atmospheric millimeter wave propagation model. *Int. J. Infrared Millimeter Waves*, **10**, 331-350, (1989).
8. G. H. Golub and C. Reinsch, Singular value decomposition and least squares solutions, *Numer. Math.*, **14**, 403-420, (1970).
9. G. E Backus and F. E. Gilbert, Uniqueness in the inversion of inaccurate gross earth data, *Phil. Trans. Roy. Soc. London*, **A266**, 123-192, (1970).
10. C. D Rodgers, The vertical resolution of remotely sounded temperature profiles with a priori statistics, *J. Atmos. Sci.* , **33**, 707-709, (1976).

Microw. Radiomet. Remote Sens. Earth's Surf. Atmosphere, pp. 137–143
P. Pampaloni and S. Paloscia (Eds)
© VSP 2000

Implications of an improved atmospheric absorption model on water vapor retrievals

CHRISTOPHER S. RUF

The Pennsylvania State University, University Park, PA 16802, USA

Abstract - Recent investigations, comparing radiosonde observations with coincident measurements by a ground based microwave radiometer/spectrometer, have resulted in a statistically significant correction to the Liebe'87 and Liebe'93 atmospheric absorption models near the 22.235 GHz water vapor line [1]. Specifically, those results indicate that the line strength of the Liebe'87 model should be increased by 6.4% and the width parameter describing pressure broadening should be increased by 6.6% for both models. These changes can significantly affect the water vapor retrievals made by radiometers operating in this frequency range. If either Liebe model is assumed when developing a retrieval algorithm, when in fact the modified model is correct, then a scale error in the retrieved water vapor can result in which both its magnitude and sign depend on the particular frequencies that are used. In addition, the variation of scale error with frequency is very different depending on which of the two Liebe models is assumed. A generalized formulation is presented for the retrieval of integrated water vapor and cloud liquid water using a ground-based two frequency radiometer. A propagation-of-errors analysis produces the spectrum of the scale error that results from this model change.

1. INTRODUCTION

A recent inter-comparison experiment addressed the accuracy of numerous models for atmospheric absorption by water vapor in the near vicinity of the 22.2 GHz water vapor absorption line [1]. In summary, an upward looking spectrometer/radiometer operating at 20.0, 20.3, 20.7, 21.5, 22.2, 22.8, 23.5, 24.0, and 31.4 GHz was deployed on the ground at US National Weather Service radiosonde launch sites. Several months of data at low to medium humidity levels were screened for: 1) clear sky conditions; 2) high radiometer calibration accuracy (via low "tipping curve" residual errors); and 3) high reliability radiosonde profiles (via absence of relative humidity "clipping" at both high and low levels). The resulting database of brightness temperatures and coincident atmospheric profiles of pressure, temperature, and water vapor density were used as constraints on a perturbation to standard atmospheric absorption models. Models based on the Gross pressure broadened water vapor line shape were found not to fit the data. Two popular models based on the Van Vleck-Weisskopf line shape could be fit to the data by small adjustments to their line strength and line width parameters. Specifically, the Liebe'87 model [2] was "best fit" to the data by an increase of 6.4% in its line strength and of 6.6% in its line width. The Liebe'93 model [3] was "best fit" by increases of 1.3% and 6.6% in the strength and width, respectively. The estimated accuracy of these corrections were $\pm 1.6\%$ for strength and $\pm 1.0\%$ for width.

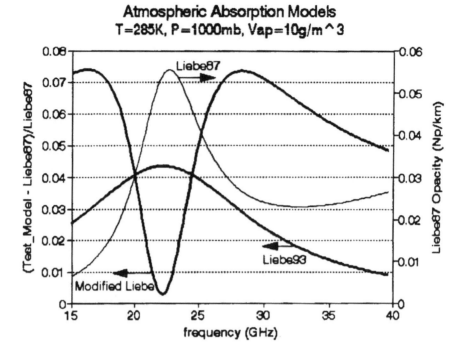

Figure 1. Spectrum of the relative change in atmospheric absorption [Np/km] predicted by the Liebe'93 and modified Liebe [1] models with respect to the Liebe'87 model, under typical clear conditions (bold lines, left vertical axis). The baseline absorption spectrum of the Liebe'87 model itself is also shown (thin line, right vertical axis). The change from Liebe'87 to Liebe'93 amounts to an increase by 5% in the sensitivity to water vapor. The change to modified Liebe involves an increase in both the strength of the water vapor line (which increases sensitivity at all frequencies) and in its width (which decreases sensitivity near line center and increases it outside of the inflection points of the pressure broadened line at ~20.7 and 23.8 GHz). The net effect is a more complicated spectral behavior.

The change in atmospheric absorption near 22 GHz from the Liebe'87 to the Liebe'93 model is principally an increase by 5% in the strength of the water vapor absorption. As such, errors in the water vapor retrieved by a microwave radiometer that resulted from assuming one of these models when the other was correct would amount to a simple 5% scale error, regardless of the particular frequencies at which brightness temperature (TB) was measured. The same is not true of the modified Liebe model. Due to the increase in line width, changes in the strength of the water vapor absorption from either Liebe model to the modified model will be frequency dependent. An illustration of the spectral behavior of the three models under consideration is shown in Figure 1 for a typical clear air atmosphere. The relative change in absorption is plotted versus frequency, with the Liebe'87 model taken as a baseline for comparison. Also shown is the absolute absorption of the Liebe'87 model for this atmosphere. Note that the relative change from Liebe'87 to Liebe'93 largely follows the shape of the Liebe'87 absolute absorption. The change from Liebe'87 to modified Liebe, on the other hand, has a more complicated frequency dependence. Changes are smallest near the 22.235 GHz line center, where the increases in line strength and width have largely offsetting effects on the total absorption. At frequencies away from line center, and especially outside of the inflection points near 20.7 and 23.8 GHz, changes in both the line strength and width will tend to increase the total absorption and their effects will add coherently.

We examine here the effect that the new absorption model has on a typical ground-based water vapor radiometer retrieval algorithm, which uses measurements of TB at two frequencies in order to correct for cloud liquid water contributions. One of the frequencies typically lies on or very near the water vapor line and the second sits in the atmospheric window just above the line. We consider 31.0 and 37.0 GHz as candidate second frequencies, which are common choices by researchers in the field [4, 5]. The choice of the first frequency is left as a free parameter in order to characterize its influence on the error.

The analysis begins by isolating only that component of the retrieval problem which is sensitive to the model change. A propagation-of-errors approach is then used to evaluate the effect of the change in the model on water vapor retrieval.

2. APPROACH - GENERALIZED SENSITIVITY OF TWO FREQUENCY RETRIEVAL ALGORITHMS

The sensitivity of the integrated atmospheric opacity to the total column of water vapor and the integrated cloud liquid water can be accurately approximated for all but the heaviest clouds by a first order series expansion of the opacity. The expansion has the form

$$\tau(f) = \tau_0(f) + a_1(f)V + a_2(f)L \qquad (1)$$

where $\tau(f)$ is the integrated opacity at frequency f, $a_1(f) = \dfrac{\partial \tau(f)}{\partial V}$, $a_2(f) = \dfrac{\partial \tau(f)}{\partial L}$, V is the integrated water vapor content, and L is the integrated cloud liquid water content. The sensitivity coefficients, a_1 and a_2, are estimated numerically in the following manner.

An artificial atmosphere is assumed, in which the temperature and pressure profiles follow the 1967 US Standard Atmosphere [6]. A relative humidity profile is superimposed that is 70% at all altitudes except 1.0-1.5 km, at which it is 100%. The water vapor density profile can then be derived. A cloud liquid water profile is also generated, by assuming that a cloud is present at altitudes with 100% relative humidity and that the liquid water density in the cloud is half the difference between the water vapor density in the cloud and at the cloud base [7]. Combined together, these conditions amount to values for V and L of 2.0 cm and 0.22 mm, respectively.

The atmospheric profiles are applied to a numerical radiative transfer routine to estimate opacity. The routine uses one of either the Liebe'87, Liebe'93, or modified Liebe absorption models for water vapor and, in all cases, the Liebe'87 model for absorption by liquid water and the Rosenkrantz'93 model for oxygen absorption [8]. This produces estimates of integrated opacity versus frequency. Additionally, a second opacity spectrum is generated in which the absolute water vapor density and cloud liquid water density are individually increased by 1% at each altitude. The partial derivatives in (1) are then estimated numerically from the perturbations to $\tau(f)$, V, and L.

In order to isolate the effects of changes in the assumed absorption model from any other possible changes in the behavior of the water vapor retrieval, the following assumptions are made: 1) The microwave radiometers produce exactly calibrated and noise free TBs. 2) The integrated opacity can be derived exactly from the TBs. This requires that the effective radiating temperature of the atmosphere is known. While it is never known exactly, this assumption essentially implies that knowledge of the radiating temperature will not be any better or worse if any particular absorption model is correct. 3) The component of opacity not due to changes in water vapor or cloud liquid, i.e. $\tau_0(f)$ in (1), is known exactly. Taken together, these conditions amount to an assumption that the retrieval algorithm for water vapor and cloud liquid will act on ideal measurements of the quantity m as defined by

$$m(f) = \tau(f) - \tau_0(f) \qquad (2)$$

Assume that the quantity m is measured at two frequencies. If the correct forward model for atmospheric absorption is the modified Liebe version, then m is given by

$$\vec{m} = \mathbf{F}_{mod}\vec{s} \qquad (3)$$

where $\vec{s}$ is a two element state vector made up of V and L, and where $\mathbf{F}_{mod}$ is the forward model given by

$$\mathbf{F} = \begin{bmatrix} a_1(f_1) & a_2(f_1) \\ a_1(f_2) & a_2(f_2) \end{bmatrix} \qquad (4)$$

in which the partial derivatives, a_i, are evaluated using the modified Liebe model. If some other model is assumed by the water vapor retrieval algorithm, then inversion of the measurements will take the form

$$\vec{s} = \mathbf{F}_{Liebe'nn}^{-1}\vec{m} \qquad (5)$$

where $\vec{s}$ is the retrieval algorithm's estimate of the atmospheric state and $\mathbf{F}_{Liebe'nn}^{-1}$ is the inverse of the forward model, as in (4) but evaluated using the Liebe'nn model. The retrieval error then follows as

$$\overset{\scriptscriptstyle \varpi}{s} - \overset{\scriptscriptstyle \varpi}{s} = (\mathbf{I} - \mathbf{F}_{\text{Liebe'nn}}^{-1}\mathbf{F}_{\text{mod}})\overset{\scriptscriptstyle \varpi}{s} = \mathbf{E}\overset{\scriptscriptstyle \varpi}{s} \tag{6}$$

where I is the 2x2 identity matrix and E as defined in (6) is the scale error matrix.

Inspection of the elements of E for both Liebe'87 and Liebe'93 and at all frequencies in the near vicinity of 22 GHz reveals that: 1) the water vapor scale error, E_{11}, is not negligible and varies significantly with frequency; 2) the coupling between vapor and liquid scale errors, as accounted for by E_{12} and E_{21}, is negligible; and 3) the cloud liquid scale error, E_{22}, is negligible. The results reported below concentrate on the water vapor scale error.

3. RESULTS - WATER VAPOR SCALE ERROR SPECTRUM

Two cases of the second frequency, f_2 in (4), are considered, namely 31 and 37 GHz. In each case, the first frequency is varied over 20-25 GHz and the water vapor scale error, E_{11} in (6), is computed for both of the Liebe'87 and Liebe'93 models. The results are shown in Figure 2 for 31 GHz and Figure 3 for 37 GHz. Note in Figure 2 that the scale error is very small for the Liebe'87 model with a 1st frequency that is close to the 22.2 GHz line center. This is consistent with the behavior of the absorption spectrum itself, as shown in Figure 1. The scale error near line center with the Liebe'93 model is much higher. The shape of the scale error versus frequency is similar for both models, since

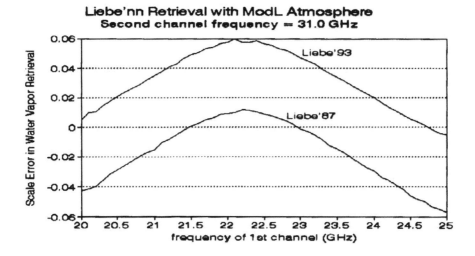

Figure 2. Scale error in the retrieval of water vapor by a radiometer operating at a variable 1st frequency plus 31 GHz, assuming that the Liebe'87 or '93 model for atmospheric absorption by water vapor is correct instead of the modified Liebe model.

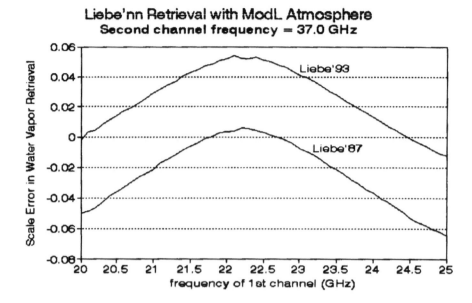

Figure 3. Similar to Figure 2 but with a second frequency of 37 GHz instead of 31 GHz.

they differ between each other by only a 5% change in sensitivity to water vapor. In fact, the difference between the two plots in Figure 2 is simply that 5% scale factor. If a similar plot were derived showing the scale error due to the assumption of Liebe'93 when Liebe'87 was correct, for example, it would amount to a constant value of 5% at all choices for the 1st frequency of the radiometer and independent of the 2nd frequency. In Figure 3, the general shape of the scale error is the same, as is the 5% offset between Liebe'87 and '93.   However, the magnitude of the error has shifted down by approximately 0.6%. This is a result of the smaller change in absorption models at 37 GHz than at 31 GHz.

4.   CONCLUSIONS

The retrieval of water vapor by a ground-based two-frequency microwave radiometer depends explicitly on the shape and strength of the water vapor absorption spectrum that is assumed. If the assumed model overestimates the strength of the absorption, then a given amount of water vapor would be expected to result in a larger integrated opacity and would produce a higher brightness temperature. Using measurements of that same TB by a radiometer, a retrieval algorithm would, then, underestimate the actual amount of water vapor present.   This is the case, for example, if the Liebe'93 model is assumed when the Liebe'87 model is correct. If both the line shape and strength are in error, then the effect this has on a retrieval algorithm is not so straightforward. The error will depend on which specific microwave frequencies are used. This is the case if either of

Table 1. Water vapor scale error due to the wrong absorption model used by the retrieval algorithm, for a microwave radiometer with a 31.0 GHz 2nd Channel

1st Radiometer Channel	Model Assumed by Retrieval	
	Liebe'87	**Liebe'93**
20.7 GHz	-2.3%	+2.6%
22.2 GHz	+1.2%	+5.8%
23.8 GHz	-2.4%	+2.5%

the Liebe'87 or Liebe'93 models is assumed when the modified Liebe model [1] is correct.

A generalized sensitivity formulation has been developed to isolate and characterize the behavior of a water vapor retrieval algorithm specifically with respect to errors in the assumed absorption model. A relative, or scale, error in the retrieved water vapor is found to result. The magnitude and sign of the scale error depend on the choice of frequencies. Specific values for the error at several popular combinations of microwave radiometer frequencies are listed in Table 1 for the case of a 1967 US Standard Atmosphere [6]. Very similar scale errors can be expected in other atmospheric conditions as well.

REFERENCES

1. S.L. Cruz Pol, C.S. Ruf, and S.J. Keihm, "Improved 20- to 32-GHz atmospheric absorption model," *Radio Science*, 33(5), 1319-1333. (1998)
2. H.J. Liebe and D.H. Layton, "Millimeter-wave properties of the atmosphere: Laboratory studies and propagation modeling," *NTIA Rep.* 87-224, Natl. Telecomm. Inf. Admin., Boulder, CO. (1987)
3. H.J. Liebe, G.A. Hufford, and M.G. Cotton, "Propagation modeling of moist air and suspended water/ice particles at frequencies below 1000 GHz," *AGARD Conf. Proc.*, 542, 3.1-3.10. (1993)
4. E.R. Westwater, "The accuracy of water vapor and cloud liquid determination by dual-frequency ground-based microwave radiometry," *Radio Science*, 13, 677-685. (1978)
5. B.L. Gary, S.J. Keihm, and M.A. Janssen, "Optimum strategies and performance for the remote sensing of path-delay using ground-based microwave radiometers," *IEEE Trans. Geosci. Remote Sens.*, 23, 479-484. (1985)
6. A.E. Cole, A. Court, and A.J. Kantor," Model atmospheres, in *Handbook of Geophysics and Space Environments*, S.L. Valley, (Ed.), Office of Aerospace Research, USAF, Cambridge Research Labs., Chapter 2. (1965)
7. M.S. Malkevich, V.S. Kosolapov, and V.I. Statskiy, "Statistical characteristics of the vertical water content structure of cumulous clouds," *Azvestiya Atmos. Ocean. Phys.*, 17, 123-136. (1981)
8. P.W. Rosenkrantz, "Absorption of microwaves by atmospheric gases," In: *Atmospheric Remote Sensing by Microwave Radiometry*, M.A. Janssen (Ed.), Chapter 2, pp. 37-79, John Wiley, New York (1993)

Microw. Radiomet. Remote Sens. Earth's Surf. Atmosphere, pp. 145–153
P. Pampaloni and S. Paloscia (Eds)
© VSP 2000

Analysis of Tip Cal Methods for Ground-based Microwave Radiometric Sensing of Water Vapor and Clouds

YONG HAN and ED R. WESTWATER

CIRES, University of Colorado/NOAA, Environmental Technology Laboratory, Boulder, Colorado, USA, 80303

Abstract -The tipping-curve calibration method has been an important calibration technique to ground-based microwave radiometers that measure atmospheric emission at low optical depth. The method calibrates a radiometer system using data taken by the radiometer at two or more viewing angles in the atmosphere. In this method, the relationship between measured brightness temperature and atmospheric opacity is used as a constraint for deriving the system calibration response. Because this method couples the system with radiative transfer theory and atmospheric conditions, evaluations of its performance have been difficult. In this paper, first a data-simulation approach is taken to isolate and analyze those influential factors in the calibration process, and effective techniques are developed to reduce calibration uncertainties. Then, these techniques are applied to experimental data. The influential factors include radiometer antenna beam width, radiometer pointing error, mean radiating temperature error, and horizontal inhomogeneity in the atmosphere, as well as some other factors of minor importance. It is demonstrated that calibration uncertainties from these error sources can be large and unacceptable. Fortunately, it was found that by using the techniques reported here, the calibration uncertainties can be largely reduced or avoided. With the suggested corrections, the tipping calibration method can provide absolute accuracy about or better than 0.5 K.

1. INTRODUCTION

Ground-based microwave radiometers (MWR) have been widely used to measure atmospheric water vapor and cloud liquid water. Frequencies on the 22.235 GHz water vapor absorption band and in the 31 GHz absorption window region are commonly used in the systems. These frequency channels differentiate in their response to water vapor and cloud liquid water and provide brightness temperature measurements from which precipitable water vapor (PWV) and integrated cloud liquid water (ICL) are derived. The absolute calibration is fundamental in determining the accuracies of these retrievals. For a dual-channel radiometer at 23.8 and 31.4 GHz, a 1.4 K calibration error will cause a 1 mm error in PWV. Such errors are usually of the bias type and can be larger than algorithm errors; e.g., errors caused by use of incorrect absorption coefficients.

The importance of the PWV measurements and, thus, the importance of the system calibration, have increased in recent years as the MWR measurements are often served as references and comparison standards for other water vapor measuring instruments, such as radiosondes, Raman water vapor lidars, and Global Positioning Systems (GPS). One of the goals of the Department of Energy's Atmospheric Radiation Program (ARM) is to evaluate various techniques for determining PWV. During Sept. 15-30, 1996, and Sept. 15- Oct. 5, 1997, Water Vapor Intensive Operating Periods (WVIOP) were conducted at the ARM Cloud And Radiation Testbed (CART) site. Various calibration uncertainties that were noticed during these WVIOPs are the motivation of our work.

Both space-borne and surface based microwave radiometers are complex systems and must be thoroughly analyzed before final calibration is applied. For example, such system factors as mismatch, ohmic losses in antenna and waveguides, frequency response characteristics, and antenna patterns, must be measured and first order corrections must be applied. Temperatures of important components be also be monitored, and a suitable radiometer equation must be developed. The techniques of our work are developed to determine a single numerical factor that determines final calibration.

Microwave radiometers that are used in space are usually calibrated using known calibration reference targets. However, it is desirable to have a target temperature close to the brightness temperatures that a MWR measures during regular observations. For the upward-looking radiometer channels considered here, the observed brightness temperature can be as low as 10 K. Two calibration methods are often applied for these channels: using a liquid nitrogen (LN$_2$) cooled blackbody target or a tipping calibration method that uses a clear atmosphere as a calibration target. If applied with care and caution, the LN$_2$ method can be useful; however, it is not practical to be applied and automated in long-term, routine operations. In addition, LN$_2$ calibration helps determine the relation between output voltage and antenna temperature; a further correction must be applied to derive the desired quantity - brightness temperature. With the tipping curve method, the radiometer takes measurements at two or more elevation angles in a horizontally-stratified atmosphere. The calibration is accomplished by adjusting a single numerical parameter that is required by the system software until the outputs of the system comply with a known physical relationship. In addition, strict quality control must be applied to the data before the calibration parameters are changed. With suitable scanning and quality control, the calibration procedure can be automated. The Environmental Technology Laboratories' and ARM's radiometer systems were all calibrated using the tipping calibration method during the experiment.

2. RADIATIVE TRANSFER AND RADIOMETER EQUATIONS

The fundamental quantity that a radiometer measures is radiative power which, in turn, is related to the specific intensity (or radiance) I_ν [1]. However, in the microwave frequency region, the intensity is usually expressed as brightness temperature, denoted as $T_{b,\nu}$. In microwave radiometry, there are two popular definitions of the brightness temperature [1]. Because of these differences, some confusion and mistakes may arise when measurements or models are compared. We will consistently use the so-called thermodynamic brightness temperature, defined as $T_{b,\nu} = B_\nu^{-1}(I_\nu)$, where B_ν^{-1} is the inverse of the Planck function $B_\nu(T_{b,\nu})$ at a temperature $T_{b,\nu}$[1]. The quantity $T_{b,\nu}$ is not linearly related to I_ν, but can be adequately approximated for $\nu \leq 300$ GHz, by the first two terms of a Taylor series expansion.

The radiometer equation relates the system output voltage to input radiation power. Usually, a MWR system measures the output voltage from two or more internal or external blackbody targets for initial calibrations. In our procedure, the final calibration depends on estimating a single calibration factor from one tipping calibration.

Each of the two ETL systems contains two independent MWRs: one operates at 20.6 or 23.87 GHz and the other at 31.65 GHz. Each radiometer has two internal blackbody loads, one at temperature T_r, near 300 K, and the other at T_h, about 100 K higher than T_r. The radiometer equation of the system [2] is given by :

$$T_{a,m} = c(\frac{T_h - T_r}{V_h - V_r}(V_s - V_r) + T_r - T_w) + T_w, \tag{1}$$

where $T_{a,m}$ is the antenna temperature [1] being measured, T_w is the temperature of the radiometer waveguide, and V_r, V_h, and V_s are voltages of a square law detector, corresponding to the radiation sources of the two internal loads and the sky, respectively. All these voltages and temperatures on the right side of the equation are measured. The unknown parameter c is the calibration factor determined through the calibration.

The ARM's system operates at 23.8 and 31.4 GHz. The system includes a noise diode injection device and an external blackbody reference target at an ambient temperature T_r. The radiometer equation of the system [3] is given by

$$T_{a,m} = \frac{qT_{nd}}{V_{r+d} - V_r}(V_s - V_r) + T_r, \tag{2}$$

where V_s is the voltage when the radiometer views the sky, V_{r+d} is the voltage when viewing the reference target with the noise diode on, V_r is the voltage when viewing the reference with the noise diode off, and T_{nd} is the noise injection temperature determined through the calibration. The constant q in the equation is a known factor counting for the radiation loss and emission of a microwave window in front of the radiometer antenna.

The radiometer equations (1) and (2) may be simplified for the sake of convenience to simulate the tipping data. We may write the parameters c in (1) as $r \cdot c_t$ and T_{nd} in (2) as $r \cdot T_{nd,t}$ where c_t and $T_{nd,t}$ are correct calibration factors, and r represents the correctness of the estimations of c or T_{nd}, with $r = 1$ representing a perfect calibration. With this consideration, (1) or (2) may be rewritten in the form as

$$T_{a,m} = r(T_a - T_g) + T_g, \tag{3}$$

where $T_{a,m}$ is an estimate of the true antenna temperature T_a in the pointing direction, $T_g = T_w$ for the ETL's systems, and $T_g = T_r$ for ARM's system. We see that the measured antenna temperature is equal to true antenna temperature T_a when a calibration is performed without error ($r = 1$). We will refer to the factor r in (3) as the calibration factor. Equation (3) provides a convenient way to simulate the radiometric measurements in which the atmospheric antenna temperature T_a may be calculated using a radiative transfer and an antenna model.

3. TIPPING CALIBRATION METHOD

The tip cal method has been commonly used throughout the microwave community. With the method, brightness temperatures are measured as a function of elevation angle θ and are then converted to optical depth τ using the mean radiating temperature approximation [4].

If the system is in calibration, then the plot of τ as a function of air mass a (= csc (θ)) will pass through the origin (a = 0); conversely, if $\tau = \tau(a) = \tau(1)a + b$ does not pass through the origin, then a parameter in the radiometer equation is adjusted (by determining r in (3) until b $\approx$ 0, usually in the least squares sense. Note that when the calibration is achieved, then the slope of the line is equal to the zenith optical depth.

4. CALIBRATION UNCERTAINTIES AND METHODS TO REDUCE THEM

Calibration uncertainties may be caused by sources from the radiometer system and the violations of the assumptions in the theory on which the calibration is based. The former include the effect of radiometer antenna pattern, radiometer pointing error, and system random noise. The latter include the uncertainty in the mean radiating temperature and the uncertainties in the fundamental relationship between the airmass and the observation angles, which can be affected by non-stratified atmospheric conditions and the earth's curvature.

We simulated these error sources and developed and tested effective techniques to reduce them. The simulations were performed for a clear-sky atmosphere by using a radiative transfer model [4], the simplified radiometer equation (3), and radiosonde pressure, temperature, and humidity profiles. A statistical ensemble of radiosonde data, referred to as *S*, with a size of 16380 soundings, were collected from five stations around the area of Oklahoma City, Oklahoma, from 1966 to 1992. We summarize the results of our simulations below, following [5]. Only air masses ≤ 4 are considered.

A. Effect of earth curvature and atmospheric refractive index

Our simulations showed that the rms differences between the exact and zero gradient are very small at all of the selected airmasses, about a few hundredths of a degree. Thus, the effect of the refractive index profile on system calibration is negligible. Earth curvature has a relatively large effect which causes airmass at an angle θ to be smaller than that of an atmosphere with a flat surface at the same angle. The differences between the two brightness temperatures with and without earth curvature are about one or two tenths of a degree at airmass 3 or 4. Although the effect is still small when compared with those from other sources, the airmass can be conveniently corrected to less than 0.05 K for air mass ≤ 4.

B. Errors caused by uncertainties in radiometer pointing angle

Our simulations showed that pointing errors could have serious impact on the performance of the tipping calibration if only one-side tipping calibration is used. The pointing angle errors can often be identified by performing tipping observations at symmetric angles. If the measurements at one angle consistently differ from those at its symmetric angle, it usually implies the existence of the pointing error (except in the situations when there is a persistent horizontal inhomogeneity in the atmosphere). Fortunately, the effect of the pointing error can often be reduced significantly by using tipping data taken on both sides under the condition that the differences among those angles are known precisely. This is due to the effect that the uncertainties in the measurements on one side due to the pointing error are partially canceled out by the uncertainties of those on the other side. Our simulations strongly suggest that tipping data should be taken in pairs on the symmetric elevation angles.

C. Effect of antenna beam width

The antenna temperature T_a of a radiometer at a specified frequency is a weighted average of incoming brightness temperature T_b (θ, ϕ) over all directions [1]. Under normal atmospheric conditions and at the weakly absorbing frequencies considered here, due to the non-linear increase of the brightness temperature when lowering the elevation angle, T_a is larger than that of the brightness temperature T_b^c at a cone-like antenna beam center direction. Based on the assumption of a gaussian beam, we derived an adjustment δT_a,

$$\delta T_a(\theta) = \frac{\theta_{1/2}^{\,2}}{16 Ln(2)} (T_{mr}(\theta) - T_{bb}) \exp(-\tau(\theta))[2 + (2 - \tau(\theta)) \tan^{-2}(\theta)] \tau(\theta), \quad (4)$$

where $\theta_{1/2}$ is the full width (in radians) at half-maximum power of the power pattern. Note that the τ in (4) is the slant path opacity at an elevation angle θ. The observed antenna temperatures should be corrected by the amount given by (4) before being used in the calibrations: $T_b^c = T_a - \delta T_a$. Note that the amount of brightness temperature adjustment is itself a function of the brightness temperature, whose correct value is unknown before the calibration. In practice, we may derive δT_a by an iteration process. However, the effect of antenna beam width is a complicated issue. In reality, the beam's side lobes may pick up radiation at low elevation angles from sources which are unpredictable. For this reason and some others discussed later, tipping observations should avoid low elevation angles. Our experiences suggest that angles with airmass larger than 3 should be avoided, especially for the 6 deg antenna used by ARM. Figure 1 shows differences between T_a and T_b and corrections made by (4).

D. Effect of mean radiating temperature

The mean radiating temperature T_{mr} plays a role in mapping the brightness temperature T_b to the opacity τ. Traditionally, T_{mr} is treated as a constant and is determined climatologically. For zenith observations, the uncertainties T_{mr} are usually not a crucial factor because the brightness temperatures are usually small. But in tipping observations, the brightness temperature can be large at a low elevation angle. Our simulations [5] suggest that by using T_{mr} that is a function of elevation angle, and is estimated by surface meteorological measurements, is an effective way of eliminating this component of error. The prediction could also be improved significantly by using remote sensor observations.

E. Errors caused by system random noise

The system random noise affects system precision. To estimate its influence to system absolute accuracy, we performed tipping calibrations using simulated tipping data from S with a 0.1 K Gaussian white noise. The calibration uncertainties are about one tenth to four tenths of a degree. We also showed that the uses of larger airmasses suffer less than the uses of smaller airmasses due to larger signal to noise ratio at large airmasses. The impact of the system noise can usually be reduced by temporal averaging.

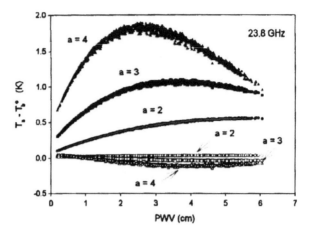

Figure 1. Differences between antenna temperature ($\theta_{1/2} = 6°$) and the brightness temperature at the beam center direction as a function of PWV. The filled symbols are those without beam effect corrections; the open symbols are those with the corrections in which the antenna temperature is adjusted by an amount given by (4). The airmasses at which the differences are calculated are indicated in the figure. Data used in the simulations are explained in the text.

F. Errors caused by uncertainty in the offset of the radiometer equation

In the analysis so far, we have assumed that the offset of the radiometer equation is known precisely. In reality, however, uncertainty may exist. To estimate the impact of the uncertainty on the system calibration, we shifted the system offset by adding $\Delta Off = 1$ K to the right side of the radiometer equation (3) and then used it to simulate tipping measurements. It would seem that a positive one-degree offset shift would cause a positive one-degree measurement error. In fact, the tipping calibration compensates for the positive shift by adjusting the calibration factor r upward (becoming large than 1) with an amount of Δr (a value about 0.003). Thus from (3), the measurement error is given by $\Delta r(T_a - T_g) + \Delta Off (< \Delta Off)$. The calibration errors (at our reference brightness temperatures) are about 0.1 K for all the four channels considered here. However, it is easy to see that the measurement error increases with the sky brightness temperature being measured. At $T_a = 100$ K, the calibration errors are about 0.3 K. Thus, as long as the offset uncertainty is less than 1 K, it will not cause serious calibration problems.

G. Errors caused by non-stratified atmospheric conditions

The airmass-angle relationship requires a horizontally stratified atmosphere. This is the reason why calibrations are usually performed under clear-sky conditions. However, even under these conditions, caution must be exercised due to spatial variations of the water vapor and temperature fields. Several instances were noted during the WVIOPs when there were significant differences and even phase shifts between these tip curves. Figure 2 shows an example when significant departures from stratification occurred.

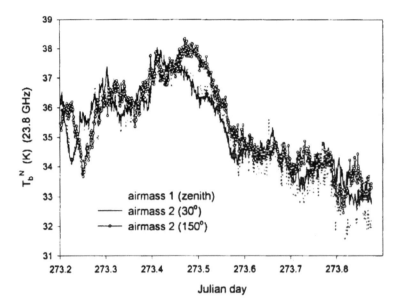

Figure 2. Brightness temperature measurements, normalized to zenith, taken during WVIOP'97. The three curves correspond to brightness temperatures observed at airmass 1 and a pair of symmetric elevation angles with airmass 2. The figure demonstrates the horizontal inhomogeneity even under clear-sky conditions, in contrast to a stratified atmosphere, which should result in an agreement among the three curves.

We simulated the effects of horizontal inhomogeneity using angular scan data from a Raman lidar and found that the magnitude of the difference reflects the degree of the horizontal inhomogeneity, and that the effects are frequently non negligible. We suggest that temporal averaging, on the order of 2-3 hrs, and careful screening of tipping data before applying them to the calibrations may reduce this important source of error.

5. DISCUSSION AND CONCLUSIONS

The tipping calibration method derives a calibration factor from a set of brightness temperature measurements at two or more observation angles. The process requires a knowledge of the fundamental relationship between the airmass and the observation angle. Horizontal inhomogeneity of the water vapor and temperature fields causes uncertainties in the relationship. The method also requires a mapping of the brightness temperature into the opacity with an estimated parameter, the mean radiating temperature T_{mr}. The mapping uncertainty becomes larger when mapping a larger brightness temperature. For this reason, the tipping calibration technique is usually not applicable to a radiometer at a frequency with large optical depth unless T_{mr} error can be reduced. Other sources we have discussed that cause calibration uncertainties include earth curvature, antenna beam width, and radiometer pointing angle errors. For each of the sources, we have developed techniques to reduce its negative effect.

We observed that calibrations using low elevation angles are more sensitive to the various error sources; however, for the same error, angles with larger separations give a mathematical advantage in least squares estimation. Thus, there is clearly a tradeoff on the choice of angles. From our experience and the earlier discussions, tipping data should not include airmass greater than 3, especially for antenna with beamwidths ≥ 6 deg.

The tipping data quality control techniques are also useful in reducing calibration errors. The technique of checking the symmetry of the tipping data taken at symmetric angles can be used to ensure a stratified atmosphere, or the correctness of system pointing angles. The standard deviation of normalized brightness temperature measurements or the airmass-opacity correlation coefficient [3] can be used to screen out erroneous tipping data. The comparisons of calibration factors derived from different combinations of airmasses can be used to check the overall performance of the calibrations.

There have been questions whether or not multiple calibration factors in a radiometer equation can be derived from the tipping data. From our experience, it is difficult to derive even two constant factors. There are two ways of looking at the tip cal procedure, assuming a single realization of an atmospheric state. In the first way, one determines the slope of the opacity vs. a curve and since the slope at $a = 1$ determines T_b, a single factor can be determined. A second way is to realize that T_b at $a=0$ is T_{bb} and again a single constant can be inferred. Theoretically, determining multiple factors depends on multiple sets of tipping calibrations that were obtained in various atmospheric states with different water vapor contents. However, the various uncertainties discussed earlier often cause the numerical solutions to be unstable. Thus, it is often better to determine one factor and estimate the remaining factors using other means, such as theoretical calculations, target calibration, or radiosonde (with RTE model) calibrations. However, calibration with a RTE model is unacceptable if one is trying to study absorption and radiative processes.

As our investigation progressed, it became clear that there were significant advantages to having nearly continuous tipping calibrations. The presence of significant horizontal inhomogeneities as they pass overhead is easily revealed by a time series of tipping calibration data. If these tipping data are done frequently and in clear conditions, a representative time series of zenith T_b is still obtained. During cloudy conditions, the off-zenith scans can be used to identify cloudy data that are not necessarily overhead. This is important because cloud ceilometers or FTIRs usually only indicate clear conditions in the zenith direction. A large initial data set is also advantageous when applying rigorous quality control methods. For example, the ARM radiometer, with its continuous scanning ability, generated roughly 3000 1-min tipping calibration scans; during the same period, the ETL radiometers generated only about 30 15-min scans.

Here, we have assumed a stable linear radiometric system whose calibration factor is independent of time for time intervals of the order of weeks. In reality, however, this assumption may not be completely valid. Although it is difficult to evaluate such situations, problems can often be identified by using strict quality control techniques of tipping data.

ACKNOWLEDGMENTS

The work presented in this paper was sponsored by the Environmental Sciences Division of the Department of Energy as a part of their Atmospheric Radiation Measurement Program.

REFERENCES

1. M. A. Janssen, An introduction to the passive microwave remote sensing of atmospheres, *Atmospheric Remote Sensing by Microwave Radiometry*, Chapter 1, 1-35, M. Janssen, ed., New York: Wiley. (1993).
2. Y. Han, J. B. Snider, E. R. Westwater, S. H. Melfi, and R. A. Ferrare, Observations of water vapor by ground-based microwave radiometers and Raman lidar, *J. Geophys. Res.*, 99, 18695-18702 (1994).
3. J. C. Liljegren, Two-channel microwave radiometer for observations of total column precipitable water vapor and cloud liquid water path, *Proc. Fifth Symposium on Global Change Studies, Amer. Meteorol. Soc.*, Nashville, TN, January 23-28, 1994, pp. 262-269.
4 E. R. Westwater, Ground-based microwave remote sensing of meteorological variables, *Atmospheric Remote Sensing by Microwave Radiometry*, Chapter 4, pp. 145-213, M.. Janssen, Ed., New York:Wiley. (1993).
5. Y. Han and E. R. Westwater, Analysis and improvement of tipping calibration for ground-based microwave radiometers, *IEEE Trans. Geosci. Remote Sensing*, (in press).

Microw. Radiomet. Remote Sens. Earth's Surf. Atmosphere, pp. 155–163
P. Pampaloni and S. Paloscia (Eds)
© VSP 2000

Observations of integrated water vapor and cloud liquid water at the SHEBA ice station

JAMES C. LILJEGREN

DOE Ames Laboratory, Ames, IA 50011, USA

Abstract-Water vapor and clouds play a critical role in moderating the Arctic climate. During the field phase of the Surface Heat Budget of the Arctic (SHEBA) project, continuous measurements of integrated water vapor and integrated cloud liquid water were obtained from a station in the Arctic ice pack over a 12-month period using a dual-channel microwave radiometer. The integrated water vapor measurements are shown to compare well with co-located radiosondes. Measured brightness temperatures support the water vapor absorption model from Liebe and Layton [7] but reveal improved agreement when the Rosenkranz [11] oxygen absorption model is substituted for that in Liebe and Layton. Comparisons of the integrated cloud liquid water with *in situ* data from the NCAR C-130 aircraft exhibit varying agreement that appears to reflect the spatial variation of the clouds.

1. INTRODUCTION

In the Arctic water vapor and clouds influence the surface radiation balance to a greater extent than at lower latitudes. Because the integrated water vapor is often less than 5 mm, substantial radiative cooling occurs in the 20 μm infrared region whereas this region is normally opaque at lower latitudes having greater water vapor amounts. The relatively thin Arctic liquid water clouds also significantly affect the surface radiation balance. Accurate measurements of water vapor and liquid water amounts are therefore necessary for understanding and modeling the surface radiation budget in the Arctic.

The Surface Heat Budget of the Arctic (SHEBA) program [1] carried out a year-long field effort to acquire comprehensive measurements of atmospheric, oceanic and sea-ice processes in order to better understand their interactions and to improve the treatment of these processes in models used to investigate potential effects of climate change. Integrated water vapor and integrated cloud liquid water were measured continuously at the SHEBA ice station over a 12-month period beginning in October 1997 with a dual-channel microwave radiometer. An overview of these measurements is presented in this paper.

2. THE MICROWAVE RADIOMETER

The microwave radiometer (MWR) deployed at SHEBA is a dual-channel instrument built by Radiometrics Corporation, Boulder, Colorado, USA that operates at 23.8 and 31.4 GHz. It was among the suite of instruments deployed at SHEBA by the U. S. Department of Energy's Atmospheric Radiation Measurement (ARM) Program [2]. It is identical to the microwave radiometers deployed by the ARM Program at its facilities in Oklahoma, Kansas, Alaska and on islands in the tropical Pacific Ocean [3].

2.1 Calibration

Beginning in December 1997, the MWR was operated in a continuous elevation angle-scanning (or "tipping") mode in order to ensure that sufficient calibration data were obtained. The range of angles scanned by the radiometer was constrained by its location on the deck of the Canadian icebreaker *Des Groseilliers* to ±45° of zenith at 5° intervals. The radiometer required two minutes to scan all 20 angles, including a second zenith measurement at the completion of the scan. The calibration algorithms applied *a posteriori* to the SHEBA microwave radiometer were essentially identical to those developed for continuous real-time calibration of other ARM microwave radiometers [4]. The calibration is based on the 2000 most recent cloud-free scans (or "tip curves"). The resulting brightness temperature calibration is believed to be accurate to ±0.2 K.

2.2 Retrieval Algorithms

The integrated water vapor (IWV) and the integrated liquid water (ILW) from clouds were determined from the microwave brightness temperatures using a statistical retrieval [5] stratified into monthly retrieval coefficients. Radiosonde data from Barrow, Alaska for 1990-1995 were used to drive the NOAA/ETL microwave radiation transfer software [6], which implements the Liebe-87 [7] absorption model, in order to generate the *a priori* data set for the retrieval. For a 0.2 K root-mean-square (RMS) error in the measured brightness temperatures, the expected RMS error in the retrieval IWV ranged from 0.1(January) to 0.25 mm (July); the RMS ILW retrieval error ranged from 0.01 mm (10 g/m^2) in January to 0.03 mm (30 g/m^2) in July.

3. GENERAL RESULTS

The hourly record of integrated water vapor (IWV), integrated liquid water (ILW), and surface air temperature at the SHEBA ice station is presented in Fig. 1. The IWV ranged from 0.1 cm during the winter to more than 2.0 cm during the summer.

The dramatic changes that occur once the surface temperature reaches 0 °C are particularly interesting. When the surface air temperature is below freezing the IWV and ILW are generally small and variations in IWV correlate well with variations in the

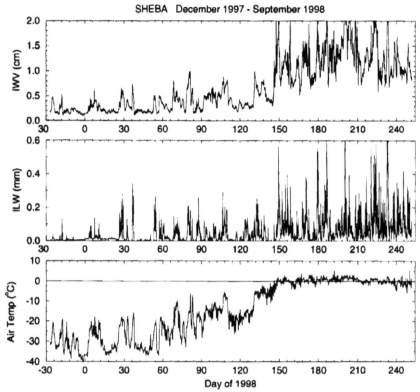

Figure 1. Hourly-averaged integrated water vapor (top), integrated liquid water (middle) and surface air temperature (bottom) for December 1997 - September 1998 at SHEBA. Note the dramatic increase in IWV and ILW when the air temperature reaches 0 °C.

surface air temperature as warmer (and more moist) air masses move through the region. However, once the surface air temperature reaches 0 °C, the net radiation no longer goes to heating the air but rather to melting snow and evaporating water. Once this happens, melt ponds begin to develop and the IWV (and ILW) dramatically increase. Subsequently they are no longer correlated with the surface air temperature. Low clouds and fog occurred frequently during this period.

It is also interesting that substantial amounts of liquid water are evident even when the surface air temperatures are substantially below freezing. The DABUL dual polarization lidar operated by NOAA/ETL confirmed that these were liquid water clouds. The occurrence of liquid water clouds is due in part to the strong surface temperature inversions that occur during the polar night (due to the radiative cooling in the 10 and 20 μm window regions). It is also likely that low concentrations of cloud condensation nuclei permitted super-cooled clouds to occur frequently.

4. WATER VAPOR COMPARISONS

4.1 Radiosonde Comparison

Radiosondes were normally launched twice per day at SHEBA. During the NASA FIRE IFO [8] periods sondes were launched four times per day. These were Vaisala RS-80 sondes with the H-Humicap relative humidity sensor. The comparison of IWV derived from radiosondes with IWV from the microwave radiometer is presented in Fig. 2 for clear sky conditions. The limited number of data points for IWV > 1.0 cm is due to the predominantly cloudy conditions that prevailed once the air temperature reached 0 °C.

The agreement between the MWR and the radiosondes is very good down to 1 mm of IWV. The calibration correction recently developed by Vaisala for the Humicap relative humidity sensors was not applied. For such low water vapor amounts the correction would not be expected to be significant [9].

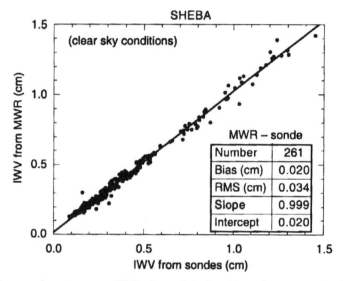

Figure 2. Integrated water vapor (IWV) from Vaisala radiosondes compared with 30-minute averaged values from the microwave radiometer (MWR) during clear sky conditions for December 1997 - September 1998.

4.2 Model Comparison

The SHEBA MWR measurements permit the examination of model performance in the limit of low water vapor where the microwave signal is dominated by oxygen emission. In Fig. 3 brightness temperatures measured with the MWR during clear-sky conditions for the period December 1997 - September 1998 are compared with calculations based on the Liebe-Layton [7] ("Liebe 87") absorption model. Although the similar values for the slopes of the regression lines suggest that the water vapor absorption model appears correct, the significant difference in the intercepts suggests that the oxygen absorption model in Liebe-87 under-predicts.

When the oxygen absorption model due to Rosenkranz [10, 11] is substituted ("Liebe 87 R93"), the calculated brightness temperatures increase such that the regression slopes are unchanged but the difference in the intercepts (MWR-model) is substantially reduced. Using the Rosenkranz oxygen model reduces the median (MWR-model) difference from 0.65 K to 0.30 K at 23.8 GHz, and from 0.75 K to 0.30 K at 31.4 GHz.

Because the IWV retrieval is based on the difference between the two channels, and

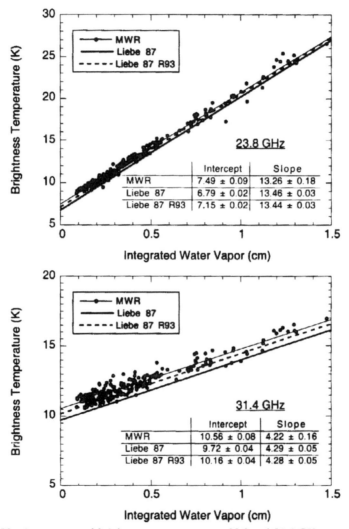

Figure 3. 30-minute averaged brightness temperatures at 23.8 and 31.4 GHz measured with the MWR compared with calculations using the Liebe-Layton ("Liebe 87") absorption model [7] for both water vapor and oxygen, and the Liebe-87 water vapor absorption model with the Rosenkranz oxygen absorption model [11] ("Liebe 87 R93"). The calculations used the same radiosondes as in Fig. 2 for clear sky conditions during December 1997 - September 1998. The integrated water vapor along the abscissa is derived from the radiosondes. The slopes and intercepts of the regression of T_B on IWV are tabulated along with their 95% confidence intervals.

because the brightness temperature biases have the same sign, they essentially cancel each other, which is why the retrieved IWV exhibits good agreement with the radiosondes despite the under prediction of oxygen absorption.

5. LIQUID WATER COMPARISONS

During the FIRE IFO periods several research aircraft flew over the SHEBA ice camp. Among them, the NCAR C-130 was equipped with both King hot-wire and Gerber particle volume monitor (PVM) liquid water content probes. Integrated liquid water from the MWR and ILW from the King and PVM probes are presented in Fig. 4 for three cases in May 1998.

For the May 4 case, the PVM indicates greater ILW than the King probe due to the presence of ice in the clouds. For the 15 and 18 May cases, the PWV and King probes give nearly identical results. The agreement between the aircraft probes and the MWR

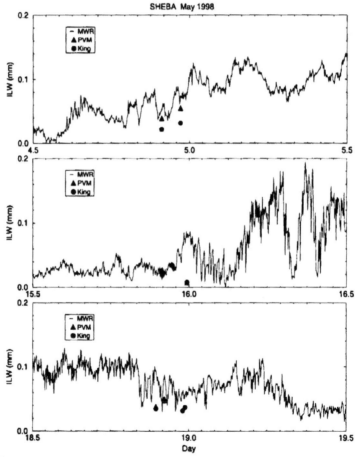

Figure 4. Time series of integrated liquid water (ILW) from the microwave radiometer for days when the NCAR C-130 aircraft with the PVM and King liquid water probes flew over the site.

vary from case to case. One possible explanation for the variable agreement is that the statistical retrieval may be inadequate. For example, the mass absorption coefficient of liquid water depends exponentially on the temperature of the liquid water. In the statistical retrieval for ILW the liquid water temperature, and thus the mass absorption coefficient, is implicitly held constant at the monthly mean value. For clouds that are warmer or colder than this mean value, the retrieved ILW would be in error. To examine this, a retrieval that explicitly accounts for the cloud temperature [12] was applied to several cases. For each case the liquid-weighted cloud temperature was calculated based on the temperature profiles from radiosondes weighted by the cloud radar reflectivity [13]. Although at times the difference in the ILW between the two retrievals exceeded 10%, during the periods when the aircraft were present the agreement between the retrievals was generally within 5%. This does not appear to explain the differences between MWR and aircraft measurements.

It appears that the agreement between the ILW from the MWR and the aircraft is better when there is less temporal variation in the ILW from the MWR, and thus less spatial variation in the clouds. Under these conditions the clouds sampled by the aircraft would be more likely to be representative of the clouds that advected through the field of view of the MWR. To examine the spatial variability and representativeness of the aircraft measurements, the difference between the median ILW measured with the MWR for 2-hour periods centered on the time of the aircraft measurement and the ILW from the King probe are plotted in Fig. 5 as a function of the inter-quartile range (75th percentile - 25th precentile) of a) the zenith ILW from the MWR, and b) the difference in ILW measured at 45° on either side of zenith over the same time. (The median and inter-quartile range statistics are used instead of the mean and standard deviation because the ILW does not generally follow a symmetric distribution; however, the trends are the same.) Figure 5a indicates that the difference in ILW measured by the MWR and King probes increases as the temporal variation in ILW increases. Figure 5b indicates that the difference between the MWR and King probe ILW also increases as the spatial variability in ILW increases. The agreement between the MWR and the aircraft generally improves as the spatial variability of ILW decreases.

6. CONCLUSIONS

The ARM microwave radiometer provided a continuous 12-month record of integrated water vapor and liquid water at SHEBA. The integrated water vapor measurements (down to 1 mm) were in very good agreement with radiosondes. Comparisons of measured brightness temperatures with model calculations suggest that the Rosenkranz absorption model for oxygen [11] is a significant improvement over that in the Liebe-87 model [7]. Integrated liquid water measurements exhibited varying agreement with the King and PVM probes on the NCAR C-130 aircraft, probably due to the spatial variations of liquid water in the clouds.

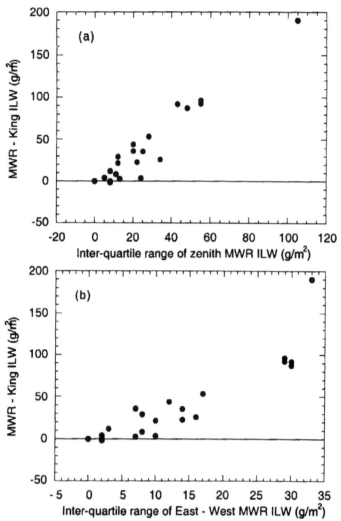

Figure 6. Differences in integrated liquid water (ILW) from the microwave radiometer (MWR) and the King probe on the NCAR C-130 plotted against (a) the temporal variation in ILW from the MWR described by the inter-quartile range (75th percentile – 25% percentile) of ILW; (b) the spatial variation in ILW described by the inter-quartile range of the difference in ILW measured at 45° angles on either side of zenith.

Acknowledgement. James Pinto (University of Colorado) provided the King and PWV probe data from the NCAR C-130 aircraft. Discussions with P. Rosenkranz regarding his oxygen absorption model were very helpful. This work was supported by the Environmental Sciences Division of the U.S. Department of Energy under the auspices of the Atmospheric Radiation Measurement (ARM) Program.

REFERENCES

1. R. E. Moritz, J. A. Curry, N. Untersteiner and A. S. Thorndike, SHEBA: a research program on the surface heat budget of the Arctic Ocean. NSF-ARCSS Tech. Rep. 3. SHEBA Project Office, Polar Science Center, University of Washington, Seattle, WA. (1993)

2. G. M. Stokes and S. E. Schwartz, "The atmospheric radiation measurement (ARM) program: programmatic background and design of the cloud and radiation test bed," *Bull. Amer. Met. Soc.*, **75**, 1201-1221 (1994)

3. J. C. Liljegren, "Observations of total column precipitable water vapor and cloud liquid water path using a dual-frequency microwave radiometer," *Microwave Radiometry and Remote Sensing of the Environment*, D. Solimini, ed. VSP Press. (1994)

4. J. C. Liljegren, "Automatic self-calibration of ARM microwave radiometers," in this volume.

5. E. R. Westwater, "Ground-based microwave remote sensing of meteorological variables," Chapter 4 in *Remote Sensing by Microwave Radiometry*, M.A. Janssen, ed. John Wiley & Sons, New York. (1993)

6. J. A. Schroeder and E. R. Westwater, "User's guide to WPL microwave radiation transfer software," NOAA Tech Memo ERL WPL-213. (1991)

7. H. J. Liebe, and D. H. Layton, "Millimeter-wave properties of the atmosphere: laboratory studies and propagation modeling," Nat. Telecom. and Inform. Admin., Boulder, CO, NTIA Rep. 87-24. (1987)

8. D. A. Randall and Co-authors, "FIRE phase III: Arctic implementation plan (a FIRE on the ice)," Tech. Rep., FIRE Project Office, NASA Langley Research Center, Hampton, VA. (1996)

9. J. C. Liljegren, B. M. Lesht, T. Van Hove, C. Rocken, "A comparison of integrated water vapor from microwave radiometer, radiosondes, and the Global Positioning System," Proceedings of the 9th Annual Meeting of the ARM Science Team, 22-27 March, 1999, San Antonio, Texas.

10. H. J. Liebe, P. W. Rosenkranz, and G. A. Hufford, "Atmospheric 60-GHz oxygen spectrum: new laboratory measurements and line parameters," *J. Quant Spectrosc. Radiat. Transfer*, **48**, 629-643, 1992.

11. P. W. Rosenkranz, "Absorption of microwaves by atmospheric gases," Chapter 2 in *Remote Sensing by Microwave Radiometry*, M.A. Janssen, ed. John Wiley & Sons, New York. (1993)

12. J. C. Liljegren, "Improved retrieval of cloud liquid water path," Preprints of the 10th Symposium on Meteorological Observations and Instruments, January 11-16, Phoenix, AZ.

13. J. C. Liljegren, "Combining microwave radiometer and millimeter cloud radar to improve integrated liquid water retrievals," Proceedings of the 9th Annual Meeting of the ARM Science Team, 22-27 March, 1999, San Antonio, Texas.

Microw. Radiomet. Remote Sens. Earth's Surf. Atmosphere, pp. 165–172
P. Pampaloni and S. Paloscia (Eds)
© VSP 2000

Measurement of plentiful water vapor and cloud liquid water around precipitation area

LIANG GU[2], LEI HENGCHI[1], LI YAN[2], SHEN ZHILAI[1], and WEI CHONG[1]

[1]*Institute of Atmospheric Physics, CAS, 100029, Beijing, China*
[2]*Meteorological Bureau of Shannxi Province, Xi'an, 710015, China*

Abstract—Ground-based microwave radiometer (GBMWRMT) is able to work automatically and continuously in monitoring atmospheric water. In this paper we focus on the estimation of the structure of precipitation cloud by using GBMWRMT. The results indicate that there is a sharp increase both in precipitable water vapor (Q) and in vertical path integrated cloud liquid water (L) values before precipitation. The precipitation will take place when the Q and L values are over respective critical value. The emerging time of the critical values is about 5 to 55 minutes ahead of rain starting. According to those observation facts, it is suggested that there is a plentiful water vapor and cloud liquid water area around the precipitation region of cloud system. Before precipitation, the passing atmospheric water above the observation site is in order of water vapor, water vapor and cloud liquid water, plentiful water vapor and cloud liquid water, and precipitation areas. The plentiful water area is the feeder of water vapor and cloud liquid water for the precipitation area. By the way, those phenomena can also be used for weather nowcasting and weather modification.

1. INTRODUCTION

Our knowledge on the cloud structure and the water field of cloud system derives from measurements. The structure of water field of cloud system, especially water vapor field is obtained using conventional radiosonde, which is made twice a day. So there are few data on microstructure of water vapor. The information of cloud structure is acquired mainly from weather radar, but the structure of water vapor field of cloud system can not be given by weather radar. The ground-based microwave radiometer (GBMWRMT) is able to work automatically and continuously as well as with high time resolution (5-minute) in monitoring atmospheric water field [1,2,3,4]. The results indicate that the GBMWRMT is a non-fungible method for getting information on atmospheric water vapor field. Recently, the GBMWRMT has been applied into many fields: atmospheric water cycle, cloud and precipitation physics, weather modification etc.. In this paper, we focus on the measurement of the temporal evolution and spatial distribution of precipitable water vapor (Q), cloud liquid water (L) and the structure of precipitation cloud by using GBMWRMT.

2. RESULTS AND DISCUSSION

The data presented in this paper are obtained by using ground-based dual-wavelength (0.86 and 1.35 cm) microwave radiometer (GBDWMWRT)(made in IAP, CAS). The main performance parameters of the GBDWMWRT are listed in Table 1. Q and L (the sampling interval is 5min.) have been measured and retrieved simultaneously and continuously. A method of empirical regressions with iteration is used to estimate cloud liquid water content and water vapor content from radiometric data [5]. The first observation was carried out at the Pacific Ocean (5°N--5°S, 125--150°E) during Sept.—Nov., 1987, and second was at Xi'an (34.3°N, 108.9°E), Shannxi Province, China, from August to November 1997.

Table 1. Parameters of GBDWMWMT

	1.35	0.85
Wavelength (cm)	1.35	0.85
Central frequency (GHz)	22.235	35.3
IF bandwidth (MHz)	±230	±210
Integration time (sec.)	3.3	1
Noise factor (dB)	9	11
Sensitivity (k)	0.5	0.3
Side lobe (dB)	≤-20	≤-17
Antenna diameter (mm)	600	368
Antenna gain (dB)	>40	>35
Width of main lobe(degree)	1.5	1.5

2.1. Common phenomena

In measured time series of precipitable water vapor and cloud liquid water, it is found that there is a sharp increase of their argument values before precipitation. The phenomenon appears in medium cloud and low cloud, cumulus and stratus, as well as over ocean and continent.

a. Over ocean

At tropical region, convective clouds play important role in exchange of heat, water vapor and kinetic energy in global atmospheric circulation. During the Third Science Investigation over the Pacific Ocean (5°N--5°S, 125--150°E) from Sept. to Nov., 1987, which was carried out by Chinese Academy of Sciences, a shipborne dual-wavelength microwave radiometer was used to detect Q and L. The results indicated that the structure of water vapor and cloud liquid water for different kind cumulus is different; for cumulus congestus (Cu cong.), Q and L increased very steeply before rainfall starting.

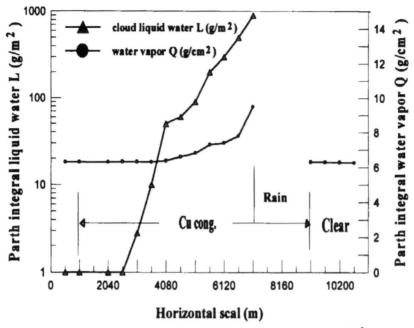

Figure 1. The distribution of vertical path integral water vapor (Q) (g/cm²) and liquid water (L) (g/m²) for cumulus congestus.

Fig.1 shows an example in which data were collected while the ship was navigating forward and across the Cu cong. area. The diameter of this cumulus is about 9km and rainfall region is nearly 2040m. At edge zone of the cumulus, the vertical path integral cloud liquid water content is very low and precipitable water vapor nearly equal to the clear region, while at interior region of cumulus before rainfall region the cloud liquid water and water vapor increase very sharply. This zone is near 4.6km width. The value of water vapor increased from 6.3(g/cm²) at clear region to 9.5(g/cm²). The value of cloud liquid water increased from nearly zero to 880(g/m²). In contrast, for cumulus humilis, the liquid water content and water vapor changes very little compared with Cu cong.. Those phenomena indicated that, at different developing stage of convective cloud, the field structure of water vapor and liquid water is different.

b. Over land

The GBDWMWRMT was fixed at the top of the experiment building of Meteorological Bureau of Shannxi Province, Xi'an City (34.3°N, 108.9°E), Shannxi Province, China. The 4 months period observation was carried out to monitor the precipitable water vapor and vertical path integral cloud liquid water content from August to November 1997. The antenna was forward the

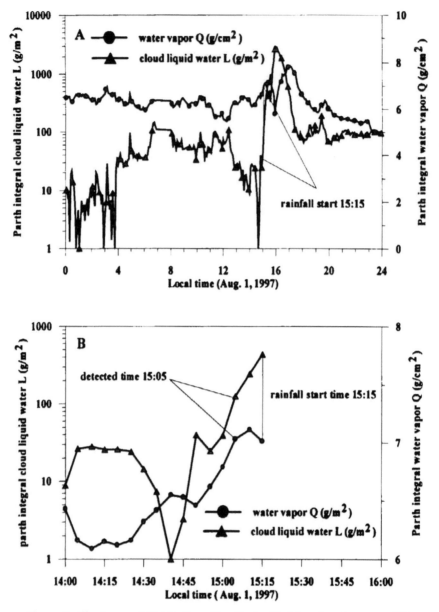

Figure 2. The temporal distribution of vertical path integral water vapor (Q) (g/cm²) and cloud liquid water (L) (g/m²) at Aug. 1, 1997. Figure 2A shows the 24hr observation results; Figure 2B is the extension of same day but only at the period of rainfall will start.

zenith to get the variation of local precipitable water vapor and cloud liquid content. That means when weather system move forward and across the observation station, the GBDWMWRT will record the time series of Q and L, which can be converted to the spacial distribution by multiplying the velocity

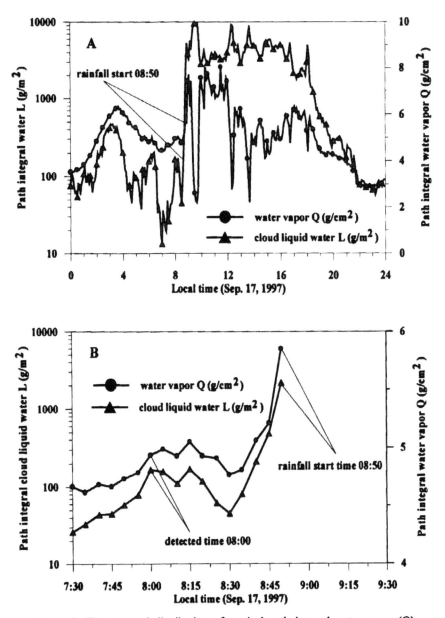

Figure 3. The temporal distribution of vertical path integral water vapor (Q) (g/cm²) and cloud liquid water (L) (g/m²) at Sep. 17, 1997. Figure 3A shows the 24hr observation results; Figure 3B is the extension of same day but only at the period of rainfall will start.

of the weather system. For example, the figure 2 (A) and (B) show a stratocumulus precipitation process that occurred at Aug. 1, 1997. Figure 2(A) is a briefly temporal structure of Q and L at that day. A detail structure of Q and L nearby the rainfall start time is given by figure 2B. At first stage, the

Liang G. Lei H. C. et al.

cloud liquid water content was very low ranging from zero to 100(g/m²). Then the value of L increased very sharply 15 minutes before rainfall starting. The water vapor content Q appears same tendency like L. The altostratus (As) (medium genera of cloud) precipitation process is presented in figure 3(A, B). The variation of Q and L is similar to the figure 2. It is a very typical case. This precipitation process occurred in Sep. 17, 1997, the middle autumn. The content of water vapor was low compared with other cases. At the early stage, the value of precipitable water varies from 2.5 to 3.0(g/cm²), late it increases very steeply when it will rain. The precipitable water vapor and cloud liquid water increased very sharply before rainfall starting: that was observed for all the 17 precipitation processes.

Table 2. Precipitation processes and their pre-forecasting test during observation period from Aug. to Nov. At Xi'an City, China

Date	detected time	Rainfall start time	Pre-forecasting time (min.)	DR* (km)	Cloud form
Aug.0 1	15:05	15:15	10	3	Ac,Sc
Aug.0 7	01:55	02:15	20	6	Ac,Sc
Aug. 15	01:00	01:15	15	5	Ac,Sc
Sept. 11	17:05	17:30	25	8	As,Fn
Sept. 17	08:00	08:50	50	15	As
Sept. 22	23:35	22:35	0	0	As
Sept. 23	21:10	22:00	50	15	As
Sept. 30	15:10	15:14	4	1	As
Oct. 02	14:25	14:30	5	1	As,Fn
Oct. 08	14:45	14:50	5	1	St
Oct. 24	21:00	21:35	35	10	Sc
Oct. 25	08:05	08:26	21	10	Sc
Nov.10	13:44	14:15	30	10	As
Nov. 11		instrument	trouble		As
Nov.25	07:05	08:00	55	18	St
Nov. 27	02:45	03:00	15	5	St,Fn
Nov. 30	19:25	19:40	15	5	St
Nov. 27	02:45	03:00	15	5	St,Fn
Nov. 30	19:25	19:40	15	5	St

DR*: Detected Range, according to the weather system moving speed.

2.2. Possible application

a. Nowcasting of precipitation

Based on the tendency of rapidly increase of precipitable water vapor and cloud liquid water simultaneously before rainfall, we defined "detected time" the beginning of sharp increasing rate of precipitable water vapor and cloud liquid water, and "rainfall start time" the beginning of rainfall time. We try to find out a threshold to predict the pre-forecasting time of precipitation occurring. In practice, the increasing rate of Q and L at the edge zone of precipitation system is great than that at none rain cloud system. Because the sampling interval of GBDWMRT was 5 min., the increasing rate of Q and L was calculated according to following equation:

$$R_X = (X_2 - X_1)/\Delta t.$$

In here, X represents Q or L respectively; subscripted label 1 represents previous record, label 2 represents present record; Δt is sampling interval (5minutes). For precipitation cloud, the range of R_Q is 0.5-0.76 (g/cm^2•min.) while RL is 100-800 (g/m^2•min.). When R_Q and R_L reached their threshold simultaneously, the "detected time" was obtained. The results were given in table 2. The "detected time" were marked in figure 2B and figure 3B also. Generally, the pre-forecasting time is 5—55 min. before rainfall starting. This advance time can be used for nowcasting of precipitation. In weather modification it can be used for seeding operation. The rate of successful prediction is 94% in the observed events. Only the event occurred on Sept. 22 was undetected.

b. Plentiful water area

The time series measurements can easily be converted in spatial information about the cloud structure. Supposing the velocity of cloud system was 20 km per hour, the corresponding range is about 1 to 18 km. It is suggested that there is an area with much plentiful water vapor and cloud liquid water around precipitation region. Before precipitation, the passing of atmospheric water above the observation site is in order of water vapor, water vapor and cloud liquid water, plentiful water vapor and cloud liquid water, and precipitation areas. The plentiful water area is the feeder of water vapor and cloud liquid water for the precipitation area, so it is important for the evolution of cloud droplets and the formation of the precipitation.

3. SUMMARY

1).   Many time series of water vapor and cloud liquid water measured by ground-based dual-wavelength (0.86 and 1.35 cm) microwave radiometer (GBDWMWRMT) indicate that there is a sharp increase in precipitable water vapor and cloud liquid water before rainfall starting. The advanced threshold of time is about 5 to 55 minutes. It could be used for nowcasting of precipitation and suggested that there is a plentiful water vapor and cloud liquid water area around the precipitation region of cloud system.

2).   The discovering of plentiful water around precipitation area promotes understanding for water field and cloud structure. It is expected that it could improve the considering of the boundary condition in numerical model simulation.

REFERENCE

1. D.C. Hogg et al., An automatic profiler of the temperature, wind and humidity in the troposphere, *Journal clim. & Appl. Meteor.*, Vol. 22, No. 5, 807-831 (1983).
2. B.L. Zhao et al., Atmospheric microwave radiation and water vapor remote sensing, *Science Letter* (in chinese), Vol. 29, No. 4, 225-227 (1984).
3. J.J. Olivero, Microwave radiometric studies of composition and structure, Ground-based techniques, *Middle Atmosphere Program Handbook for MAP*, Vol. 13, 43-55 (1984)
4. C. Wei et al., Microwave remote sensing for precipitable water vapor and cloud liquid water over West Pacific Ocean, *Scientia Atmospherica Sinica* (in chinese), Vol. 13, No. 1, 101-107 (1989).
5. C. Wei et al., A Comparison of Several Radiometric Method of Deducing Path- Integrated Cloud Liquid Water, *J. Atmos. Oceanic Technol*, Vol.6, No.6, 1001- 1012 (1989).

Microw. Radiomet. Remote Sens. Earth's Surf. Atmosphere, pp. 173–182
P. Pampaloni and S. Paloscia (Eds)
© VSP 2000

Characteristics of atmospheric water on measurements of ground-based microwave radiometer and rain recorder

WEI CHONG, WU YUXIA, WANG PUCAI and XUAN YUEJIAN

LAGEO, Institute of Atmospheric Physics, Chinese Academy of Sciences, 100029, Beijing, China

Abstract—Precipitable water vapor (Q), vertically path-integrated cloud liquid water content (L) and 5-minute averaged rain rate (R) are measured simultaneously and continuously by a dual-wavelength (0.86 and 1.35 cm) ground-based microwave radiometer and automatic rain measurement system at a fixed point (2°S, 158°E) in the tropical Western Pacific "Warm Pool" (WPWP) during TOGA COARE Intensive Observation Period (IOP) and in Beijing (39°N, 116.1°E) during June - September 1996, respectively. In addition, the time series of Q and L in WPWP in every fall from 1986 to 1989 as well as the series of Q in Beijing for the four seasons of 1982 are also obtained by the radiometer. This paper focuses on the statistical characteristic of the three parameters of atmospheric water and their spatial and temporal variations based on the observed data. The main aspects are 1) Diurnal, seasonal, annual and interannual variation rates of Q and their geographical differences. 2) The relationship between surface humidity and Q and its specific character of tropical ocean. 3) Seasonal statistical distribution of Q or L and their middle latitude land/tropical ocean similarity and difference. 4) Intraseasonal and Interannual variations of the statistical averages of L over WPWP and their association with the 30-60-day Oscillation (or Madden-Julian Oscillation) or *El Nino* event. 5) Quantitative relationship among the three variables of atmospheric water. The results promote our understanding of atmospheric water's behaviors, which is essential in the researches of climate and global change.

1. INTRODUCTION

Atmospheric water plays a central role in many processes influencing climate and global variation. Up to now, our knowledge of atmospheric water's behaviors is still very limited. We are faced with a challenge of detecting of atmospheric water with high resolution, high accuracy, and under all weather conditions. Among many means of observation, ground-based microwave radiometer is a unique tool to monitor atmospheric water with its merits, such as high time resolution (comparing to radiosonde), high accuracy (comparing to space-born instrument), long-term unmanned operation, and synchronous taking samples of gaseous and liquid variables of atmospheric water.

We have been involved in the Project "Tropical Ocean and Global Atmosphere" (TOGA) for many years. The observation area is in the Western Pacific Warm Pool (WPWP). From 1986 to 1989, every autumn there were meteorological and oceanic surveys, sponsored by the Chinese Academy of Sciences, in the area ranging over (0°-5°N, 125°-145°E) that covers most part of the WPWP. A made-by-own dual-wavelength (0.86 & 1.35 cm) microwave radiometer (RMT) was placed in a Chinese surveying ship

"Shiyan-3" to monitor atmospheric precipitable water vapor (Q) and vertically path-integrated cloud liquid water content (L). During TOGA COARE Intensive Observation Period (IOP) (Nov. 1992 - Feb. 1993), as one of the join-observation ships, the 'Shiyan-3' moored at a fixed point (2°S, 158°E) within the Intensive Flux Arrays of this campaign. In addition to the radiometer system for measuring Q and L, an automatic rain rate measurement system (ARRMS) of US Integrated Sounding System (ISS) is placed in the ship, so that three parameters (Q, L and rain rate R) of atmospheric water were measured simultaneously and continuously. In order to compare the similarities and differences of atmospheric water between Tropical Ocean and middle latitude land, an observation was made at Beijing (BJ) from June 3 to September 9, 1996. The same radiometer was used and a tipping bucket rain rate recorder (TBRRR) replaced the ARRMS used during TOGA COARE IOP. Besides, the water vapor measurement was also made by a single wavelength (1.35cm) microwave radiometer in Beijing in the four seasons of 1982.

This paper presents the analysis of above observation data. The analysis focuses on quantitative estimation of distribution and variation of atmospheric water and their geographical differences.

2. MEASUREMENTS AND THEIR ACCURACY

For all observations mentioned above, the sampling rates of radiometer are set in 5 minute. The fifth method in reference [1] is used for the retrieval of clear and non-precipitation cloudy measurements. The precipitable water vapor measured by the radiometer system was systematically compared with that by *in situ* released radiosonde (over WPWP, during 1986-1989), that by *in situ* OMEGA radiosonde system of ISS (during TOGA COARE IOP), and that by conventional meteorological radiosonde (in Beijing), respectively. The statistical relative deviations, namely, the standard deviation vs. the average are 12.5%, 12.9% and 14% for 87, 230 and 80 cases, respectively. Taking account of the difference of sampling manner between radiosonde and radiometer as well as the measurement errors of 3%-10% of radiosonde itself, it can be considered that the accuracy of the microwave radiometer can be compared with that of radiosonde.

As for the radiometer data in rainy day, two kinds of correction are made for the original measurements. The first kind is known as rain correction, through which the rain-layer microwave radiation is removed from the RMT-received total microwave radiation of rainy atmosphere. The second one is the instrumental correction, through which the best part of the microwave radiation from the rain drops on the RMT's antenna can be removed. Consequently, the retrieved results can be much improved, so that most corrected measurements still can be used for the purpose of statistical analysis and some parameters relevant to the precipitation process can be obtained. The statistical comparison between the retrieved values of Q from the corrected measurement and that by radiosonde shows that the statistical relative deviations are 18.5% for the BJ's 41 cases and 15.1% for the WPWP's 31 cases under the condition of light and middle rain ($R<5$ mm/h), which accounts to more than 97% of the total rainfall hours. Therefore, the data in nearly all weather can be used for the purpose of statistical analysis. Please consult reference [2] for further details of the correction method.

· Both the ARRMS' raw data of rain rate and those of the TBRRR's are assimilated into the precipitation data with 5-min time resolution by averaging the raw data over every 5 min in order to be compatible with those of water vapor and cloud liquid water.

Finally, representative time series of atmospheric water are obtained over WPWP (winter of 1992) and Beijing (summer of 1996), known as WPWP and BJ series hereafter. Each of the series includes three synchronous variables: precipitable water vapor Q, vertically path-integrated cloud liquid water content L and 5-minute-averaged rain rate R. The data are characterized by high time resolution, high accuracy, nearly all weather conditions, good continuity and lasting for more than three months. It is these time series that provide a sound basis for further research on characteristics of atmospheric water over the observing sites.

3. CHARACTERISTICS OF PRECIPITABLE WATER VAPOR Q

3.1. Diurnal Variation of Q

Fig. 1a) and b) show the statistical average of diurnal relative departure of Q in clear, cloudy and rainy days over WPWP and BJ, respectively. It can be seen that the diurnal departures of Q over WPWP in every kind of weather all are quite small, within ±4%, while that over BJ are much larger, even more than 20% for cloudy days.

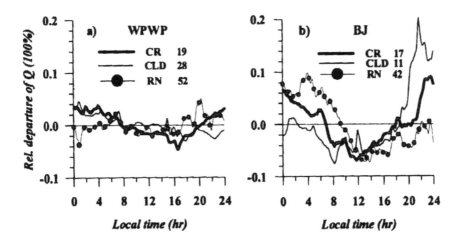

Figure 1. Statistical averages of diurnal relative departure of Q in clear (CR), cloudy (CLD) and rainy (RN) days, respectively. The values in the legends are the numbers of statistical cases.

3.2. Seasonal Statistical Distribution and Variation Rate of Q

Fig.2 shows the seasonal statistical distribution of Q. The solid lines in the figure are the fitting curves for WPWP's and BJ's distributions, respectively, and they both obey the normal distribution,

$$ P(Q) = \frac{1}{\sqrt{2\pi} \cdot \sigma} \exp\left[-\frac{(Q-Q^*)^2}{2\sigma^2} \right] , \qquad (1) $$

where Q^* is mathematical expectation (average of Q), σ is standard deviation. The values of Q^* are 5.63 and 4.26cm and those of σ are 0.73 and 1.56cm for WPWP and BJ series, respectively. Now we use the ratio σ/Q^* to represent seasonal relative variation of Q and their respective values are 13% and 37%. Similar distribution form is obtained for other courses of observations in WPWP and the distribution parameters are shown in table 1, from which we can see that the σ/Q^* for every course of observation is relatively small with the maximum of 14%. The Table also shows that these seasonal averages of water vapor in WPWP are plentiful and range from 5.31 to 6.08cm. They are equivalent to about 53,000 to 60,800 tons of water in a vertical atmospheric column with a bottle area of one square kilometer. Let us use the average value 4.26 cm of BJ series to represent the typical value of Beijing's humid summer, it can be calculated that the Autumn- average of atmos-

Table 1.
Seasonal statistical distribution parameters of Q and L at different years in WPWP.

Year	Season	No. of cases	Seasonal Average Q^*(cm)	Standard deviation σ (cm)	σ/Q^*	seasonal Average L^* (g/m²)
1986	Autumn	3,047	6.08	0.75	0.12	21.0
1987	Autumn	7,643	6.01	0.85	0.14	36.8
1988	Autumn	5,174	5.71	0.61	0.11	98.7
1989	Autumn	6,347	5.31	0.60	0.11	97.9
1992	Winter	23,300	5.63	0.73	0.13	360.0

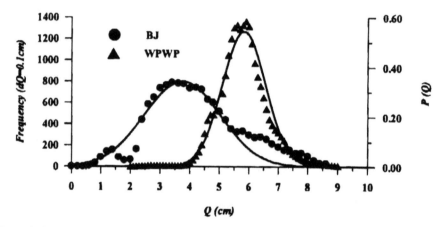

Figure 2. Seasonal statistical distribution of precipitable water vapor Q over WPWP and Beijing. *P(Q) is* the probability distribution *of Q.*

pheric water vapor storage (about 5.7cm) over WPWP is about 1.4cm more than that in Beijing's humid summer. It also should be noted that the WPWP's standard deviation is only half of that of BJ's. Consequently, WPWP's seasonal relative variation rate is only one-third of that of BJ's.

3.3. Annual Variation of Q

The largest contrast among every seasonal averages of Q during one year is used to represent the annual variation of Q and the largest contrast vs. the annual average of Q is used to represent the annual relative variation rate. Four typical seasonal averages of Q are calculated based on the BJ's radiometer measurements in 1982. The minimum is around 0.7 cm for the winter, which is about 5 cm less than the above mentioned autumn-average of Q^* in WPWP, and the maximum is about 3.5 cm in the summer, from which the annual variation rate in Beijing can be accessed and it is more than 130%. As for annual variation over WPWP, the seasons with largest contrast of Q during one year will be different from those in middle latitude and autumn's vs. winter's is expected, as the annual variation of sunshine at Tropical area is different from that in middle latitude area. Although there is not radiometer's observation through one year in the tropical area, the character that the annual variation of Q^* is quite small over WPWP can be expected by comparing the value in the last line of table 1 with those in other lines, which shows that the maximum relative deviation of Q^* between the two seasons in different years is no more than 8%.

3.4. Interannual Variation of Statistical Distribution of Q

From Table 1, it can be seen that the absolute variations of seasonal average Q^* among different years are around 0.4 to 0.77 cm, and the largest interannual variation rate of Q^* is calculated by using the ratio of the maximum of these absolute variations to their average and it is about 14%.

All above-mentioned variation rates of Q on different time scales over WPWP and Beijing are summarized in Table 2 and will be used in the Summary (final section of this paper).

Table 2.
The variation rate of precipitable water vapor Q at different time scales.

Time scale	Variation rate of Q	
	WPWP	Beijing
Diurnal	$\leqslant\pm4\%$	$\pm8\% - \pm20\%$
Seasonal	$\pm11\% - \pm14\%$	$\sim\pm37\%$
Annual	Small	$\sim130\%$
Interannual	$\leqslant14\%$	

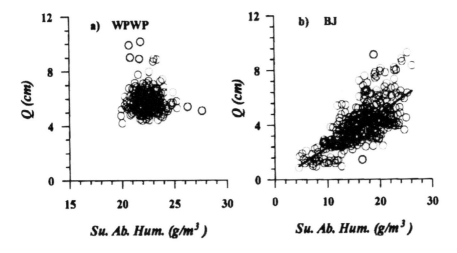

Figure 3. The relationship between surface absolute humidity and precipitable water vapor Q.

3.5. Relationship between Q and Surface Absolute Humidity

Fig. 3a) and b) show the scattering diagrams between the RMT-measured Q and the surface absolute humidity ρ by ventilated psychrometer with sampling frequency of 4 sets per day over WPWP during TOGA COARE IOP and over Beijing in summer of 1996, respectively. It can be seen that there is little linear correlation between the two variables in WPWP, while significant correlation exists over the land with correlation coefficient more than 0.7. Similar scattering diagrams are made for every observation course both on the sea and on the land, such as 'the course WPWP 1988 autumn', 'the course BJ 1982 winter-summer', and so on. The same characteristics as those shown in Fig.3 are manifested themselves in every diagram. Furthermore, the historical radiosonde data typical of seasons are used and calculated for Q and ρ, in Yap (9.6°N, 138°E), trust territory of the United States, which is so small an island in the tropical western Pacific ocean that its radiosonde still has the tropical oceanic representation. The results further confirm the characteristics of boundary layer over tropical sea. It should be demonstrated that this specific characteristic of boundary layer of Tropical Ocean has emerged evidently in several scattering diagrams (Fig.5 –Fig.8) of the reference [3], where monthly summaries of atmospheric soundings taken over 17 years from 49 mid-ocean stations at small islands and weather ships distributed from equator to 66°N or 54°S over major oceans are used to make the scattering diagrams between Q and ρ for each ocean. In those diagrams a common tendency shows that if the precipitable water vapor Q is greater than 4cm, which generally represents the samples over tropical oceans, the relation between Q and ρ shows little correlation, while significant linear correlation exists between the two variables when Q is less than 4cm. Although at that paper the author did not discuss this character or the cause of the phenomenon, those diagrams suggest the same fact as shown here in our paper, in spite that the sampling frequency and sampling method in that paper are quite different from ours. It means that in Tropical Ocean on the sea surface there always exists a boundary layer with high humidity, which is scarcely influenced by above atmospheric condition.

4 CHARACTERISTICS OF VERTICALLY PATH-INTEGRATED CLOUD LIQUID WATER CONTENT L

4.1. Seasonal Statistical Distribution of L

Fig. 4 shows the accumulated seasonal statistical distribution of vertically path-integrated cloud liquid water content L both for WPWP and BJ. It can be seen that both of the two distributions obey negative exponential law, that is

$$f_i (L \geq L_0) = c \bullet \exp (-d \, L_0) \qquad (L_0 > 0) \qquad (2)$$
$$f_i (L = 0) = 1$$

where f_i is normalized accumulative distribution function of L, c and d are the distribution parameters and their values are given in Table 3, from which it can be seen that the parameters over WPWP and BJ are significantly different and the winter's average of cloud liquid water content of WPWP is two times more than that of the summer's of BJ.

The same distribution form as Equations (2) with different distribution parameters was also obtained based on a set of 12,403 individual radiometer measurements taken in Shearwater, Nova Scotia, Canada during the 52-day period from 21 January to 13 March 1986 as a part of Canadian Atlantic Storms Program (CASP). Please see Figure 3 of reference [1] for the details of the distribution. The sampling geographical position (45°N, 61°W) and the sampling season (best part of winter) for CASP data are obviously different from those of WPWP and BJ, which suggests that the seasonal distribution law shown here may be applicable for many places and other seasons. However, the distribution parameters are variable, depending on geographical positions and seasons.

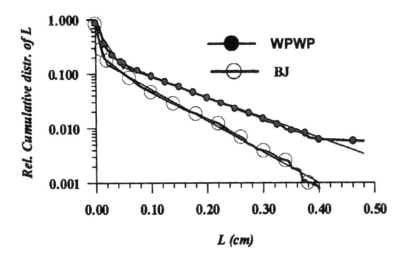

Figure 4. Accumulated statistical distribution of vertically path-integrated cloud liquid water content L.

Table 3.
Seasonal statistical distribution parameters of L

Place	Character point L_i (cm)	$L \leq L_i$		$L > L_i$	
		c	d	c	d
WPWP	0.05	0.943	39.3	0.209	8.38
BEIJING	0.02	0.889	79.6	0.233	13.2

4.2. Statistical Average of L in Several Synoptic Processes Related to MJO Convective Action over WPWP

"Tropical convection and circulation features undergo changes on several temporal and spatial scales. Within the intraseasonal time-scale the 30-60-day or Madden-Julian Oscillation (MJO) is a prominent large-scale tropical convection and circulation phenomenon. The equatorial cloud complexes that propagate eastward from the Western Indian Ocean to the central Pacific during the 'wet phase' of MJO are frequently composed for a hierarchy of cloud clusters consisting of eastward-propagating super cloud clusters (SCC) with time-scales of 7-20 days and spatial scales on the order of 10^3 km and westward-propagating cloud clusters (CC) whose scale are 10^2 km and whose lifetimes are only 1-3 days," Andres Fink, et al., stated in reference [4]. Three typical synoptic processes related to above mentioned convective actions are chosen from the WPWP series. The first is a westerly burst process related to the action of SCC lasted 16 days from 20 December 1992 to 5 January 1993. The second is a strong convection process related to the action of CC from 17 to 22 January 1993. The third is a steady synoptic process with weak convective action from 30 November to 8 December 1992. The averages of L for these processes are 532, 299 and 23 g/m² (or 0.0532, 0.0299, and 0.0023 cm), respectively. It can be seen that the averages of L for different synoptic processes are quit different. The convective action in the westerly burst process is strongest and its average of L exceeds that of the strong convection process by a factor of 1.8 and that of the steady synoptic process by a factor of 23.

4.3. Interannual Variation of Seasonal Averages of L

The seasonal statistical averages of L at different years in WPWP are listed in the last column of Table 1, from which we can see that the statistical averages show significant interannual variation in WPWP and the largest variation rate can go beyond several times or more. Moreover, the seasonal averages of L in 1986 and 1987 are significantly lower than that in other years. Considering the fact that an *El Nino* event occurs from 1986 through 1987, and that the convective action in WPWP during *El Nino* event is weaker than that during the period when no *El Nino* event occurs [5], we suggest that the low value is a sign of *El Nino* event.

5. QUANTITATIVE RELATIONSHIPS AMONG THREE VARIABLES OF ATMOSPHERIC WATER

5.1. The Ratios of Seasonal Averages between Q, L and R

Table 4 shows the seasonal average of precipitable water vapor Q^*, the seasonal average of vertically path-integrated cloud liquid water L^*, the total amount of precipitation Sr, and the ratios between them in WPWP and BJ series, respectively. It can be seen that on the average, the cloud liquid water is about six thousandths of the water vapor storage and every day about one fifth of the water vapor storage falls into the sea in precipitation over Western Pacific Warm Pool during TOGA COARE IOP. We are only aware of one experimental assessment of the percentage of the atmospheric liquid water vs. the gaseous water, namely, that reported by Obukhrov, A.M., et al., in reference [6]. They used the measurements over oceans by radiometers with 1.35 cm and 0.8cm wavelengths on the satellite "Cosmos-243" during continuous 23 courses around the earth and assessed that the liquid water is about 1% of total of the water vapor in the atmosphere. Considering that the liquid water in Obukhrov's experimental assessment includes rain water, while only cloud liquid water is accounted for the value of L^* in Table 4 of this paper, we think our results are of the reasonable order of magnitude. These experimental assesses will be of great value to us in understanding the characteristics of atmospheric water cycle.

Table 4.
Seasonal statistics of Q, L and R and their ratios for WPWP and BJ series

Place	Duration (day)	Q^*(cm)	L^*(cm)	Sr (cm)	L^*/Q^*	Sr/Q^*(day⁻¹)
WPWP	85.7	5.63	0.036	97.64	6‰	20%
BJ	99.0	4.26	0.018	72.17	4‰	17%

5.2. Parameters Relevant to Precipitation Process

The time series of atmospheric water are divided into some sub-series according to the time periods when the observation site is covered by cloud system relevant to corresponding synoptic process. Each of the sub-series is known as a synoptic process or a cloud system. All periods of precipitation included in one synoptic process are known as a rain event. Respective 28 and 26 typical processes are selected from Beijing and WPWP series. Two parameters of process, cloud water cycle parameter C_{yc}, as well as process-amount precipitation P_p and average precipitation intensity I are obtained for each process. Based on these parameters two regressive relationships are obtained as follows,

$$C_{yc} = e\, P_p^f, \tag{3}$$

and

$$E_{ff} = g\, I^h, \tag{4}$$

where e, f, g and h are regressive coefficients, the values of which show obviously geographical difference. Please consult reference [7] for further details.

6. SUMMARY AND DISCUSSIONS

Based on the time series measured by microwave radiometer and rain rate recorder, following characteristics of atmospheric water are shown.

1) From the view-point of short-term climate (seasonal average), Western Pacific "Warm Pool" is a huge storage of atmospheric water: in a vertical atmospheric column with a bottle area of one square kilometer there is about 53,000 to 60,800 tons of water vapor, which is about 14,000 tons more than that of BJ's in humid summer and about 50,000 tons more than that of BJ's in dry winter. This storage seems fairly stable: all the variation rates of atmospheric water stores at different time scales (diurnal, seasonal, annual and interannual) are small, no more than ±14%, though their absolute variation can not be neglected. On average, every day about one-fifth of the storage falls into the sea in rainfall. It can be inferred that equivalent water will escape from the Ocean to the atmosphere in evaporation and in this way the huge storage keeps its balance. A specific character of the Tropical Oceanic boundary is revealed that little correlation exists between the sea surface humidity and the total water vapor Q. It means that on the sea surface there always exists a layer of air with high humidity, which is scarcely influenced by above atmospheric condition. However, the precipitable water vapor over middle latitude land (BJ) may have much large variation rates at all time scales, even more than 130% on annual scale. The land-surface humidity shows significant correlation with the total water vapor content in a vertically atmospheric column.

2) On seasonal average, the amount of liquid water in the form of cloud drops in atmosphere is very small, only several thousandths of atmospheric water vapor storage (e.g., 6‰ over WPWP during TOGA COARE IOP and 4‰ over Beijing in summer of 1996). The cloud liquid water content is variable. Case study shows that the average of cloud liquid water over WPWP in a westerly burst process, which is associated with the action of Supper Cloud Cluster related to the 30-60-day Oscillation, can be 1.8 times as great as that of strong convection process and 23 times as that of steady synoptic process. Over WPWP, the seasonal averages of cloud liquid water in the period of *El Nino* event (autumns of 1986 and 1987) are obviously lower than those when no *El Nino* event occurs (autumns of 1988 and 1989). These results suggest that the averages of vertically path-integrated cloud liquid water contents on different time scales can be used as the good indexes of tropical convection actions on corresponding scales.

3) On seasonal time scale over WPWP and BJ, several statistical distribution functions are obtained. The distributions of precipitable water vapor obey normal statistical distribution law and those of vertically path-integrated cloud liquid water content obey negative exponential law. Both effectiveness of precipitation vs. average of precipitation intensity of rain event and cloud water cycle parameter vs. process-amount precipitation obey the power law. The parameters describing these laws are significantly different between WPWP and BJ, showing obvious geographical features.

4) It is indicated that ground-based microwave radiometer serves as a unique tool for monitoring atmospheric water. Provided the observations are conducted in the areas typical of climate, our understanding of the behaviors of atmospheric water in meteorology, hydrology and climatology will be improved significantly.

ACKNOWLEDGEMENT

This paper is supported by National Nature Scientific Foundation of China (Grant No. 4957245).

REFERENCES

1. Chong Wei, H. G. Leighton and R. R. Rogers, A comparison of several radiometric methods of deducting path-integrated cloud liquid water, *J. Atmos. Oceanic Technol.*, 6(6), 1001-1012, Dec. (1989).

2. Wei Chong, Wang Pucai, Wu Yuxia and Xuan Yuejiang, The corrections of measurements on rainy days by a dual-wavelength ground-based microwave radiometer, In: *Research on Detection of Global Environment and climate Change*, pp. 130-136, Meteorology, Beijing (1997).

3. Liu. W. T., Statistical relation between monthly precipitable water and surface-level humidity over global ocean, *Mon. Wea. Rev.*, 114, 1591-1602, (1986).

4. Andreas Fink and Peter Speth, Some potential forcing mechanisms of the year-to-year variability of the tropical convection and its intraseasonal (25-27-day) variability, *International Journal of climatology*, 17, 1513-1534, (1997).

5. Huang Ronghui and Wu Yifang, The influence of ENSO on the summer climate change in China and its mechanism. *Advances in Atmospheric Sciences*, 6 (1), 21-33, (1989).

6. Obkhrov,A.M., et al., Research of atmosphere by using the microwave radiometric measurement on the man-made satellite "Cosmos-243", *Cosmic Research (in Russia)*, 9 (1), 66-76, (1971).

7. Wei Chong ,Wu Yuxia, Wang Pucai and Xuan Yuejian, On measurement of probability and effectiveness of precipitation by combined ground-based microwave radiometer and rain recorder, *Proceedings on Seventh WMO Scientific Conference on Weather Modification*, Chiang Mai, Thailand, 17-19 Feb., (1999), WMO/TD-No. 936, Volume 2, pp. 569-572.

Microw. Radiomet. Remote Sens. Earth's Surf. Atmosphere, pp. 183–191
P. Pampaloni and S. Paloscia (Eds)
© VSP 2000

Using a micro-rain radar to assess the editing of ground – based microwave radiometer data

LUBOMIR GRADINARSKY, GUNNAR ELGERED, AND YAWEI XUE

Onsala Space Observatory, Chalmers University of Technology
SE-439 92 Onsala, Sweden; e-mail: geo@oso.chalmers.se

Abstract - A microwave radiometer measuring the radiation at 21.0 and 31.4 GHz, provides data from which it is possible to retrieve simultaneously both the integrated amount of water vapor and the cloud liquid in the atmosphere. The algorithm used for modeling the emission from the liquid water is based on the assumption that it is inversely proportional to the frequency squared. This is valid as long as the size of the liquid particles in the atmosphere is less than a few tenths of a millimeter. When independent information on the liquid water is not available the normal data reduction procedure is to screen and edit the radiometer data using its own estimates of the liquid water to decide if the quality is acceptable. Typically data acquired in presence of more than 0.7 mm of integrated liquid water are discarded. A disadvantage of this method is that the poor data quality may affect the editing process. We have installed a micro-rain radar in order to measure rain events. It transmits a frequency modulated continuous wave (FM-CW) at 24.1 GHz with a power of 50 mW, which results in an operating range of 1.4 km. The Doppler information is used in order to detect when the liquid drops are large enough to fall. Then we will have a strong indication that our radiometer data are of poor quality. We report from our first tests and evaluation of using the radar data in the radiometer data editing. We compare the results with those obtained from the standard editing and find an agreement, in terms of the number of discarded data points, when using a threshold value of 0.4 mm of liquid water.

1. INTRODUCTION

The ground-based radiometer has the ability to estimate the atmospheric water vapor content with a measurement uncertainty of approximately 1 mm [1]. We call this instrument a Water Vapor Radiometer (WVR). It is useful both as a stand-alone instrument as well as a tool to assess the accuracy of other remote sensing methods. One such example is the recently established networks of continuously operating GPS receivers (see, e.g., [2]). We have used WVR data as ground truth for comparisons with GPS data [3] and are therefore concerned about the quality of the WVR data. In general the increased importance of accurate measurements of water vapor for climate studies and meteorology stimulate further improvement of the WVR data quality. The estimation of water vapor is based on modeling the emission from the sky due to the liquid water as inversely proportional to the frequency squared. This assumption is not true when large drops appear (the drop size must be much smaller than the wavelength of the observed radiation, in our case $1/10\ \lambda \approx 1$ mm). The present analysis uses the liquid water (LW) estimation based on the measurements of the sky brightness temperature in the two radiometer channels. When the obtained value of the integrated liquid water content exceeds a certain threshold LW_{max} ($LW > LW_{max}$) the data are discarded.

In the following sections we will discuss the influence of the value of the liquid water threshold on the WVR data reduction. We will compare results obtained when using the WVR data themselves to indicate the presence of large drops of liquid water, with results obtained when we instead use information from an independent instrument, namely the Micro-Rain Radar (MRR). We start with a brief introduction of the instruments used, their basic principles, the data acquisition and the data quality. Next the presently used LW algorithm is described and the method of editing the WVR data using the MRR data is introduced. Finally the results of the comparison and the effect of the threshold value on the data reduction is presented followed by the main conclusions of the work.

2. EQUIPMENT AND DATA ACQUISITION

The Onsala Space Observatory is equipped with a dual frequency WVR (Figure 1) which by measuring the sky emission is able to retrieve the propagation delay of radio waves due to water vapor in the troposphere. This delay can easily be used to infer the integrated water vapor content [4]. In the following discussion we will concentrate on the wet delay. Table 1 presents the main Onsala WVR specifications. The radiometer is continuously scanning the sky providing estimates for all directions in azimuth and for elevation angles exceeding $15°$. More information on the WVR and its calibration is presented in [5].

For the purpose of this study, periods of WVR data acquired in 1997 and 1998 were selected. The presence of rainy periods in the data was desired in order to compare the data editing based on WVR estimates of liquid water versus the data editing using MRR data. This means that the data set used is not statistically representative for the average weather at the site.

The Micro-Rain Radar (Figure 2) used in the experiment is commercially available. It uses a Frequency Modulated - Continuous Wave (FM-CW) carrier. A block diagram of the device is shown in Figure 3. The carrier frequency is continuously modulated with a saw tooth function. The transmitted and the received signals are mixed in the mixing diode and the delay difference between them is then proportional to the target range. After

Figure 1. The water vapor radiometer operating at 21.0 and 31.4 GHz at the Onsala site.

Table 1. Main specifications of the Onsala WVR.

Frequencies	21.0 GHz 31.4 GHz
Antenna beam width	$6°$ for both channels
Slew rate (AZ and EL)	$1.8 °/s$
Reference loads temperature	313 K / 370 K

detection, the Fast Fourier Transform (FFT) is applied to obtain the Doppler spectrum. This range resolved Doppler spectrum is then used to estimate the liquid water density and the rain rate averaged over different height intervals. The estimation of the rain rate is based on a relation between terminal-fall velocity of the drops and a certain drop diameter.

We will use the rain-rate data only as an indication of the presence of rain, i.e., a positive value of rain rate will indicate that the WVR wet delay data are of poor quality. We will not pay attention on the absolute values of the rain rate. Table 2 shows the main rain-radar specifications. It is interesting to note that the integration time can be varied from 10 to 3600 s. We have used an integration time of 60 s. This will give us a sampling rate comparable to that of the WVR. The vertical resolution can be varied between 10 and 200 m. We use a 200 m range resolution and the rain rate estimate from the lowest height interval which is 200 m.

Figure 2. The FM-CW Micro-Rain Radar (MRR) operating at 24.1 GHz at the Onsala site.

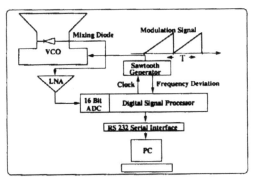

Figure 3. The block diagram of the MRR

Table 2. Specifications of the MRR

Beam width	$3°$
Frequency	24.1 GHz
Averaging time	10 - 3600 s
Range gates	number - 6 (0 to 1400 m)
Range resolution	10 - 200 m
Transmitted power	50 mW

3. WVR DATA EDITING ALGORITHM

The wet delays are calculated and edited in a number of steps (see Figure 4). First the "raw" sky brightness temperature measurements from both WVR channels are combined with the ground based meteorological data and converted to wet path delay using the inversion algorithm described in [6].

At this stage also estimates of the liquid water are calculated based on the algorithm presented in Figure 5. This liquid water algorithm was derived using 3 years of radiosonde data which provide height profiles of pressure $P(h)$, temperature $T(h)$, and humidity $\rho_v(h)$.

We assume that the relative humidity exceeding a certain value indicates a presence of clouds and using a relation between the cloud thickness and the liquid water density we can obtain the liquid water profile $\rho_L(h)$ [7]. The attenuation coefficients due to water vapor, liquid water, and oxygen, are used in the equation of the radiative transfer to calculate the primary WVR observables – the sky brightness temperature at the two channels - T_{2l} and T_{3l}. They are then related to the coefficients of the liquid water model (a_1, a_2, a_3) through the equation $L = a_1 + a_2 T_{2l} + a_3 T_{3l}$ and are derived using the standard method of least squares. Based on those coefficients we estimate the liquid water content from the WVR observed sky brightness temperatures, used later for wet delay data editing.

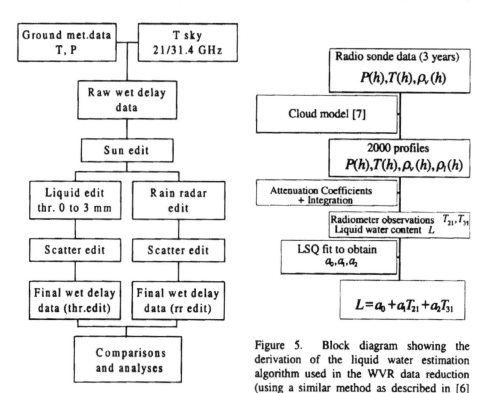

Figure 4. Block diagram of the WVR data reduction – the analysis using the standard procedure is the left branch and the analysis using rain radar is the right branch.

Figure 5. Block diagram showing the derivation of the liquid water estimation algorithm used in the WVR data reduction (using a similar method as described in [6] for the wet delay).

The wet delay data reduction continues by removing observations made within a great circle distance of 15 degrees of the sun (Figure 4). After this step, we have applied two different editing techniques. The first one is using the standard WVR editing procedure where a variable threshold value from 0 to 3 mm was applied (the threshold value for our standard editing procedure is 0.7). The second one uses the rain detection information coming from the rain radar. The last stage of the editing process common for both methods above consists of removing the outlier points (the "scatter edit" box in Figure 4). The resulting time series from the two different (thr. edit and rr edit) liquid water editing procedures are finally compared.

4. RESULTS

We have analyzed a total number of 21 days of WVR data from selected rainy periods during tha fall of 1997 and the winter of 1998. In total the "raw" time series consisted of

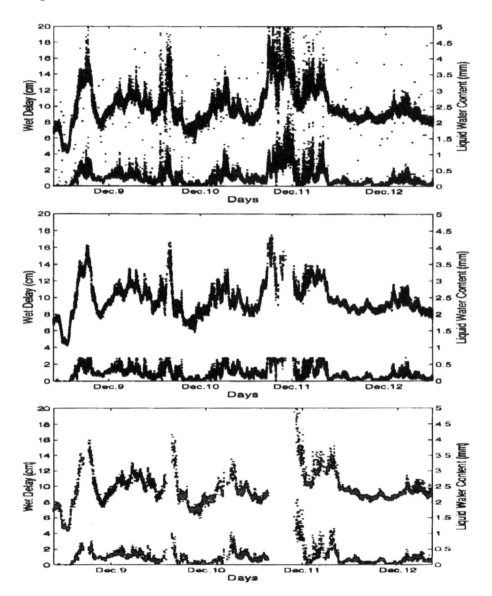

Figure 6. Examples of "raw" wet delay and liquid water data (top frame, 14481 data points) and the results from the two editing techniques; the upper curve in each frame shows the wet delay (left scale) and the lower curve shows the liquid water content (right scale); (middle frame) wet delay data using the standard liquid water threshold of 0.7 mm (2362 removed data points); (bottom frame) wet delay data after editing using the micro-rain radar data (3363 removed data points).

112 600 data points. Figure 6 presents examples of wet delay and liquid water data: (top) "raw" unedited data, (middle) data edited by using the standard value of liquid water threshold of 0.7 mm, and (bottom) data edited by using the MRR data. Studying the radar edited time series we note that both the wet delay and the liquid water content are larger after a rain event compared to before. Liquid water drops remaining on the WVR feed system for some time after the rain event could be the cause of this. We have not been able to study this effect further but we note that this is an important task for future work.

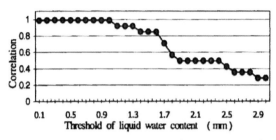

Figure 7. The correlation coefficient between the results from the standard editing procedure using different threshold values for the liquid water content and the results from the rain radar edited data. The analyses is based on all 21 days.

To evaluate the threshold value of the standard editing method we compare the number of removed points and the correlation coefficients of the obtained wet delay time series.

Table 3. Results from the comparison of the two editing procedures using the 21 days of data containing originally 112 600 data points.

Threshold value of liquid water [mm]	Number of removed data points [x 1000]	Number of data points removed by threshold editing, normalized to the number of removed points by radar editing	Correlation coefficient between the time series of threshold and radar editing	Percentage of removed data points using the threshold method
0.0	112.6	3.45	0.952	100.0
0.1	77.8	2.46	0.985	69.1
0.2	52.1	1.67	0.990	48.5
0.3	37.5	1.18	0.991	35.4
0.4	27.3	0.85	0.992	26.3
0.5	20.7	0.64	0.991	20.1
0.6	16.3	0.50	0.990	15.7
0.7	13.1	0.39	0.990	12.3
0.8	10.5	0.31	0.990	9.5
0.9	8.4	0.24	0.989	7.4
1.0	6.9	0.19	0.989	5.9
1.5	2.8	0.13	0.847	3.8
2.0	1.6	0.06	0.494	1.2
2.5	1.0	0.04	0.423	0.8
3.0	0.7	0.04	0.282	0.6

The sequences, which are compared, are edited using a liquid water threshold value from 0 to 3 mm and then correlated with the rain-radar-edited data. It is worth pointing out the difficulty in comparing the two, based on the correlation coefficient. We have several possibilities to handle the gaps due to removed data in the different time series. One option we tried was to insert zeros instead of the missing points, but this had a strong effect on the correlation coefficient results and the correlation dropped to low values. One other method is to linearly interpolate the data between the neighboring points. This resulted in high values of the correlation coefficient and thus presented a difficulty to find an optimum value of the threshold. Figure 7 displays the results based on all data sets and using the linear interpolation method. As seen in Figure 7 and Table 3 large values of correlation are obtained for low threshold values with a slightly larger value at the 0.4 mm level. As already indicated above this result depends highly on the interpolation method used for the periods of removed points. In Table 3 we highlighted the values where the number of removed points using the two strategies is of comparable size and where the maximum correlation coefficient is obtained. We see that the two methods agree best for a threshold value of 0.4 mm. On the other hand it is only for threshold values above 1 mm were the correlation decreases significantly.

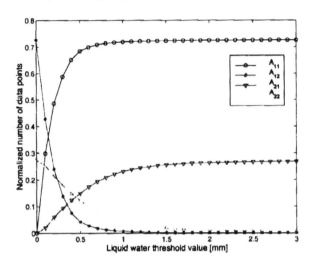

Figure 8. Results for the coefficients of the "confusion matrix" (Note that the sum of the four elements always equals to one, see text).

We also calculated a 2 x 2 "confusion matrix". The matrix elements are calculated for all threshold values from 0 to 3 mm using an increment step of 0.1 mm of LW, thus producing 31 x 4 values. Each value is normalized by the total number of data points before the editing. Figure 8 illustrates the results. The corresponding elements of the matrix are defined as follows:

- Element A_{11} is the normalized number of accepted data points using the threshold editing and at the same time accepted using the radar editing. When the threshold increases and more data are accepted by the threshold editing, A_{11} will approach the total number of accepted data points using radar editing

- Element A_{12} is the normalized number of data points which are accepted by the radar editing and at the same time discarded by the threshold editing. For low threshold values A_{12} approaches the total number of accepted data points using radar editing, where for large values of the threshold $A_{12} \rightarrow 0$.

- Element A_{21} is the normalized number of data points which are discarded by the radar editing and at the same time accepted by the threshold editing. For zero threshold where no points are accepted by the threshold editing, $A_{21} \rightarrow 0$. For high values of the threshold (data are discarded by the threshold editing), A_{21} reaches the total number of data discarded by the radar.

- Element A_{22} is the normalized number of data points discarded by both methods. For 0 threshold, A_{22} equals the number of discarded by radar points and decreases (as the threshold increases) approaching $A_{22} \rightarrow 0$. This corresponds to the fact that the threshold editing does not remove any data points for large thresholds.

In Figure 8, it is interesting to point out that the intersection of the coefficients A_{12} and A_{21}, occurs around the 0.4 mm threshold level indicating the level where the same amount data are accepted by one and discarded by the other method. If both methods were in a perfect agreement A_{12} would drop to 0 and A_{21} would start increasing (from 0) at the "correct" threshold value corresponding to the radar editing result. This is not true in our case and one obvious reason is the limited resolution of the radar data. On the other hand we do not expect a well defined "correct" threshold value because a few large (falling) drops can correspond to the same liquid water content as many small drops. The point where A_{12} and A_{21} cross indicates a threshold where the amount of data accepted by the radar method and discarded by the threshold method is the same as the amount of data accepted by the threshold method and discarded by the radar method. If the threshold is larger/smaller one of the techniques will be discarding more/less data than the other. At this optimum threshold level the difference is about 10%. From the saturation level of A_{21} we observe that radar is discarding about 27 % of the data. We finally note that this large percentage is a consequence of the fact that for the purpose of this study we used selected periods with presence of rain.

5. CONCLUSIONS AND FUTURE WORK

Based on the analyzed rainy periods of 21 days of WVR data, we find that the micro-rain radar observations are useful for data editing. Based on the radar data, we find an optimum threshold value of 0.4 -0.5 mm in the liquid water content. On the other hand we see no dramatic decrease in the data quality for threshold values of up to 1 mm.

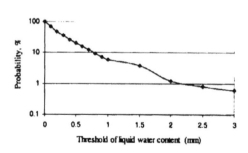

Figure 9. Cumulative distribution function for exceeding a certain threshold value of liquid water based on the WVR measurements.

One additional comparison could be based on the probability of exceeding a certain threshold of liquid water based on the WVR liquid water estimates and the probability of rain based on the rain radar estimates. Figure 9 displays the cumulative distribution function (CDF) of the liquid water for the selected rainy periods. Such CDF could also be calculated based on a longer period of WVR data including all kinds of weather conditions. This result shall then be consistent with the average probability of rain on the West Coast of Sweden. This statistical method, which we have not yet evaluated, for an appropriate large data set, would offer an alternative way to assess the standard WVR data editing using a simple threshold.

Finally we mention that future work must include studies of the effects of water drops remaining on the radiometer feed system after rain events.

ACKNOWLEDGEMENTS

We thank the two anonymous reviewers for constructive criticism including the suggestion to present the "confusion matrix". This work has been carried out as part of the WAVEFRONT Project, which is funded by the European Commission Environment and Climate Program   (EC Contract ENV4-CT96-0301). The Project is a collaboration between the IESSG, University of Nottingham (UK), Onsala Space Observatory, Chalmers University of Technology (Sweden), ETH, Swiss Federal Institute of Technology (Switzerland), and CSIC, Institut d'Estudis Espacials de Catalunya (Spain), in association with the Astronomical Institute at the University of Berne (Switzerland), the Meteorological Office (UK) and the Danish Meteorological Institute (Denmark).

REFERENCES

1. E.R. Westwater. Ground-Based Microwave Remote Sensing of Meteorological Variables in Atmospheric Remote Sensing by Microwave Radiometry.
    Ed. M. Janssen, Wiley & Sons, pp. 145-213, New York, (1993).
2. H. Tsuji, Y. Hatanaka, T. Sagiya, M. Hasimoto. Coseismic crustal deformation from the 1994 Hokkaido-Toho-Oki earthquake monitored by a nationwide continuous array in Japan. *Geophys. Res. Lett.*, **22**, 1669-1672, (1995).
3. T.R. Emardson, G. Elgered, J.M. Johansson. Three months of continuous monitoring of atmospheric water vapor with a network of Global Positioning Systems receivers. *Journal of Geophysical Research*, **103**, 1807-1820, (1998).
4. T.R. Emardson, H.J.P. Derks. On the relation between the wet delay and the integrated precipitable water vapor in the European atmosphere. *Meteorological Applications*, accepted (1999).
5. G. Elgered and P.O.J. Jarlemark. Ground-based microwave radiometry and long-term observations of atmospheric water vapor. *Radio Science*, **33**, 707-717, (1998).
6. J.M. Johansson, G. Elgered, J.L. Davis. Wet Path Delay Algorithms Using Microwave Radiometer Data. *Contributions of Space Geodesy to Geodynamics*: *Technology*, eds. D.E. Smith and D.L. Turcotte, AGU Geodynamics Series, **25**, (1993).
7. E.R. Weswater, M.T. Decker, F.O. Guiraud. Feasibility of atmospheric temperature sensing from ocean data buoys by microwave radiometry. *NOAA Technical report ERL 375 - WPL 48*, National Tech. Inform. Serv. Springfield, VA (1996).

Microw. Radiomet. Remote Sens. Earth's Surf. Atmosphere, pp. 193–201
P. Pampaloni and S. Paloscia (Eds)
© VSP 2000

Microwave Ground Based Unattended System for Cloud Parameters Monitoring

A. KOLDAEV[1], E. KADYGROV[1], A. MIRONOV[1]

[1]*Central Aerological Observatory, 141700, Dolgoprudny, Russia*

Abstract: This paper describes a new remote instrument for "real-time" retrieval of cloud parameters such as Liquid Water Path (LWP) and the average temperature of the liquid- water layer. These parameters are very important for diagnosing winter clouds which are potentially dangerous as they may cause aircraft in-flight icing. This system consists of two microwave radiometers operating at 85 GHz and 37 GHz. The main feature of this system is that it can operate in an unattended mode within a very wide range of weather conditions such as heavy rain, freezing rain, freezing drizzle, and heavy snowfall at temperatures varying from -40C^0 to +10C^0. The automated data quality control as well as internal calibration provided make it possible to obtain fully processed data in "real time" without the need of microwave data flow estimation by an expert. Preliminary tests of this system under field winter conditions are also discussed.

1. INTRODUCTION

The current high air traffic intensity in the main highly industrialized countries has made the problem of aircraft in-flight icing very pressing. It is well known that during the last three winter seasons at least one airplane crash with many damaged and killed is reported in the USA and Canada each winter.

It is 50 years ago (Lewis, 1951) that aircraft icing in clouds was first accounted for by the presence of small supercooled water drops in clouds. The main parameters that determine icing intensity are: Liquid Water Content (LWC) of clouds, mean temperature of water drops, airplane speed and shape.

To measure LWC, a variety of "in-situ" sensors have been developed and used. One of the most recent devices is described in (A. Nevzorov, 1996). When the diagnostics of icing is performed from board an airplane using "in-situ" sensors, the information obtained is most appropriate from a scientific point of view as then one knows LWC and temperature at the flight level, and the speed of the airplane. But as to the practical use of this information, it seems too late when the pilot only learns about the airplane icing when flying through a dangerous cloud. For this reason, remote sensing of cloud zones where icing is likely to occur, is of much more use, because the information about the current conditions can be used preventively to avoid such zones.

Unfortunately, weather radar can not detect such zones containing small water drops, and the only tool for remote sensing as applied to the problem concerned is microwave passive radiometry. It is well-known that microwave ground-based systems provide data on the integral LWC (E. Westwater, 1980). But as it has been stated above, the knowledge of LWC is not enough. With this in mind, a method of cloud temperature retrieval by two-wavelength sounding has been developed (Khaikin, 1994). The results of the field tests conducted are reported in (Koldaev, 1998).

This paper describes the first attempt to create a commercial model of a two-wavelength sounding system capable of operating in an unattended mode, with self- calibration and data quality control provided. The final goal is to make a routine use of this system at civil airports by the local personnel, without any need for additional assistance from experts in running microwave instrumentation.

2. LWH REMOTE SENSING TECHNIQUE

As stated above, the possibility of LWP remote determination is well-known (Hogg, 1983). That is why we shall dwell upon the retrieval of the mean cloud drop temperature and, therefore, on the possibility of Liquid Water Height (LWH) determination.

The physical concept of LWH microwave remote sensing builds upon liquid water property to change its absorption coefficient depending on thermodynamic temperature. The cloud brightness temperature T_b, measured from the zenith direction by ground-based microwave radiometer in the absence of liquid precipitation can be given by the following equation (Khaikin,1994):

$$T_b(f) = T_a\{1-exp[-\tau(f)]\} \tag{1}$$

where T_a is the mean atmospheric temperature and $\tau(f)$ is the total atmospheric opacity at the sounding frequency f. Far away from the absorption lines of atmospheric gases, the opacity τ, can be presented as the integrated absorption of liquid water, water vapor and molecular oxygen:

$$\tau(f) = B(T,f)W + C(f)Q + \tau_{o_2}(f) \tag{2}$$

where $B(T,f)$ is the absorption coefficient of liquid water drops at temperature T and sounding frequency f, W is the integral liquid water path (LWP), $C(f)$ is the absorption coefficient of water vapor at frequency f, Q is the integral water vapor content and $\tau_{o_2}(f)$ is molecular oxygen absorption.

If measurements are made with a two-wavelength system, a scatter plot in $\tau(f1) \times \tau(f2)$ coordinates can be drawn for any set of data. During "clear sky" conditions when W=0 and only integral water vapor is varied, the points of the scatter plot will be approximated by a straight line with the tangent of the inclination angle equal to $C(f1)/C(f2)$, because oxygen absorption is almost constant.

Under cloudy conditions, LWP variation is much faster and liquid water absorption is much higher, so the straight line fitting the scatter plot best will have the tangent of the inclination angle equal to $B(T,f1)/B(T,f2)$. Thus, the tangent of the inclination angle of the best fitting line in $\tau(f1) \times \tau(f2)$ coordinates will only depend on the temperature of liquid water drops.

It has been shown (Khaikin, 1994) that for 3 mm and 8 mm wavelengths the function $F(T)=B(T,f1)/B(T,f2)$ is almost linear with a deviation less than 3% within a $-20^0C/+5^0C$ cloud drop temperature range.

As soon as we are able to get the mean temperature of cloud drops, the height of the liquid water layer can be estimated using standard radiosonde data on the vertical atmospheric temperature profile. This procedure is described in Fig 1. The method and algorithm of LWP and LWH retrieval employed is reported in detail in (Koldaev, 1996).

Experimental tests of the LWH retrieval technique were conducted during three field projects in Canada. The results of comparisons with "in-situ" aircraft measurements (Koldaev, 1996) as well as the results of statistical comparisons with radiosonde data (Koldaev, 1998) have shown a good agreement between the retrieved and directly measured LWH.

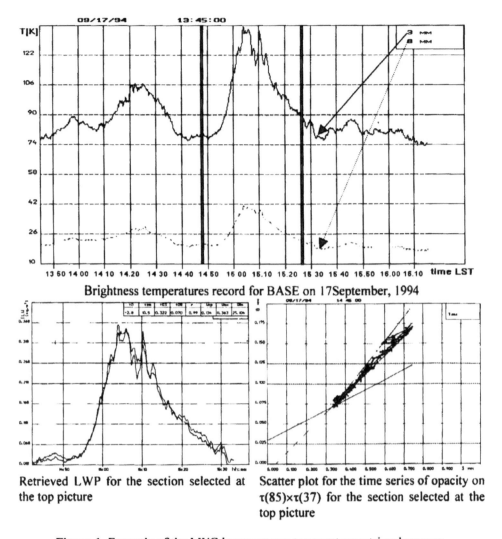

Retrieved LWP for the section selected at the top picture

Scatter plot for the time series of opacity on $\tau(85) \times \tau(37)$ for the section selected at the top picture

Figure 1. Example of the LWC layer average temperature retrieval process

3. MICROWAVE UNATTENDED SYSTEM

3.1. Technical requirements

During the field experiments mentioned above, two wavelength ground-based microwave systems were used. The most advanced measuring system configuration had been designed for the Canadian Freezing Drizzle Experiment III which was carried out in Ottawa region during December 1997-February 1998. It was a trailer housing with thermo-stabilizing system, in which two microwave radiometers with 85 GHz and 37 GHz working frequencies were installed. Atmospheric radiation was received through the radio-transparent window after its reflection by a flat metal sheet. The calibration was performed by using a "black body" and "clear sky" technique. For this purpose, an absorber was placed beneath the flat reflector. Then, the radiometers received signals from the absorber when the reflector was looking at the nadir direction. In fact, systems of this kind mounted in a trailer to make long-term measurements are commonly used (Westwater, 1986). The system was run by skilled personnel because almost all the procedures involved (such as calibration) required manual operation or at least supervision (data quality and radiometer operability control). The experience gained both from these operations and previously allows formulating basic limitations of most recent ground-based microwave systems and, therefore, developing technical requirements for an unattended system. The main problems and their possible solutions are presented in Table 1 in terms of comparison.

PROBLEM	SOLUTION
Precipitation problem for antenna system	Automated cleaning system adjusted for specific precipitation type
"Black body" quality control: termal equilibrium with the environment, precipitation, direct sun and windshield factor	Black body in weatherproof housing
Termo-stabilization of the microwave system's operation volume.	Automated control and maintaining the whole system's stable temperature
Manual calibration during "clear sky" conditions.	Automated regular calibration procedure under any weather conditions with a built-in "black body"
Expert control of environmental conditions acceptable for proper calibrations	Automated monitoring of "clear sky" conditions
Commonly, the systems are not portable	Portable, easy-to- install system with a possibility of autonomous power supply and data transmission via radio modem.

Table 1. Problems, to be resolved for creating an unattended system

3.2 .Construction of the system

The drawing and the general view of the system are presented in Fig. 2.
The main features of the system are as follows:

I) Weather protection: each radiometer is installed in a separate housing. The housing has the shape of a tube 14" in diameter and 32" long. The tube consists of two parts - a stationary and a rotating one. The stationary part includes a microwave radiometer with an antenna, a thermo-stabilizing system, a steping motor, two electromagnetic clicks, a microprocessor control board, and a set of sensors. The rotating part includes a turning flat metal reflector with one side used as a reflector and the other covered by an absorber. Atmospheric radiation is received through a radio-transparent window at the top of the rotating part. The connection between the rotating and stationary parts is waterproof.
II) All-weather operation: the shape of the weather protection housing prevents snow and water accumulation. The devices are installed in separate housings . To protect the radio-transparent window from water drops or snow, the rotating part is turned each 5-20 minutes and the windshield cleaners, which are solid with the stationary part of the housing, clean it. Two thermo-stabilizing systems are used to provide a $+40^0C/-40^0C$ range of operation. The first one is inside the radiometer. It maintains $+50^0C+/-0.5^0C$

temperature inside the radiometer box. The second one is in the stationary part of the housing. It does not provide such an accurate termo-stabilization, but makes it possible to control the inner temperature by using a heating/cooling system.

III) Self-calibration: when the rotating part is turning to clean the window, one of the electromagnetic clicks releases the flat mirror for it to turn over and expose the absorber. The other click fixes the mirror's position. The rotating part is not thermally insulated from the environment but has good thermal insulation from the stationary part and the absorber being at ambient temperature.

As the ambient temperature changes at least by 5^0C every 24 hours, one needs to get information about the recent temperature value and the output code of the radiometer to continuously update the calibration data set.

IV) Unattended operation: unattended mode of operation as well as "real-time" data processing and transmission via radio modem is performed by the in-built microprocessor board. A 12 V automobile accumulator is used as power supply.

Specification
1. Radiometer
2. Microprocessor card
3. Stepping motor
4. Mirror
5. Sensor of antenna position
6. Magnet
7. Termistor
8. Teflon
9. Fans
10. Mirror control units
11. "Black body"
12. Cleaning unit
13. Precipitation sensor with heater
14. Cleaning heater

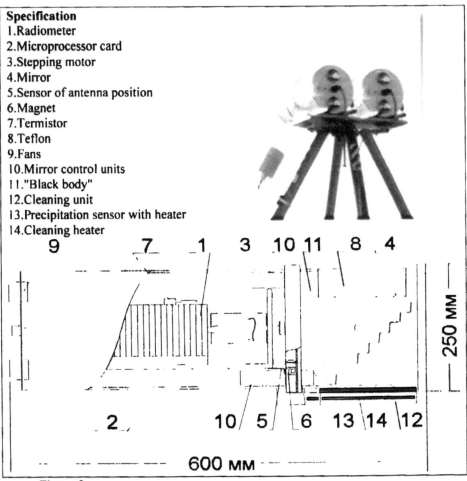

Figure 2.

The given system includes 4 functional blocks or units: the rotating and stationary parts of the weather protection housing, a commercial computer with a radio modem and a user's terminal (Fig. 3).

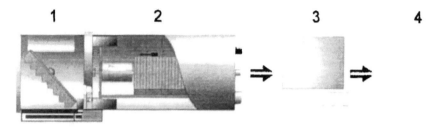

Figure 3. Functional diagram

The rotating and stationary parts of the housing are connected only mechanically (without any wires or rotating contacts). The computer and weather protection housing are connected with a special cable enabling operation at ambient temperatures up to $-60^{0}C$. The user's terminal and computer are connected via radio modem line with RS232 exchange protocol. Each part of the system fulfills specific operational functions:

The rotating part (No. 1 in Fig. 3):
– Precipitation Control
– Cleaning radio-transparent window
– Angular scanning
– Calibration using turnable "black body"

The stationary part (No. 2 in Fig. 3):
– Data receipt/transmission via RS232
– Program management using microprocessor
– Program execution control
– ON/OFF for all build-in sensors and units
– Scanning management

The industrial computer (No. 3 in Fig. 3)
– Scanning protocol management depending on precipitation sensor data
– "Clear sky" calibration management
– "Black body" calibration management
– Control and management of the measuring system
– Data quality control
– Calibration quality control
– Real-time computation of LWP and LWH
– Data transmission via radio modem
– Database support

The user's terminal (No. 4 in Fig. 3)
– Short-term forecasting of aircraft in-flight icing ("Nowcasting")
– Indication of precipitation type with rough intensity estimate

3.3. Mode of operation

The system's operational mode is illustrated by Fig. 4. It presents a time-series diagram of different sensors or procedures depending on current weather conditions.

For "clear sky" the rotating part is looking at the zenith direction for 20 minutes. After that the rotating part is turned by 360^0. The precipitation sensor is requested and if the signal is "0", the situation is identified as "clear sky" and the same procedure is repeated within the next 20 minutes. After three 20-min. periods a "clear sky" confirmation is obtained by two angle measurements. For this purpose, the radiometers first receive radiation at 60^0 relative to the zenith direction during 2 min. and then at 71^0 during another 2- min. period. After that, the rotating part is turned at the nadir direction, and the electromagnetic click releases the mirror so as to expose the receiver to "black body" signals. The computer is run to provide simultaneously the ratio of atmospheric absorption measured at the zenith direction and at a 60^0 angle as well as the ratio of absorption for $60^0/71^0$. If these ratios prove equal, the confirmation of "clear sky" situation is stored.

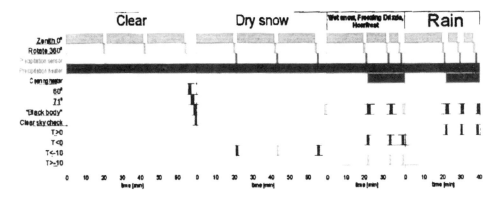

Figure 4. Time series diagram of the system operation

For "dry snow" the procedure is run in the same fashion during 20 minutes first. But after the first check of the precipitation sensor when the rotating part is turned by 360^0, we get signal "1". Immediately after that, the ambient temperature sensor is checked. If the ambient temperature is below -10^0C, the situation is classified as "dry snow".

Nevertheless, at the end of an hour's period the system checks "clear sky" conditions and performs "black body" calibrations. A disagreement between the two absorption ratios would testify to the existence of precipitable clouds.

For "wet snow", "freezing drizzle" and "Hoarfrost" conditions, the procedure is the same as for "dry snow", but the ambient temperature will be within the range of $0^0C/-10^0C$. In this case the cleaner's heaters will be activated to prevent icing of the rotating part. No "clear sky" check in this situation is done. Precipitation sensor data are processed in a parallel mode, and if the intensity of precipitation is more than 1 mm per hour, the period of the rotation of the moving part is decreased from 20 to 10 minutes. At a 5-mm or higher precipitation values the period of rotation is once again decreased two-fold. For "rain" the situation is similar to the previous one, with the only difference in the ambient temperature sensor's readings. If the temperature is above 0^0C, precipitation is recognized as rain.

4. PRELIMINARY TESTS AND DISCUSSION

A prototype of the system described has been designed and manufactured by the "Atmospheric Technology" company (Russia). This system was run at a field site in Dolgoprudny, Moscow Region, during the winter 1998/99. All the design parameters of the system have been validated. The "wet snow" and "rain" precipitation records obtained are exemplified in Fig. 5. It can be inferred from the precipitation sensor records that "snow" and "rain" situations could even be distinguished after analyzing signal shape.

The reliability of the system during "freezing drizzle" was subject to special investigation. Unfortunately, this kind of precipitation is not frequently observed in Moscow Region, and thus it had to be simulated by way of spraying water aerosol at $-10^0C/-17^0C$ temperatures under anticyclone clear-sky conditions. Specifically, this experiment has made it possible to establish a 20-min. optimum time period for the rotation. It has been found out that the first layer of small water drops is glaciated, approximately, within this time interval, while the following layers are transformed into ice crystals much faster.

The authors realize that the reliability of the unattended microwave system can only be verified through comparisons of the output data bases from regular (manually serviced) and unattended systems. Similar comparisons are usually made during field experiments, with 24-hour observations conducted throughout winter. Such a comparison is planned to be accomplished within the framework of the Canadian Freezing Drizzle Experiment during 1999/2000 winter season.

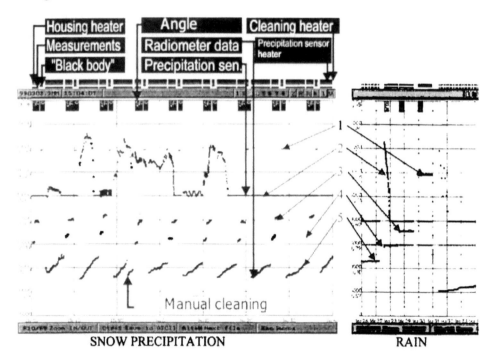

Figure 5. Time series diagram of the system operation
1. signal from "black body"
2. signal from precipitation sensor
3. signal from 71^0 angle
4. signal from 60^0 angle
5. zenith signal

ACKNOWLEDGEMENTS

The authors are very thankful to Walter Strapp for his kind assistance in the experimental work with attended microwave systems during a number of field projects conducted by the Atmospheric Environment Service of the Environment Canada, and supported by Transport Canada.

We also thank Arkadi Troitcky for his permanent attention to this work and for his very useful recommendations.

REFERENCES

1. A.R. Lewis. Physical and operational aspects of aircraft icing. *Compendium of Meteorology. Amer. Meteor. Soc.*, p.1190, (1951).
2. A.N. Nevzorov. CAO aircraft instrumentation for cloud physics, *Proc. of 12th International Conference on Clouds and Precipitation*, v.1 pp.371-375, Zurich, Swizerland, (1996).
3. E.R. Westwater, F.O. Guirand. Ground-based microwave radiometric retrieval of precipitable water vapour in the presence of clouds with high liquid content, *Rad.Sci.*, v15,No15, 947-957 (1980).
4. M.N. Khaikin, A.V. Koldaev. Remote sensing of aircraft icing zones parameters, *TECO-94 WMO/TD-No 588 Report No 57*, 413-417 (1994).
5. A.V. Koldaev, E.N. Kadygrov, A.V. Troitsky. Remote sensing of average LWC layer height inn winter clouds, *Proceedings of the eighth atmospheric radiation measurement (ARM) science team meeting*, 381-384 (1998).
6. D.C. Hogg, F.O. Guiraud, J.B. Snider, M.T. Decker and E.R. Westwater. Asteerable dual-channel microwave radiometer for measurement of water vapor and liquid in the troposphere. *J.Climate Appl. Meteor.*, 22, 789-806, (1983).
7. A.V. Koldaev, A.V. Troitsky, J.W. Strapp and B. Sheppard. Ground based microwave measurements of integral liquid water content of Arctic clouds during BASE. *Proc. of 12th International Conference on Clouds and Precipitation*, v.1 pp.442-445, Zurich, Swizerland, (1996).

Microw. Radiomet. Remote Sens. Earth's Surf. Atmosphere, pp. 203–211
P. Pampaloni and S. Paloscia (Eds)
© VSP 2000

Statistical analysis of the tropospheric radio-path delay using radiosonde and ground-based radiometric data

ERMANNO FIONDA, FRANCESCO BARBALISCIA and PIER GIORGIO MASULLO

Fondazione Ugo Bordoni, viale Europa 190, 00144 Roma, Italy.

Abstract - The total radio-path delay plays a key-role for many scientific fields such as space geodesy, radar altimetry, satellite-based positioning system (Global Positioning System: GPS), and radio-astronomy (VLBI). The radio-path delay wet component, due to the tropospheric water vapour, is in particular the main source of inaccuracy. As well known, the tropospheric atmospheric emission allows the retrieval of the corresponding atmospheric integrated contents of water vapour and cloud liquid, strongly variable in both time and space. Consequently, the wet radio-path term is highly variable and accurate analysis is required of its features to give guidance for the above mentioned technical applications. In this paper, the troposheric radio-path delay is analysed by means of radiosonde profiles and radiometric measurements, both observed for the Roma area, Italy, pointing out a good agreement between the two separate sources. Statistical trends are also analysed of both the total and the wet path delay, showing seasonal dependence and a rather limited year-to-year variability, while no significant daily-time dependence has been found.

1. INTRODUCTION

For many application fields concerning geophysical topics, navigational satellites and radio astronomy in particular, a correct estimate is needed of the tropospheric radio-path delay. Different methods can be utilised to that purpose, the choice depending on both the availability of predictors and on the required accuracy. From radiosonde data, theoretical values of total radio-path delay can be extracted directly from the refractivity definition, as well as the separate dry and wet components can also be calculated. Through prediction algorithms, climatic data measured at ground surface level can be used, as predictors, to estimate the radio-path delay. The thermal downwelling emission from the atmosphere, measured by a dual-channel microwave radiometer can be utilised for determining radio-path delay, also using water vapour as an alternative predictor. In this paper, the availability of 14 years of radiosonde profiles and of 4 years of radiometric experimental observations, at 23.8 and 31.4 GHz, allows a twofold analysis of the radio-path delay and the assessment of its statistical features.

2. ALGORITHMS FOR ESTIMATING RADIO-PATH DELAY

The total path-delay ΔH_T (in cm), intended as the difference between the electrical and the geometrical length of the radiowave path, is defined as by the Bean and Dutton equation [1]:

$$\Delta H_T = 10^{-6} \int_0^H N(z, P, T, e)\,dz \qquad (1)$$

In Eq (1), $N(z) \equiv N(z,P,T,e)$ is the atmospheric refractivity depending on the spatial position (z in cm), on the air pressure (P in hPa), on the temperature (T in K), and on the partial pressure of water vapour (e in hPa). In Eqn. (1) the $10^{-6} N(z)$ factor is a convenient expression of the atmospheric index of refraction (n). The integration of (1) ranges from the ground surface level to the maximum altitude (H in km) at which the atmosphere can affect the path-delay. A high degree of accuracy of the (P), (T), and (e) along the atmospheric path provides a high degree of accuracy of $N(z)$ with a consequently high precision in the computation of the ΔH_T. The analytic equation of the $N(z)$, commonly used, is the well known approximate formula given by [1]:

$$N(z) = 77.6 \left(\frac{P(z)}{T(z)} \right) + 3.73 * 10^5 \left(\frac{e(z)}{T(z)^2} \right).$$

(2)

The tropospheric refractivity $N(z)$ is the sum of two separate components: the first one due to the atmospheric gases, called the "dry" (or hydrostatic) component $N_D(z)$ and the second one due to the atmospheric water, called the "wet" component $N_W(z)$, which results from precipitable water vapour (V). The corresponding total path-delay can be written as:

$$\Delta H_T = 10^{-6} \left(\int_0^H N_D(z)dz + \int_0^H N_W(z)dz \right) =$$

$$= 10^{-6} \left(\int_0^H N_D(z)dz + \int_0^H N_V(z)dz \right) =$$

(3)

$$= \Delta H_D + \Delta H_W = \Delta H_D + \Delta H_V.$$

The dry path-delay term (ΔH_D) can be calculated from the air pressure measured at the ground surface level, the accuracy being strongly dependent on the instrumental accuracy of the predictor. The ΔH_D which is predominant in the determination of the total path-delay, can range from 220 to 240 cm at sea level. Accurate studies have shown that the variability of the dry term increases with the decreasing of the latitude of sites, and is limited to a few centimetres. Nevertheless the dry mean values have shown long-term stable features [2]. On the contrary, the troposphere wet path-delay term (ΔH_W) is much more difficult to compute using climatic parameters observed at the ground surface level as predictors. The primary source of the ΔH_W comes from the atmospheric water vapour term, which is quite critical mainly because of the complexity in rebuilding, with high accuracy, the actual non-uniform vertical distribution of the atmospheric water vapour. The ΔH_V component shows also a large variability due to the high space-time variability of water vapour. The presence in the atmosphere of suspended cloud drops can add an additional contribution to the wet path-delay (ΔH_L), which depends on the integrated amount of cloud-liquid (L) and on cloud temperature. However, ΔH_L is typically small, and can be neglected.

Ground-based microwave radiometric measurements of the atmospheric brightness temperature ($T_b(f_i)$ in K) give the possibility to provide values of the cloud component to the wet path delay [2], [3].

Radio-path delay from refractivity
Theoretical dry and water vapour path-delays, and therefore the total path-delay, can be estimated by applying a straightforward algorithm on radiosonde (RAOB) profiles of air pressure, temperature, and humidity using the well known equation [1]:

$$\Delta H_T = \Delta H_D + \Delta H_V = 77.6*10^{-6}\int_0^H \left(\frac{P}{T}\right)dz + \int_0^H 3.73*10^{-1}\int_0^H \left(\frac{e}{T^2}\right)dz \qquad (4)$$

Here, theoretical delay values, at the radiometric channels, are calculated by applying the Eqn. (4) on a representative data set of 14-year RAOB profiles collected in Rome. Below, the values of path delay from the refractivity are assumed as the "true" values.

Statistical inversion algorithm
For a non-scattering atmosphere in local thermodynamic equilibrium, the $T_b(f_i)$ measured at the ground at two frequencies, the first one (f_1) around the water vapour absorption peak (22.2 GHz), and a second one (f_2) within the liquid window-band (30-40 GHz), is given by [4]:

$$T_b(f_i) = T_C e^{-\tau(0,\infty)} + \int_0^\infty T(z)\alpha(z)e^{-\tau(0,z)}dz \qquad (5)$$

where τ is the optical depth (in Np), T_C is the cosmic background temperature; T(z) is the absolute physical air temperature (in K), z (in km) is the spatial position of the emitting air volume, $\alpha(z)$ is the atmospheric volume absorption coefficient (Np/km). Eqn. (5) states that the zenith sky brightness temperature is the sum of the cosmic background term and of the contributions of the infinite single atmospheric layers along the observed path, both reduced by the absorption of the atmosphere underneath. $T_b(f_i)$ depends on the vertical profile of the temperature and absorption coefficients; the latter, in turn, corresponding to the vertical distribution and physical properties of the relevant components.
The corresponding opacities $\tau(f_1)$ and $\tau(f_2)$ (in Np) can be estimated by the definition of the mean radiating temperature $T_m(f_i)$ (in K) [5]:

$$\tau(f_i) = ln[(T_m(f_i) - T_C)/(T_m(f_i) - T_b(f_i))] \qquad (6)$$

This equation is commonly used to convert the radiometrically-measured brightness temperature into the path attenuation. Eqn. (6) should be used, however, only for low values of path attenuation. As the latter increases, in fact, the radiometer measures values of brightness temperature closer and closer to the atmospheric effective radiating temperature, thus leading to a saturation effect. The opacity τ takes into account the spectral absorption by oxygen, water vapour and cloud liquid contents into the volume of air observed by the radiometer-antenna. The use of the radiometric opacities as predictors allows one to derive integrated atmospheric vapour (V) and liquid (L) [5], [6]. Since the wet and dry path-delays show a linear relationship with V and L [1], [2], the total path-delay is expected to be determined from a linear combination of the radiometric opacity $\tau(f_1)$ and $\tau(f_2)$, [3]:

$$\Delta H_T = a_0 + a_1\tau(f_1) + a_2\tau(f_2) \tag{7}$$

where the a_i are the linear statistical inversion coefficients at the two operating-frequencies (f_i) of the radiometer. The inversion coefficients, which depend on the frequency and on the physical properties of the atmosphere, are estimated from RAOB profiles on a statistical basis by using theoretical models. For the computation, a large data set of RAOBs collected in Rome, twice a day from the Italian Air Force Service from 1981 to 1993 is used. Appropriate atmospheric absorption/emission models and radiative transfer equation model have been applied on RAOB data, simulating the theoretical "true" brightness temperature values [7]. The inversion coefficients of (7) have been estimated by minimizing the sum of square of the differences between the values of ΔH_T, calculated by (3) and the corresponding values predicted by (7), when simulated opacity values $\tau(f_i)$ from the same RAOBs are used. In order to compute inversion coefficients of Eqn. (7), instrumental errors for each predictor have been assumed by adding a zero mean Gaussian error with a standard deviation of 0.5 K on $T_b(f_i)$ and 5 K on $T_m(f_i)$. Moreover in this paper, the Decker criterion has been adopted to provide the simulation of cloud-liquid contents [7]. Since measurements of cloud-liquid are not reliable even with radiosonde sensors, this parameter must be found with great caution.

Similarly, through the same approach the wet path-delay (ΔH_W) can be derived from the radiometrically measured τ's using similar inversion coefficients b_i's:

$$\Delta H_W = b_0 + b_1\tau(f_1) + b_2\tau(f_2) \tag{8}$$

As from the application of the above procedures, the inversion coefficients for both ΔH_T and ΔH_W are reported in Table 1a and Table 1b, respectively.

Table 1a.
Linear regression parameters for the total path-delay as a function τ's, Eqn. (7). (Rome–RAOBs data 1981-1993).

	a_0, [cm]	a_1, [cm/Np]	a_2, [cm/Np]
ΔH_T	232.205	27.338	-15.987

Table 1b.
Linear regression parameters for the wet path-delay as a function τ's, Eqn. (8). (Rome–RAOBs data 1981-1993).

	b_0, [cm]	b_1, [cm/Np]	b_2, [cm/Np]
ΔH_W	0.643	31.171	-17.694

A second possibility is also offered through radiometric measurements for calculating the ΔH_W component, which can be related to the total V by a linear regression:

$$\Delta H_W = c_0 + c_1 V \tag{9}$$

directly evaluating the wet delay from V, for those cases when the value of V is given from other sources (i.e. RAOB, maps, etc.). Table 2 shows the c's linear regression parameters, taking into account different sky conditions.

Table 2.
Linear regression parameters for the wet path-delay as a function of the total V. (Rome–RAOBs data 1981-1993).

Parameters	Clear Sky	Cloudy Sky	Clear+Cloudy Sky
c_1, [1/cm]	6.293	6.323	6.345
c_0, [cm]	0.293	0.459	0.301
Rms, [cm]	0.165	0.199	0.207
Corr. Coeff.	0.9994	0.9991	0.9992
N. of samples	3691	2822	6513

3. OBSERVED LONG-TERM PATH-DELAY FEATURES

Ground-based radiometric measurements of the sky brightness temperature have been carried out in Italy, continuously for 4 years, from September 1992 to August 1996, at the very south side of Rome, in a temperate zone, at 15 km from the sea and subject to marine breeze regime and urban environment influence. The radiometer is a dual-channel, total power, portable type, the model WVR-1000 manufactured by the Radiometrics Corporation, Boulder, Colorado (USA). The radiometer operates at 23.8 GHz, near the weak vapour resonant line, and at 31.4 GHz, within the window region almost transparent to water vapour. Radiometric observations, acquired at a 145-s sample rate, have been edited in different ways to check reliability. Although the major part of the data have satisfactory quality, several natural or instrumental types of outliers might occur, which can arise from the presence of liquid, ice, or snow on the antenna window, spurious electromagnetic signals, calibration drift in the receiver or errors due to the data transmission system [8], [9]. Records showing outliers at either channel have been entirely disregarded, as well as those acquired during rainy events, identified by a tipping-bucket rain gauge operated close to the radiometer. Values of $T_m(f_i)$ of 276.39 and 272.98 K, at 23.8 and 31.4 GHz, respectively, are used, calculated by applying theoretical absorption models on long-term RAOB data [9].

Statistical analysis of radio-delay has been carried out for the entire 4-year period (1992-1996) of measured emission levels of the ground-based radiometer, as well as for seasons, and daily-time, by applying the linear statistical inversion as from Eqn. (7) to derive the radiometric total path-delay and using RAOB data (via Eqn.4) to calculate the "true" correspondent total path delay. In Figure 1 the cumulative distribution functions (CDF) of the above values are reported, the radiometric one in terms of individual years and of total period average. Radiometric and radiosonde total path delay show good agreement, being the maximum discrepancy less than 1 cm, for less than 10% of the time. The year-to-year variability keeps below 2%, with the total mean and rms values limited to only 3 cm and 1 cm, respectively. This variability seems indeed due totally to the wet component of path delay, calculated according to Eqn.(8) and reported in Figures 2 and 3, the latter showing seasonal CDFs.

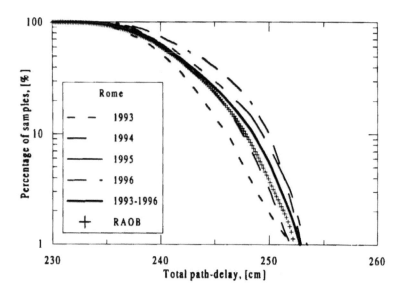

Figure 1. Yearly cumulative distribution functions of the zenithal total path, together with the predicted true values.

The wet path-delay variability is, in fact, appreciably higher, about 20 % in average, among years, while the seasonal variability is much pronounced for the wet component, as expected, reaching in summer time values as higher as twice with respect to the cold season.

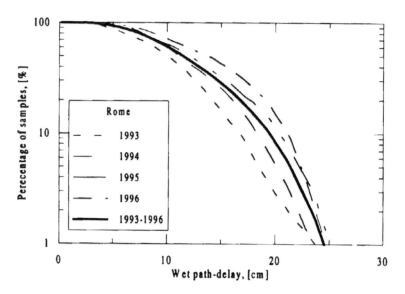

Figure 2. Yearly cumulative distribution functions of the zenithal wet path-delay.

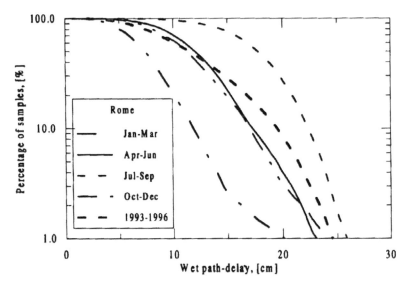

Figure 3. Three-monthly cumulative distribution functions of the zenithal wet path-delay.

Figure 4 shows the monthly trends of the wet path-delay throughout 4 years, for percentile levels from 90% to 2 %. Clear thermal-seasonal cycle dependence can be noted, with the highest peak values, at each level, recorded in August. From the analysis of percentiles behaviour along the hours of the day, no significant differences can been noted.

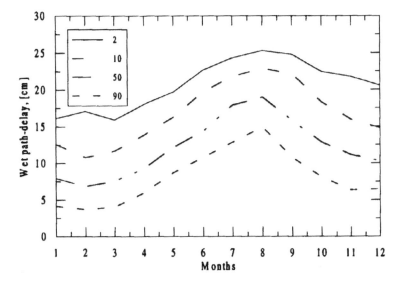

Figure 4. Monthly percentile levels of the zenithal wet path-delay.

The comparison is offered in Figure 5 of the CDFs of the wet delay calculated through the τ_s (Eqn.8) and the total vapour content V (Eqn.9) from radiometric measurements. The same curve calculated with the Elgered model by using the RAOB data is also reported

[2]. It is worthwhile to notice that the difference between the delay calculated via the τs and that from V keeps low, some 1 cm, that is less than 5%. This allows the use of V, which is more readily available than the τs for calculating the path delay.

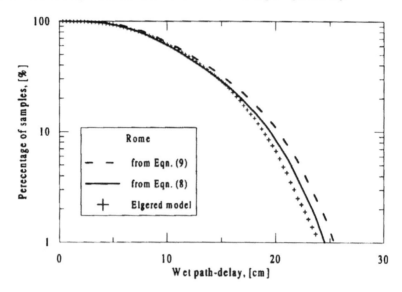

Figure 5.- Cumulative distribution functions of the wet path delay calculated from opacities τ (Eqn.8), total vapour V (Eqn.9) and by the Elgered model.

4.   CONCLUSIONS

The paper reports the statistic features of the tropospheric zenithal radio-path delay, at 23.8 and 31.4 GHz, for Roma, Italy, located in a temperate climatic zone. The process for retrieving the radio-path delay was based on a representative historical data base of radiosonde profiles, and on four years of downwelling radiometric emission measurements. Applying a radiative transfer model on radiosonde profiles has provided a calculated data base of atmospheric parameters including brightness temperature and radio-path delay, either total and wet. The analysis carried out has shown that radiometrically derived path delay is in good agreement with the radiosonde calculated one, as well as an estimate can be obtained directly through the total vapour content, available from the radiosonde or maps. Statistical trends are also analysed of the wet path delay, showing marked dependence upon the seasonal climatic station features, while the year-to-year variability of either total and wet path-delay appears rather limited.

Acknowledgements   The authors thank the Italian Air Force for radiosonde data as well as A. Capitanio and F. Consalvi for assistance in instrumental keeping. Grateful thanks are also due to Gunnar Elgered for the help given to apply his model. The authors also thank the two anonymous reviewers for useful comments and suggestions in improving the original manuscript.

REFERENCES
1.   Bean B. R., E. J. Dutton, *Radio Meteorology*, National Bureau of Standard Monograph, Department of Commerce, Washington D. C., USA (1966).

2.   Elgered G., Troposheric radio-path delay from ground-based microwave radiometry, Chapter n.5 of the *Atmospheric Remote Sensing by Microwave Radiometry*, Edited by M. Janssen, John Wiley & Sons.(1993).

3.   Ciotti P., Basili P., D'Auria G., Marzano F.S., Pierdicca N., Microwave radiometry of the atmosphere: an experiment from a sea-based platform during ERS-1 altimeter calibration, *Int. J. Remote sensing*. **16**(13), 2341-2356(1995).

4   Ulaby F.T., R. K. Moore, and A. K. Fung, *Microwave remote sensing: active and passive. Volume I: Microwave remote sensing fundamental and radiometry*, London Addison-Wesley edition (1981).

5.   Westwater E. R., The accuracy of water vapour and cloud liquid determination by dual-frequency ground-based microwave radiometry, *Radio Science*, **13**(4), 677-685, 1978.

6.   Basili P., P. Ciotti, and E. Fionda, Comparison of algorithms for the retrieval of water vapour, cloud liquid and atmospheric attenuation by microwave radiometry, in *Proceedings of PIERS 1994*, pp. 571-574, Euro. Space Agency, ESTEC, Noordwijk, Netherlands(1994).

7.   Schroeder J., E. R. Westwater, Users' guide to WPL microwave radiative transfer software, *NOAA Technical Memorandum ERL WPL-213*(1991).

8.   Fionda E., M. J. Falls, and E. R. Westwater, Attenuation statistics at 20.6, 31.65, and 52.85 GHz derived from emission measurements by ground-based microwave radiometers, IEE Proc., Part H, **138**(1) 46-50, (1991).

9.   Barbaliscia F., E. Fionda, and P. G. Masullo, Ground-based radiometric measurements of atmospheric brightness temperature and water contents in Italy, *Radio Science*, **33**(3), 697-706(1998).

Microw. Radiomet. Remote Sens. Earth's Surf. Atmosphere, pp. 213–220
P. Pampaloni and S. Paloscia (Eds)
© VSP 2000

First results of tropospheric transmission measurements at 94 and 212 GHz

A. LÜDI*, L. MARTIN*, A. MAGUN*, C. MÄTZLER*, N. KÄMPFER*,
P. KAUFMANN[†], J.E.R. COSTA[†], C. G. GIMENEZ DE CASTRO[†], M. ROVIRA[‡]
and H. LEVATO[§]

* : *Institute of Applied Physics, University of Bern, Switzerland*
[†] : *CRAAE, Instituto Presbiteriano Mackenzie, S. Paulo, Brazil*
[‡] : *IAFE, Buenos Aires, Argentina*
[§] : *CASLEO, San Juan, Argentina*

Abstract—In order to investigate the microwave characteristics of the troposphere, simultaneous transmission measurements at 94 and 212 GHz were carried out. The main objectives were to study attenuation and scintillation effects in the same air volume under different atmospheric conditions.

The temporal variation of the atmospheric attenuation over an extended period of time was recorded and compared with calculations performed with the Millimeter-Wave Propagation Model (MPM'93) of Liebe. At 94 GHz data have been continuously collected in 1-minute intervals since June 1998. At 212 GHz the measurements were carried out in October, November 1998 and January 1999. It was found that MPM'93 computes the atmospheric absorption at both frequencies with an accuracy better than 5% for clear and rainy atmospheres.

Scintillation measurements have been made at 212 GHz with the new Solar Sub-mm Telescope (SST). It features a multi-beam antenna what allows to determine the angle of arrival of the incoming radiation with an accuracy better than 1 arc second at millisecond time resolution. Depending on atmospheric conditions the angle of arrival was scattered between 1 and 10 arcseconds corresponding to an estimated refractivity turbulence structure constant C_n of $0.5 \cdot 10^{-7} - 8 \cdot 10^{-7}\ m^{-1/3}$. Also small anisotropy in the angle of arrival was observed.

The measuring principle and first results are presented.

1. INTRODUCTION

For the investigation of the microwave characteristics of the troposphere, transmission measurements at 94 and 212 GHz were carried out. The atmospheric absorption over long time scales (minutes to hours) was measured in order to verify the Millimeter-Wave Propagation Model (MPM '93) [7], [8] and its applicability to horizontal test ranges. Transmission measurements are rarely used to verify the atmospheric attenuation, although proposed e.g. by COST Action 712 [6]. In the atmospheric windows around 90 GHz and 210 GHz, this seems to be the first experiment of that kind.

Additionally, line-of-sight propagation measurements were carried out at 212 GHz where short time scale variation of the atmospheric influence are of interest. In this frequency range no similar measurements have been available, so far. This investigation is done to quantify the influence of the atmosphere on radioastronomical observations in the mm-range and also to compare the results with fluctuation theory.

The investigation is still in progress, and we present first results.

2.   TEST RANGE AND MEASUREMENT EQUIPMENT

The slightly tilted test range at Bern, Switzerland with a length of 7.5 km crosses a mix of urban, suburban, rural and forest terrain at heights from 580 to 990 m above sea level. Next to the transmitter and the receiver weather stations are operated which measure the meteorological data (air temperature, relative humidity, air pressure) necessary for the MPM. The transmitters at 94 and 212 GHz were placed on a telecommunication tower 50 m above ground to minimize multi-path propagation.

At 94 GHz the transmitter had a HPBW of 9° which is large enough to cause multipath propagation due to reflection on the ground. In order to average out possible reflections, the signal frequency is periodically changed over a bandwidth of 100 MHz with a modulation frequency of 60 Hz, and the received signal is integrated over 5 seconds. Data are taken every minute.

At 212 GHz the transmitter operates at a constant frequency with a HPBW of 0.8°. The Solar Sub-mm Telescope (SST) [1], [2] which will be finally used for the observation of solar flares in the Argentinean Andes, was used as detecting system. It is a multi-beam and multi-frequency system with four antenna beams at 212 GHz and two at 405 GHz. The main reflector has a diameter of 1.5 meter providing half power widths of 4 and 2 arc minutes at 212 and 405 GHz respectively. Three beams at 212 GHz are intersecting near their half-power values for the localization of solar burst sources. The angle of arrival of the incident radiation can be measured with arc second precision (see section 4.).

3.   LONG TIME MEASUREMENTS COMPARED WITH MPM

Transmission measurements at 94 GHz have been collected continuously since June 1998 and intermittently at 212 GHz during winter 1998/1999. The measurements were compared with the MPM by Liebe [7],[8] for the clear atmosphere and rainfall events. Good agreement between measurements and the MPM was found when the test range was free of fog, haze and clouds. A typical example of such a day is shown in Figure 1. The agreement between measurements (solid line) and modelling results (dashed line) is better than 5% on all clear days. This small difference is of the same order as the errors in the MPM-results caused by the limited accuracy of the used meteo parameters.

The attenuation on rainy days is shown in Figure 2. Generally, when rainfall occurs over the whole testrange, MPM models the absorption quite well (left panel of Figure 2 between 14:00 and 16:00 UT and between 17:00 and 19:00 UT). However, rainfall is often locally concentrated and then the computation fails as can be seen between 6:00 and 9:00 UT. In the right panel of Figure 2 measurements at both frequencies are represented. Between 12:00 and 16:00 the absorption ratio of the two frequencies is approximately a factor 5 (i.e. frequency squared) and is due to vater vapor. However, the additional increase of absorption due to rainfall (16:30 UT) is at both frequencies almost the same.

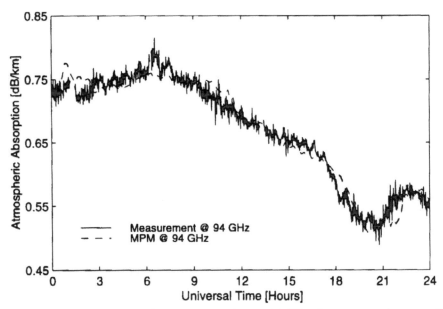

Figure 1: Absolute attenuation of the clear atmosphere at 94 GHz. The agreement between MPM-calculation and measurement is better than 5%. (Date: 5. Aug. 1998)

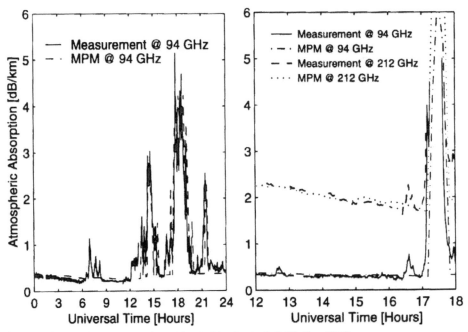

Figure 2: Atmospheric attenuation with rainfall: On the left (30. Oct. 1998), measurements compared with MPM-calculations cover a full day. The right panel (26. Oct. 1998) shows the atmospheric absorption at both frequencies.

4. SCINTILLATION MEASUREMENTS

In the following section we consider only measurements at 212 GHz with the main interest in short time scale variations of the received signal due to atmospheric air motion and scintillation. These effects were measured through the angle of arrival of the radiation from the transmitter.

4.1 Measuring principle

For determining the angle of arrival we use the method ([3]) that has been succesfully applied to the precise location of solar bursts at 48 GHz with the multibeam system at Itapetinga, Brazil [4],[5].

For a point source at the angular positon (ϑ_s, φ_s), the measured signal I_i in each of the three channels is proportional to the power pattern $P_i(\vartheta_s, \varphi_s)$ of the individual antenna beams (i).

$$I_i \propto P_i(\vartheta_s, \varphi_s)$$

With at least 3 antenna beams that point to slightly different directions and with known antenna pattern, the angular position of the source (i.e. the angle of arrival of radiation) can be reconstructed.

$$\frac{I_i}{I_{j \neq i}} = \frac{P_i(\vartheta_s, \varphi_s)}{P_{j \neq i}(\vartheta_s, \varphi_s)} \qquad i, j = 1...3 \qquad (1)$$

This equation is solved numerically with the measured farfield pattern to obtain the angle of arrival (ϑ_s, φ_s). The accuracy of the measured angular positions, discussed in this paper, is better than 1 arc second.

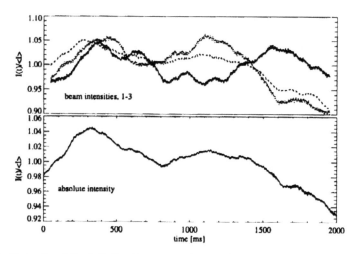

Figure 3: Example of the individually measured antenna beam intensities and the corresponding reconstructed absolute intensity (bottom).

As the angle of arrival can be determined it is possible to reconstruct the absolut intensity fluctuations due to atmospheric absorption for each instant of time, irrespective of the angle of arrival. Therefore we can distinguish between fluctuations due to variation in the angle of arrival and in absorption. An example for the temporal behaviour of the normalized individually measured beam intensities and the corresponding reconstructed intensity is shown in Figure 3. The beam intensities do not vary synchronously because the angle of arrival of the radiation is changing. The reconstructed intensity fluctuates less than the beam intensities because the fluctuations due to the variations in angle of arrival are eliminated.

4.2 Preliminary Results

Because the instrumental sampling rate is high (1 ms) it is possible to see the movement of the angle of arrival almost continuously. In Figure 4 a typical measurement of the angle over 30 s is shown. Its trace over time which looks like a random walk is in fact found to be a continuous line.

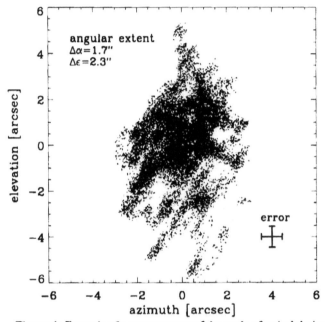

Figure 4: Example of a measurement of the angle of arrival during 30 s.

For a statistical analysis the angular extent of angle of arrival is represented by its standard deviation over periods of 10 minutes. Figure 4 shows that the measured angular extent in elevation ($\Delta\epsilon$) is larger than in azimuth ($\Delta\alpha$). This small anisotropy is substantiated in Figure 5 for another 27 measurements, as the fitted line (least-square) with a slope of 1.5, that significantly lies above one. This anisotropy can be explained by a probable vertical stratification of the atmosphere.

From turbulence theory (e.g. [10],[12]) it is known that the amount of the angular extent depends mostly on the atmospheric vertical gradient of the potential refractive index. In the mm-range the refractivity N depends considerably on the frequency and was calculated with MPM based on the meteo information from the two stations at different altitude. We

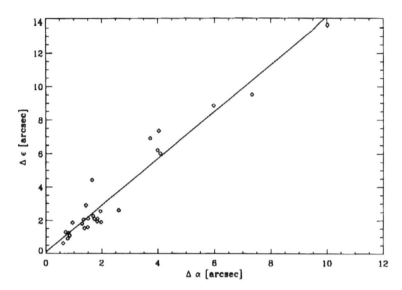

Figure 5: Angular extent in elevation ($\Delta\epsilon$) versus angular extent in azimuth ($\Delta\alpha$), showing a typical $\Delta\epsilon/\Delta\alpha \approx 1.5$.

assume that the vertical gradient of the potential refractivity is approximately equal to the measured one, $\Delta N/\Delta z$. In Figure 6 the dependence between the angular extent and the vertical gradient of the refractivity is shown, where its correlation is obvious. With the assumption of homogeneous turbulence and with the approximation of smooth perturbation theory for a spherical wave[1],the mean square fluctuation in the angle of arrival $(\Delta\phi)^2$ is given by [10],[11]

$$(\Delta\phi)^2 = \frac{1}{3} 1.4 C_n^2 L d^{-1/3} \qquad l_o < d < \sqrt{\lambda L} , \tag{2}$$

where d is the diameter of the reflector of the receiver, L the length of the test range, l_o the inner scale of turbulence, λ the wavelength of the radiation and C_n the refractivity turbulence structure constant that is calculated and also shown in figure 6. C_n is found to be between $0.5 \cdot 10^{-7}$ and $8 \cdot 10^{-7}$ $m^{-1/3}$, similar order of magnitude is found by other authors (e.g. [10],[13]).

With the calculated C_n, an estimate of the outer scale length L_o of the turbulence spectrum is feasible. We use the equation from [10]

$$C_n^2 = a^2 \alpha L_o^{4/3} (\frac{\Delta N}{\Delta z})^2 \tag{3}$$

where a^2 is a universal constant which is taken as 2.8 [13] , α is the ratio of thermal to momentum diffusivity and varies slightly with atmospheric stability but is taken as unity.

---

[1]If most of the random medium between the transmitter and the receiver is located in the far zone of the transmitter, then the radiation can be well approximated by a spherical wave. $(\Delta\phi)^2_{spherical} \approx (\Delta\phi)^2_{plane}/3$

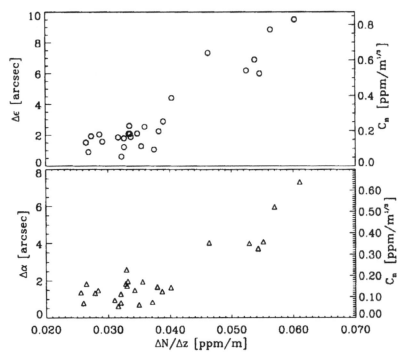

Figure 6: Angular extent in Elevation ($\Delta\epsilon$) and in Azimuth ($\Delta\alpha$) versa $\Delta N/\Delta z$ (z-axis points downwards). The right vertical axis shows the corresponding C_n.

With the equations 2 and 3, $\langle L_o \rangle$ was found to be approximately 40 m and 20 m in elevation and in azimuth, respectively. This result is in good agreement with other autors who estimate $L_o \approx 10 - 100\, m$ ([9],[10],[12],[13]).

5. CONCLUSIONS AND OUTLOOK

5.1 Long time measurements

Our results show that MPM computes the atmospheric absorption at mm-wave frequencies for clear days with an accuracy better than 5%. With respect to the resolution of the meteorological data, also the calculated absorption for rainy days is quite good.

In a next step the investigation of cloud water effects is planned together with a statistical evaluation of the comparison between modelling and measurements.

5.2 Scintillation measurements

The determination of the angle of arrival of the radiation can be done with very high precision ($< 1''$) at high sampling rates. Because of the high measurement accuracy of the angular extent the refractivity turbulence structure constant C_n ($= 0.5 \cdot 10^{-7} - 8 \cdot 10^{-7}\, m^{-1/3}$) is determined quite accurately. From simultaneous measurements of weather parameters it is also possible to estimate the outer scale of turbulence L_o. It was found that L_o dominates in elevation by a factor 2 in respect to azimuth (approx. 40 m and 20 m respectively).

The evaluation of our measurements including absorption will be extended to the spectra of temporal variations and also new measurements are planned in the Argentinean Andes.

Acknowledgements. The SST project was funded by Brazil's S. Paulo State Foundation for Support of Research (FAPESP, Contract 93/3321-7) and also supported by the Swiss National Science Foundation (contract 2000-049391.96/1) and Argentina institutes (IAFE and CASLEO). CRAAE is a joint center between Brazil institutes Mackenzie, INPE, USP and UNICAMP.

REFERENCES

1. N. Kämpfer et al., Proceeding of the 6th Int. Symposium on Recent Advances in Microwave Technology, *Beijing, Aug., 3-6, 1997.*

2. P. Kaufmann et al., Coronal Physics from Radio and Space Observations, *Lecture Notes in Physics, Springer, 202-206, 1996.*

3. V.A. Efanov, I.G. Moiseev, The Observational Method of Solar Radio Emission Bursts by Sharp-Directional Aerials. *Russian Academy of Science, Moscow 1967.*

4. R. Herrmann et al., A multibeam antenna for solar mw-wave burst observations with high spatial and temporal resolution *Solar Physics, 142, 157-170, 1992.*

5. J.E.R. Costa et al., A Method for Arc-Second Determination of Solar Burst Emission Centers with High Time Resolution and Sensitivity at 48 GHz,*Solar Physics, 159, 157-171, 1995.*

6. C. Mätzler (ed.), Development of Radiative Transfer Models, *Report from review Workshop of Prject 1 of COST Action 712, held at EUMETSAT, Darmstadt, Germany, April 8-10,1997.*

7. H.J. Liebe, MPM-An Atmospheric Millimeter-Wave Propagation Model, *Int. J. of Infrared and Millim. Waves, 10, 631-650, 1989.*

8. H.J. Liebe et al., Propagation Modelling of Moist Air and Suspended Water/Ice Particles at Frequencies below 1000 GHz, *AGARD 52nd Specialists' Meeting of the Electromagnetic Wave Propagation Panel, Palma De Mallorca, Spain, 17-21 May, 1993.*

9. T.E. VanZandt et al., Vertical profiles of refractivitz turbulence structure constant: Comparison of observations by the Sunset radar with a new theoretical model, *Radio Sci.,* 13, 819-829, 1978.

10. V.I. Tatarskii, The effects of the turbulent atmosphere on wave propagation, *Israel Program for Scientific Translations Ltd., 1971.*

11. A.I. Kon and V.I. Tatarskii, Parameter fluctuations of a space-limited light beam in a turbulent atmosphere, *Izv. VUZ Radiofizika,* 8, 870-875, 1965.

12. A. Ishimaru, Wave Propagation and Scattering in Random Media, *vol 2, academic press, 1978.*

13. D. Narayana Rao, et al., Studies on refractivity structure constant, eddy dissipation rate, and momentum flux at a tropical latitude, *Radio Sci.,*32, 1375-1389, 1997.

2. Remote sensing of atmosphere

2.2 Satellite observations

Microw. Radiomet. Remote Sens. Earth's Surf. Atmosphere, pp. 223–233
P. Pampaloni and S. Paloscia (Eds)
© VSP 2000

Influence of cloud temperature on the retrieval of integrated cloud liquid water content from space radiometry

ALBERT GUISSARD

Université Catholique de Louvain, Louvain-la-Neuve, B1348, Belgium

A simple physically-based retrieval procedure developed previously by the author is used for determining the influence of cloud temperature (or cloud mean altitude) on the retrieved cloud liquid water content from a spaceborne radiometer. The importance of this effect is illustrated with SSM/I data from 1993.

1. INTRODUCTION

With the launch of the ESMR instrument (Electrically Scanned Microwave Radiometer) on-board of Nimbus-5, in December 1972, both IR and microwave observations of the Earth surface and of the atmosphere became possible. In the next years a number of algorithms were developed for the retrieval of atmospheric integrated water vapor (WV) and integrated cloud liquid water (CLW) from the brightness temperature measurements provided by the Nimbus-5 nadir looking NEMS (Nimbus-E Microwave Spectrometer) (e.a. [1-2]), and later by Nimbus-7 and SEASAT SMMR (Scanning Multichannel Microwave Radiometer) both launched in 1978 (e.a. [3]).

Continuity of passive microwave observations has been provided by the SSM/I instrument (Special Sensor Microwave Imager). A full description of this instrument has been provided by Hollinger & al. [4], and retrieval algorithms for WV and CLW have been proposed by e.a. Alishouse & al. [5-6], Petty [7], Greenwald & al. [8], Liu and Curry [9], Gerard and Eymard [10]. Most of the proposed algorithms are based on statistical regressions constructed either on a measurements data base, or on a simulated data base, with in some cases physical inputs used for determining the analytical form of the regression relationships.

A more simple and very efficient approach is based on the inversion of a physical radiative transfer model. In a previous paper [11], we proposed such a procedure for the evaluation of atmospheric integrated water vapor and cloud liquid water contents from radiometric measurements at an incidence around 50°. The method was applied to SSM/I data. An extended version of the method was presented later [12] for measurements at both 50° and nadir, and applied to SSM/I and TOPEX data. The procedure is based on the inversion of a simplified radiative transfer model. The main assumptions are that (1) a simple model for the emissivity of the rough ocean surface is adequate, (2) a simple model for the atmospheric gases attenuations as taken from CCIR [13] can be used, (3) the same average sea surface temperature is taken for all inversions and (4) the atmospheric temperature at ground level is assumed equal to the average sea surface temperature. The results were quite encouraging, showing the advantage of this approach, as compared to regressions that require heavy and costly measurement campaigns and/or time consuming computing.

In the case of SSM/I, only vertically polarised brightness temperatures were used at the incidence angle of 53.1°, for which the influence of the wind is negligible, and a flat surface can be assumed. In the case of TOPEX on the other hand, the look angle is to the nadir and the wind has some influence on the observed brightness temperatures. It was pointed out that the retrieval of cloud liquid water is highly dependent on the cloud temperature, or on the cloud mean height above sea level. Greenwald et al. [8] also pointed out the problem and considered that a temperature of 6°C lower than the surface temperature would be representative of most marine boundary layer clouds. In our previous inversions, we assumed an average height of 2 Km above sea level. With a temperature lapse rate of −6.5°C/Km, this leaded to a temperature of 13°C below the surface temperature.

The purpose of this paper is to examine more carefully the influence of the cloud temperature on the atmospheric retrievals. This will be done using SSM/I brightness temperatures at vertical polarisation, where the influence of the wind is the lowest.

2. ATMOSPHERIC ATTENUATION

In this section, we summarize the approximate formulas for the atmospheric gases attenuation as given by CCIR [13] and for cloud attenuation as given in [14]. The atmospheric transmittivity along a path at angle θ from the vertical is given by $\mathcal{T} = \exp[-\tau/\cos\theta]$ assuming a flat Earth (not valid for grazing incidences). The opacity τ is related to the specific attenuation $\alpha(z)$ (Neper/m) at level z by $\tau = \int \alpha(z)\, dz$.

Let us first consider attenuation by oxygen. Various models were compared in [15, section 4.2]. A more exact model has been developed by Liebe [16] and was approximated by Greenwald et al. [8] at 19 and 37 GHz by cubic regressions with respect to the temperature. The various models are similar at 19 GHz and not very different at 37 GHz. Since oxygen attenuation is not the dominant term at these two frequencies, the CCIR model is considered as satisfactory. An exponential attenuation profile with a 5.5 km characteristic height is assumed.

Let us next consider the attenuation induced by atmospheric water vapor. In the approximate expression of the CCIR, the specific attenuation is proportional to the water vapor content ρ_v (g/m^3). This allows to express the total attenuation on a vertical path (the opacity) in a simple form. Let us define the factor η as the ratio of the characteristic heights h_v and h_{av} of the vertical profiles of WV content and of WV attenuation respectively, assuming exponential profiles. This factor is a function of frequency and of h_v [17, Vol.2, fig.2.1]. Assuming an average height

$h_v = 2$ Km, we find the following values : $\eta = 0.95, 1.25, 0.85$ for the frequencies $f = 19, 22.2, 37$ GHz respectively. Then the opacity reads

$$\tau_v = \eta A V \tag{1}$$

where $V = \rho_v(0) \times h_v$ is the total columnar water vapor (V in Kg/m^2, h_v in Km, $\rho_v(0)$ is the ground level value in g/m^3) and A depends on frequency and temperature. This leads to a simple inversion scheme, since V can be extracted without the knowledge of the detailed vertical distribution.

More recent results for atmospheric WV attenuation have been proposed by Liebe in several publications (see e.a. [16]. CCIR expression and Liebe's more exact expression differ mainly by the dependence on the water vapor content. In the CCIR expression the dependence is linear, and this allows to combine ρ_v and h_v in the single parameter V, as stated above. The inversion is then linear with respect to both V and L. On the other hand, Liebe's expression contains a quadratic term in the continuum for the water vapor and the inversion with a simple linear model is no more possible. The quadratic term should be included, if a more exact calculation of T_B is required in the window around 37 GHz, for high values of water vapor. However, since we want to stress in this paper the importance of cloud temperature, we shall use the linear approximation of the CCIR.

The specific attenuation by clouds is calculated using the Rayleigh approximation, valid up to 100 GHz. Within this approximation, the attenuation is proportional to cloud liquid water content ρ_ℓ (g/m^3), so that the product with the cloud thickness H_c (Km) provides a total attenuation proportional to the integrated cloud liquid water content L (Kg/m^2). Attenuation by cloud liquid water is discussed in several papers and reference books. It can be put into the following form [14]

$$\gamma = 24.5\, \phi(\lambda, t)\, \rho_\ell / \lambda^2 \qquad \text{(dB/Km)} \tag{2}$$

where λ is the wavelength (cm), t is the temperature ($^\circ$C) and the function ϕ is

$$\phi(\lambda, t) = \lambda \Im(\epsilon_r) / |\epsilon_r + 2|^2 \tag{3}$$

Here ϵ_r is the complex relative permittivity of water and $\Im$ stands for the imaginary part. This leads to a total attenuation (or opacity) for clouds, along a vertical path

$$\tau = B \times L \qquad \text{(Neper)} \tag{4}$$

where (with the frequency f in GHz)

$$B = 6.27 \times 10^{-3} \phi(\lambda, t)\, f^2 \qquad \text{(Neper/Kgm}^{-2}\text{)} \tag{5}$$

A good approximation of the function ϕ is as follows

$$\phi(\lambda, t) = (a + bt) / 100 \tag{6}$$

where $a = 3.19$ and $b = -0.067$. This approximation is valid, with an accuracy of 4.5%, for -8°C$\leq t \leq 18^\circ$C and 15GHz$\leq f \leq 35$GHz. For temperatures between 0 and 22°, it differs by at most 4.5% from the model of Petty [7], as approximated by a cubic regression in [8].

The ϕ function is drawn in figure 1 for 18, 21 and 37 GHz, as a function of temperature, together with the linear approximation (6). This figure clearly shows the importance of cloud temperature : for an increase from 0 to 20 degrees, the attenuation (proportional to ϕ) decreases in the ratio $\cong 3.2/1.85 = 1.73$ or 2.4 dB. The retrieved CLW value will increase in the same ratio. This leads to a very large uncertainty in the CLW inversion, if no information is available on the cloud altitude and thickness, as will be illustrated below.

3. INVERSION PROCEDURE

The inversion procedure is summarized in this section. With the linear approximation (1) for water vapor opacity, the total opacity has the following expression at frequency j

$$\tau_j = \eta_j A_j V + B_j L + C_j \tag{7}$$

where C_j represents the oxygen contribution. We then get the following linear model

$$\ln[T_a - T_{Bj}(\hat{s})] = -\frac{2}{\cos\theta_s}(\eta_j A_j V + B_j L + C_j) + \ln[(1 - e_j(\theta_s))T_b] \tag{8}$$

where T_{Bj} is the brightness temperature for channel j, s is the unit vector in the observation direction, θ_s is the corresponding incidence angle and $e(\theta_s)$ is the sea surface emissivity at angle θ_s. The temperatures T_a and T_b are corrected ground level atmospheric temperatures as follows

$$\begin{aligned} T_a &= T_t(0) - (1 - e(\theta_s)T)\Delta T_1 \\ T_b &= T_t(0) - \Delta T_2 - T_c \end{aligned} \tag{9}$$

where $T_t(0)$ is the ground level atmospheric temperature (t stands for troposphere), T_c is the cosmic background brightness temperature and ΔT_1, ΔT_2 are corrections for the atmospheric temperature lapse rate.

If we have several channels (different frequencies and polarisations), we can write in matrix form

$$y = Hp + d \tag{10}$$

where y is the column matrix with elements $y_j = \ln[T_a - T_{Bj}(\hat{s})]$, p is the column matrix of unknown parameters $p = (V\,L)^T$, H is a rectangular matrix and d a column matrix that can be deduced directly from (8).

We apply a maximum likelihood estimate to the linear model (10). Let V_n be the covariance matrix of the measurements noise. One finds for the estimates $\hat{p}$ of the parameters vector p

$$\hat{p} = M(y - d) \tag{11}$$

where the matrix M is given by

$$M = (H^T V_n^{-1} H)^{-1} H^T V_n^{-1} \tag{12}$$

Here we shall assume that all channels are affected by the same noise with variance σ_n^2, so that the matrix $V_n = \sigma_n^2 \times U$, where U is the unit matrix. A short discusssion of the error budget can be found in [11].

4. APPLICATION TO SSM/I DATA

The above inversion procedure has been applied to SSM/I data of 10/10, 29/10 and 13/11/1993, above the Atlantic Ocean, from 60° latitude North down to the Equator. The incidence angle is $\theta_s = 53.1°$. The sea surface temperature (SST) ranges between 275 K and 303 K and the wind speed from almost 0 to more than 20 m/s, according to collocated data from the European Center for Medium-Range Weather Forecasts (ECMWF). Figure 2 displays the geographical locations of the measurements, and the values of SST and wind speed at these locations.

We are using the brightness temperatures in vertical polarisation for the three frequencies : 19.35, 22.235 and 37.0 GHz. Greenwald et al. [8, p.18,474] have found that the 37 GHz brightness temperature measurements would have a bias of -3.58 K. We thus introduced a correction of +3.6 K to the 37 GHz channel. According to the same authors [8, p.18,487], "errors in the retrieval (of cloud liquid water) associated with differences in the footprint sizes of the 19.35- and 37-GHz channels are shown to be smaller than errors caused by imprecise knowledge of the state of the atmosphere and sea surface". Therefore these differences have been ignored in the present work.

For vertical polarisation around 50° incidence, the wind has only a small influence on the brightness temperature. From previous simulations [15], we have found that a wind speed change from 0 to 20 m/s is producing a change of the vertically polarised brightness temperatures of the order of ±1.5 K, in the frequency range of interest. This allows us to approximate the emissivities by the flat surface values.

We note from figure 2 that there is a continous grid of sampling points. The only missing points are near the coast of Europe or Groenland and all remaining points are above the ocean. If rain is present, what we retrieve is an *equivalent* liquid water content, because scattering by the larger rain drops is ignored. The maximum average rain rate within the antenna footprint (37x28 km [4]) at the highest frequency used here (37 GHz) is roughly $R \cong 5$ mm/h for a cell of maximum rain rate equal to 100 mm/h [15]. Such a rain rate has a probability of 0.02% (i.e. 100 minutes in one year, or less than 1 point on the total number of 2,387 data points used in this analysis) for a highly rainy region such as Miami [18]. The value $R = 5$ mm/h corresponds to a liquid water content of 0.35 g/m^3 [17, Vol.2, p.1.19], or 1.65 Kg/m^2 for a rain height of 5 km.

The other parameters are as follows. The ground level atmospheric temperature $T_t(0)$ is taken as equal to the average sea surface temperature, and the temperature lapse rate corrections are evaluated as $\Delta T_1 \cong \Delta T_2 = 5$. The oxygen and water vapor attenuations are evaluated for a temperature at level z : $T_t(z) = T_t(0) - \Delta T_{Ov}$, where z is taken as approximately 1 Km, leading to $\Delta T_{Ov} \cong 6$ K.

The cloud temperature is given by $T_{cl} = T_t(0) - \Delta T_3$ and is the variable parameter in the following inversions. It is assumed that cloud average altitude ranges between 0 and 3 Km with 500 m steps, corresponding to 7 values of ΔT_3 from 0 to 19.5 K. For

	slope	CC	bias	sdif
UCL	1.02	0.95	-1.77	3.84
Petty	1.10	0.95	-2.35	4.15

Table 1: Global results for UCL and Petty water vapor algorithms

	slope	CC	bias	sdif
UCL	1.07	0.96	0.01	0.037
Alishouse	1.29	0.81	0.076	0.108
NRL	1.09	0.86	-0.10	0.073

Table 2: Global results for UCL, Alishouse and NRL cloud liquid water algorithms

comparison we use as "ground truth" the ECMWF values for water vapor, and the output of Gérard algorithm [19] for liquid water, since ECMWF does not provide liquid water data and we do not have other ground truth data (Note that SSM/I data have been introduced in ECMWF forecasts only from June 26, 1998 [J.F. Mahfouf, priv. comm.]).

Figure 3 (left pannel) shows an example of integrated water vapor retrieved from 10/10/93 data, with $\Delta T_3 = 13$ K (clouds at 2 Km altitude) : the inversion points are shown (+), together with the regression line. The inversion of integrated water vapor is not affected by the cloud temperature. For comparison, the right pannel displays the results obtained with Petty algorithm [7]. In table 1, we compare the slopes of the regression lines (slope), the correlation coefficients (CC), the bias (intersection with the vertical axis) and the rms differences (sdif). It can be seen that the correlation coefficients are the same, but that the UCL algorithm has better slope, and lower bias and sdif.

Figure 4 (upper pannel) shows an example of cloud liquid water contents retrieved from 10/10/93 data, for the same cloud altitude. Here again the inversion points are shown (+), together with the regression line. For comparison, the lower left and right pannels are the results obtained with Alishouse algorithm [6] and with the NRL algorithm [20, p.A.37] respectively (we have taken the NRL mid-latitude Spring-Fall climate code for cloud liquid water over the ocean). It can be seen that UCL and Alishouse have small negative values, while NRL has negative values up to -0.23 Kg/m^2 (not represented on the figure). There is more dispersion of the data for Alishouse and NRL algorithms. Table 2 provides the comparison of global results, in a similar form as for water vapor. The UCL algorithm has the best slope and correlation coefficient, the lowest bias and standard difference.

Figure 5 displays the regression lines of cloud liquid water retrieval for 10/10/93 data for changing cloud temperatures. The seven regression lines correspond to a cloud average altitude from 0 to 3 km in steps of 500 m. Figure 6 (upper pannel) gives the corresponding bar charts for the slope of the regression lines, the bias and the rms difference. Cases 1 to 7 correspond to the seven altitudes from 0 to 3 Km. The correlation coefficient remains constant at the value 0.96. The best slope is obtained for case 6 (altitude 2.5 Km), while the bias and rms difference decrease continuously. However the results are somewhat different for the two other dates, as appears from figure 6 (middle and lower pannel) which display similar bar charts for 29/10/93 and 13/11/93 data respectively. For the second date (figure 6-middle), the correlation coefficient equals 0.984, the best slope is obtained for case 4 (altitude 1.5 Km), and the lowest rms difference for cases 4 and 5. For the last date, 13/11/93 (figure 6-lower), the correlation coefficient equals 0.98, the best slope corresponds to case 5 and the lowest rms difference to case 6.

These comparisons illustrate the fact that the cloud temperature has an important influence on the retrieved cloud liquid water content, with negligible effect on the retrieved water vapor content. We must stress here again that, for cloud liquid water content, we are comparing our results to the inversion outputs of Gérard's algorithm and not to ground truth values, since the latter were not available. Therefore what appears to be the best case, does not necessarily would be the best case when compared to ground truth. What these results clearly show however is that a knowledge of the cloud altitude or temperature is a requirement if one wants to improve the retrieval of cloud liquid water.

5. CONCLUSION

In this paper, we are considering the influence of cloud temperature on integrated water vapor and cloud liquid water retrievals from spaceborne radiometers. It appears that water vapor retrieved values are practically independent of cloud temperature. However, cloud temperature has a large effect on cloud liquid water retrieval. Therefore, in order to increase the performances of the retrieval, the inversion should be coupled with other instruments data such as infrared data, that would allow to estimate the cloud altitude, and therefrom their average temperature.

REFERENCES

1. N.C. Grody. Remote sensing of atmospheric water content from satellite using microwave radiometry. *IEEE Trans. Antennas and Propag.*, AP-24(2):155–162, 1976.

2. D.H. Staelin, K.F. Kunzi, R.L. Pettyjohn, R.K.L. Poon, and R.W. Wilcox. Remote sensing of atmospheric water vapor and liquid water with the NIMBUS 5 microwave spectrometer. *Journal of Applied Meteorology*, 15:1204–1214, 1976.

3. E.G. Njoku, J.M. Stacey, and F.T. Barath. The SEASAT scanning multichannel microwave radiometer (SMMR) : instrument description and performance. *IEEE J. Oceanic Eng.*, OE-5(2):100–115, April 1980.

4. J.P. Hollinger, J.L.Pierce, and G.A. Poe. SSM/I instrument calibration. *IEEE Trans. Geosc, Remote Sensing*, 28(5):781–790, September 1990.

5. J.C. Alishouse, S.A. Snyder, J. Vongsathorn, and R.R. Ferraro. Determination of oceanic total precipitable water from the SSM/I. *IEEE Trans. Geosc. Remote Sensing*, 28(5):811–816, September 1990.

6. J.C. Alishouse, J.B. Snider, E.R. Westwater, C.T. Swift, C.S. Ruf, S.A. Snyder, J. Vongsathorn, and R.R. Ferraro. Determination of cloud liquid water content using the SSM/I. *IEEE Trans. Geosc. Remote Sensing*, 28(5):817–822, September 1990.

7. G.W. Petty. On the response of the special sensor microwave/imager to the marine environment - implications for atmospheric parameter retrievals. Technical report NASA Grant NAG5-943, Univ. of Washington, Seattle, December 1, 1990.

8. Th.J. Greenwald, G.L. Stephens, Th.H. Vonder Haar, and D.L. Jackson. A physical retrieval of cloud liquid water over the global oceans using special sensor microwave/imager (SSM/I) observations. *J. Geophys. Res.*, 98(D10):18,471–18,488, October 20 1993.

9. G. Liu and J.A. Curry. Determination of characteristic features of cloud liquid water from satellite microwave measurements. *J. Geophys. Res.*, 98(D3):5,069–5,092, March 20 1993.

10. E. Gérard and L. Eymard. Remote sensing of integrated cloud liquid water : development of algorithms and quality control. *Radio Science*, 33(2):433–447, March-April 1998.

11. A. Guissard. The retrieval of atmospheric water vapor and cloud liquid water over the oceans from a simple radiative transfer model : Application to SSM/I data. *IEEE Trans. Geosc. and Remote Sensing*, 36(1):328–332, January 1998.

12. A. Guissard. A simple physically-based retrieval procedure for atmospheric water vapor and cloud liquid water, applied to SSM/I and TOPEX data over the oceans. In *Progress in Electromagnetics Research Symposium (PIERS), July 7-11, 1997, Cambridge, MA, USA*. Schlumberger Doll Research, Ridgefield, CO, USA, 1997.

13. CCIR. *Recommandations et rapports du CCIR*. Union Internationale des Télécommunications, Genève, 1986.

14. A. Guissard. *Propagation et Rayonnement (Vol. 1 et 2)*. Diffusion Universitaire CIACO, Louvain-la-Neuve, 1993.

15. A. Guissard. Study of the influence of the atmosphere on the performance of an imaging microwave radiometer. Report for ESA contract n. 4124/79, Université Catholique de Louvain, Laboratoire de Télécommunications et d'Hyperfréquences, Place du Levant,3, B1348 Louvain-La-Neuve, 1980.

16. Liebe H.J. An updated model for millimeter wave propagation in moist air. *Radio Science*, 20:1069–1089, 1985.

17. A. Guissard and M. De Coster. Influence of the atmosphere on the performance of spaceborne imaging microwave radiometers - distance measurements correction. Report for ESA contract n. 4702/81, Université Catholique de Louvain, Laboratoire de Télécommunications et d'Hyperfréquences, Louvain-La-Neuve, Belgium, 1982.

18. L. Boithias. *Propagation des ondes radioélectriques dans l'environnement terrestre*. Dunod, Paris, 1983.

19. E. Gérard. Restitution de l'eau liquide nuageuse par radiométrie hyperfréquence. Thèse de doctorat, Université Paris VI, Paris, 1996.

20. J. Hollinger, R. Lo, G. Poe, R. Savage, and J. Pierce. Special sensor microwave/imager user's guide. Technical report, Naval Research Laboratory, Washington, D.C., 1987.

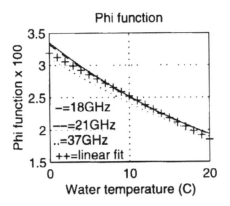

Figure 1: The ϕ function for cloud liquid water attenuation (equation (3))

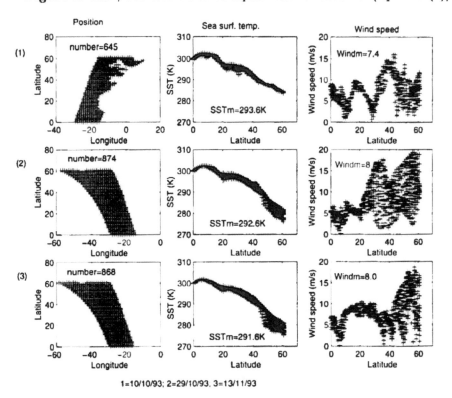

1=10/10/93; 2=29/10/93, 3=13/11/93

Figure 2: Sampling points, sea surface temperature and wind speed.

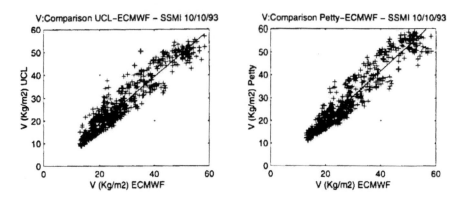

Figure 3: Detailed results of water vapor retrieval for 10/10/93 data. Algorithms of UCL (left) and Petty (right).

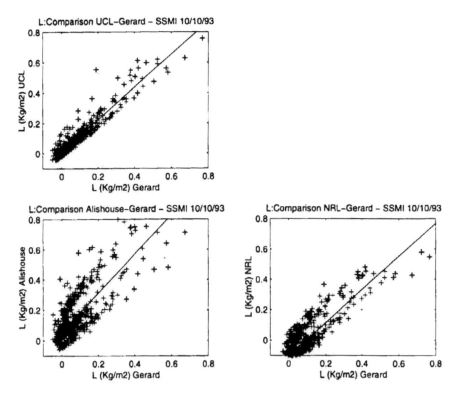

Figure 4: Detailed results of cloud liquid water retrieval for 10/10/93 data. Algorithms of UCL (upper), of Alishouse (lower left) and of NRL (lower right).

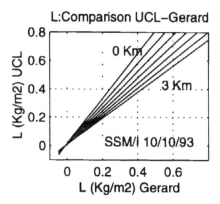

Figure 5: Effect of cloud temperature on the retrieval of CLW for 10/10/93 data : regression lines for altitudes from 0 to 3 Km, in steps of 500 m.

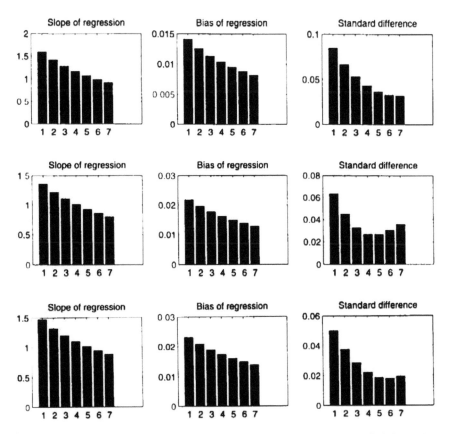

Figure 6: Effect of cloud temperature on the retrieval of LW for 10/10/93 (upper), 29/10/93 (middle), 13/11/93 (lower) : bar charts for the slope of the regression line, the bias and the rms difference.

Microw. Radiomet. Remote Sens. Earth's Surf. Atmosphere, pp. 235-245
P. Pampaloni and S. Paloscia (Eds)
© VSP 2000

Analysis of cloud liquid water content characteristics from SSMI and a ship-borne radiometer

LAURENCE EYMARD

CETP/UVSQ, 10-12 avenue de l'Europe 78140 Vélizy Villacoublay, France

Abstract- A systematic comparaison of the retrieved liquid water content from the shipborne radiometer and SSMI was performed in the framework of CLOREVAL, using data obtained during the FASTEX experiment. Comparisons between the retrieved water vapour, liquid water and surface wind (directly measured on the ship) were made. Despite the different sampling from the ship and SSM/l, there is a rather good agreement between retrieved liquid water contents, but the standard deviations of both data set differ by an order of magnitude. In a second step, clouds were individually identified in both data sets, and their mean characteristics were analyzed. Three properties were found similar for the ship and the SSMI : the power law of the size distribution, a linear relation between the cloud heterogeneity, quantified as the standard deviation of liquid water inside the cloud versus the mean liquid water content, and a correlation between the variabilities in liquid water and water vapour inside each cloud.

1. INTRODUCTION

The European program Cloud Retrieval validation experiment (CLOREVAL, Offiler et al, 1998), cooperation between the UK Met. Office, CRS (Bristol, UK), ECMWF, IFM (Kiel, Germany), and CETP, was aimed at the quantitative validation of the cloud liquid water retrieval from spaceborne microwave radiometers. Several methods were evaluated by comparison to direct aircraft measurements of the C130 in stratiform clouds. Details can be found in [1]. In addition to this comparison of SSMI retrievals, two shipborne radiometers (IFM, CETP) were used.

A systematic comparaison of the retrieved liquid water content from the shipborne radiometer and SSMI was performed in the framework of CLOREVAL using our algorithms, developed with the same method for the shipborne radiometer and SSMI. Data obtained during the FASTEX experiment (Jan. - Feb., 1997, North Atlantic) were used. We then undertook a statistical analysis of clouds, as depicted by microwave radiometry.

In the following, we present the retrieval method for liquid water content, then the FASTEX experiment. In the fifth section, we detail the method and the results of the comparison SSMI/shipborne radiometer, and the sixth section presents the cloud analysis.

2. RETRIEVAL ALGORITHMS

The retrieval method is physico-statistical, based on a multilinear regression on simulated brightness temperatures using a data base and a radiative transfer (RT) model. The database consists of ECMWF global 36-hour forecast fields in order to get a good balance of the hydrological cycle [2]. The cloud water is given by a prognostic scheme in the model, providing a better distribution of the cloud liquid water [3].

The RT model, based on [4], has been improved by [5]. It includes :

- The atmosphere transmission model of Liebe'93 MPM [6]
- The surface electromagnetic model of [7], derived from [8]
- The surface roughness described using the [9] slope probability density
- The sea water permittivity formulation of [10] and [11], based on laboratory measurements)
- The foam cover described following [12], the emissivity being given by [13].

Algorithms are loglinear combinations of the brightness temperatures. A first set was obtained by [14], using low frequency channels (18 to 37 GHz). A new set was created with the addition of the 85 GHz channels. The algorithms for SSMI and ship radiometer liquid water are given hereafter :

SSMI : LWP =0.728+0.0275Ln(280.-T19V)+1.143Ln(280.-T19H)+0.083Ln(280.-T22V)-0.402Ln(280.-T37V)-1.063Ln(280.-T37H)+0.0183Ln(280.-T85V)-0.035Ln(280.-T85H)

ship : LWP =16.364+1.824Ln(280.-T23.8)-4.764Ln(280.-T36.5)

3.VALIDATION OF SSMI AND SHIP RADIOMETER RETRIEVALS

The CETP radiometer is a zenith-viewing instrument with two channels at 23.8 and 36.5GHz. The antenna is a corrugated horn, and the antenna aperture is a twelve degree cone. It is thermally controlled in order to ensure stability of the gain.

During FASTEX, the ship-borne radiometer worked continuously, but the high variations in air temperature occasionally caused problems for the internal temperature stabilisation. For this reason, is was necessary to check the calibration, and a comparison was made with simulated brightness temperatures computed using the RT model and coincident radiosoundings performed on board the ship in clear atmosphere. It led to an adjustment of the internal absorption coefficients by a few %. Once re-calibrated, the retrieved water vapour has been validated by comparison with the global set of radiosoundings (about 200), showing no significant bias, and the liquid water in clear air was verified to range within the retrieval noise. In addition, the CETP and IFM radiometers were compared over a three-day period, showing very consistent measurements, as well for brightness temperatures as for water vapour and liquid water retrievals (mean bias in LWP : 0.025 kgm^{-2}). Data, sampled at a 1s rate, were averaged over 1minute intervals before computing the water vapour and liquid water contents.

The SSMI algorithm was tested, among a set of 19 retrieval methods, including statistical methods, neural network retrievals, and assimilation in the ECMWF model on SSMI data got for two experimental periods (CLOREVAL experiment, Nov. 1996, and ACE-2, June 1997) during which the C130 of the UK Met Office was flown in stratus and strato-cumulus clouds. For the above algorithm, the mean bias versus the aircraft measurement was found to be 0.012 kgm^{-2}, with an RMS error of 0.046 kgm^{-2}, similarly to the 6 other best retrievals [1].

4. FASTEX

The FASTEX (Front and Atlantic Storm Track EXperiment) experiment was devoted to the study of strong atmospheric perturbations in the North Atlantic during winter (January - Februry, 1997). Several aircraft and ships participated in the field campaign. Among them, the R/V Le Suroit covered the area 30-45W- 40-50N, performing measurements in the atmosphere and the ocean. Atmospheric data consisted of radiosoundings, UHF radar wind profiles, microwave retrievals of water vapour and liquid water contents, low atmosphere mean meteorological parameters (sampling rate 10s) and flux measurements on an instrumented mast (16 m above sea level). All meteorological ship data were checked after the experiment, then archived after averaging over 1 minute intervals.

The weather varied from calm, anticyclonic conditions up to very strong storms. The atmosphere was often dry (ranging from 0.5 to 2.5 gcm^{-2}), and clouds were of all types (cirrus, deep stratus, strato-cumulus, as well as cumulus, with sometimes snow), the low atmosphere temperature ranging between 0 and 18°C.

SSM/I data from F10, F11 and F13 in the experimental area (30 - 45 W, 40 – 55 N) were processed and water vapour and liquid water algorithms were systematically applied. Alishouse's algorithm [15] was used to retrieve the water vapour content, and Goodberlet's one [16] the surface wind. The liquid water content was retrieved using the above algorithm.

5. COMPARISON OF SSMI AND SHIP DATA

The first step was to select collocated data. As the ship samples the meteorological events which overpass it, and the satellite provides horizontal integrations of instantaneous phenomena, they cannot be directly compared. However, the consistency of the two data sets can be checked at various scales using various data sampling. To get a sufficient statistical significance, SSMI data were taken over 1 square degree, and compared to time samples over one hour, to stay within the expected atmosphere stationarity.

Thus, for every SSM/I overpass, data falling in a square degree centred on the ship location were kept. We obtained 190 images for the entire FASTEX period. Time series of the ship surface wind (to check the wind retrieval) and of the radiometer were extracted for one hour centred on the satellite overpass. The CETP radiometer data were filtered to eliminate the maximum of rain-contaminated data.

The time distribution of ship data in the 1-hour interval and the space distribution in the square degree of SSM/I data, both cumulated over the 190 comparison cases, were compared in histograms (Figure 1).

We find (after normalisation of the distribution to each other) comparable distributions for the water vapour, the liquid water and surface wind. However, the wind distributions are shifted with respect to each other (SSM/I overestimate), and the liquid water distribution shows that there are more SSM/I data for smaller values (0–0.1kg.m^{-2}) than from the ship, whereas the highest values are under-represented in the SSM/I dataset.

Each collocated case was then examined by comparing space and time averages, standard deviations, together with the minimum and maximum values found in each sample. This comparison was made by filtering out some data for which the time and space variations are too different (ship standard deviation greater than twice the SSM/I one, plus 0.1kg.m^{-2}). These correspond to cases for which the ship and the satellite did not "see" the same scene. Statistical results are summarized in Table 1.

We report in this table the mean bias (ship-SSMI), the corresponding standard deviation, as well as similar parameters obtained by selecting the minimum value (respectively the maximum value) in each sampling cell (in the 1° box for SSMI, and within the 1-hour time sample for the ship).

The water vapour show similar standard deviations for ship time series and SSM/I images, and the minimum and maximum values correspond well to each other.

The surface wind presents higher scatter, but again standard deviations, representative of the local variability, are of comparable magnitudes. Surprisingly, the maximum values fit better than the mean and minimum values, but there is a mean bias of about 1.6ms[-1] between SSMI and the ship direct measurements, which cannot be explained by the small difference in reference height (10m for Goorberlet's algorithm, 16m for the ship data). Simulations of the air flow distortions on this ship showed that the wind is overestimated by less than 10%. Ship measurement errors can therefore not be responsible of this discrepancy. However, [17] show that the SSMI wind compares well with ship measurements during two other experiments : Semaphore (Oct. -Nov. 1993, near the Azores) and Toga/Coare (Equatorial Pacific, winter 1992-1993). The difference between FASTEX and them is the very rough sea state, with both wind waves and high swell, which might affect the brightness temperature – wind relationship due to the change in emissivity, and thus could bias the SSMI algorithm The liquid water comparison (Fig. 2) shows more discrepancy between the ship and the satellite. The minimum values range from zero to 0.2kg.m^{-2}.

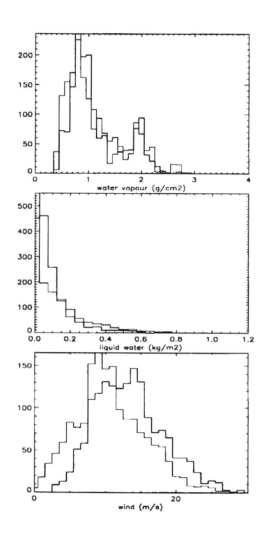

Figure 1. Distributions of water vapour (upper panel), liquid water content (middle panel), and surface wind (bottom panel) from SSMI (thick line) and the ship (thin line).

The maximum values show that high contents are measured from the ship (up to 1kg.m^{-2}), but are generally not retrieved by SSM/I. Moreover, the standard deviations within the sample obtained with the ship are of an order of magnitude greater than for the SSM/I (overimposed on the left and right panels for the SSMI and the ship, respectively). However, the comparison of average values shows a surprisingly good consistency, since the mean bias is 0.02kg.m^{-2}, with a correlation coefficient of 0.75 and a standard deviation of 0.07 kg.m^{-2}. These values suggest that most of the average values (for liquid water contents below 0.2 kg.m^{-2},agree with each other.

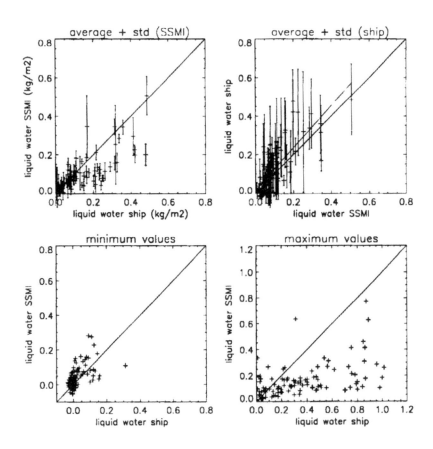

Figure 2. Liquid water plots. Upper left : SSMI versus the ship, with the SSMI standard deviation for each individual comparison sample superimposed; upper right : ship versus SSMI, with the ship standard deviation superimposed; bottom left : plot of the minimum values encountered in each comparison sample; bottom right : plot of the maximum values encountered in each comparison sample.

In addition, we made a "local" comparison, taking only the SSM/I pixel located over the ship, and a one hour ship window. Fig. 3 shows the results for the water vapour, liquid water and surface wind (the 1-hour window corresponds to 36km for a $10 m.s^{-1}$ wind).

As previously, we find a very good agreement for water vapour (mean bias $-0.047 gcm^{-2}$, standard deviation 0.179 gcm^{-2}), an average consistency between liquid water retrievals (mean bias 0.019 kgm^{-2}, standard deviation 0.065 kgm^{-2}), and a significant bias for the surface wind (mean bias -1.55 ms^{-1}, standard deviation 2.47 ms^{-1}).

	Water vapour $(kg.m^{2})$	Liquid water $(kg.m^{2})$	Wind speed (ms^{-1})
Bias (ship-SSM/I)	-0.019	0.024	-1.61
Standard deviation of the difference	0.175	0.074	1.99
bias of the minimum values	-0.034	-0.022	-1.95
Standard deviation for minimum values	0.182	0.055	2.08
bias of maximum values	0.03	0.203	-1.57
Standard deviation of maximum values	0.238	0.257	2.45

Table 1. Statistical comparison between SSM/I (taken in a 1° square box) and ship borne data (taken over 1 hour centred on the SSM/I overpass).

6. CLOUD STATISTICAL ANALYSIS

The discrepancy between the liquid water retrievals from the ship and the SSMI led us to perform a specific study of the cloud characteristics, as observed by the two instruments. The method used consists of selecting every cloud then computing its mean properties of size, water vapour and liquid water content (respectively noted WV and LWC in the following). To select clouds as well from the time series as for the SSMI images, a simple threshold of 0.05 kgm^{-2} was taken, corresponding to the estimated error on the retrieval. The cloud size was obtained by calculating the number of contiguous points. Mean values inside each cloud of LWC and WV are noted hereafter mLW and mWV hereafter, and the cloud heterogeneity is evaluated by computing the standard deviation of LWC and WV (sdLW and sdWV respectively).

Assuming that there is no significant evolution of the mean cloudiness properties within the two months of January and February, all data from the ship and SSMI were considered as a global data set. Moreover, a second assumption is that there is no significant horizontal variation inside the 15°x15° SSMI box (and ship cruise domain). The SSMI and ship data are processed independently, without looking for any collocation. However, as the cloud size does not have the same definition when considered along the wind (ship time sampling), and spatially, two additional cloud data set were created by considering one-dimension SSMI samples, selected as the lines along track and across track crossing at the ship location in each image. This sampling of SSMI can be assumed to be equivalent to the time analysis, for a mean wind 10ms^{-1} (the mean wind measured during FASTEX) if there is no preferred wind direction. The global characteristics of all cloud data sets are summarized hereafter:
- Shipborne radiometer : time cloud characteristics - (minimum duration 1 min ~ 0.6 km)
- Shipborne radiometer : selection of clouds larger than the SSMI sampling (i.e. about 40 minutes)
- SSMI : surface analysis surface (size = number of pixels x 25*25 km^{2})
- SSMI : linear analysis along lines and column crossing over the ship location (linear cloud size : minimum 25 km).

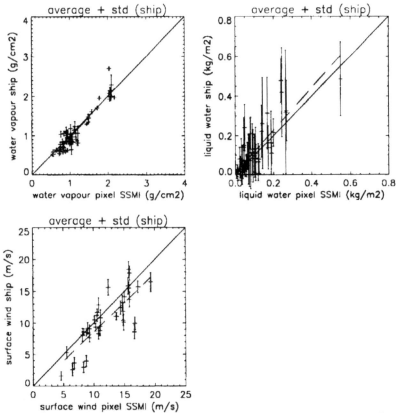

Figure 3. Plot of the ship average data (time interval 1 hour) awith the corresponding standard deviation superimposed versus the SSMI closest point. Upper left : water vapour; upper right : liquid water ; bottom : surface wind.

Results of the statistical analysis are gathered in Table 2. The two linear analyses of SSMI data are rather similar to each other, and only results of the so-called "columns" are shown, because the mean length of each line is greater (about 85 points instead of 64), allowing to sample larger clouds. The size distribution, plotted in logarithmic coordinates, allowed us to evaluate the size power law. The size average has not a high significance, because the distribution is not gaussian. It simply shows that the two SSMI analyses are consistent. Fig. 4 and 5 show respectively the cloud characteristics obtained with the ship and SSMI (surface analysis). In addition of looking at mean distributions, plots of sdL versus mLW, sdL versus mLW and sdW versus mWV are shown. A correlation appears between sdL and mLW, as well as between sdL and sdW. The corresponding mean slopes are given in Table 2. No other significant correlation could be found, when looking for relations sdW - W, mLW - mWV, mLW - cloud size.

	ship	ship (clouds> 25km)	SSMI linear analysis	SSMI surface analysis
mWV (kgm^{-2})	11.6	13.3	14.0	12.8
mLW (kgm^{-2})	0.18	0.26	0.13	0.09
max. mLW	0.85	0.6	0.5	0.45
slope sdL/mLW	0.8	0.7	0.7	0.9
slope sdL/sdW	0.13	~ 0.1	~ 0.06	~ 0.09
size characteristics: decreasing distribution law of type N(size) = a x size ^{b}				
power b	-1.6		-1.5	-1.5
average size (km)	~ 11	~ 70	~ 250	~ 280 (square cloud)

Table 2. Mean statistical characteristics of clouds observed using the shipborne radiometer (first and second columns "ship"), and the SSMI, for lines along the track (col. 3) and in surface analysis (col. 4).

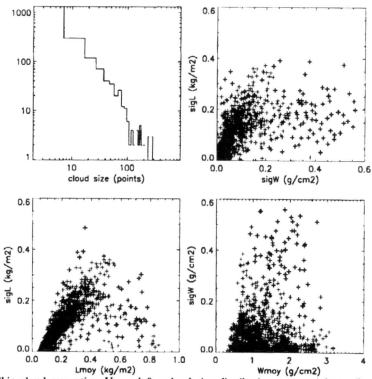

Figure 4. Ship cloud properties. Upper left : cloud size distribution; upper right : sdL versus sdW; bottom left : sdL versus mLW; bottom right : sdW versus mWV.

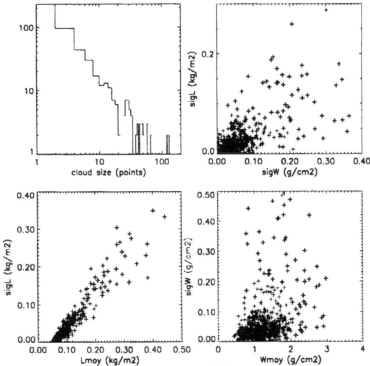

Figure 5. Same as Fig.4, but for SSMI surface cloud analysis

These results show that SSMI and ship cloud data sets have similar behaviours, despite the difference in sampling. The linear relationship of the cloud liquid water variability (sdL) and the variability in water vapour (sdW) can be explained using a simple model [18]: in this model established for stratocumulus, the LWC is proportional to the difference between the base and top specific humidities (q_{base} - q_{top}). Any cloud depth variation within the cloud will therefore induce a change in q_{top}, thus in water vapour. We note that this relation is found more accurate for ship when taking all small clouds, i.e. when considering all boundary layer clouds, and not mainly large stratiform clouds associated with frontal perturbations. Frontal systems might also contribute to this correlation, since they are characterized by simultaneous variatins of water vapour and cloudiness. However, these events were partly suppressedby filtering out raining occrences.

The second linear relationship (sdL versus mLW) is more surprising, since there is no evidence in the literature (at our knowledge) of such behaviour. Moreover, the slope of the mean line is nearly the same, whatever be the data set. Another common property is the power law of the size distribution (about -1.5). Analyses of cloud size distribution have been made for a long time using visible satellite images of various resolutions (AVHRR, Landsat, SPOT, mainly). For example, such a distribution power law was also found by [19] who interpreted it as an invariance property using fractal analysis. It might be possible that the LWC law obtained with microwave radiometry be also interpreted in this manner.

7. CONCLUSION

Data of a shipborne radiometer were obtained for a two-month period in winter 1997, in the NW Atlantic ocean, providing a continuous survey of the water vapour and liquid water content in clouds over the ship. They were compareD to SSMI data, selected in the 15°x15°

experimental domain. The question to answer was "Is it significantand meaningful to compare instantaneous data sampled over $25 \times 25 \text{km}^2$ with time series sampled every minute?"

Comparisons between the retrieved water vapour, liquid water and surface wind (directly measured on the ship) were made. Despite the different sampling from the ship and SSM/I, there is a very good agreement between the retrieved water vapours using Alishouse's algorithm for SSMI, but an overestimate of the SSMI wind using [16] with respect to the direct ship measurement (1.6 ms^{-1}), and a 2 ms^{-1} standard deviation. The average agreement between retrieved liquid water contents in the range 0–0.6kg.m^{-2} is within the retrieval error (bias 0.02kg.m^{-2}), except for some particular cases, but the standard deviations of both data set differ by an order of magnitude. A possible reason could be that many of the observed clouds are convective in the cold sector of a depression. These have scales of a few kilometres, and sometimes high liquid-water contents. Such clouds are smoothed within each SSM/I pixel. Averaging the ship borne radiometer data over 2 hours led to the same results, and the comparison ship (1-hour) - SSMI individual pixel does not differ significantly from the comparison with the 1° square box.

To investigate the causes of this difference in liquid water contents, clouds were individually identified in both data sets, and their mean characteristics were analyzed. Three properties were found similar for the ship and the SSMI : the power law of the size distribution (about - 1.5), a linear relation between the cloud heterogeneity, quantified as the standard deviation of liquid water inside the cloud versus the mean liquid water content (slope ~0.8), and a correlation between the variabilities in liquid water and water vapour inside each cloud. The proportionality of liquid water variability to the average liquid water content could partly explain the results obtained in comparing SSMI and the ship data : low values (below 0.2 kgm^{-2}) with low variability agree better than high values, which are found greater in average from the ship than SSMI, and thus associated to a greater variability.

These preliminary results show the interest of microwave radiometry, as a complement of visible/infrared images for analyzing cloud properties. Here, all clouds were considered without paying attention to the meteorological situation, and often clouds were artificially joined, since the cloud selection was only based on a threshold in liquid water. The FASTEX data base will allow us to improve the interpretation of results, by distinguishing between the various cloud types.

Acknowledgments : The author is grateful to the crew and the scientific teams who participated in the FASTEX/CATCH cruise. The SSMI data were obtained through the CLOREVAL group activity, under support of the EU/Environment and Climate programme. The reviewers are thanked for their valuable comments.

This work was supported by INSU/CNRS and EU/Environment and Climate programme.

REFERENCES

1. Offiler, D., L. Eymard, D. Kilham, E. Gérard, H. Gaeng, "Cloud Retrieval Validation Experiment (CLOREVAL) Final Report". *Report for the European Commission, programme Environment and Climate*. Edité par le UK Met. Office. (1998)
2. Gérard, "Restitution de l'eau liquide nuageuse par radiométrie hyperfréquence". *Thèse de l'Université Paris-VI*, 290 pp (1996)
3. Tiedtke, M., "Representation of clouds in large-scale models". *Monthly Weather Review*, **121**, 3040–3061. (1993)
4. Guillou C., "Etude du transfert radiatif dans l'atmosphère, en microondes, à partir d'observations radiométriques aéroportées", *Thèse de doctorat de l'Université Paris VII*, (1994)
5. Boukabara, S.A., "Couplage des mesures hyperfréquences actives et passives". *Thèse de l'Université Paris-VII*, (1997)

6. Liebe, H.J., G.A. Hufford and M.G. Cotton, "Propagation modelling of moist air and suspended water/ice at frequencies below 100GHz". In *Proceedings AGARD 52nd Specialists' Meeting of the Electromagnetic Wave Propagation Panel*, Palma De Mallorca, Spain, 3-1-3-10. (1993)

7. Guillou, C., S.J. English, C. Prigent and D.C. Jones, "Passive microwave airborne measurements of the sea surface response at 89 and 157 GHz". *J. Geophys. Res.*, **101**(C5), 3775-3788. (1996)

8. Wilheit, T. T., "A model for the microwave emissivity of the ocean's surface as a function of wind speed", *IEEE Trans. Geosci. Electron.*, GE-17, 244-249, (1979).

9. Cox, C., and W. Munk, "Measurements of sea surface from photographs of the sun's glitter". *Journal of the Optical Society of America*, **44**, 838–850. (1954)

10. Ellison, W., A. Balana, G. Delbos, K. Lamkaouchi, L. Eymard, C. Guillou et C. Prigent, "New permittivity measurements of sea water". *Radio Science*, **33**, 639-648, (1998).

11. Guillou, C., W. Ellison, L. Eymard, K. Lamkaouchi, C. Prigent, G. Delbos, G. Balana and S.A. Boukabara, "Impact of new permittivity measurements on sea surface emissivity modelling in microwaves". *Radio Science*, **33**, 649–667. (1998)

12. Monahan, E. C., and M. Lu, "Acoustically relevant bubble assemblages and their dependance on meteorological parameters", *IEEE J. Ocean. Eng.*, **15**, 340-349, (1990).

13. Droppleman, J. D., "Apparent microwave emissivity of sea foam", *J. Geophys. Res.*, **75**, 696-698, (1970).

14. Gérard, E. and L. Eymard, "Remote Sensing of integrated cloud liquid water: development of algorithms and quality control". *Radio Science*, **33**, 433–447. (1998)

15. Alishouse, J.C., J.B. Snider, E.R. Westwater, C.T. Swift, C.S. Ruf, S.A. Snyder, J. Vongsathorn and R.R. Ferraro, "Determination of cloud liquid water-content using the SSM/1". *IEEE Transactions on Geoscience and Remote Sensing*, **28**, 5, 817–822. (1990)

16. Goodberlet, M.A. and C.T. Swift, "Improved retrievals from the DMSP wind speed algorithm under adverse weather conditions". *IEEE Transactions on Geoscience and Remote Sensing*, **30**, 1076–1077. (1992)

17. Bourras, D., L. Eymard and C. Thomas, "A Comparison of the errors associated with the retrieval of the latent heat flux from individual SSM/I measurements, this issue pag. 47

18. Slobin, S.D., "Microwave noise temperature and attenuation of clouds : statistics of these effects at various sites in the United States, Alaska ans Hawaii". *Radio Science*, **17**, 1443 - 1454. (1982)

19. Calahan and Joseph, "Fractal statistics of cloud fields". *Monthly Weather Rev.*, **117**, 261-272 (1989)

Microw. Radiomet. Remote Sens. Earth's Surf. Atmosphere, pp. 247–254
P. Pampaloni and S. Paloscia (Eds)
© VSP 2000

Estimation of water vapor vertical distribution over the sea from Meteosat and SSM/I observations

FEDERICO PORCÚ[1], DAVIDE CAPACCI[1] and FRANCO PRODI[1,2]

[1]*Physics Dept. University of Ferrara, I-44100 Ferrara, Italy*
[2]*ISAO-CNR, Clouds and Precipitation Group, I-40129 Bologna, Italy*

Abstract – Object of this work is to explore the capabilities of multi-sensor water vapor (WV) observations for identification and classification of fronts and air masses in northern Atlantic and Mediterranean areas. We used data from the 6.3 μm channel of the European geostationary satellite Meteosat: we retrieved the distribution of WV mean content in the layer between 600 and 300 hPa for cloudless areas. Multifrequency data from Special Sensor Microwave/Imager (SSM/I) are used to estimate: 1) the distribution of WV mean content in lowest 500 m of the troposphere, and 2) the distribution of total WV content in the troposphere. The retrieval is performed over marine areas and outside heavy precipitation areas. We combined the three WV retrievals and we estimated the vertical WV profile at three tropospheric levels: the lowest one below 500 m (1000-960 hPa, from the SSM/I), the layer between 600 and 300 hPa (from the Meteosat) and the layer between 500 m and 600 hPa (as difference between these two fields and the total columnar content as from SSM/I). The performances of the three techniques are evaluated by comparison with European Center for Medium Range Weather Forecast (ECMWF) analysis: good agreement is found for both SSM/I retrievals (percentage of error between 15 and 25%) while for the 6.3 μm retrieval higher values are reported (about 45%). The combined approach is used to estimate vertical profiles of WV content with an accuracy suitable for semi-quantitative analysis of the moisture structure. Vertical cross sections of WV fields are obtained along frontal surfaces and discussed for one case study.

1. INTRODUCTION

Recent climatological studies [1] pointed out that most of the flooding in Europe over the last years are related to some degree of cyclogenesis and are often linked to frontal system development. Frontal cloud systems are the main mechanisms for producing heavy and long lasting precipitation, especially in autumn/winter, while in spring/summer the frontal lines act as synoptic or mesoscale forcing mechanisms, which are able to organize convection and trigger for large convective complexes. In particular, frontal surfaces often develop over the sea (North Atlantic and Western Mediterranean) and affect the coastal areas [2].

On the other side, there is a lack in the knowledge of these systems especially in the Mediterranean. The need of a better understanding of these systems and the development of new and more adequate conceptual models has been recently pointed out as a preliminary result of the COST 78 Action [3].

In this work we focus our attention on the spatial distribution of the water vapor: due to the scarcity of detailed ground-based observations of water vapor fields over the sea, we use a remote sensing approach. Moreover, microwave water vapor retrievals have proved to be an effective tool in identifying and classifying frontal areas over the sea [4].

2. ALGORITHMS AND DATASET

2.1 Meteosat retrieval

The algorithm for retrieval of Upper Tropospheric Humidity (UTH) from Meteosat water vapor image data is based on radiative transfer calculations for a cloudless atmosphere. It uses an approximated integral solution of Schwarzchild's equation for the absorption of radiation between 5.7 and 7.1 μm by water vapor. The integral solution is obtained by neglecting the surface contribution to the outgoing radiance, dividing the troposphere into a number of homogeneous layers (between 14 and 20 layers), knowing the temperature and the relative humidity of these tropospheric layers and using an empirical water vapor transmittance function [5]. Ninety percent of the computed radiance is determined by the water vapor content in the 600-300 hPa layer.

The radiance measured by the satellite can be simulated using the above calculations by introducing the radiometer filter function in the spectral interval [5.7, 7.1] μm and the satellite zenith angle. The temperature profile is also needed for the radiative transfer calculations, and we derived such profiles from ECMWF analysis. For each clear-sky pixel in the image, the operative model [6] simulates six possible radiance measures by fixing six values of relative humidity into the 600-300 hPa layer, and interpolates the true measure on the six simulated measures in order to extrapolate the value of correspondent relative humidity. The water vapor content in the 600-300 hPa layer (hereafter referred as Upper Tropospheric WV Content, UTWVC) is calculated using relative humidity and temperature.

2.2 SSM/I retrievals

For the microwave frequencies, radiative transfer calculations cannot be carried out for each pixel in the current image, because it would require a description of the transmission and of the surface emissivity for the actual atmospheric/surface conditions. For this reason, the algorithms for retrieval of Integrated Water Vapor Content (IWVC) and Boundary Layer Water Vapor Content (BLWVC) are based on regression equations resulting from appropriate radiative transfer calculations for a representative set of atmospheric situations. This regression equations have been subsequently validated by the application to SSM/I data and by the comparison with corresponding measurements from radiosonde ascents.

For the BLWVC retrieval, a radiative transfer scheme [7] has been set up for the simulation of the SSM/I radiometer signals at given atmospheric and oceanic parameters, but scattering is not treated. For that reason, areas covered by clouds whose liquid-water content higher than 0.03 g/cm^2, or heavy precipitation are excluded. The representative set of atmospheric situations entered into the radiative-transfer model are defined by radiosonde and surface measurements, and by introduction of randomly distributed liquid-water contents in the range 0-0.09 g/cm^2.

The 19 and 22 GHz (both at vertical polarization), provide the main predictors for BLWVC, the use of the 37 GHz at vertical polarization channel is helpful for correcting liquid-water, and the 19 GHz channel at horizontal polarization is used to account for wind-speed variations. The regression equation used in this work is:

$$BLWVC = -5.9339 + 0.03697 \times T19v - 0.02390 \times T19h + 0.01559 \times T22v - 0.00497 \times T37v$$

A sophisticated radiative transfer model has been used to simulate the SSM/I radiometer signals for the IWVC estimates [8]. The model takes into account multiple scattering as well as polarization. The set of atmospheric situations entered into model is defined by radiosonde measurements and by the random introduction of clouds and rain. The separation of the radiative simulations into cloud-free, cloud-covered, and raining scenes is used to provide separate retrieval algorithms for each scene type. For cloud free areas the equation we used in the present work is:

$$IWVC = 33.92 - 4.943 \times \ln(300 - T22v) + 2.406 \times \ln(300 - T85v)$$
$$-3.295 \times \ln(300 - T85h) - 0.03093 \times T37h$$

2.3 Dataset used

We carried out this study for the period 1-22 November 1992, when DMSP f10 and Meteosat 4 data were available. The working area was selected between 35 and 55 N and 55 W and 25 E in order to consider both Atlantic and Mediterranean situations. Satellite data have been remapped onto a gaussian grid of $1/8^{th}$ of degree in latitude and longitude, resulting in a pixel of about 130 km^2.

We used ECMWF analysis data for initializing the Meteosat radiative transfer scheme and to validate the retrievals. The data were available on a 0.5 × 0.5 gaussian grid and 12 pressure levels were used (surface, 1000, 925, 850, 700, 500, 400, 300, 250, 200, 150 and 100 hPa). For the Meteosat retrieval a minimum of 14 levels was required and we got the levels at 800, 600, 450 and 350 hPa by logarithmic interpolation.

3. ALGORITHMS VALIDATION

The first step in our work was to validate separately the individual retrieval algorithms, comparing the retrieved WVC in the different layers with the ECMWF analysis at the same layers. In order to minimize errors due to time differences between the observations from different sensors and synoptic time, we only considered satellite data closer than 15 minutes to the ECMWF analysis. Reference time for the satellite data was assigned considering the scanning time of the 45°N latitude.

As the Meteosat retrieval algorithm makes use of the ECMWF analysis for the UTWVC estimations, the Meteosat data were averaged onto the ECMWF grid of 0.5°×0.5°. Comparison between satellite-derived and analysis-derived water vapor content were carried out on the same grid.

For UTWVC retrieval, we screened out cloudy pixels making use of infrared images: the pixels with an infrared brightness temperature lower than the 850 hPa level are not considered in the water vapor retrieval.

For the algorithms validation, we selected seven nearly-simultaneous ECMWF-SSM/I and ECMWF-Meteosat couples and inside the satellite observed areas, we selected about thirty subareas of about 200 pixel in size. The subareas were selected looking for marine regions with the highest number of cloud-free pixels (for Meteosat) and rain-free pixels (for SSM/I). For these areas we retrieved UTWVC, BLWVC and IWVC fields by the algorithms previously described. We also computed temperature and specific humidity using the ECMWF data for the same tropospheric layers. We integrated the boundary layer content from the surface to 960 hPa, the upper tropospheric content between 600 and 300 hPa and the total tropospheric content integrating from the surface to 100 hPa. Pressure levels which weren't available directly from ECMWF data (e.g. 960 and 600 hPa) were obtained by logarithmic interpolation.

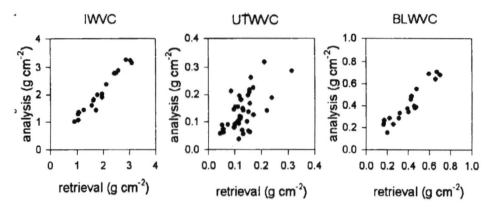

Figure 1 Comparison between satellite retrieval and analysis for IWVC (left), UTWVC (center) and BLWVC (right).

In figure 1 the scatterplots for the three algorithms are shown: the SSM/I shows good performances for both algorithms, while Meteosat shows poor correlation (r^2=0.41). Best performances are reached by the IWVC algorithm (r^2=0.96), while slightly higher errors are found in the BLWVC (r^2=0.88).

4. COMBINED USE OF MULTI-SENSOR RETRIEVALS

The vertical distribution of water vapor at three different layers is estimated as follows (see fig. 2a): BLWVC and UTWVC are directly computed by the retrieval algorithms from SSM/I and Meteosat data respectively, while the Lower Tropospheric Water Vapor Content (LTWVC) is computed by subtracting from the IWVC (retrieved by SSM/I data) the boundary layer and upper tropospheric contributions.

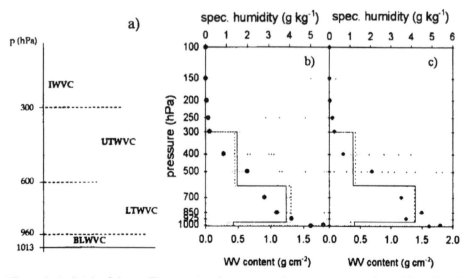

Figure 2 a) sketch of the profile reconstruction scheme. b) and c) two sample profiles: ECMWF derived profiles (solid), satellite retrieved profiles (dotted) and specific humidity ECMWF data (black circles) are plotted

The contribution of the layer above 300 hPa is neglected, being around 10^{-2} g cm^{-2} and lower. The resulting LTWVC is an estimate of the water vapor content between 960 and 600 hPa.

In figs. 2b and 2c we plotted the reconstructed vertical moisture profiles for two different clear sky marine locations: dryer (fig. 2b) and moister (fig. 2c). The retrieved layers are compared to profiles derived from ECMWF analysis; the values of specific humidity for all available pressure levels are reported too.

The discrepancies between retrieved and observed profiles are rather low and the characteristics of the two profiles are correctly estimated by the reconstructed profiles: the two vertical structures mainly differ in the lower tropospheric layer, while both BLWVC and UTWVC are similar, as confirmed by ECMWF data.

For a more complete evaluation of the errors, we compared LTWVC maps with ECMWF analysis for a number of subareas where both satellites were available at synoptic times. For each k subarea, for the N_k cloud-free or rain-free pixels, we computed the Percentage Error (PE) as follows:

$$PE_k = \frac{2}{\overline{A}_k + \overline{S}_k} \sqrt{\frac{1}{N_k} \sum_{i=1}^{N_k} (A_i - S_i)^2}$$

where A_i and S_i are the water vapor content (g/cm^2) computed by ECMWF analysis and satellite retrieval, respectively, while $\overline{A}_k$ and $\overline{S}_k$ are the subarea averaged values.

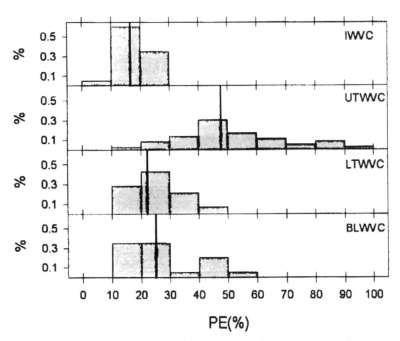

Figure 3 PE values distribution for the different fields: from the top to the bottom: IWVC, UTWVC, LTWVC and BLWVC. Solid black lines show the PE averaged values

F. Porcú et al.

In figure 3 the frequency distribution of PE values for the different tropospheric layers is shown, the averaged PE were also reported for each histogram. The averaged PE for LTWVC is slightly higher than the one found for IWVC: this is due to the relatively low values of UTWVC and BLWVC (below 0.4 and 1.0 g cm^{-2}, respectively), while IWVC ranges between 1 and 4 g cm^{-2} and it is the main contributor to LTWVC an its error.

5. APPLICATIONS

As an application of the combined retrievals, we studied two moisture structures over North Atlantic observed by both satellites on November, 4 and 5 1992 (see figure 4). On 4th November a dry slot ahead of a cold front was in clear sky and all the retrievals were possible, while on 5th of November no particular structure was evident under clear sky.

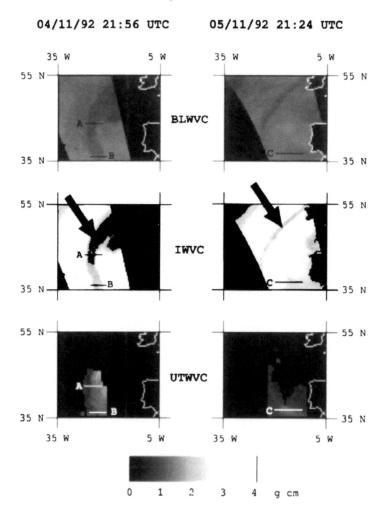

Figure 4. Water vapor maps as obtained from satellite data for the two cases. Black arrows indicate the position of the dry slots.

In fig 5 we plotted water vapor content cross sections as indicated by the labeled lines in fig. 4. For 4th of November case, the sections cross the dry area at two different latitudes, as it is clearly shown by the steep gradient in the IWVC plot (figs. 5a and 5b). The behaviour of BLWVC is similar to the IWVC one along both sections, while the section of UTWVC is rather different: in section A (fig. 5a) increases to the west across the dry IWVC area, while it is almost constant alongside section B. In section C, on the contrary, only weak structures can be noted in the IWVC and BLWVC patterns, while the UTWVC shows a marked peak in correspondence with a weak minimum in IWVC. The different coupling of WV content in lower and upper troposphere can be interpreted in terms of vertical structure of the frontal region and linked to correspondent conceptual models.

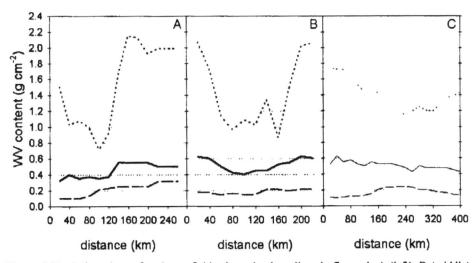

Figure 5 Vertical sections of moisture fields along the three lines in figure 4: A (left), B (middle) and C (right). IWVC (dotted), BLWVC (solid) and UTWVC (dashed) are plotted.

6. CONCLUSIONS

Our multi sensor approach allows an estimate of the vertical profile of water vapor in the troposphere in three layers over the sea under clear sky conditions. The accuracy of such an estimate makes possible a semi-quantitative analysis of the moisture field, especially in presence of frontal structures when WV gradients are present and marked dry intrusion can be identified as important part of the cyclonic structure. In one case study we shown that there is the capability of classify different regions along a cold front, by means of different vertical WV distribution.

Vertical moisture sections can be useful in identifying and classifying frontal structures with two clear limitations: Meteosat retrievals work only over cloud free areas and SSM/I works only over the sea. Further work will include: the use of a broader database to analyze frontal structures in the spring/summer time and to determine the correlation between IWVC and UTWVC over dry areas. Also the integration with other sensors' data (e.g. SSM/T2) is envisaged, in particular looking forward to the use of the 7.2 μm channel on board the Meteosat Second Generation.

AKNOWLEDGEMENTS

This work was supported by the Commission of the European Communities, Environment and Climate Program, under contract ENV-CT96-0281 (MEFFE). Daniel Taurat (at the Max Plank Institute for Meteorology) provided the SSM/I retrievals. The help of Marianne König and Johannes Schmez (both at EUMETSAT) in implementing and evaluating the Meteosat retrievals is gratefully acknowledged. Access to ECMWF analysis was ensured by the Italian "Servizio Meteorologico dell'Aeronautica Militare" (SMAM) in the framework of an agreement with the CNR.

REFERENCES

1. F. Porcú, F. Prodi, S. Franceschetti and S. Pasetti, Short term climatology of cloud systems leading to flood events in Europe (1991-1996), In *Proc. 1997 EUMETSAT Meteorological Satellite Data Users' Conference*, 461-466, EUMETSAT, Darmstadt, EUM P 21. (1997)
2. E.C. Barrett and J. Michell. Satellite remote sensing of natural hazards and disaster in the Mediterranean region, In: *Current topics in remote sensing* (Barrett, E. C., Brown, K. A. & Michallef, A., editors), pp 51-67. Gordon and Breach. (1991)
3. B.J Conway, L. Gerard, J. Labrousse, E. Liljas, S. Senesi, J. Sunde, and V. Zwatz-Meise, *Nowcasting. A survey of current knowledge, techniques and practice*, EC-DG XII, Brussels. (1996)
4. K.B. Katsaros, I. Bhatti, L.A. McMurdie, and G.W. Petty, Identification of atmospheric fronts over the ocean with microwave measurements of water vapor and rain. *Weather and Forecasting*, 4, 449-460. (1989)
5. W. Wiscombe and J. Evans, Exponential-sum fitting of radiative transmission functions. *J. Comput. Phys.*, 24, 416-444. (1977)
6. J. Schmetz, and O.M. Turpeinen, Estimation of the Upper Tropospheric Relative Umidity Field from Meteosat Water Vapour Image Data, *J. Appl. Meteor.*, 27, 889-899. (1988)
7. J. Schulz, P. Schlüssel, and H. Grassl, Water vapour in the atmospheric boundary layer over oceans from SSM/I measurements. *Int. J. Remote Sensing.*, 14, 2773-2789. (1993)
8. P. Schlüssel and P. Bauer, Rainfall, total water, ice water, and water vapor over sea from polarized microwave simulations and Special Sensor Microwave/Imager data, *J. Geophys. Res.*, 98, 20737-20759. (1993)

Microw. Radiomet. Remote Sens. Earth's Surf. Atmosphere, pp. 255–262
P. Pampaloni and S. Paloscia (Eds)
© VSP 2000

Effects of AMSU-A cross track asymmetry of brightness temperatures on retrieval of atmospheric and surface parameters

FUZHONG WENG, RALPH R. FERRARO and NORMAN C. GRODY

NOAA/NESDIS/ORA, 5200 Auth Road Camp Springs, MD USA 21042

Abstract - The AMSU-A instrument on board NOAA-K provides global information on atmospheric temperature sounding, water vapor, clouds, precipitation , snow cover and sea ice concentration. However, brightness temperatures for the window channels exhibit a cross-track asymmetry which can significantly impact the retrieval of AMSU non-sounding products. As an example, the cloud liquid water along each scan line shows a pronounced asymmetry, having higher amount on one side of nadir than the other. A polynomial scheme is developed to correct the brightness temperature asymmetry along the scan line. It is shown that the quality of cloud liquid water distribution is significantly improved after the asymmetry correction is applied.

1. INTRODUCTION

In 1998, National Oceanic and Atmospheric Administration (NOAA) began launching the next generation polar-orbiting operational environmental satellites (NOAA-K, L, M and N). The Advanced Microwave Sounding Unit-A (AMSU-A) is one of the most important instruments on board these satellites. The instrument was designed primarily for retrieving atmospheric temperature profiles from the surface to 50 km. However, it is also utilized to derive non-sounding products such as column water vapor, cloud liquid water, precipitation rate, sea ice concentration, and snow cover. We are currently studying the quality of these non-sounding products and comparing with those derived from other satellite sensors as well as from ground-based measurements.

While the AMSU-A instrument was calibrated before the satellite was launched into space [1], actual measurements exhibit a pronounced asymmetry along scan lines, especially at window channels. This study first presents the asymmetric characteristics by comparing the observed with simulated brightness temperatures. Then, the effects of the asymmetry on the retrievals of non-sounding products is demonstrated. While the causes of the asymmetry remain unknown, a procedure is developed to correct the asymmetry using a high order polynomial which adjusts the brightness temperatures to those predicted from the radiative transfer model. The significance of this correction is illustrated by comparing the global cloud liquid water distribution before and after the asymmetry correction is applied.

2. AMSU-A OBSERVATIONS

The AMSU-A is composed of two separate modules, A1 and A2. The A1 module has 12 channels in the oxygen band (3-15) (see Table 1) and provides information on the atmospheric temperature profile, whereas the A2 module has two channels (1-2) which are primarily used to derive non-sounding products.

Table 1.

AMSU-A instrument parameters specification

Channel Number	Center Frequency (GHz)	Number of Pass Bands	Band width (MHz)	3-dB Beamwidth (deg)	NEΔT (K)
1	23.80	1	251	3.53	0.21
2	31.40	1	161	3.41	0.26
3	50.30	1	161	3.76	0.22
4	52.80	1	380	3.72	0.14
5	53.59 ± 0.115	2	168	3.70	0.15
6	54.40	1	380	3.68	0.15
7	54.94	1	380	3.61	0.13
8	55.50	1	310	3.63	0.14
9	57.29 = fo	1	310		0.24
10	fo ± 0.217	2	76	3.51	0.25
11	fo ± 0.322 ± 0.048	4	34		0.28
12	fo ± 0.322 ± 0.022	4	15		0.40
13	fo ± 0.322 ± 0.010	4	8		0.54
14	fo ± 0.322 ± 0.004	4	3		0.91
15	89.00	1	2000	3.80	0.17

The AMSU-A is a cross track instrument, scanning between ±47.85 degree off nadir and obtains 30 measurements along each scan line, covering nearly a 2300 km swath on Earth. Beam position 1 and 30 are the extreme scan positions of the Earth views, while beam position 15 and 16 are at $1^0 40'$ and $-1^0 40'$, respectively, from the nadir. Since the antenna beamwidth is about $3.3^0 \pm 10\%$, the AMSU-A IFOV on the surface is nearly the same for all frequencies at a given beam position; however the IFOV increases as the scan angle increases. Onboard calibration is performed every 8 seconds by viewing an internal blackbody target at 180^0 from nadir and the cold space in one of four possible viewing positions located between -76^0 and -84^0 from nadir. However, the antenna temperature is different from the brightness temperature (seeing within the main lobe of the antenna) due to the antenna sidelobes which can view cold space and the satellite platform [1].

Brightness temperatures, T_b are empirically derived from antenna temperatures, T_a using the following relationship [1]

$$T_a = \alpha T_b + \beta T_c + \gamma T_s, \qquad (1)$$

where $\alpha + \beta + \gamma = 1$; T_c and T_s are the cold space and satellite platform temperatures, respectively. The coefficients, α, β, and γ are the fraction of each object viewed by the instrument. The last two terms in Eq. (1) are negligibly small [1] compared to the first term so that $T_b = T_a / \alpha$. Since α decreases as the scan angle increases, the difference between T_a and T_b increases with an increase of scan angle.

The nominal correction of the antenna to brightness temperatures based on Eq. (1) is less than 3 K, as shown in Fig. 1. Under clear atmospheres, monthly mean satellite brightness temperatures, T_b(sat) are warmer than antenna temperatures, T_a(sat) with larger corrections being made at larger scan angles.

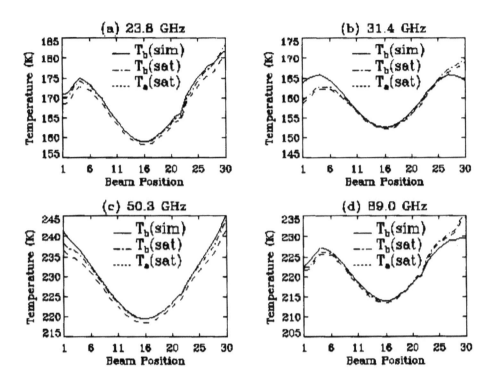

Figure 1. Global monthly (March 1999) mean satellite antenna, brightness temperatures and simulated brightness temperature at AMSU-A 23.8, 31.4, 50.3 and 89 GHz over ocean

To further identify the characteristics of the brightness temperature along each scan, we simulate the brightness temperatures at 23.8, 31.4, 50.3 and 89 GHz over ocean corresponding to each AMSU-A beam position. The radiative transfer modeling is performed under clear atmospheric conditions. Sea surface temperature (SST), wind vector and temperature/moisture profiles are obtained from the NCEP global data assimilation system (GDAS) and are used as inputs to the radiative transfer model. The SST is generated using a blended technique of satellite and conventional observations [2]. Surface wind data are analyzed through assimilating available microwave satellite estimates from the SSM/I with buoy measurements. The GDAS global analyses have a resolution of one degree in latitude and longitude and are produced four times a day. As shown in Fig. 1, simulated brightness temperatures apparently have discrepancy from observed brightness temperatures. In particular, simulations at 31.4 GHz are warmer than observations at beam position 1-6 and colder from beam position 25 to higher.

The difference between simulations and observations is further illustrated in Fig. 2. It appears that the bias is asymmetric relative to the nadir. For example, at 31.4 GHz (Fig. 2b), a bias is positive for the beam position from 1 to 14 and negative for the beam position from 15 to 30.

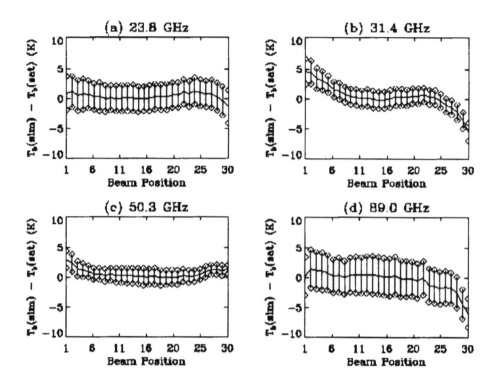

Figure 2. Mean bias of simulated brightness temperatures from observed under clear atmosphere over oceans. Vertical bars show the standard deviation of the biases corresponding to each beam position.

While the measurements at 23.8 and 31.4 GHz are obtained from the same AMSU-A2 module, the bias seems to be smaller at 23.8 GHz (Fig. 2a) than that at 31.4 GHz (Fig. 2b). Also, the bias at 50.3 GHz (Fig. 2c) is smaller than that at 89 GHz (Fig. 2d) although both channels are situated on the AMSU-A1 module. The asymmetry appears the worst at 31.4 GHz where the atmosphere is the most transparent.

The asymmetric brightness temperature along each scan line can result in an asymmetry for the AMSU-A products. This is illustrated by the cloud liquid water distribution over oceans as shown in Fig. 3. The cloud liquid water algorithm is described in detail in the appendix. It is a physical retrieval similar to that developed for the SSM/I [3]. The algorithm requires three auxiliary parameters, sea surface temperature, surface wind speed, and cloud layer temperature. These parameters are not directly measured by AMSU but can be obtained using a prior information or auxiliary data. For example, the surface temperature and wind speed can be obtained from the GDAS analyses. Also, infrared measurements at 11 μm from the Advanced Very High Resolution Radiometer (AVHRR), also aboard the NOAA-K, can be used to estimate the cloud layer temperature[4,5].

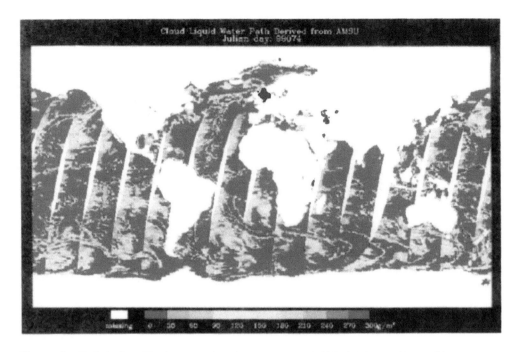

Figure 3. Global cloud liquid water distribution using AMSU-A 23.8 and 31.4 GHz

It is very obvious that the cloud liquid water along each scan line is asymmetric. Higher amounts occur on the one side of nadir than the other side. This distribution is not attributed to the natural variability of cloud liquid water. It does rather suggest that there may be some errors in the instrument calibration or there may be a lack of understanding of the radiative transfer process which may require more details to account for actual satellite scan geometry.

3. CORRECTIONS OF AMSU-A CROSS-TRACK ASYMMETRY OF BRIGHTNESS TEMPERATURES

The asymmetric brightness temperatures shown in Fig. 2 is corrected using the following polynomials.

$$\Delta T_b = A_0 \exp\left\{-\tfrac{1}{2}\left[(\theta - A_1)/A_2\right]^2\right\} + A_3 + A_4\theta + A_5\theta^2 , \qquad (2)$$

where the coefficients, A_{0-5} are given in Table 2. This correction provide an additional adjustment to the brightness temperature given Eq. (1). The adjustment varies from 3.0 to - 6.0 K (see Fig. 2). The largest correction is required for the 31.4 GHz channel.

Table 2.
Coefficients used to correct the AMSU-A cross-track asymmetry of radiance

A_0	A_1	A_2	A_3	A_4	A_5
-8.18554	76.1569	6.2002	0.0568309	0.0154784	0.000578763
-37.7958	100.615	30.6613	0.134141	0.0328018	0.00203639
1.27105	-56.2647	3.67390	-0.0328722	-0.00237711	0.000481114
-3.40744	56.9120	4.01828	0.0722392	-0.0282945	-0.000311757

After the bias correction is applied to the AMSU-A brightness temperature using Eq. (2), the asymmetry of the cloud liquid water disappears as shown in Fig. 4.

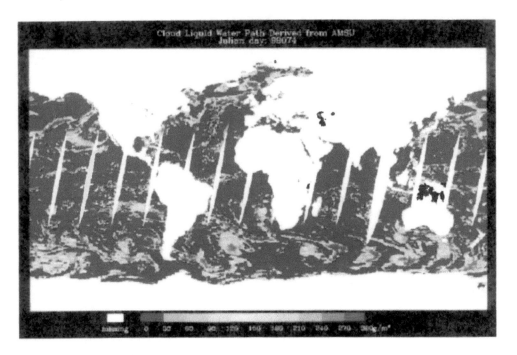

Figure 4. Global cloud liquid water distribution after the asymmetry correction

4. CONCLUSIONS

The AMSU-A instrument clearly exhibits a cross-track asymmetry of brightness temperature, especially for the window channel at 31.4 GHz. The asymmetry is identified by comparing simulated brightness temperatures with observed brightness temperatures along the AMSU-A scan lines. For lower beam positions (see Fig. 2), the simulations tend to be warmer than observations, whereas the opposite is true for the high beam positions. At this time, the actual cause of the asymmetry is not fully understood.

Appendix: Physical Retrievals of Cloud Liquid Water and Water Vapor

For a scattering-free atmosphere, brightness temperatures at microwave frequencies may be approximated as

$$T_b = T_s \left[1 - (1 - \varepsilon)\Im^2 \right], \qquad (A.1)$$

when transmittance, $\Im$ is close to unity. Here, ε is the surface emissivity and T_s is the surface temperature.

Atmospheric transmittance is further related to

$$\Im = \exp\left[-(\tau_0 + \tau_v + \tau_1)/\mu \right], \qquad (A.2)$$

where τ_0, τ_v and τ_1 are the optical thickness for oxygen, water vapor and cloud liquid water, respectively and μ is the cosine value of local zenith angle.

In the following derivations, we establish a simple parametric model for the transmittance in terms of water vapor path and liquid water path. First, water vapor optical thickness is related to

$$\tau_v = \kappa_v V, \qquad (A.3)$$

where κ_v is the water vapor mass absorption coefficient and $V = \int_0^\infty \rho_v dz$, which is the vertically integrated water vapor. κ_v can be derived from the slope of the relationship between τ_v and V. Using 1400 radiosonde station data having temperature and moisture profiles, we derive κ_v for four AMSU-A window channels as shown in Table A.1

Cloud optical thickness may be derived in a similar manner as

$$\tau_L = \kappa_L V \qquad (A.4)$$

where κ_L is the cloud liquid water mass absorption coefficient and $L = \int_0^\infty \rho_L dz$ is vertically integrated cloud liquid water . Using Rayleigh's approximation, one can express κ_L in terms of cloud layer temperature, T_L as

$$\kappa_L = a_L + b_L T_L + c_L T_L^2. \qquad (A.5)$$

Oxygen optical thickness is parameterized as a function of sea surface temperature[3] through

$$\tau_0 = a_0 + b_0 T_S . \qquad (A.6)$$

where a_0 and b_0 are shown in Table A.1

Table A.1.
The parameters calculated at four AMSU-A channels, which are used in water vapor and cloud liquid water algorithms

	23.8	31.4	50.3	89
K_v	4.80423E-3	1.93241E-3	3.76950E-3	1.15839E-2
$K_l - a_l$	1.18201E-1	1.98774E-1	4.53967E-3	1.03486E00
$K_l - b_l$	-3.48761E-3	-5.45692E-3	-9.68548E-3	-9.71510E-3
$K_l - c_l$	5.01301E-5	7.18339E-5	8.57815E-5	-6.59140E-5
$\tau_o - a_o$	3.21410E-2	5.34214E-2	6.26545E-1	1.08333E-1
$\tau_o - b_o$	-6.31860E-5	-1.04835E-4	-1.09961E-3	-2.21042E-4

Substituting Eqs. (A.2) and (A.3) into (A.1) results in

$$\kappa_L V + \kappa_L L = -\frac{\mu}{2}\left\{\ln(T_s - T_b) - \ln\left[T_s(1-\varepsilon)\tau_o^2\right]\right\}. \tag{A.7}$$

Using two properly channeled measurements, one may solve for V and L, simultaneously, as has been demonstrated in previous studies [4]. For AMUS-A, measurements at 23.8 GHz are near a weak water vapor line and provide sensitivity to water vapor and those at 31.4 GHz are more affected by cloud liquid absorption. Thus,

$$L = a_0\left\{\ln\left[T_s - T_b(31.4)\right] - a_1\ln\left[T_s - T_b(23.8)\right] - a_2\right\}, \tag{A.8}$$
$$V = b_0\left\{\ln\left[T_s - T_b(31.4)\right] - b_1\ln\left[T_s - T_b(23.8)\right] - b_2\right\}. \tag{A.9}$$

where coefficients a_{0-2} and b_{0-2} are mainly a function to water vapor and cloud mass absorption coefficients.

REFERENCES

1. T, Mo,: AMSU-A Antenna pattern correction. *IEEE Trans. Geosci. . Remote Sensing* , (37), 103-112, (1999)
2. R. W. Reynold, and T. M. Smith, Improved global sea surface temperature analyses using optimum interpolation, *J. Climate*, (7), 929-948, (1994).
3. Weng, and N. C. Grody, Retrieval of cloud liquid water using the special sensor micowave imager (SSM/I), *J. Geophys. Res.*, (99}, 535 - 25,551, (1999).
4. T. J. Greenwald, G. L. Stephens, T. H. Vonder Haar, A physical retrieval of cloud liquid water over the global oceans using special sensor microwave/imager (SSM/I) observations, *J. Geophys. Res.*, (98), 18471 - 18488, (1993).
5. P. C. Pandey, E. G. Njoku, and J. W. Waters, Inferences of cloud temperature and thickness by microwave radiometry, *J. Clim. Appl. Meteorol.*, (22), 1894-1898, (1983).

Microw. Radiomet. Remote Sens. Earth's Surf. Atmosphere, pp. 263–270
P. Pampaloni and S. Paloscia (Eds)
© VSP 2000

Comparison of Advanced Microwave Sounding Unit observations with global atmospheric model temperature and humidity profiles and their impact on the accuracy of numerical weather prediction

CAROLINE POULSEN, STEPHEN ENGLISH, ANDREW SMITH, RICHARD RENSHAW, PAUL DIBBEN, PETER RAYER

Meteorological Office, Bracknell, RG12 2SZ, United Kingdom

Abstract—The Advanced Microwave Sounding Unit was launched on NOAA-15 in May 1998. This represented a very significant improvement in the information available from meteorological polar orbiting satellites over the previous Microwave Sounding Unit, particularly for humidity and vertical resolution of temperature in cloudy areas. In preparation for assimilation of the observations in to a three dimensional analysis of atmospheric temperature and humidity, the observations have been compared with calculated top of atmosphere brightness temperatures using a radiative transfer model and six hour forecast temperature and humidity profiles. This showed that the observations could be predicted to within 0.2K for some tropospheric channels. Early in 1999 a series of experiments testing the impact of assimilating AMSU observations on the accuracy of numerical weather prediction forecasts have been completed. These show that the use of microwave observations reduce forecast errors by well over 20% in the southern hemisphere and over 5% in the northern hemisphere. Further improvements have also been found from assimilating more data over the land.

1. INTRODUCTION

Between 1979 and 1998 the only microwave temperature sounding information on the civil polar orbiting satellites came from the Microwave Sounding Unit (MSU) which utilised the strong absorption by oxygen between 50 and 60 GHz to provide some information on atmospheric temperature in very deep layers. A similar instrument (SSM/T) flew on the Defence Meteorological Satellite Program. Despite the coarse resolution MSU has been and still is an important source of information in cloudy regions, especially in the southern hemisphere were there are few conventional soundings (radiosondes). In 1998 the Advanced MSU (AMSU) was launched on NOAA-15 which increased (with respect to MSU) the number of temperature sounding channels peaking below 50 hPa from 3 to 6. AMSU has 5 channels peaking above 50 hPa so it also takes over the function of the Stratospheric Sounding Unit (SSU); an infra-red pressure modulated radiometer that has operated on successive polar orbiting satellites since 1979. Furthermore a humidity sounding instrument, AMSU-B, was added which uses the water vapour line at 183.31 GHz to provide information on humidity structure, although problems with radio-frequency interference have prevented the full exploitation of AMSU-B [1]. MSU, together with the SSU and the High Resolution Infra-red Radiation

Communicating author (senglish@meto.gov.uk)

264   *C. Poulsen* et al.

Sounder (HIRS), formed a suite of instruments known as TOVS. The new suite, with AMSU replacing MSU and SSU, is known as Advanced TOVS (ATOVS).

The channel frequencies, noise characteristics and polarisation are listed in Table 1. For comparison the standard deviation of the differences between the observation "O" and the calculated brightness temperature "B" using a six hour short range forecast are listed for each channel ("O-B" – see section 3 for more details). Note that for many channels the O-B standard deviation is only marginally higher than the NEΔT. The results for channels 12-14 are however influenced by the use of a stratospheric extrapolation based on the observations themselves in the construction of the first guess. These O-B statistics are for single scan positions following quality control and correction of the AMSU-B radio-frequency interference problem. Note that the height of peak sensitivity for AMSU-B channels is highly variable with only channel 18 never peaking at the surface, although in very dry conditions even this channel will have a surface contribution to the measured brightness temperatures. This makes humidity retrievals from AMSU very sensitive to emissivity error, as reported by [2].

Channel	Channel Frequency MHz	Band width MHz	Measured NEΔT K	Clear air "O-B" Standard deviation	Angle of Polarisation	Typical peak altitude hPa
1	23,800±72.5	125	0.21	4.5	90-θ	Surface
2	31,000±50	80	0.26	3.0	90-θ	Surface
3	50,300±50	80	0.22	2.5	90-θ	Surface
4	52,800±105	190	0.14	0.4	90-θ	950
5	53,596±115	170	0.15	0.2	θ	750
6	54,440±105	190	0.15	0.2	θ	400
7	54,960±105	190	0.13	0.3	90-θ	250
8	55,500±87.5	155	0.14	0.5	θ	150
9	57,290.34±87.5	155	0.24	0.5	θ	85
10	57,290.34±217	78	0.25	0.4	θ	50
11	57,290.34±322.2±48	36	0.28	0.5	θ	25
12	57,290.34±322.2±22	16	0.40	0.4	θ	10
13	57,290.34±322.2±10	8	0.54	0.3	θ	6
14	57,290.34±322.2±4.5	3	0.91	0.1	θ	2.5
15	89,000±1,000	1,000	0.17	5.0	90-θ	Surface
16	89,000±900	1,000	0.40	6.0	90-θ	Surface
17	150,000±900	1,000	0.76	5.0	90-θ	Sfc-850
18	183,310±1,000	500	1.12	3.0	90-θ	800-300
19	183,310±3,000	1,000	0.73	4.0	90-θ	Sfc-500
20	183,310±7,000	2,000	0.94	4.0	90-θ	Sfc-650

Table 1. AMSU channel characteristics and typical altitude of peak sensitivity. The NEΔT and channel specifications are taken from [3].

A major application of MSU and AMSU observations is assimilation into numerical weather prediction (NWP) models in order to produce more accurate weather forecasts. There are three alternative methods to assimilate the information: 1. NWP-independent retrievals; 2. NWP-dependent retrievals; 3. Direct radiance assimilation. All three options are currently being used at different global NWP centres. At the UK Meteorological

Office the second option is currently operational. The method is described by [4]. The technique is a one dimension variational analysis (1D-var).

2. DESCRIPTION OF 1D-VAR

1D-var retrieves a new profile of temperature and humidity given the observed radiances and the current best guess of all the variables to which the observed radiances are sensitive. Most of these are available from a six hour NWP forecast, the `background`). It is also essential to have an estimate of the observation error, the background error and the error in the model which relates observed radiances to meteorological parameters. This retrieval is optimal in a mathematical sense [5]. However the analysis created following assimilation of the retrieval will be sub-optimal unless the correlation between the errors in the background and the retrieval are properly accounted for.

The calculation of top of atmosphere radiances is achieved using a fast radiative transfer model (RTTOV) based on work by [6-7]. For the microwave channels the fast model reproduces the brightness temperatures which would have been calculated using a line by line model for water vapour [8] and oxygen [9]. A fast ocean emissivity model is used [10] but fixed values are used for land (0.95) and sea ice (0.92).

The retrieved profiles are then used in a three dimensional variational assimilation to create a new analysis. Following initialisation this is then used to generate a numerical weather forecast out to 144 hours using the global atmospheric model [11].

3. COMPARISON OF OBSERVATIONS WITH MODEL BACKGROUND

As part of 1D-var it is necessary to carry out a forward calculation from the space of the temperature and humidity profiles to the space of the observation. Thus we get a difference between observations (O) and background (B) in observation space. To allow for observation bias this O-B difference is averaged over a long time period to calculate the mean bias at each scan position which is then subtracted from each observation to ensure that the ensemble of observations is unbiased with respect to the background. For some channels this correction varies with airmass [12]. Typical values of the standard deviation of O-B for the observations (over the sea only) which pass quality control are given in Table 1.

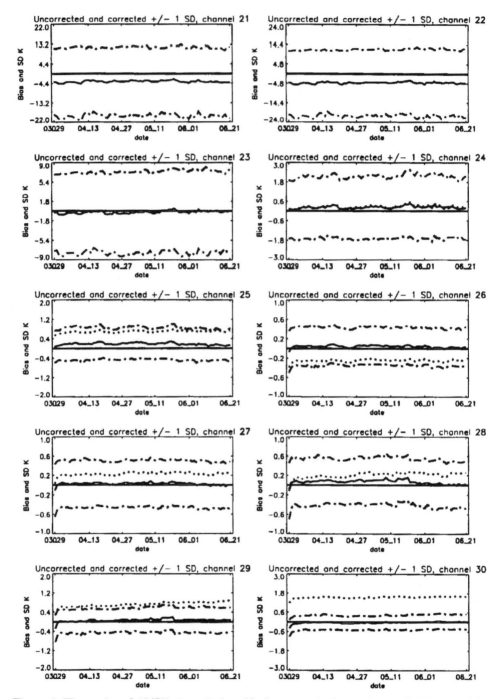

Figure 1. Time series of AMSU channels 1 to 10 observation-background from 29 March to 21 June 1999. Channels are defined in Table 1. Dotted line: O-B before bias correction; continuous line: O-B after bias correction; dash-dot line: one standard deviation around the mean bias. For channels with no dotted line no bias correction is applied.

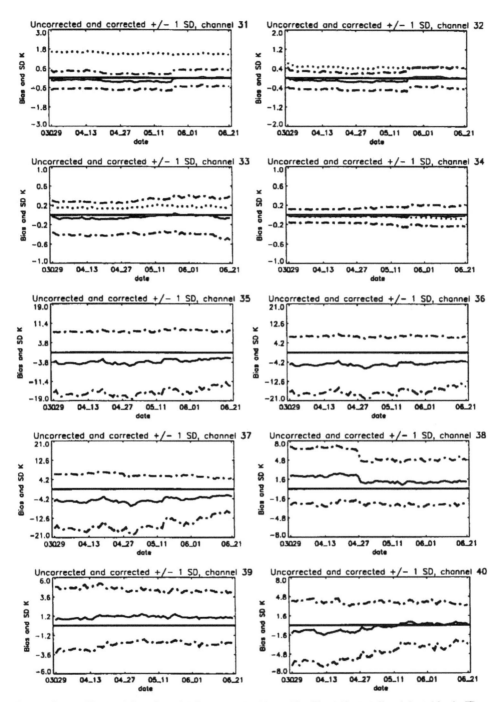

Figure 2. As Figure 1 but for AMSU channels 11 to 20. Channels are listed in table 1. The improvements for the AMSU-B channels (16-20) arise from the application of corrections to the RF interference provided by Met. Office Remote Sensing and NASA (N. Atkinson, pers. comm).

Figures 1 and 2 show the global mean O-B before and after bias correction and the standard deviation of O-B after bias correction for the period 29 March to 21 June 1999 for a complete set of AMSU channels and including land and sea observations. The standard deviations are much higher than those listed in Table 1 because of the inclusion of land data where surface emissivity and skin temperature are both less well known and there is no reliable quality control for cloud and rain. Only AMSU channels 4-14 are used operationally and over land no information is assimilated below 400 hPa. The other channels are not used either because they are very sensitive to cloud liquid water (which is not analysed) or suffer from radio-frequency interference. The assimilation of AMSU observations began on 29 March 1999 and a change in both the mean and random difference between the model background and the observations is evident in the first 2-3 days. Following that the global mean and standard deviation are very stable and very low, as low as 0.4K for several channels and if high land is excluded this falls to 0.2K (not shown). Typically the standard deviation is only slightly higher than the pre-launch measurements of instrument noise. This indicates that forward model noise is also very low and that the combined instrument and forward model noise (the total observation error as seen by the data assimilation system) is comparable with the errors in the NWP background. Clearly the lower the observation error compared to the background error the higher the probability that the observation will have positive impact on numerical forecast accuracy. If observation error is high compared to the background error (whether this error comes from the instrument or the forward model) the observation is unlikely to have a positive impact. The fact that background errors for temperature are so low imposes a severe discipline on the use of the data. Very small sources of additional error (e.g. from a sub-optimal retrieval method, poorly handled non-linearities or retrievals with complex error characteristics not easily handled by the assimilation system) would limit the usefulness of the observations. In the next section experiments are described which determine the impact of TOVS and ATOVS observations using the current two stage (and slightly sub-optimal) assimilation technique.

4.   OBSERVATION IMPACT EXPERIMENTS

ATOVS data was assimilated for AMSU channels 4 (52.8 GHz) to 14 (57.29 GHz) and for the same HIRS channels used for TOVS [4]. Quality control was applied to the AMSU data to prevent assimilation of the two lowest tropospheric channels when significant liquid water or precipitation was present [13]. The infra-red channels were quality controlled to prevent the use of all tropospheric channels when cloud was present [14]. 1D-var retrieved profiles were assimilated for all heights over the sea but only above 400 hPa over land.

Three global impact experiments were completed: 1. Replacing the TOVS observations from NOAA-11 with ATOVS observations from NOAA-15 (TOVS data from NOAA-14 was assimilated in both control and experiment); 2. Not using any TOVS or ATOVS radiances; 3. Using only SSU, MSU and AMSU radiances (i.e. no HIRS). All the experiments were run from 12 January to 9 February 1999. Bias corrections were recalculated after nine days.

The impact of the new microwave observations was to reduce forecast errors by 5-20%. By contrast only a small positive impact was found when HIRS data was added (1-2%).

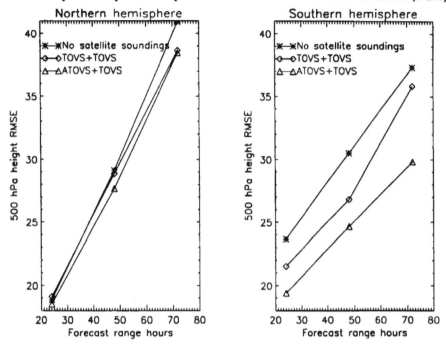

Figure 3. Impact of 1D-var (A)TOVS retrieval assimilation on 500 hPa height RMSE error in the northern hemisphere 24 to 72 hour forecast range measured against a control in which no sounding radiance information was assimilated.

Therefore the impact of the microwave observations is much larger than the infra-red observations and the overall impact of TOVS and ATOVS on the accuracy of NWP is very large.

Figure 3 shows the root mean square error for 500 hPa height in the first three days of the forecast for the northern and southern hemispheres respectively. The total impact of TOVS and ATOVS in the southern hemisphere is in excess of one day of forecast range and in the northern hemisphere around one quarter of a day. The impact of having one ATOVS and one TOVS is at least twice that of having two TOVS. The impact of ATOVS is largest in the southern hemisphere. There are two reasons for this. Firstly no data is used below 400 hPa over land; this restricts its use in large regions of the northern hemisphere. Secondly the forecast background error is larger in the southern hemisphere because there are fewer other observations.

A fourth experiment showed a 1% reduction in forecast error over the whole northern hemisphere if data was assimilated to the surface over Asia in a region 30E-130E 50N-70N. For short range forecasts (1-2 days) the impact over Siberia and the NW Pacific was 5-10% reduction in forecast root mean square error. This result shows that the total impact of TOVS and ATOVS observations in the northern hemisphere could be increased by less cautious use of the data.

5. CONCLUSIONS

On the basis of these experiments AMSU observations have been assimilated operationally at the UK Meteorological Office since 29 March 1999. The overall positive impact of the sounding radiances on the two current operational satellites (NOAA-14 and NOAA-15) is around 5% in the northern hemisphere and 20% in the southern hemisphere in terms of root mean square error. The most important contribution to positive impact is from the microwave sounding channels between 50 and 60 GHz. It is anticipated that the impact would be larger if the radiances were assimilated directly and used more extensively below 400 hPa over land. The latter has been confirmed by assimilating over land in a selected region of Asia.

6. REFERENCES

1. N.C. Atkinson and S. Mclellan. Initial evaluation of AMSU-B in orbit. *Microwave Remote Sensing of the Environment, Proceedings of SPIE, Beijing, China*, **3503**, 276-287 (1998).
2. S.J. English. Estimation of temperature and humidity profile information from microwave radiances over different surface types. Accepted by *J. Appl. Meteorol.* [in press] (1999).
3. R.W. Saunders. A note on the Advanced Microwave Sounding Unit. *Bulletin of the American Meteorol. Soc.*, **74**, 2211-2212 (1993).
4. A.J. Gadd, B.R. Barwell, S.J. Cox and R.J. Renshaw. Global processing of sounding radiances in a numerical weather prediction system. *Q. J. Royal Meteorol. Soc.*, **121**, 615-630 (1995).
5. C.D. Rodgers. Retrieval of atmospheric temperature and composition from remote measurements of thermal radiation. *Revs. of Geophys. and Space Phys.*, **14**, 609-624 (1976).
6. J.R. Eyre. A fast radiative transfer model for satellite sounding systems. *ECMWF Tech Memo.* **176**, (1991).
7. R.W. Saunders, M. Matricardi and P. Brunel. An improved fast radiative transfer model for assimilation of satellite radiance observations. *Q. J. Royal Meteorol. Soc.*, **125**, 1407-1426 (1999).
8. H.J. Liebe. MPM – An atmospheric millimetre-wave propogation model. *Int. J. Infrared Millimeter Waves*, **10**, 631-650 (1989).
9. P.W. Rosenkranz. Absorption of microwave by atmospheric gases. *Atmospheric remote sensing by microwave radiometry, M.A. Janssen, Ed., Wiley-Interscience*, (1992).
10. S.J. English and T.J. Hewison. A fast generic millimetre-wave emissivity model. *In microwave Remote Sensing of the Atmosphere and Environment, Proceedings of SPIE, Beijing, China*, **3503**, 288-300 (1998).
11. M.J.P. Cullen. The unified forecast/climate model. *Met. Mag.* 122, 81-94 (1993).
12. J.R. Eyre. A bias correction scheme for simulated TOVS brightness temperatures. *ECMWF Tech. Memo.* **186**, (1992).
13. S.J. English, R.J. Renshaw, J.R. Eyre, P.C. Dibben. The AAPP module for identifying precipitation, ice cloud, liquid water and surface type on the AMSU-A grid. *Tech. Proc. 9th International TOVS Study Conference, Igls, Austria, 20-26 Feb 1997; Ed. J.R. Eyre; Published by ECMWF, Reading, UK.* 119-130 (1997).
14. S.J. English, J.R. Eyre and J.A. Smith. A cloud detection scheme for use with satellite sounding radiances on the context of data assimilation for numerical weather prediction. Accepted by *Q. J. Royal Meteorol. Soc.* [in press] (1999).

Microw. Radiomet. Remote Sens. Earth's Surf. Atmosphere, pp. 271–279
P. Pampaloni and S. Paloscia (Eds)
© VSP 2000

Radiometric measurements in the 10/50 GHz band for application to radio wave propagation

CARLO CAPSONI[1], ADA V. BOSISIO[1] and M. MAURI[2]

[1] Politecnico di Milano, Dipartimento di Elettronica e Informazione, Milan, Italy,
[2] Centro Studi per le Telecomunicazioni Spaziali, CNR, Milan, Italy

Abstract - Since 1992 some radiometric equipment have been operating in the framework of SHF and EHF propagation experiments carried out with geosinchronous satellites at the experimental station located at Spino d'Adda, 30 km east of Milan. After a resume of the techniques used to validate the radiometric data both by using sounding profiles and by applying techniques of mutual consistency among the different brightness temperatures ad hoc developed, the work presents the statistical results of the analyses carried out on the data base obtained so far, like the cumulative distribution function (cdf) of brightness temperature T_b at the five radiometric frequencies and the cdfs of the cumulative integrated water vapor (V) and liquid water (L) contents. Moreover, the retrieval of V and L performed with the classic pair of brightness temperatures at 20 and 30 GHz has been compared with the one obtained by using the 20/50 GHz pair. The results have shown that also this last one allows the correct retrieval of the two quantities of interest.

1 INTRODUCTION

The experimental station at Spino d'Adda is well equipped with a lot of apparata, such as meteorological instruments, radar, satellite beacon receivers and radiometers. As far as radiometers are concerned, two of them, one single band at 13 GHz and one dual band at 23.8/31.65 GHz were installed in 1992 to operate with a co-located receiving station for the Olympus satellite, and a third three-band equipment at 22.2, 31.65 and 50 GHz was added in the middle of 1994 to operate with a co-located receiver station for the Italsat satellite. Since September 1993 all radiometers were pointed towards Italsat satellite, which means a propagation slant path with an elevation angle equal to 37.8°.

The brightness temperatures T_b measured by the equipment along the satellite paths were mainly used to infer extra attenuation due to gases and suspended liquid at the satellite frequencies. In this respect, radiometers are invaluable tools in radio propagation studies as they allow the evaluation of attenuation along a path, with respect to the free space, in non rainy conditions which is a quantity not obtainable by direct beacon measurements, but of basic importance for the design of low margin telecommunication systems. That means, radiometric measurements let us evaluate clear sky attenuation that we use to define a «zero dB» level and consequently to calibrate the satellite beacon receivers.

Since radiometric measurements are indirect ones, a very accurate calibration of the equipment is needed to obtain the desired precision. As defined during the Olympus satellite experiment [1,2], if frequencies are properly chosen, clear sky fade contribution can be calculated within 0.1 dB, if radiometers measure brightness temperature with an accuracy better than 1 K. The characteristics of our equipment, developed by Electronik Centralen and Farran Technology, fulfill this requirement as shown in Table 1.

To accomplish this objective, tip curves are performed and correction to data are then introduced, off-line, in a semi automatic way. A preliminary campaign, devoted to assess knowledge and confidence in tip curve calibration technique has helped us in deciding how often it has to be performed. As a conclusion, a weekly calibration was found to be sufficient for radio propagation studies.

Table 1.
Radiometers main characteristics. (EC) = Elektronic Centralen; (F) = Farran Technology.

Frequency (GHz)	13 (EC)	22.2 (F)	23.8(EC)	31.65 (EC & F)	50.2 (F)
Integration time (s)	1-32	1-32	1-32	1-32	1-32
resolution (K)	0.5	0.5	0.5	0.5	0.5
accuracy (K)	1	1	1	1	1
bandwidth (MHz)	150	150	150	200	250
beamwidth (deg)	1.8	2	2	2	2.05

2. DATA ANALYSIS

As briefly explained in the introduction, the results presented hereinafter have been collected in the framework of a measurement campaign aimed at studying radio wave propagation at frequencies above 20 GHz. The objective is to achieve a deeper knowledge about each frequency behavior and the relationship among them. Moreover, we are interested in combining measurements performed by various equipment to extract as much information as possible.

Because of the measurement technique, the impact of the weather condition and of all the troubles that may happen while recording, it is necessary to figure out some procedures to analyze raw data. In principle, they must allow to filter out artifacts and anything else wrong and to calculate the time duration of every period correctly measured. In this way, we can then deduce automatically the cumulative distribution functions of any parameter of interest by including only good quality data and by referring them to the proper time basis.

In this paragraph, we show the two steps-procedure that we have adopted to transform raw data in values that can be used without screening. The first step consists in eliminating all those data affected by problems which may derive from undue «obstacles» or phenomena along the propagation path. The second one is related to the removal of measurements affected by mechanical or electrical problems of the radiometric equipment.

2.1 Data preprocessing

The data preprocessing step appears to be mandatory since some anomalies and artifacts are evident in the recorded data stream. A list of the typical problems is shown in Table 2, together with the kind of procedure used to face them. They are spikes, because of people or birds crossing the propagation path, or the sun, which is a very «hot target», and, because of the pointing direction, is in the path twice a year. Moreover, correction to data has to be introduced in order to take into account changes in the calibration curve.

Table 2. Example of possible anomalies in data and removing techniques.

artifacts	detection & filtering techniques
spikes	Automatic
sun activity	Automatic
tip curve calibration	semi automatic
slow drift	Manual

2.2 Data validation

Provided that all undue measurements have been identified and withdrawn from the data stream, a second step, called data validation, is then recommended. In fact, some subtle artifacts may affect measurements as it happens, for instance, in some meteorological situations when a dew layer on the antenna feed or on the dish can occur. In this case, after the rainy event, there is some delay before the radiometers, or particular frequencies, start again to measure in a correct way, that is to follow temperature changes in a proper way. Data validation aims at revealing those kind of problems by checking if actual measurements are within the proper confidence range and if the brightness temperatures at different frequencies are consistent each other. From one side sounding data, combined with the Liebe's [3] propagation model, have been used to calculate statistically meaningful data bases of synthetic brightness temperatures at various frequencies and V and L values corresponding to different sky conditions [4].

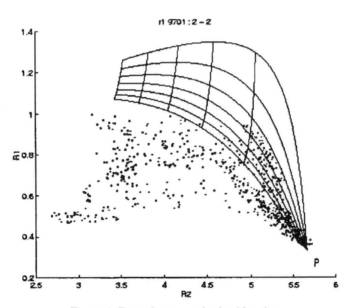

Figure 1 -Examples: cross check with ratios

From the other side theoretical evaluation of the relationships among the quantities of interest are used to cross-check the measurements. Figure 1 shows an application of temperature ratio which helps us in detecting wrong measurement.

The grid shown in the figure on the plane R1 R2 which are ratios between brightness temperatures (see eq. 2) represents the theoretical geometric loci of constant path integrated water vapor (V) and liquid water (L) content. The expressions of each line is derived from the radiometric expression of brightness temperature written as a function of atmospheric attenuation:

$$T_b = T_m - (T_m - T_c) \cdot 10^{-(a_v V + a_l L + a_o)/10}$$ (1)

where a_v, a_l and a_o are frequency dependent coefficients derived from statistical regressions over a lot of sounding data. The point P in figure represents the lowest bound corresponding to an atmosphere with V and L equal to zero.

It follows that measurements have to fall inside the grid for every physical situation, corresponding to an actual pair of V , L values. We have chosen the ratio between brightness temperatures as an indicator more suitable than the temperature itself. In fact, we have observed that a linear relationship exists between temperature measured at different frequencies in clear sky condition [4,5]. Since we are observing the same medium, the atmosphere, at frequencies which present a different sensitivity to different atmospheric phenomena, we can discriminate, in principle, the medium condition with this analysis. Starting from this observation, a set of ratios has been defined (see eq 2), where k_1 and k_2 are constants introduced for a better grid drawing and they are fixed for a given pair of frequencies:

$$R_1 = \frac{T_{30} + k_1}{T_{20}}; \quad R_2 = \frac{T_{30} + k_2}{T_{13}}$$ (2)

Therefore, if data had a physical sounding the cloud of points in Fig 1 should be inside the grid. On the contrary , we may appreciate a relative rotation between the grid and the cloud of points, suggesting that something wrong is happening at 30 GHz.

Hence, measurements corresponding to this event should be withdrawn from the database, unless a correction based on physical basis in applied.

3. STATISTICAL RESULTS

In this section, some cumulative distribution functions are shown concerning brightness temperature and the inferred quantities as V and L. Radiometers' elevation angle is 37.8°. Data are measured at 1 second rate. Cumulative distributions are normalized to total time.

3.1 Brightness temperature

Fig 2 shows the cumulative distribution functions (cdf) of brightness temperature collected during 1997, which is our best year since all radiometers were working together for the longest period. The observed year to year variation of cdfs (not shown for conciseness) is in agreement with the climate of our region. For instance, at 20/30 GHz, it is about some tens of kelvin, which corresponds to up to 2 dB in attenuation values. We may also appreciate that the 30 GHz and 50 GHz curves show a similar behavior; in fact, in the 100 - 2 % range the two lines have a similar behavior with a knee around the 10 % probability level which is not present on the other lines. The two frequencies seem to react analogously to the same physical phenomena, at least in the above mentioned probability range. This observation suggested us the opportunity of applying the classical dual frequency algorithm for V and L retrieval [6] using the 50 GHz data instead of the 30 GHz ones (see §3.2).

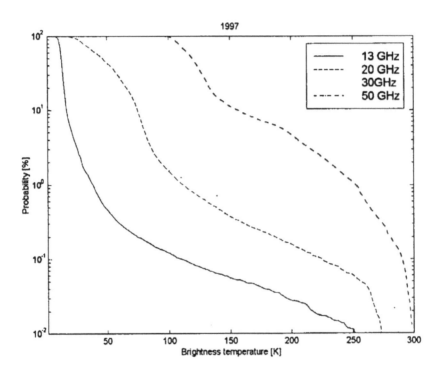

Fig 2 Cumulative distribution function of brightness temperature measured during 1997.

3.2 V and L retrieval

Applying the V and L retrieval algorithm to measurements, without any knowledge about weather condition introduces errors particularly during rainy events: V and L are then overestimated. In order to understand what happens, we have calculated the ratio between the instantaneous values of retrieved V and L. Fig 3 shows the corresponding cdf.

Except for the low level probabilities corresponding to rainy condition, we may appreciate the linear relationship between V and L. This facts warn us not to apply the retrieval algorithm to the highest temperature values, let say to those greater than 90 K, which corresponds to a probability of some percent according to the frequency.

As explained in the previous paragraph, V and L are retrieved with the classical linear algorithm based on two frequencies in the vapor and liquid water absorption bands [6]. The cumulative distributions are presented in Figs 4 and 5. Values are those calculated using brightness temperatures below 90 K for both frequencies:

Besides the 20 and 30 GHz lines, we also tested another pair of frequency (20/50). It is a preliminary test on the capability of this pair, and we applied this algorithm only over a subset of data, namely the first half of the April 1997. The coefficients are reported in Table 3 and the scatterplot of V retrieved with the 20/30 versus the 20/50 algorithm is shown in Fig, 6. The regression line build over those data gives us a correlation coefficient equal to 0.96, which represents a quite good agreement between the two approaches.

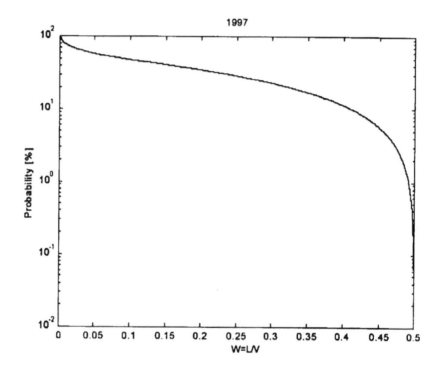

Fig 3 – Cdf of the ratio obtained by the instantaneous values of V and L retrieved by the whole stream of data over one year (1997)

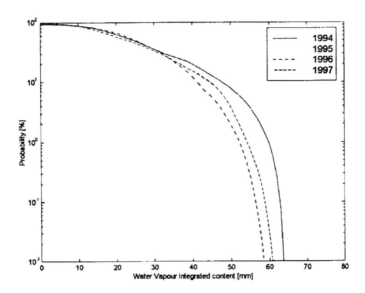

Fig 4 - Cumulative distribution function of Integrated Water Vapor content calculated over brightness temperature lower than 90 K.

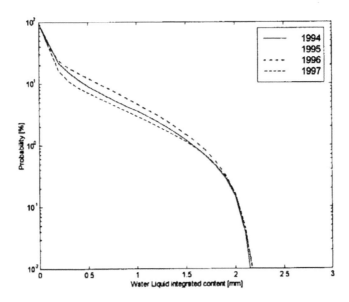

Fig. 5 - Cumulative distribution function of Integrated Liquid Water content calculated over temperature brightness lower than 90 K.

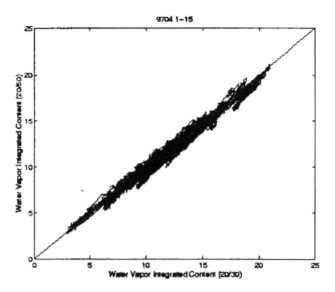

Fig. 6 - Comparison between Water Vapor Integrated Content calculated using 20 / 30 and 20 /50 GHz. The correlation coefficient is equal to 0.96.

Table 3 - Coefficients used in calculation of V and L starting from the classical dual frequency algorithm:

$$V = Cv_{20} * A_{20} + Cv_{30} * A_{30} + Dv$$
$$L = Cl_{20} * A_{20} + Cl_{30} * A_{30} + Dl$$

	A(23.1)	A(31.6)	D
V[mm]	50.27 [mm/dB]	-28.93 [mm/dB]	-1.06 [mm]
L[mm]	-0.72 [mm/dB]	0.09 [mm/dB]	-0.29 [mm]

4. CONCLUSIONS

Radiometers are a very powerful tool in radio wave propagation applications, in instances such as total path attenuation estimation due to oxygen, water vapor, liquid and rain along the path for which frequencies lower that 13 GHz are the most suitable [7], clear sky investigations and V,L retrieval.

As we have shown, data preprocessing is mandatory if we want to detect and filter out anomalies and artifacts in the measurements. Data validation, which is the subsequent step, is very useful to discriminate errors due to malfunctioning of the equipment, that is to say, non systematic errors.

Once the data have been processed, we are allowed a straightforward calculation of cumulative distribution functions. The last and critical step concerns V and L retrieval. We show a linear relationship between these quantities, limited to cloudy condition only. The retrieval of V and L is performed on a new pair of frequencies. The results are in agreement with those obtained at 20 and 30 GHz. This points to the 50 GHz line being very promising. The next step will be to apply the algorithm to the whole set of data.

REFERENCES

1. Opex, Second Workshop of the Olympus Propagation Experiments, Vol.3, *Reference Book on Radiometry and Meteorological Measurements*, Noordwijk, November 1994.

2. Mallet C., J. Lavergnat, «Beacon calibration with multi frequency radiometer, *Radio Science*, Vol.27, Nr 5, pp 661-680, 1992.

3 Liebe H.J., G. A. Hufford, M.G. Cotton, «Propagation modeling of moist air and suspended water/ice particles at frequencies 1000 GHz», *AGARD 52nd Specialist's Meeting of EM Wave Propagation Panel*, pp 3-1, 3-10, Advis. Group foe Aerosp. Res. And Dev., Brussels, Belgium, 1993

4. Bosisio A.V., C. Capsoni, «Effectiveness of brightness temperature ratios as indicators of the atmospheric path conditions», in *Microwave Radiometry and Remote Sensing of the Environment*, VSP International, ISBN 90-6764-189-8, 1995

5. Bosisio A.V., C. Mallet, «Influence of cloud temperature on brightness temperature and consequences for water retrieval«, *Radio Science*, Vol. 33, Nr 4, pp 929-939, 1998.

6. Westwater E.R, The accuracy of water vapor and cloud liquid determination by dual frequency ground-based microwave radiometry, *Radio Science*, Vol.13, N.4, pp 677-685, 1978.

7. Brussaard G., Radiometry. A useful prediction tool? ESA SP-1071 April 1985

3. Remote sensing of clouds and precipitation

3.1 Radiative transfer models

Microw. Radiomet. Remote Sens. Earth's Surf. Atmosphere, pp. 283–290
P. Pampaloni and S. Paloscia (Eds)
© VSP 2000

Effects of aspherical ice and liquid hydrometeors on microwave brightness temperature

E. MOREAU, C. MALLET, C. KLAPISZ

CETP, 10-12 Avenue de l'Europe, 78140 Vélizy, France

Abstract-. The objective of our activities is to develop a new algorithm for precipitation over ocean, with the use of a neural network model based on brightness temperature simulations. The quality of the simulated data base that we have built is determining to obtain a retrieval algorithm without bias. We therefore focus on the radiative transfer model and in particular on the hydrometeor modelization before performing inversion. Scattering by nonspherical particles of rain and ice at the TRMM Microwave Imager (TMI) frequencies is investigated in order to study its impact in term of brightness temperatures. The scattering calculations have been carried out with Waterman's Tmatrix approach. The brightness temperatures have been simulated using a plane radiative transfer model based on a discrete ordinate method. The atmosphere is supposed to be composed of plane parallel homogeneous layers. Our main interest is to quantify in terms of brightness temperatures, the impact of spheroid particles which lead to a more realistic modeling of natural collections of scattering particles. The rain drops modeled are oblate spheroids. The modeled results show that the difference with the brightness temperatures calculated with equivolumic spherical particles is less than 2 K for both 37 and 85.5 GHz for realistic rain rates. The ice layer is composed of spheroids particles randomly deformed. Different ice density are investigated. Brightness temperature depends strongly on the ice density. At 85.5 GHz, significant depolarizations are found. The implications of the results for the retrieval of precipitation are outlined.

1. INTRODUCTION

This study fits into a project which aims to find new retrieval algorithms to estimate the atmospheric water vapor, total water path, ice path, and surface rain rate from TMI measurements by simulating radiative transfer through cloudy and precipitating atmospheres. A statistical approach (using Neural Network, NN) has been selected for the retrieval. Indeed, this methodology has already showed great efficiency in the case of water vapor and cloud liquid water path parameters retrieval [1]. The ability of NN methods to retrieve atmospheric quantities for very different meteorological situations is connected to the representativeness of the data base [2]. A great care must be taken to build these simulated data so as no biases are introduced when the NN will be applied to real data. In this context, sensitivity studies about the microphysics must be done, especially on the influence of liquid and ice particles shape on the radiative transfer. In the millimeter and the centimeter wavelength range, the radiometric signal is related to cloud and rain microphysical properties. Thus the development of exact radiative transfer algorithms including reliable assumptions about atmospheric and cloud structures is crucial for the understanding in the radiometric signals. In the context of passive microwave observation from satellites, many investigators have presented sensitivity studies relating variable cloudy and rainy conditions to simulated brightness temperatures ([3] among others). They showed the dependence of the brightness temperatures and polarizations on increasing water paths and rain rates. Two effects are influencing the brightness temperatures in an opposite way: the absorption by the particles leads to an increase of the brightness

284   *E. Moreau et al.*

temperatures, while scattering has the opposite influence. This last effect increases with the size of the particles, and is more pronounced for ice than for liquid hydrometeors. In all the cases, those effects depend on the microphysics of the hydrometeors in the atmosphere.

The originality of this study with respect to previous ones ([4] among others) is the use of the Waterman's T-matrix approach [5] combined with an exact radiative-transfer algorithms as opposed to Eddington's [3] method for instance. A detailed influence of study of three different ice shape, with different bulk densities have been performed at high frequencies [6]. However, the results are only given for a particular case with fixed ice water content. The present study tries to quantify the different effects on brightness temperatures due to the bad knowledge of the microphysics of hydrometeors. Because we have a data base generation in mind, this sensitivity study will be, a the end, interpreted in terms of representativity.

The paper is organized as follow. Section 2 shortly describes the polarized radiative transfer model used to simulate brightness temperatures. Next, section 3 discusses the characteristics of the modeled liquid and ice hydrometeors. Section 4, presents the obtained results on simulated brightness temperatures. In section 5, an approach to take into account the results of this study in the data base is presented.

2. POLARIZED MICROWAVE RADIATIVE TRANSFER MODEL

This model considers the radiative transfer of polarized radiation through a horizontally stratified atmosphere, also called one dimensional plane-parallel model. The radiation intensities and the polarization state are described by the modified Stokes vector $I(z,\mu) = (I_v, I_h, U, V)$. The radiations considered here are natural, so $U \approx V \approx 0$. At microwave frequencies, the Rayleigh-Jeans approximation is used, and the intensity of radiation is expressed in term of brightness temperature $T_B = (\lambda^2/k)I$, where λ is the wavelength and k the Boltzmann's constant. The radiative transfer equation is written [7]:

$$\mu\frac{dT_B(z,\mu)}{dz} = -k_e(z,\mu)T_B(z,\mu) + k_a T(z) + \int_{-1}^{1} M(\mu,\mu')T_B(z,\mu')d\mu' \qquad (1)$$

where z is the height; μ', μ are the cosines of the incident and emergent zenith angles, respectively; $k_e(\mu)$ is the extinction matrix; k_a is the absorption vector and T(z) the environmental temperature. The matrix $M(\mu, \mu')$, describes the angle-dependent scattering of each vector $T_B(z, \mu')$ in the direction of μ.

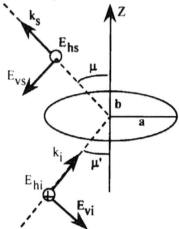

Figure 1. Schematic diagram of the scattering by an oblate drop. k represent the direction of the propagation of the wave. Ev and Eh the electric field components in the vertical and the horizontal polarization, respectively. Indicies i and s are for the incident and the scattered field components, respectively.

The components of the scattering matrix **M** are computed using the Mie method for spherical particles and using the Waterman's T-matrix method for nonspherical particles. The model uses the discrete ordinate-eigenanalysis method [8] to solve the radiative transfer equation. In this method the continuum of propagation direction is discretized into a finite number of directions (for these calculation, six angles per hemisphere were used) so that the integro-differential equations are converted into a system of ordinary differential equations with constant coefficients, the solution of which can be calculated by eigenanalysis.

The water vapor and oxygen absorption are computed with Liebe's model developed in 93 [9] and the cloud liquid absorption model uses the Rayleigh approximation. For modeling sea surface emissivity the parametrization of Hollinger [10] as a function of temperature and near-surface wind speed was implemented.

3. MICROPHYSICS OF LIQUID AND ICE HYDROMETEORS

The liquid hydrometeors are assumed to be oblate spheroids (Fig. 1). The canting angle of the particle depends on the drop size and increases with the size of the drop. Nevertheless the mean value of the canting angle distribution is near zero degrees [11]. It is assumed that liquid hydrometeors are oriented so that their larger dimensions are horizontal. The axial ratio defined like the ratio of the minor to the major axis (b/a) of the particle depends on the drop size, generally expressed as an equivolumetric radius ($\bar{r} = (b^2a)^{1/3}$). The axial ratio decreases with the increase of the size of the drop to take into account the more flattened shape of the bigger drops. The Morrison&Cross [12] relationship is used for simplicity reason. In fact, we have found that the computed brightness temperatures depend weakly on the choice of this relationship. An exponential shape drop size distribution has been chosen with the form,

$$N(D) = N_0 \exp(-\lambda D) \tag{2}$$

where N_0 and λ are the distribution parameters set as in Joss [13]. The values of N_0 and λ have been chosen because the concentration of the bigger drops is relatively higher than for other drop size distributions. Because the bigger drops have an axial ratio lower than the smaller drops, the use of the Joss drop size distribution maximise the non-spherical particles effects.

Ts	300 K	(surface temperature)
V	0 m/s	(near surface wind speed)
θ	52.8 °	(incident angle)

Height [km]	Temperature [K]	Liquid Rr [mm/h]	Ice Rr [mm/h]	H [%]	CLW [g/m³]
1.6	291.20	X	0	98	0.1
3.2	282.20	X	0	98	0.1
4.8	273.20	X	0	98	0.1
6.4	264.30	0	X	98	0.0
8.0	255.33	0	X	98	0.0
9.6	246.36	0	X	98	0.0

Tab 1. Atmospheric parameters used for microwave radiative transfer modeling. H is the relative humidity. CLW is the cloud liquid water content. Rr is the equivalent rain rate for liquid and ice respectively. X varies between 1 and 30 mm/h.

A wide variety of ice particles happens in the atmosphere. The most striking feature of the ice phase in clouds is the extreme diversity and complexity of the crystal habits [14] which leads to some uncertainty in their morphological and aerodynamical properties. So in order to elucidate the impact of the ice phase in terms of brightness temperatures, unavoidable but necessary assumptions on the shape and on the density of ice must be done. It is assumed that ice hydrometeors are oblate spheroid in shape and oriented so that their larger dimensions are horizontal [14]. To take into account the wide variety of shapes, the axial ratio for each drop size is randomly set between 0.5 and 1.0. The ice layers are so modeled by oblate spheroids randomly deformed. Two different bulk densities (0.91 and 0.4 g/cm^3) that corresponds to pure ice and graupel respectively, were used. The ice particles were assumed to be randomly oriented in the horizontal plane. In the atmosphere, the drop size distribution for ice crystals varies according the type of the ice. For simplicity reason, an exponential type of drop size distribution is assumed with the Marshall Palmer coefficients [15]. And in order to preserve the mass between the rain and the different type of ice, the intercept parameter of the drop size distribution was normalized by the bulk density [16]. The size distribution was actually made up of thirty radius intervals from 0.02 to 3.0 mm. For the low-density particles the index of refraction is reduced according to the Lorentz-Lorentz mixing rule.

4. RESULTS

In the first test the effects of aspherical rain drops without any additional ice layer is considered. Precipitating systems are considered over ocean. Table I details the atmospheric properties. Fig 2 shows the upwelling brightness temperatures at 37 and 85.5 GHz as a function of the surface rain rate for oblate spheroids and spheres of equivalent volume. The brightness temperatures in the vertical polarization for the sphere and the spheroid assumption are very close, the maximum difference is less than 1 K. The brightness temperature difference between the two shape assumptions is less than 2.5 K for the horizontal polarization. In the case of the classical Marshall Palmer drop size distribution the difference between spheroid and sphere is less than 2 K for both polarization. For the horizontal polarization, the brightness temperatures for oblate spheroids are always smaller than for the corresponding spheres which indicates that the scattering cross section is greater for spheroid that for sphere drops. At incident angle of 52.8°, spheroid drops exhibit larger geometrical cross section than sphere drops. It is interesting to notice that for 37 and 85.5 GHz the depolarization (Tbv -Tbh) is bigger for the spheroids than for spheres. At 37 GHz , the depolarization is about 6 K for high rain rate. Modeling using liquid spheres shows none of the interesting polarization characteristics that oriented liquid particles presents.

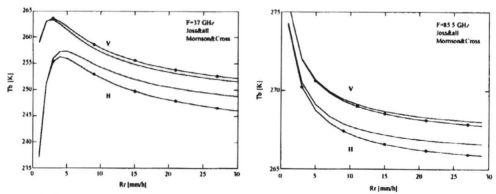

Figure 2. Vertically and horizontally polarized simulated brightness temperatures at 37 (left) and 85.5 GHz (right) versus the surface rain rate. Solid lines stand for the sphere case and solid lines with dots stand for the spheroid case.

In the second test the effects of aspherical ice particles is considered. We consider one rain cell between 0 and 5 km and one ice cell just above. In the rain cell, sphere particles are used. The atmospheric properties are detailed Table 1. For the lower frequency channels negligible effects are observed. Fig 3 shows the upwelling brightness temperatures at 85.5 GHz as a function of the surface rain rate for oblate spheroids and spheres of equivalent volume with the two bulk ice densities. The brightness temperatures computed with oblate spheroids are lower than those computed with spheres for both polarizations. These differences are less than 3 K for the vertical polarization and less than 15 K for the horizontal polarization, for both ice densities. The biggest difference in the horizontal polarization is due to the fact that the ice particles are horizontally oriented. For horizontally oriented particles the depolarization is positive, indicating more vertically polarized than horizontally polarized radiation. The depolarization is quite important, even for relatively small surface rain rate. Results are similar to the liquid case but with stronger effects.

There is a strong dependence of the brightness temperature depression on the particle bulk density. The refractive index of ice varies with density, its real part falling almost linearly from a value of 1.78 for a density of 0.91 g/cm^3 to 1.32 for a density of 0.40 g/cm^3. In all cases the imaginary part remains very small.

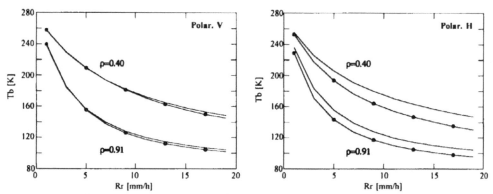

Figure 3. Vertically (left) and horizontally (right) polarized simulated brightness temperatures at 85.5 GHz for two bulk densities versus the surface rain rate. Solid lines stand for the sphere case and solid lines with dots stand for the spheroid case. ρ corresponds to the bulk density of the ice particles in g/cm^3.

5. DISCUSSION AND CONCLUSION

Our aim is to build a simulated data base in which the actual variability of the rain situation is kept. Because of the lack of knowledge concerning the hydrometeors morphological properties, necessary assumptions must be done. However, these different assumptions must not introduce a bias nor limit the representativeness of simulated data.

In this paper we assume a plan parallel homogenous atmosphere. However, following this study, another one is performed on the effects of horizontal heterogeneous atmosphere on simulated brightness temperatures [17].

A wide variety of vertical structures of atmospheres, clouds, and rains must be simulated to enable global application. These data are provided by the global circulation model (GCM) of the European Center for Medium Range Weather Forecasts (ECMWF) on a grid of 0.5deg$_x$0.5deg. The use of this kind of profiles makes us certain of a great variability of atmospheric vertical structures. Because precipitating quantities are diagnosed, we don't have any information concerning microphysical descriptions of hydrometeors.

For liquid rain particles, sphere assumption is quite good for weak rain rate, but for rain rate higher than 10 mm/h error is about 2.5 K. In these last cases, non spherical particles are used. Concerning the density of ice particles, the effects are very important. So the ice density is set for each of the simulated atmosphere at a random value between 0.9 and 0.1 g/cm^3. The so created database is representative of many different meteorological cases.

Concerning the shape of ice particles, the difference between the sphere and the spheroid assumption can lead to 15 K for the horizontal polarization for the 85.5 GHz channel. However ice can have many different shapes. If we modelize ice with a particular shape, the data base won't be representative of the natural variability of ice particles shape. Because of the lack of information and the difficulty encountered to modelize different ice particles shape, we modelized ice particles by spheres.

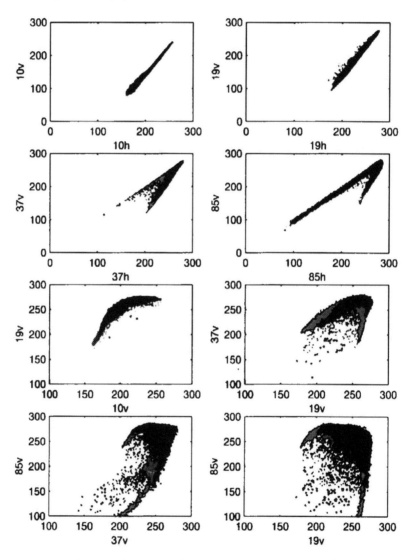

Figure 4. Scatterplots of the simulated upwelling brightness temperatures of the data set (grey points) and TMI data (black points) which corresponds to a mesoscale convective system over the central pacifique ocean (Orbit 1033, 98/02/01).

Because the data base will be used for a neural network type retrieval method, its representativity in terms of ice particles shape is improved with the addition of a random noise on the 85.5 GHz channel. This random noise varies between 0 and 3 K for the vertical polarization and between 0 and 15 K for the horizontal polarization. It's seems paradoxical to take into account the non spherical raindrops, for high frequency (85.5 GHz) when at the same time a random noise is added to take into account errors made on the ice particles shape. However, at 37 GHz, the effects of the ice particles shape are negligible, and the non-sphericity effect of rain drops must be taken into account. In fact to generate the data base, the microphysics of hydrometeors is the same for all the channels.

The conclusions of this study have been used to simulate a wide data base including 3000 different precipitating situations. A comparison with measured TMI brightness temperatures has shown a good agreement (Fig. 4), no bias has been observed, and the variability of the brightness temperatures is realistic. The brightness temperature of the observed scene exist in the data base which indicates the good representativity of the later. This type of verification will be performed extensively in the future before performing the retrieval with a neural network. This kind of retrieval method is highly sensitive to "gaps" in the data base.

REFERENCES

1. E. Moreau, C. Mallet, and C. Klapisz. Determination of integrated cloud liquid water and total precipitable water using a neural network algorithm. *Proc. of SPIE First Specialist Meeting of Microwave Remote Sensing of the Atmosphere and Environment*, **3503**, 135-143 (1998).

2. D. Tsintikidis, J.L. Haferman, E.N. Anagnostou, W.F. Krajewski and T.F. Smith. A neural network approach to estimating rainfall from spaceborne microwave data. *IEEE Trans. Geosci. Remote Sens.*, **35**(5), 1079-1093 (1997).

3. C. Kummerow and J.A. Weiman. Determining microwave brightness temperature from precipitating horizontally finite and vertically structured clouds. *J. Geophys. Res.*, **93**(D4), 3720-3728 (1988).

4. R. Wu and J.A. Wieman. Microwave radiances from precipitating clouds containing aspherical ice, combined phase, and liquid hydrometeors. *J. Geophys. Res.*, **89**(D5), 7170-7178 (1984).

5. P.C. Waterman. Matrix formulation of electromagnetic scattering. *Proc. IEEE*, 53, 8, 805-812 (1965).

6. K.F. Evans and J. Vivekanandan. Multiparameter radar and microwave radiative transfer modeling of nonspherical atmospheric ice particles. *IEEE Trans. Geosci. Remote Sens.*, **28**(4), 423-437 (1990).

7. S. Chandrasekhar, *Radiative Transfer*, Dover Publications, Inc., New York (1960).

8. L. Tsang J.A. Kong, R.T. Shin. *Theory of Microwave Remote Sensing*. Wiley, New-York (1985).

9. H.J. Liebe G.A. Hufford, and M.G. Coton. Propagation modeling of moist air and suspendede water/ice particles below 1000GHz. *Proc. of AGARD Fifty-Second Specialists Meeting of Panel on Electromagnetic Wave Propagation*, 3-1-3-10 (1993).

10. J.P. Hollinger. Passive microwave measurement of sea surface roughness. *IEEE Trans. Geosci. Electron.* 9, 165-169 (1971).

11. J. Tan, A.R. Holt and D.H.O Bebbington. Radar estimation of apparent canting angle in raindrops. *Elect. Lett.*, **28**, 944-945 (1992).

12. J.A. Morrison, and M.J. Cross. Scattering of a plane electromagnetic wave by axi-symmetric raindrops. *Bell Syst. Tech. J.*, 52, 955-1019 (1974).

13. J. Joss, J.C. Thams, and A. Waldvogel. The variation of raindrop size at Locarno. *Proc. Int. Conf. on Cloud Physics*.369-373 (1968).

14. H.R. Pruppacher, and J.D. Klett. *Microphysics of Clouds and Precipitation*. Boston, Mass.: D.Reidel (1978).

15. J.S. Marshall and W.M.K. Palmer. The distribution of rain drops with size. *J. Meteorol.*, 5, 165-166 (1948).
16. J. Vivekanandan, J. Turk and N. Bringi. Ice water path Estimation and characterisation using passive microwave radiometry. JAM, 30, 1407-1421 (1991).)
17. C. Mallet, C. Klapisz, N. Viltard. Effects of heterogeneous precipitating atmospheres on simulated brightness temperatures. *Sixty Specialist Meeting on Microwave Radiometry and Remote Sensing of the environment.* 15-16 March 1999, Firenze, Italy.

Microw. Radiomet. Remote Sens. Earth's Surf. Atmosphere, pp. 291–298
P. Pampaloni and S. Paloscia (Eds)

Effects of heterogeneous precipitating atmospheres on simulated brightness temperatures

CECILE MALLET, CLAUDE KLAPISZ and NICOLAS VILTARD

CETP, 10-12 Avenue de l'Europe, 78140 Vélizy, France

Abstract The three-dimensional (3-D) structure of raining clouds deeply affect the simulated brightness temperatures at microwave frequencies. Difficulties that arise considering the view angle (53°) of conical scanning satellite radiometers are here analysed by means of a radiative transfer model and a cloud model. The radiative transfer model is based on Eddington's approximation used in a 3-D geometry corresponding to the TRMM Microwave Imager (TMI) . The cloud model is a three-dimensional mesoscale non-hydrostatic cloud model (MESO-NH). The proper treatment of heterogeneous atmospheres effects is fundamental to generate a reliable cloud-radiation database for the development of retrieval algorithm in using a neural network approach. In particular the effects of the downward directed radiation paths has to be taken into account for the lower frequency channels.

1. INTRODUCTION

Satellite-borne microwave radiometers are useful tools in remote sensing of atmospheric variables, especially over oceans where ground-based measurements are scarce. In particular the amount of precipitation can be obtained from different sensors such as the Scanning Multichannel Microwave Radiometer (SMMR), more recently the Special Sensor Microwave/Imager (SSM/I) and the TMI; coming next is the Advanced Microwave sensing Radiometer (AMSR). Our aim is to develop new retrieved algorithms in using a neural network approach. Artificial neural network must be taught how to process inputs before they can be used in an application. That's why, the main point in this approach is the need of an important database. This database has to contain the multifrequencies brightness temperatures on the one hand and the geophysical parameters to be retrieve on the other hand. Owing to the lack of in situ measurements over ocean we have decided to use simulated data. It is necessary to simulate brightness temperatures to study their sensivity to the geophysical quantities we wish to retrieve.

In this study we simulate brightness temperatures corresponding to TMI measurements. A description of the atmospheric variables that may arise from a cloud model simulation is necessary: vertical profiles of pressure, temperature, water vapor, liquid cloud content, precipitating and/or non-precipitating ice content and rain are necessary. Surface wind speed used in surface emissivity models and vertical temperature profiles are also needed. However, as explained in Kummerow and Weinman [1], one major problem is the extreme spatial variability of rain. To account for the horizontal as well as the vertical variability over scale lengths which are often smaller than the radiometer footprint size, the cloud model should describe the atmospheric variations in vertical and horizontal directions. The radiative transfer model is adapted to correspond to TMI characteristics, so the same view angle, frequencies, footprint sizes are used.

The contribution of the downward radiation reflected by the surface is studied on a particular case that corresponds to a tropical squall line. The influence of the integrated water content along the downward and the upward radiation paths is considered for different TMI frequencies. Different positions of the satellite in relation to the rain cells are considered.

In section 2, the 3-D model used for the simulations is presented. In section 3, the 3-D geometry of the problem is described. The results obtained with the radiative transfer model are presented and discussed in section 4. Section 5 gives the conclusion of the present study.

2.   SIMULATION AND CASE STUDY

The cloud model used in this study, (Meso NH Atmospheric Simulation System) is a joint effort of the Centre National de Recherches Météorologiques (Météo-France) and the Laboratoire d'Aérologie (CNRS) [2]. It comprises a numerical model able to simulate atmospheric motions, ranging from the large meso-scale down to the microscale, with a comprehensive physical package.

The horizontal simulation domain extends 150 km along the latitude axis and 100 km along the longitude axis. A mesh size of 1.25 km is chosen in each of the horizontal directions. Vertically, a stretched grid is used, varying from 70 m at ground level to 700 m at the top level (20 km). The basic atmospheric variables are temperature, pressure, density of moist air and various mixing ratios. Some of these ratios are used here, after conversion into contents $(g\ m^{-3})$

The case we have selected for our study represents a tropical squall line observed on the 22 February 1993 during TOGA-COARE (Tropical Ocean Global Atmosphere-Coupled Ocean Atmosphere Response Experiment). This case was an eastward fast propaging oceanic squall line characterized by a 100 km long convex north-south leading edge. The leading edge is oriented, during its more linear stage, roughly perpendicular to the low level wind shear [3].

The TMI brightness temperatures corresponding to the meteorological situation previously described are thus computed. Figure 1 shows the footprints of TMI channel at 19.35 GHz on a schematic view of the cloud model. The line AB materialize the satellite line of sight. Figure 2 shows the microphysical output of the model along the line AB, depicted in Figure 1.

For different positions of the satellite in relation to the cloud model (see section 3.) we simulate the TMI brightness temperatures. For these simulations we have modified a plane-parallel model using Eddington's second approximation of radiative transfer [1,4,5] to obtain a modified 1-D model. In the modified 1-D model, we performed two 1-D calculations using clouds of different vertical structure. For the calculation of the upward directed path (UDP) radiation, the vertical profile of the hypothetical cloud is given by an intersection of the line of sight from the radiometer to the surface through the real cloud. For the calculation of the downward directed path (DDP) radiation, which is reflected in the surface, a horizontally homogeneous cloud is assumed. The vertical profile of the hypothetical cloud is defined by the intersection of the specular reflected line of sight through the real cloud. A part of this DDP radiation is reflected and it contributes to the

UDP. These two hypothetical clouds are represented in Liu et al. [6] for simplified cloud geometry.

The large differences between a plane-parallel model and 3-D radiative transfer calculations for off-nadir observations are mainly due to the effects of geometry. A modified 1-D model approximated 3-D conditions by slant-path simulations for both downwelling and upwelling directions. To a large extent the 3-D effects of radiative transfer can be accounted for by much less time -consuming modified 1-D modeling. For the original cloud model resolution, early studies [6,7] have shown that the result of this modified 1-D model is much closer to the 3-D model than the plane parallel model. In fact, for a radiometer resolution like TMI or SSM/I the obtained difference between the spatially averaged radiation temperatures from modified 1-D and 3-D modeling is only 1 to 3 K.

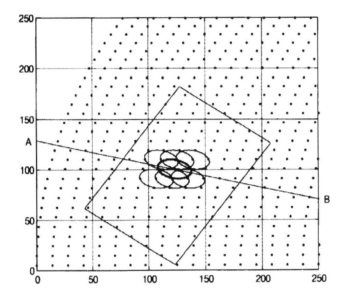

Figure 1. Schematic view of the MESO-NH domain (box), ground sampling (dots) of the TMI moving from right to left along a ground track coincident with x axis, and footprints of TMI channel at 19.35 GHz (ellipses). The straight line AB is the ground plan of the line between the central TMI pixel and the satellite.TRMM The axes are in km.

3. GEOMETRY OF THE PROBLEM

The radiative transfer model has been adapted to correspond to TMI particular view angle and frequencies [8]. In Figure 2, the vertical cross sections are standing in the plane containing the straight line AB of Figure 1. The distances on the x axis of the four pictures are calculated along the straight line and the distance 0 km corresponds to the center of the pixel represented by the thick ellipse in Figure 1. The top left panel shows the ice content as computed by the MESO-NH model. The precipitating ice is located mostly in the tailing stratiform region of the system. In the top right part of Figure 2, the liquid cloud presence corresponds to the updrafts associated with the leading convective region. The cell is moving from the left hand side to the right hand side of the picture. The bottom left picture shows the rain. The maximum rain is located at 20 km from the center and is part of a

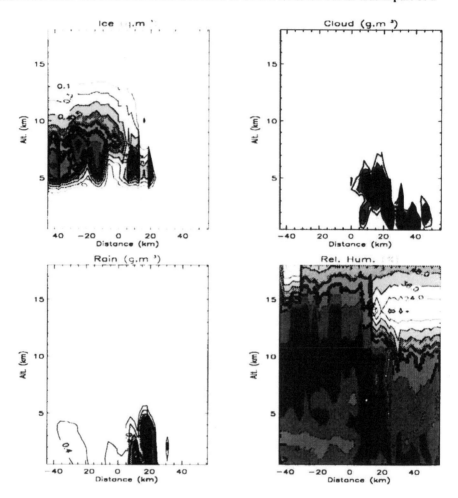

Figure 2. Vertical cross-section in MESO-NH domain simulation.

convective cell. The statiform region extends from 10 km to -40 km with light rain due to the ice particles falling from above. The last picture is for relative humidity (%). Where the relative humidity is equal or above 100% the cloud model produces liquid cloud below the freezing level (4.7 km) and ice cloud above it. All the following calculations are made in the vertical plane of this cross section.

The purpose of Figure 2 is to present the structure of the different fields of interest. First, it is noticeable that ice precipitation (upper left panel) is maximum in the stratiform part (up to 1 g.m-3). On the other hand, the cloud liquid water (upper right panel) is maximum in the convective region (up to 3 g.m-3). The model doesn't produce any liquid cloud in the stratiform region which is consistent with observations. The heavy liquid precipitation (bottom right panel) is concentrated also in the convective region (up to 5 g.m^{-3}). On the other hand, falling ice gives light precipitation in the stratiform region, with some signs of evaporation. Finally, the relative humidity (bottom right panel) exhibits saturation everywhere in the precipitation and cloudy regions.

4. RESULTS AND DISCUSSION

In the modified 1-D model, for each viewing position we performed two 1-D calculations using clouds of different vertical structure. One vertical structure corresponds to the DDP and the other one to the UDP. These two hypothetical clouds correspond to different integrated water contents. For the same altitude, the UDP and DDP do not

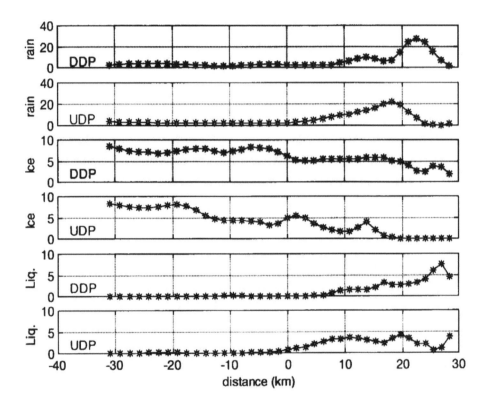

Figure 3. Integrated liquid precipitating water (rain), integrated precipitating ice (Ice) and integrated cloud liquid water (Liq.) along the downward directed path (DDP) and upward directed path (UDP). The integrated quantities are in kg m^{-2}.

interact with the same quantities of water. The integrated water contents (kg m^{-2}) along the UDP and DDP are thus different as we can see in Figure 3. The variations of the integrated contents agree with the vertical cross sections presented in Figure 2.

The variation of horizontally polarized brightness temperatures for a modified 1-D model and for a plane-parallel model are shown in Figure 4 for 10.6 GHz and 85.5 GHz

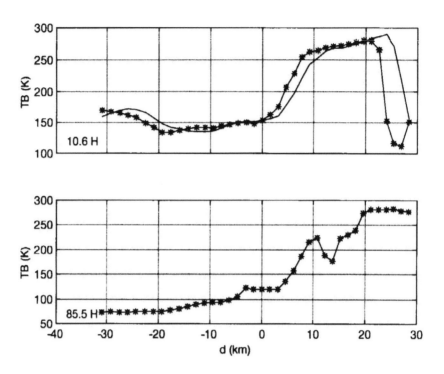

Figure 4. Simulated, horizontally polarized brightness temperature for a modified 1-D model (-)
and for a plane parallel model (-*)

at the original cloud model resolution. As explained in section 2, so the two models used
the same approximations concerning the radiative transfer equation, the differences are
only due to the effects of geometry .In fact, for the plane parallel model, only the
atmospheric profile that corresponds to the UDP is considered, when for the modified 1D
model the profile used to compute the downward radiation is modified. Because the
higher-frequency channels are more sensitive to the ice contents, within the higher layers,
the contribution of DDP radiation is negligible. That's why the two curves are
superimposed in the bottom picture. On the contrary, the contribution of the DDP
radiation to the total signal becomes important at low frequencies, especially when the
UDP goes through clear air layers.

The differences of horizontally polarized brightness temperatures between the two
models have been also computed after application of antenna patterns. These were
approximated by Gaussian functions determined by the 3 dB footprint size corresponding
to TMI characteristics [8]. The results, not presented here, lead to same qualitative
conclusions. Two combined effects lead to frequency dependent conclusions. In fact, the
low frequencies are sensitive to DDP radiation and they corresponds to footprint wider
than the rain cell. For example for the 10.7 GHz channel the along-track footprint size is
63 km and it has to be compared with the horizontal extend of the structure represented in
Figure 2. The obtained difference due to geometric effects is more than 40 K. On the
contrary, the 85.5 GHz channel isn't sensitive to DDP radiation and corresponds to an
along-track footprint size of 7 km. No differences are observed.

5. CONCLUSION

The measured brightness temperatures depend on the integrated atmospheric variables contained within the antenna beam. Because the lowest frequencies are more sensitive to emission effects and have a larger footprint size, they may be dramatically affected by the downard radiation contribution. This effect becomes even more critical in convective situations which exhibit a high spatial heterogeneity.

When retrieving the surface rainrate from radiometric measurements, the results may be strongly biased. Corrections that are applied to take the beamfilling errors into account [7] usually correct for the clear-air region contribution but neglect the downward radiation contribution.

So as, to take all these problems into account in future rain retrieval techniques, it seems necessary to develop both strongly heterogeneous convective situations and a radiative transfer scheme accounting for finite cloud geometries in the used database.

REFERENCES

1. C. Kummerow and J.A. Weinman. Determining microwave brightness temperatures from precipitating horizontally finite and vertically structured clouds. *J. Geophys. Res.*, **93**, (D4), 3720-3728 (1987).

2. J-P Lafore, J. Stein, N. Asencio, P. Bougeault, V. Ducrocq, J. Duron, C. Fisher, P. Héreil, P. Mascart, J-P Pinty, J-L Redelsberger, E. Richard, and J. Vilà-Guéreau de Arellano. The Meso-NH Atmospheric Simulation System. Part I: Adiabatic formulation and control simulation. *Ann. Geophys.* **18**, 90-109 (1998).

3. T. Montmerle. Validation et initialisation d'un modèle tri-dimensionnel méso-échelle non-hydrostatique par des données expérimentales issues de TOGA-COARE. *Thèse de l'Université Paris 6* (1998).

4. C. Kummerow. On the accuracy of the Eddington approximation for radiative transfer at microwave frequencies. *J. Geophys. Res.*, **98**, 2757-2765.(1993).

5. L. Roberti , J. Haferman, C. Kummerow. Microwave radiative transfer through horizontally inhomogeneous precipitating clouds. *J. Geophys. Res.*, **99**, 707-718 (1994).

6. Q. Liu, C. Simmer, and E. Ruprecht. Three-dimensional radiative transfer effects of clouds in the microwave spectral range. *J. Geophys. Res.*, **101**, 4289-4298, (1996).

7. P. Bauer, L. Schanz and L. Roberti. Correction of three-dimensional effects for passive microwave remote sensing of convective clouds. *J. Appl. Meteor.*, **37**, 1619-1632 (1998).

8. C. Kummerow, C. Barnes, T. Kozu, J. Shiue and J. Simpson. The Tropical Rainfall Measuring Mission (TRMM) sensor package. *J. Atmos.Oceanic Technol.*, **15**, 809-817 (1998)

9. C. Klapisz, N. Viltard and V. Marécal. Effects of field of view inhomogeneities on the rain estimation from microwave radiometry. *Int. J. Remote Sensing*, **19**, 211-215 (1998).

Microw. Radiomet. Remote Sens. Earth's Surf. Atmosphere, pp. 299–311
P. Pampaloni and S. Paloscia (Eds)
© VSP 2000

Determination of cloud structure from spaceborne microwave observations

IRENE G. RUBINSTEIN

Earth Observations Laboratory, Centre for Research in Earth and Space Technology, York University Campus, 4850 Keele St., Toronto, Ontario, Canada, M3J 3K1

Abstract - Microwave remote sensing has become an important tool for monitoring the atmospheres and surfaces of planetary objects, with increased applications in meteorology. The increasing need for the atmospheric structure information to provide initial conditions for numerical weather predictions and for near-real time applications has made it necessary to develop methodology for retrieval of cloud structure parameters. The retrieval of these parameters from the observed microwave radiances involves the use of inversion methods and/or statistically derived classification rules. To circumvent the problem of ill-conditioning associated with inverse problems, constraints on the solution are introduced. These constraints are derived using statistical database or using forward radiative transfer modelling. Simulated brightness temperatures for a broad range of meteorological conditions were generated using scattering, absorption, and emission properties of atmospheric constituents. This database was used to generate retrieval algorithms for the precipitation and possible cloud ice contents associated with convective and stratiform cloud systems. In addition, simulations of winter precipitation over snow covered land were carried out. Test algorithms for detection of winter precipitation were evaluated using SSM/I data for several winter storm events in Canada. One of the main objectives in carrying out the analysis was to determine what type of information is required to optimize the use of spaceborne passive microwave observations.

1. INTRODUCTION

The first in the series of 7 Special Sensor Microwave Imager (SSM/I) aboard a Defense Meteorological Satellite Program (DMSP) polar orbiting satellite was launched in 1987 and the SSM/I data delivery will probably continue through the year 2001. At that time it will be replaced by a combined imager/sounder SSM/IS. The SSM/I sensor measures radiances at 19.35, 22.23, 37.0 and 85.5 GHz frequencies. All frequencies except the water vapour absorption channel (22.23 GHz) are dual polarized. The footprint size varies with frequencies and range from approximately 48 km at 19.35 GHz to 12 km at 85.5 GHz. The availability of well calibrated data led to very active developments of retrieval algorithms. Where as retrieval of oceanic geophysical parameters, e.g. sea ice cover parameters, ocean surface winds speeds, quickly found its utilization in weather forecasting centres and generation of near-real time operational sea ice analysis charts, the use of precipitation over land and land surface parameters, e.g., snow water equivalent, is still at the development stage. Main reason for this lag is that over land the emissivity variability due to surface type and vegetation is very complex. The high emissivity of the most of land surfaces makes it difficult to separate atmospheric contribution from the total signal observed by a spaceborne sensor. Typical emissivities over land range (at SSM/I frequencies) from 0.85 to 0.95. If a pixel within the instruments field of view contain lakes, wetland, or rivers, the emissivity will vary as a function of fraction of water covered surface. Because of the coarse resolution of the SSM/I , most of the pixels over land will contain a broad range of land surface types. Unless the exact composition of the land surface types within each pixel is known, detection of the rain-free clouds and low precipitation rate clouds is not always possible.

Microwave radiation reaching the Earth-viewing reflector is a resultant of the emitting and reflecting properties of the surface, attenuating and emitting properties of the atmosphere as well as a receiving antenna geometry. This implies that a number of assumptions and simplifications have to be made in order to uncouple surface and atmospheric information. While at higher frequencies propagation of microwave radiation through the atmosphere is affected by the interactions with ice particles, at lower frequencies thermal emission from liquid water droplets dominates the atmospheric effects. The emission and scattering processes depend on many characteristics of the cloud particles such as type (i.e. liquid or frozen), size, shape, and vertical distribution.. The success of cloud parameter retrieval is also a function of the surface within the field of view characteristics. Since microwave emissivity of the ocean surface is relatively low, emission from clouds represents a significant contribution to the measured radiation intensity. The snow free or dry land surface emissivity is high (~0.9) at all frequencies, thus detection of emission contribution from clouds may not be possible. Moist and snow covered soil will appear to be relatively cold and emission from clouds becomes an important contribution to the observed signal. Models that relate emissivity to physical parameters are discussed in detail in [10] and [11].

For severe weather events, when large liquid drops and ice particles are present, or when land surface is snow covered, detection of precipitation can be more successful.

In this paper we will limit our discussion to analysis of precipitating systems. There are two types of precipitating systems that are relevant to forecasting of environmental emergency. Hail storm events that cause severe damage to vegetation and winter/spring precipitation occurring over snow covered land.. It was pointed in several research papers that SSM/I observations are well suited for detection and quantifying hailstorm and high rain rate events [4, 15]. In this work we will present confirmation of this capabilities and evaluate the potential of SSM/I utilization for the winter precipitation analysis. Radiative transfer models and the weighting functions developed by Weinman [3] were used to reassure that for convective systems retrieval of the precipitation rates , or to be more exact, the information about the amount of precipitation size liquid droplets within the atmosphere and the amount of frozen types of hydrometeors can be carried out.

Radiative transfer models were also utilized to carry out evaluation of the SSM/I precipitation retrieval algorithms for winter (snow) storms applications.

2. RADIATIVE TRANSFER MODEL

Finite difference method for azimutally isotropic plane-parallel cloud structure was used to find the solution of the radiative transfer model [3]. Using Eddington approximation [19] a he radiation intensity measured at a height z ($I(z, \theta, \varphi) = I_o(z) + I_1(z, \theta)$) then can be de calculated numerically. The contribution from a thin layer (extinction coefficient $k_{extf}(z)$) to the radiation intensity measured at a level ($z = h$) is:

$$\int_0^h J(z, \theta) \mu(z) \exp[-\int_z^h \mu(z')dz']dz \quad (1)$$

where: $\mu(z') = k_{extf}(z')/\cos\theta.$

For a layer at temperature T and total scattering albedo, ω, $J(z,\theta)$ (source function) is defined as :

$$J(z,\theta) = (1-\omega)T + \omega(I_o + I_l \cos\theta)$$

The integrand in Equation 1 is known as the weighting function $W(z)$. Profile of $W(z)$ shows contributions of different atmospheric layers to the radiation at the top of the atmosphere. The usefulness of this type of function is illustrated in Figures 1 a and 1b. The sensitivity of the 37 GHz and 19 GHz channels to the precipitation size liquid droplets in the atmosphere is evident from the changes in the weighting function values. Since the precipitation rate in this case study was greater than 5 mm/hr , the 85 GHz weighting function is the same in both figures. Radiative transfer model was used to test sensitivities of top of atmosphere (TOA) brightness temperatures to the presence of different types of hydrometeors. One very important results that should be mentioned is the warming effects at 37 GHz and 85 GHz of liquid droplets presence at altitudes above the precipitation layer . The changes in TOA brightness temperatures were estimated to be about 5° at 37 GHz and 10° at 85 GHz corresponding to the changes in supercooled liquid droplet amount of 0.25 g/m^3 [4].

The accuracy of using the finite difference solution was evaluated by comparing the results of simulations with published results of the Multi-stream approach [15]. Several reported hailstorm events were used to calculate the scattering index and the precipitation rates.

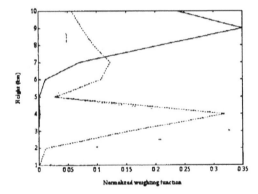 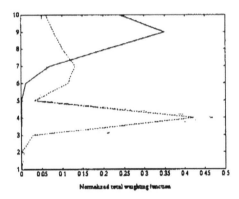

Figure 1a. Normalized weighting functions for a convective cell. Symbols used: Dash line –19 GHz, dot –line for 37 GHz, and solid line for 85 GHz. Vertical profiles of hydrometeor types (rain droplets, cloud particles, graupel, and ice) and distribution are described in [4]. Note that at 85 GHz there is no contribution to theTOA brightness temperatures below 5 km.

Figure 1b. Normalized weighting functions for a convective cell. The amount of rain size particles is double the amount used to generate Figure 1. Symbols used: Dash line –19 GHz, dot –line for 37 GHz, and solid line for 85 GHz.

3. SUMMER STORM EVENTS

The rainfall and cloud structure algorithms development was based on results of two stream radiative transfer modelling with pre-selected cloud microphysics and Eddington approximation for microwave single particle interaction parameters [1]. Airborne Multi - channel Microwave Radiometer (AMPR) data, acquired during COMHEX campaigns [2], were used to evaluate the radiative model performance. Translation from airborne scene simulations to scales compatible with SSM/I resolution required the use of rainfall intensity areal distribution as a function of rain rates [2]. Standard cloud model vertical profiles of hydrometeors [12] were used in addition to profiles described in [13] .

In order to determine the main contributions to the TOA brightness temperature from different layers of the precipitating atmosphere, airborne passive microwave measurements were compared with the space and time coincident weather radar returns.

The traditional retrieval of precipitation is to use the intensity of the weather radar reflectivity to determine the precipitation rates and the height of the precipitating cloud system. The sensitivity of the TOA brightness temperatures to hydrometeors vertical distribution was compared with the weather radar observations carried out simultaneously with the measurements by a multi-frequency radiometer during CaPE [16]. The correlation coefficients of the TOA brightness temperatures and the radar returns from different layers within the precipitation system are presented in Figures 2a and 2b. The data were partitioned into two groups one group with RR<5mm/hr (stratiform rain cases) and second with RR>5mm/hr (convective systems). For the first set of case studies, at 19 GHz highest correlation (0.55) was with the return at 2 km altitude. The highest correlation (0.8) at 37 GHz was with the returns at 1 km . This could be attributed to the maximum of the cloud liquid water amount at that height. For 85 GHz correlation coefficients were between 0.6 and 0.95 for the returns above 6 km and below 14 km, i.e., cloud ice distribution. For the case studies with rain rates above 5 mm/hr both 37 GHz and 85 GHz had correlation coefficients >0.6 with the radar returns above 8 km. The correlation coefficients with the returns below 8 km were less than 0.6 for all three frequencies. This is in agreement with the forward radiative transfer calculations [1, 3,4]. These calculations show that for precipitation rates greater than 5 mm/hr, the contribution s to TOA are mainly from layers above the precipitation layer.

The analysis of numerous weighting functions [1,4] generated for the case studies listed in Table 2 indicates that emission type of algorithms can be used to derive columnar amounts of rain droplets (translated in equivalent Marshall-Palmer rain rates), while scattering index type of algorithms can be used to derive the amounts of cloud ice. The use of this method is illustrated in Figures 3a and 3b. These figures show calculations along a transect from 40 N to 50 N (at 79-80 W), for April 20, 1996. Two very intense hailstorm events (at 44 N and 46 N) were captured during SSM/I overpass (10 minutes of the reported hailstorm events). The same scattering index (SI) type of algorithm ($SI=A - B*$ V85; A=500, B=2) was also used to identify pixels containing cloud ice for extreme events such as hurricanes (Figures 4 and 5). Since at 85 GHz the contribution from the surface to the TOA brightness temperature is negligible, the same type of algorithm can be used for land and ocean surfaces with regionally optimized A and B coefficients . A database of Doppler weather radar and SSM/I observations for hailstorm events will be used to derive coefficients suitable for Ontario.

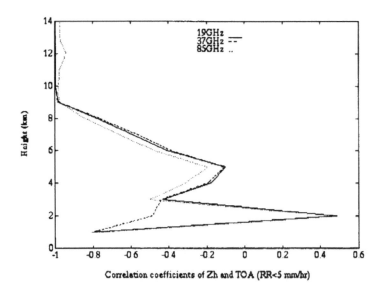

Figure 2a. Correlation coefficients of the AMPR TOA brightness temperatures measured during COHMEX 1991 with the weather radar returns from different atmospheric layers. Stratiform sections of the precipitating systems were used in this analysis.

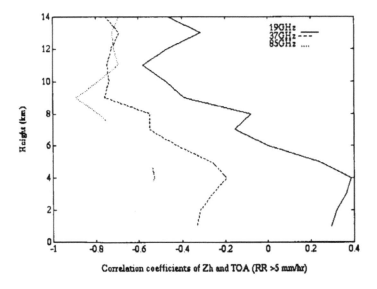

Figure 2b. Correlation coefficients of the AMPR measured brightness temperatures and the radar returns for convective type of precipitating systems (precipitation rates >5mm/hr) . COHMEX 1991 AMPR and weather radar data were used in this analysis.

Table 1. Radiative transfer model comparison for a land surface. The TOA brightness temperatures were calculated using the finite difference method (FDM) and compared with multi-stream (MM) calculations [15]. The simulations were carried out for a zenith angle of 49° for three frequencies and three Marshall-Palmer rain rates (RR).

Rain and ice Freq. (GHz)	RR mm/hr	TB (K) FDM	FDM-MM Difference (K)	Rain no ice TB (K) FDM	FDM-MM Difference (K)
19	2	278.54	0.40	278.73	0.27
	10	274.21	-0.74	274.21	-3.18
	50	257.56	0.36	267.39	-0.91
37	2	269.55	0.30	276.84	4.60
	10	242.45	0.40	260.21	-0.66
	50	193.00	3.59	259.08	5.40
85	2	240.68	-2.25	264.21	0.17
	10	190.38	-0.56	259.66	1.00
	50	137.56	-1.14	259.27	-0.21

Table 2. List of meteorological case studies used for database generation.

Case #	Meteorological scenario type	Source of atmos. profile information	Reference #
1	Convective cells, July 11, 1986 storm	Cloud model and radar	[13],[18]
2	June 23, 1986 , COHMEX storm	Radar	[16]
3	August 12, 1991, Cape storm	Radar	[17]
4	Stratiform rain with ice layer	Cloud model	[18]
5	Rain/snow and rain/freezing rain	Cloud model	[4]
6	Cloud over snow covered land	Cloud model	[4]

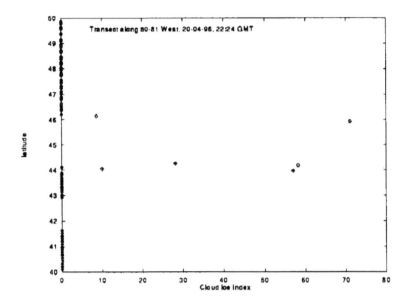

Figure 3a. Estimated cloud index calculated for SSM/I data taken across a severe weather system. The satellite overpass was within 10 minutes from the time of reported hailstorms was at 44 North and 46 North (~79W).

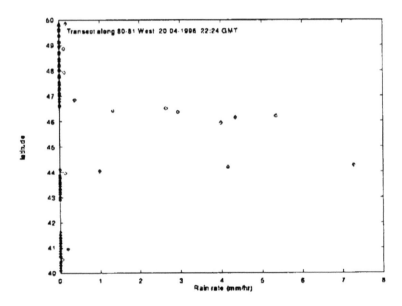

Figure 3b. Transect along a severe weather system in Ontario. Tornadoes, heavy downpour, and hailstorm events were reported at several locations within the study area. The SSM/I overpass was within 10 minutes of reported severe events.

Figure 4. Hurricane Andrew, cloud ice distribution as determined using 85.5 GHz SSM/I channel. Data from satellite overpass 25-08-1992, 22:52 GMT.

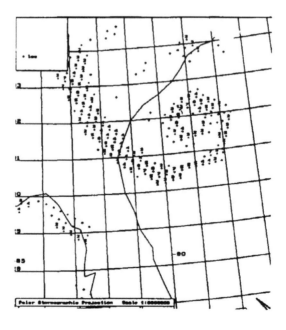

Figure 5. Hurricane Hugo cloud ice distribution calculated using 85 GHz SSM/I data from a satellite overpass on 21-08-1988, 23:32 GMT.

4. WINTER STORM EVENTS

The warming effects of the cloud cover on the brightness temperatures at 37 GHz and 85 GHz can be as high as 8° and 12° even for moderate thickness of cloud cover and 20 cm snow cover on land [11, 5]. For snow covered land and non-precipitating cloud systems with varied amounts of cloud liquid water [11, 4] theoretical calculation generate following conditions for no rain: (precipitation criteria for snow covered land, PCS) TB85- TB19 >0 as well as TB37-TB19>0 . Figure 6 generated using SSM/I data from snow covered land and fully overcast, but not precipitating sky (according to weather forecast charts), confirms the theoretical predictions. To test if scattering type of algorithm can be applied to winter storms systems, we compared the precipitation detection criteria described in [18] and 85 GHz vs 37 GHz scatterplot for SSM/I data acquired over snow and mixed snow/rain March 1998, March 1999 events (Figure 7b). Most of the points are below the AFD line, defined as one of the criteria for detection of precipitation. The importance of using no precipitation criteria (PCS) is illustrated in Figure 7a. Using AFD criteria by itself could result in false identification of precipitation.

Transects of SSM/I data across reported snow storms, Figures 8a and 8b, provide additional confirmation that winter type precipitation can be detected over land covered with snow using TB85-TB19 and TB37-TB19 criteria. Database of winter storm events and forward calculations is currently being compiled in order to carry out a quantitative analysis of the traditional algorithms and derivation of more physically based algorithms.

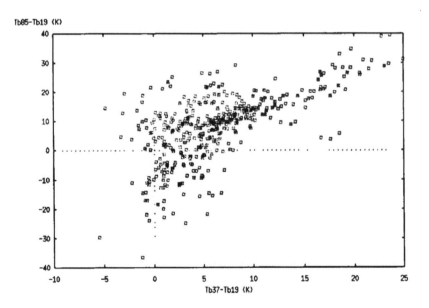

Figure 6. Scatter plot for the SSM/I pixels taken from a cloud covered area within a winter storm system. The algorithm identified these pixels as rain free. Land surface within this study area was snow covered.

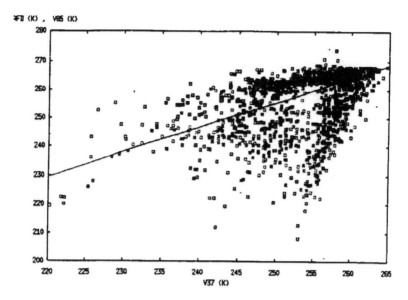

Figure 7a. Evaluation of the precipitation detection criteria with the rain discrimination function AFG derived in [18]. SSM/I data were extracted from a transect across a non-precipitating system. No correction for the snow cover presence.

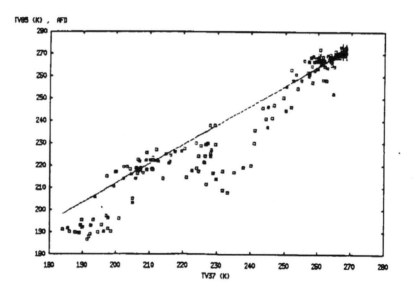

Figure 7b. Comparison of the precipitating detection criteria with the rain discrimination function AFG derived in [18]. SSM/I data were extracted from a transect across a (snow/rain) precipitating system. Note that most of the points (75%) are located below the AFD line.

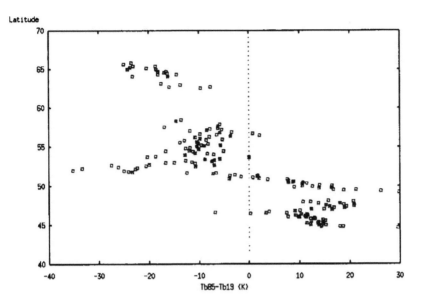

Figure 8a. Transect along 69 West, across a snow storm. All region from 40 North to 55 North was under a heavy cloud cover. Precipitation was at isolated locations for pixels south of 50 North.

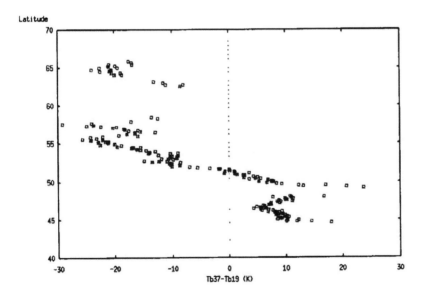

Figure 8b. Difference of brightness temperatures 37 GHz and 19 GHz (vertical polarizations) for a transect along 69 West, across a heavy winter storm system. Very wet snow was observed at latitudes below 50 North. Land surface was already snow covered.

5. CONCLUSION

This paper has focused on two types of precipitation scenarios over land relevant for environmental emergency management. Forward radiative transfer calculations were carried out for these types of events. This allowed development of retrieval algorithms that use relative channel weight factors. These weight factors are predetermined by identifying the type of meteorological scenario were evaluated. At this stage the identification has to be made by the user. The methodology for identifying the presence of convective type of clouds within the field of view of the sensor and calculation of probable precipitation rates and total columnar ice amount was developed and applied to a hailstorm event. Second type of weather events that were considered correspond to winter/spring snow storm events. Microwave emissivities corresponding to snow covered land surfaces were used in forward computations. The retrieval of the winter precipitation over snow covered land is possible because the contribution to the signal from the atmosphere is relatively large. A simple algorithm utilizing 85 GHz channel combined with the CAL/VAL precipitation over land algorithm identified winter storm locations. The quantitative comparison of the forecast and the precipitation type and intensities estimated using SSM/I is one of the current research activities at CRESTech and the Environment Canada cloud physics specialist.

One of the main objectives of this work was to determine which additional parameters are required for flagging particular type of event. Since temperature information for the surface as well as the cloud type is very crucial, the use of AVHRR type of data acquired at the same time as SSM/I will be beneficial. In addition, for correct identification of the stage in precipitation development, information about the weather prior to the time of satellite overpass must be known. The prior weather effects on the land surface emissivity (e.g., changes in soil moisture or snow deposit on land surface) introduce changes of the same order of magnitude as the precipitation.

REFERENCES

1.   Kummerrow , C., "On the accuracy of the Eddington approximation for radiative transfer in the microwave frequencies", *J. Geophys. Res.*, 98, 2757-2765, (1993).

2.   Atlas, D., D. Rosenfeld, and D.A. Short, "The estimation of convective rainfall by area integrals. Part I: The theoretical and empirical basis". *Conf. On Mesoscale Precipitation*, MIT, Cambridge, MA. , (1988).

3.   Wu., R., and J.a. Weinman, "Microwave radiances from precipitating clouds containing aspherical ice, combined phase, and liquid hydrometeors", *J. Geophys. Res.*, 89, 7170-7168, (1984).

4.   Rubinstein, I.G., "Effects of cloud microphysical properties on space-based passive microwave remote sensing". *PhD Thesis, York University, Toronto, Canada*, (1995).

5.   Heymsfield, A.J., "Precipitation development in stratiform clouds: A microphysical and dynamics study". *J. Atmos. Sci.*, 34, 367-381, (1977).

6.   Rosenfeld, D., D. Atlas, and D.A. Short, "The estimation of convective rainfall by area integrals. Part II: The height area rainfall threshold method". *Conf. On Mesoscale Precipitation*, MIT, Cambridge, MA., (1988).

7.   Heymsfield, A.J, "A comparative study of the rates of potential graupel and hail embryos in High Plains storm". *J. Atmos. Sci.*, 39, 2847-2866, (1982).

8.   Fujiyashi, Y., "Melting snowflakes", *J. Atmos . Sci.*, 43, 307-311, (1986).

9. Heymsfield, A.J. and A. G. Palmer, "Relationships for deriving thunderstorm anvil ice mass for CCOPE storm water budget estimates". *J. Clim. Appl. Meteor.*, 25, 691-702, (1986).

10. Ulaby, F.T., R.K. Moore, and A.K. Fung, *Microwave Remote Sensing*, Vol. I., Addison-Wesley Publishing Company, Reading, MA, USA, (1981).

11. Ulaby, F.T., R.K. Moore, and A.K. Fung, *Microwave Remote Sensing*, Vol. II and III., Artech House Pub., Norwood, MA, USA, (1983).

12. House, R. A., *Cloud dynamics*, Academic Press, Harcourt Brace Jovanovich Publ., (1993).

13. Smith, E. A., A. Mugnai, H.J. Cooper, G.J. Tripoli, and X. Xiang, "Foundation for statistical –physical precipitation retrieval from passive microwave satellite measurements. Part I: Brightness temperature properties of a time-dependent cloud radiation model", *J. Appl. Met.*, 31, 506-531, (1992).

14. Tsang, L., J. A. Kong, and R.T. Shin, *Theory of microwave remote sensing*, J. Wiley and sons, New York, N.Y, (1985).

15. Evans, K.F., and G.L. Stephens, "Polarized radiative transfer modeling: An application to microwave remote sensing of precipitation". *Tech. Report, p. 78., Dept. of Atmos. Sci.* , Colorado State University, Fort Collins, CO. (1990).

16. Vivekanadan, J.J., Turk, G.L., Stephens, and V.N. Bringi, "Microwave radiative transfer studies using multispectral radar and radiometric measurements during COHMEX", *J. Appl. Meteor.*, 29, pp. 561-585. (1990).

17. Vivekanadan, J.J., Turk, G.L., and V.N. Bringi, *Advanced Microwave Precipitation Radiometer (AMPR) and multispectral radar comparison*, preprint. (1992).

18. Smith, D. M., D. R. Kniveton and E. Barrett, "A statistical method approach to passive microwave rainfall retrieval", *J. of Appl. Meteor.*, 37, pp.135-154, (1998).

Microw. Radiomet. Remote Sens. Earth's Surf. Atmosphere, pp. 313–323
P. Pampaloni and S. Paloscia (Eds)
© VSP 2000

Modelling and measurements of Stokes vector microwave emission and scattering for a precipitating atmosphere

ACHIM HORNBOSTEL[1], ARNO SCHROTH[1], ANDREY SOBACHKIN[1],
BORIS KUTUZA[2], ANDREY EVTUSHENKO[2] and GENNADIJ ZAGORIN[2]

[1]*Deutsches Zentrum fuer Luft- und Raumfahrt, D-82230 Oberpfaffenhofen, Germany*
[2]*Russian Academy of Sciences, Moscow, Russia*

Abstract – Passive and active fully polarimetric ground-based measurements of rain signatures have been carried out. The obtained results, in particular for the third Stokes parameter and the orientation angle of polarization are compared with theoretical calculations. The principles and special features of the used radiometer and radar instruments are described.

1. INTRODUCTION

Dual polarization measurements have been successfully utilized in radar meteorology and passive remote sensing to derive information about the rain structure, particle size distribution, discrimination of different particle types, and finally, to improve rain rate estimation [1,2]. The dual polarization technique makes use of the first and second Stokes parameter and is now well established.

Model calculations [1,3] show that also the third Stokes parameter of rain emission can reach values up to 3 K and is in particular sensitive to the mean canting angle of the rain drops. In order to validate these simulations, polarimetric radiometer measurements at 13.3 GHz, which allow to derive the third Stokes parameter, were performed and are presented here.

In contrast to passive measurements, which are integral values, active measurements provide a spatial resolution of the medium. Principally, they have the same sensitivity to the particle properties as passive measurements, i. e. the second Stokes parameter is sensitive to the shape of particles and the third is sensitive to the canting of particles. However, the Stokes vector of the backscattered signal depends on the polarization of the transmitted signal. The Stokes vectors of the received and transmitted signals are related to each other by the 4x4 Mueller matrix, which describes the polarization properties of the scattering medium. Measurements with the C-band polarimetric weather radar of DLR allow the construction of the Mueller matrix elements either by coherent measurements or incoherently utilising combinations of different polarization bases for reception and transmission. Both methods have been tested and are discussed.

2. STOKES VECTOR PRESENTATION AND RELATION TO PARTICLE PROPERTIES

The Stokes vector expressed in a horizontal-vertical linear polarization basis (h, v-basis) is built from the electrical field components E_h and E_v The Stokes vector components have the dimension of intensity. It can be shown that the third and fourth component are the difference between the intensities in 45 and 135 degree linear polarization (d,x) and between left-hand and right-hand circular polarization (l,r).

$$[S] = \begin{bmatrix} I \\ Q \\ U \\ V \end{bmatrix} = \begin{bmatrix} I_h + I_v \\ I_h - I_v \\ I_d - I_x \\ I_l - I_r \end{bmatrix} = \frac{1}{\eta} \begin{bmatrix} \langle E_h E_h^* \rangle + \langle E_v E_v^* \rangle \\ \langle E_h E_h^* \rangle - \langle E_v E_v^* \rangle \\ \langle 2\,\mathrm{Re}(E_h E_v^*) \rangle \\ \langle 2\,\mathrm{Im}(E_h E_v^*) \rangle \end{bmatrix}, \tag{1}$$

where η is the wave impedance and $\langle...\rangle$ is the expected value. Eq. (1) is valid for active and passive measurements, where the field components refer either to the emitted or to the transmitted and received field. The Stokes vector can be decomposed in an unpolarized and polarized vector:

$$[S] = [S]_{unp} + [S]_p = [I - \sqrt{Q^2 + U^2 + V^2},\ 0,\ 0,\ 0]^T + [\sqrt{Q^2 + U^2 + V^2},\ Q,\ U,\ V]^T \tag{2}$$

The subscript ^{T} denotes the transpose of a vector. The Stokes parameters are related to the orientation angle φ and ellipticity angle ψ of the polarization ellipse, which describes only the polarized part, by the following equations:

$$\varphi = 0.5\arctan\left(\frac{U}{Q}\right). \tag{3}$$

$$\psi = 0.5\arcsin\left(\frac{V}{\sqrt{Q^2 + U^2 + V^2}}\right) \tag{4}$$

With the degree of polarization

$$m = \frac{\sqrt{Q^2 + U^2 + V^2}}{I} \tag{5}$$

and with the aid of Eqs. (3) and (4) one obtains [4]:

$$V = m \cdot I\, \sin(2\psi), \tag{6}$$

$$\sqrt{Q^2 + U^2} = m \cdot I\, \cos(2\psi), \tag{7}$$

$$Q = m \cdot I \, \cos(2\psi) \, \cos(2\varphi), \tag{8}$$

$$U = m \cdot I \, \cos(2\psi) \, \sin(2\varphi). \tag{9}$$

Note, that Eq. (7) defines the total linear polarization intensity in the plane of polarization.

The first Stokes parameter is a measure for the rain intensity. The second Stokes parameter is sensitive to the shape of particles and, therefore, allows to derive the drop size distribution [1,2]. The third Stokes parameter is sensitive to the azimuthal asymmetry of the volume source conditioned by the non-sphericity of rain drops and the preference orientation of their symmetry axis [1].

For passive measurements it is common to use the modified Stokes vector $[S_{mod}] = [I_h, I_v, U, V]^T$; it is usually expressed as a brightness temperature vector applying the Rayleigh-Jeans approximation:

$$[T_h, T_v, T_U, T_V]^T = \frac{\lambda^2}{k_B}[I_h, I_v, U, V]^T, \tag{10}$$

with k_B being the Boltzmann constant and λ being the wavelength. In the following we define $T_Q = T_h - T_v$ and $T_I = 0.5(T_h + T_v)$. Note, that T_I is defined here as the halfsum of the components T_h and T_v in contrast to Eq. (1), where the intensity I is the sum of the components I_h and I_v.

In [1] a relation for the orientation angle of the polarisation ellipse was derived for downwelling radiation and small mean canting angles β of rain drops:

$$\varphi = 0.5 \arctan\left(\frac{T_U}{T_Q}\right) \approx 0.5 \arctan\left(2\frac{\sin 2\beta}{\sqrt{1-\mu^2}}\sin\phi\right), \tag{11}$$

where μ is the cosine of the direction of incidence relative to the zenith, β is the orientation angle of the drop symmetry axis with respect to the vertical, and ϕ is the azimuth direction of canting relative to the direction of propagation.

Fig. 1 presents simulation results for T_Q and T_U for ground-based observation of rain brightness temperatures under an elevation of 30° at 13.3 GHz. The rain drops were modelled as oblate spheroids with an axial ratio

$$\frac{a}{b} = 1 - 0.091\bar{a}, \tag{12}$$

where $\bar{a}$ is the equivolumetric drop radius [5]. The drop size distribution was described by the Marshall-Palmer law. A plane-parallel homogeneous rain layer with 4 km height and a constant thermodynamical temperature of 293° K was assumed. For the simulation the vector radiative transfer equation was solved analytically including first-order of scattering and separation into zero-, first-, and second-order azimuthal harmonics. For further details refer to [1].

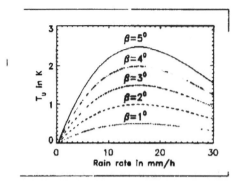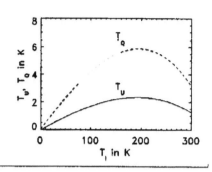

Figure 1. Second and third Stokes parameter for downwelling radiation from a homogeneous rain layer with 4 km height for passive ground-based observation at 13.3 GHz under an elevation of 30°. Left panel: Third Stokes parameter T_U versus rain rate in dependence of mean drop canting angle β. Right panel: Second and third Stokes parameter T_Q and T_U versus first Stokes parameter T_I for $\beta=5°$.

3. PASSIVE MEASUREMENTS

In order to validate the simulations, passive polarimetric measurements were carried out with a special designed radiometer at 13.3 GHz. The block diagram is shown in Fig. 2. The main difference to a conventional dual polarization radiometer is a voltage controlled polarizator unit $\varphi(u)$ which is automatically adjusted to the polarization plane of the incoming radiation. The polarizator unit consists of two Faraday ferrite polarizators. The first polarizator rotates the polarization of the incoming signal into a plane of polarization, which is determined by the control voltage u. At the output of this polarizator two orthogonal linearly polarized components parallel and perpendicular to this plane T_1 and T_2 are detected. The second polarizator rotates the polarization plane by further 45 degrees. The two linearly polarized orthogonal components parallel and perpendicular to this plane are again detected. By the feed back loop the difference between these two latter components is minimized. Because the second polarization basis is rotated by 45° to the first basis, the linear polarization difference at the output of the first polarizator $\Delta T = T_1 - T_2$ is maximized, i.e. the polarization basis is adjusted to the orientation angle φ of the incoming radiation and the control voltage u is proportional to this angle.

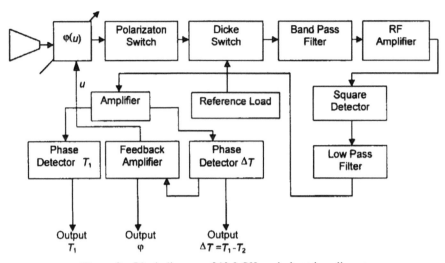

Figure 2. Block diagram of 13.3 GHz polarimetric radiometer.

Emission from rain is a partially polarized noise signal. The two orthogonal components with brightness temperatures T_1 and T_2 contain equal parts of unpolarized and circular polarized radiation. Since the measurement basis is adjusted to the plane of polarization, T_1 contains all linearly polarized radiation and T_2 no linear polarization, i.e. the linear polarization difference $\Delta T = T_1 - T_2$ is equal to the linearly polarized part of radiation as defined in Eq. (7). The first three Stokes parameters in h, v -basis follow then from Eqs. (8), (9) and (10)

$$T_I = \frac{T_1 + T_2}{2} = T_1 - \frac{\Delta T}{2} ,$$ (13)

$$T_Q = \Delta T \cos(2\varphi) ,$$ (14)

$$T_U = \Delta T \sin(2\varphi) ,$$ (15)

where φ is the angle of the polarization plane (orientation angle) with respect to the horizontal axis.

The phase detectors in the block diagram of Fig. 2 measure only phases of internal signals, but for the incoming signal only powers are detected. Therefore, the ellipticity angle ψ and the fourth Stokes parameter cannot be measured with this instrument. However, according to our simulations both are negligible small for rain emission. The calibration of the radiometer was performed with a black body, clear sky temperatures and an external polarization grid.

Fig. 3 shows the main component T_1, the linear polarization difference ΔT, and the double orientation angle 2φ measured during a convective rain event under an elevation of 30^0. Fig. 4 presents the corresponding second and third Stokes parameters T_Q and T_U, and the double orientation angle 2φ in a different scale. The maximum tilt angle of the polarization ellipse φ from horizon is about 7 degree.

By comparison of Figs. 3 and 4 a decrease of T_U and T_Q for high values of T_I is visible, which is in agreement with the simulated results of Fig. 1. This effect can be also seen in Fig. 5, where the second and third Stokes parameter are plotted versus the first Stokes parameter. Fig. 5 contains additionally some theoretical points of T_Q and T_U for rain drops with a canting angle of 3^0 perpendicular to the direction of propagation, i.e. $\beta=3^0$ and $\phi = 90^0$ in   Eq. (11), which were calculated for rain rates of 2.5, 5.0 and 12.5 mm/h and are taken from Fig.1.

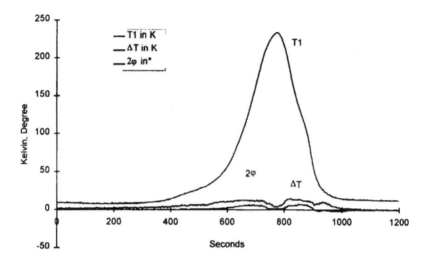

Figure 3. Main component T_1, polarization difference ΔT and double orientation angle 2φ of down-welling radiation, elevation 30^0 (Oberpfaffenhofen, Germany, 30th July 1998).

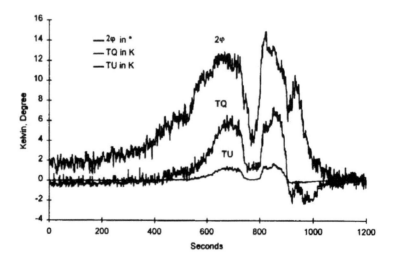

Figure 4. Second, third Stokes parameter and double orientation angle of polarization ellipse.

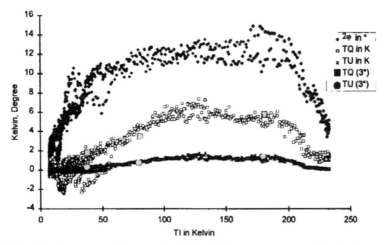

Figure 5. Double orientation angle, second and third Stokes parameter vs. first Stokes parameter in comparison with model calculations for rain drops with 3^0 canting angle.

4. ACTIVE MEASUREMENTS

For active measurements the Stokes vector of the backscattered signal $[S]_{rec}$ at the location of the radar antenna is:

$$[S]_{rec} = \frac{1}{R^2}[M][S]_{tr} = \frac{V_{eff}}{R^2}[M'][S]_{tr},\qquad(16)$$

where $[M]$ is the 4x4 Mueller matrix, R the distance of the scattering volume to the radar antenna and $[S]_{tr}$ is the Stokes vector of the transmitted signal. Here $[M']$ denotes the specific Mueller matrix per unit volume and V_{eff} is the effective scattering volume.

For the following discussion we introduce the modified Stokes vector $[S_{mod}]=[I_h,I_v,U,V]^T$. The corresponding modified specific Mueller matrix $[M_{mod}']$ is built from the complex volume backscattering amplitudes f_{ij}, where i and j denote the receive and transmit polarization, respectively.

$$[M_{mod}']=\frac{1}{V_{eff}}\begin{bmatrix} \langle|f_{hh}|^2\rangle & \langle|f_{hv}|^2\rangle & \langle\mathrm{Re}(f_{hh}f_{hv}^*)\rangle & \langle\mathrm{Im}(f_{hh}f_{hv}^*)\rangle \\ \langle|f_{vh}|^2\rangle & \langle|f_{vv}|^2\rangle & \langle\mathrm{Re}(f_{vh}f_{vv}^*)\rangle & \langle\mathrm{Im}(f_{vh}f_{vv}^*)\rangle \\ \langle2\mathrm{Re}(f_{hh}f_{vh}^*)\rangle & \langle2\mathrm{Re}(f_{hv}f_{vv}^*)\rangle & \langle\mathrm{Re}(f_{hh}f_{vv}^*+f_{hv}f_{vh}^*)\rangle & \langle\mathrm{Im}(f_{hh}f_{vv}^*-f_{hv}f_{vh}^*)\rangle \\ -\langle2\mathrm{Im}(f_{hh}f_{vh}^*)\rangle & -\langle2\mathrm{Im}(f_{hv}f_{vv}^*)\rangle & -\langle\mathrm{Im}(f_{hh}f_{vv}^*+f_{hv}f_{vh}^*)\rangle & \langle\mathrm{Re}(f_{hh}f_{vv}^*-f_{hv}f_{vh}^*)\rangle \end{bmatrix}\quad(17)$$

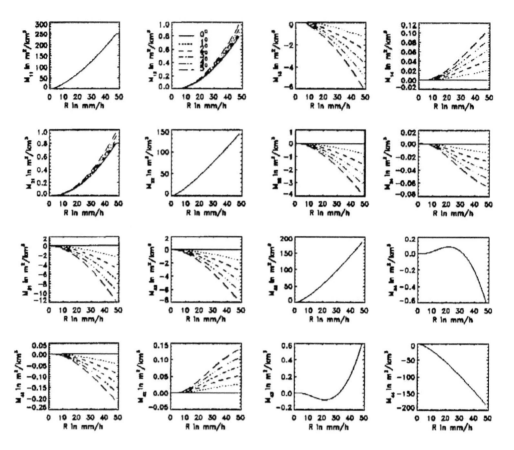

Figure 6. Specific Mueller matrix [M$_{mod}$'] for backscattering in dependence of mean drop canting angle β and rain rate (Elevation: 3°, frequency 5.5 GHz, standard deviation of canting angle: 10°). The corresponding canting angles for different line types are given in the legend of element M$_{12}$.

Fig. 6 shows a simulation of the elements of [M$_{mod}$'] for non-spherical oriented rain drops. For the orientation angles of the drops a Gaussian distribution with 10° standard deviation and a mean canting angle β perpendicular to the direction of propagation was assumed. Curve parameter in the diagrams is the mean canting angle β = 0,1, ...5°.

The backscattering amplitudes can be measured directly with a coherent dual polarization radar receiver [6]. Fig. 7 shows a modified Stokes vector, which was derived by multiplication of a measured modified Mueller matrix with the normalized modified Stokes vector for horizontal transmit polarization [S$_{mod}$]$_{tr}$= [1, 0 ,0 ,0], i.e. the elements of the resulting vector correspond to the first column of the measured Mueller matrix [M$_{mod}$']. Additionally shown are the orientation and ellipticity angle of the backscattered signal calculated with Eqs. (3) and (4) and $Q=I_h\text{-}I_v$. The x- and y-axis denote the distance from the radar and the height above ground.

Alternatively, the elements of $[M_{mod}']$ can be obtained by incoherent measurements applying 6 different combinations of h, v, 45°, 135° (d,x), and left-hand, right-hand circular (l,r) polarizations for receive and transmit, e. g. transmit in h,v-basis and receive in d,x-basis [7]. The advantage of this method is that no phases must be measured and logarithmic receivers with higher dynamic range can be used. However, the six polarizations must be calibrated carefully. For such measurements we used a ray-to-ray switching scheme instead of a scan-to-scan switching scheme to minimize the decorrelation in time. The polarization was changed from one radar ray to the following one and groups of six successive rays were combined to „super-rays" for which the Mueller matrix elements were computed in the data processing. Each radar RHI-scan consists then of several such super-rays.

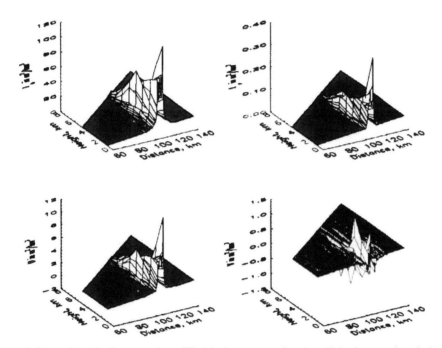

Figure 7. Normalized backscattered modified Stokes vector of a rain cell for h-transmit polarization.

With both methods low cross-polar powers must be measured which come to the limit of the radar channel isolation of about -30 dB. As a special case the normalized Stokes vector for an (assumed) unpolarized input signal can be obtained from only co-polar incoherent measurements:

$$[S']_{rec} = \frac{[S]_{rec}}{V_{eff}/R^2} = [M'] \begin{bmatrix} 1 \\ 0 \\ 0 \\ 0 \end{bmatrix} = \begin{bmatrix} \langle |f_{hh}|^2 \rangle + \langle |f_{vv}|^2 \rangle \\ \langle |f_{hh}|^2 \rangle - \langle |f_{vv}|^2 \rangle \\ \langle |f_{dd}|^2 \rangle - \langle |f_{xx}|^2 \rangle \\ \langle |f_{ll}|^2 \rangle - \langle |f_{rr}|^2 \rangle \end{bmatrix} \tag{18}$$

Here we have used again the unmodified Stokes vector $[S]=[I, Q, U, V]^T$. The normalized received Stokes vector $[S']_{rec}$ corresponds then to the first column of the specific unmodified Mueller matrix $[M']$.

An example for the normalized Stokes vector obtained from such incoherent co-polar measurements is presented in Fig. 8. The figure includes also the orientation and ellipticity angles according to Eqs. (3) and (4), which were calculated only, if the second Stokes parameter Q exceeded a threshold of 4 m²/km³ in order to eliminate the influence of measurement noise. In the special backscattering case of Eq. (18) the orientation angle φ for low elevation angles is equal to the effective drop canting angle β in the plane perpendicular to the direction of propagation. The peak values of β=3–4° are in good agreement with the results from passive measurements (compare angle 2φ in Figs. 4 and 5 and relation of β to φ for passive measurements according to Eq.(11)).

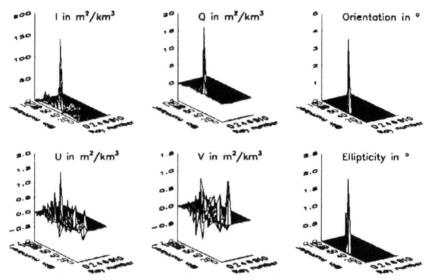

Figure 8. Normalized Stokes vector and polarization angles of a backscattered unpolarized signal derived from incoherent radar measurements in different polarizations. (The ray numbers correspond to groups of 6 successive rays with different polarization bases. The bin numbers refer to the 300 m radar resolution cells).

5. CONCLUSIONS

The third Stokes parameter reaches values of about 1% of the first Stokes parameter in passive measurements of rain. The measured orientation angles of the polarization ellipse are in the range of 4 - 7 degrees. This corresponds to effective mean canting angles of rain drops between 2 and 3.5 degrees. With respect to other applications, i.e. measurements of the upwelling Stokes vector emission from sea surface, these results mean that rain can significantly disturb such measurements.

Both coherent and incoherent radar measurements of the Mueller matrix have been performed, but presently no clear preference can be given. However, the results for the orientation angle of the polarization ellipse and the corresponding canting angles of rain drops obtained from incoherent radar measurements are in good agreement with the passive measurements.

REFERENCES

1. B. G. Kutuza, G. K. Zagorin, A. Hornbostel and A. Schroth, "Physical Modeling of Passive Polarimetric Microwave Observations of the Atmosphere with Respect to the Third Stokes Parameter," *Radio Science*, 33, 677-695 (1998).
2. A. Hornbostel, A. Schroth, B. G. Kutuza and A. Evtuchenko, "Dual Polarization and Multifrequency Measurements of Rain Rate and Drop Size Distribution by Ground-Based Radar and Radiometers," *Proc. of IGARSS'97*, Singapore, (1997).
3. A. Schroth, A. Hornbostel, B. G. Kutuza and G. K. Zagorin, "Utilization of the First Three Stokes Parameters for the Determination of Precipitation Characteristics," *Proc. IGARSS'98* , Seattle, pp. 141-143, (1998).
4. L. Tsang, Kong, J. A., and Shin, R. T., *Theory of Microwave Remote Sensing*, 121ff., John Wiley and Sons, New York (1985).
5. T. Oguchi, „Scattering from Hydrometeors: A Survey," *Radio Science*, 16, 691-730 (1981).
6. A. Schroth, M. Chandra and P. Meischner, *Coherent Polarimetric Radar Techniques for Microwave Propagation and Cloud Physics Research*, DFVLR-FB 88-47, DFVLR, Cologne (1988).
7. O. Kobayashi , Y. Matsuzaka and H. Hirosawa, "Measurements of Modified Mueller Matrices of Backscatter from Random Media Using a Power-Measuring Scatterometer," *IEEE Trans. on Geosc. and Rem. Sens.*, 28, pp. 438-442 (1990).

Microw. Radiomet. Remote Sens. Earth's Surf. Atmosphere, pp. 325–335
P. Pampaloni and S. Paloscia (Eds)
© VSP 2000

Polarimetric radiometry of rain events: theoretical prediction and experimental results

A. CAMPS[1], M. VALL-LLOSSERA[1], N. DUFFO[1], F. TORRES[1], J. BARÁ[1], I. CORBELLA[1], and J. CAPDEVILA[1,2]

[1] *Universitat Politècnica de Catalunya, Campus Nord, D3, 08034 Barcelona, SPAIN*
[2] *University of Massachusetts, Amherst, 01003 MA, USA (since January 1999)*

Abstract—During the last several years a number of theoretical research studies and experimental measurements have focused on polarimetric radiometry, most of them devoted to the polarimetric emission behavior of natural surfaces, such as the sea surface roughened by the wind. More recently, some theoretical results have also been published on the polarimetric emission behavior of rain, under some hypothesis [1]. In this paper, known results on rain attenuation and depolarization due to rain scattering and raindrop shapes are incorporated into a radiative transfer model, which is used to predict the down-welling polarimetric emission behavior of rain events. Because the four-element Stokes vector depends on the canting angle of the raindrops, their dependence several frequencies and rain rates is evaluated. This model indicates that the non-zero third and fourth Stokes parameters are mainly due to the depolarization induced by differential phase constant and attenuation induced by canted raindrops. Experimental results obtained with an X-band polarimetric radiometer developed at the Universitat Politècnica de Catalunya (UPC) [2] are presented. They range widely in rain intensity, from typical intense Mediterranean storms to light rain. Measurements are discussed and compared to rain intensities and wind measurements collected by the Barcelona gauge network and by the meteorological station ESCRA (Environmental Radioactivity Service, Energy Techniques Institute, UPC).

1 INTRODUCTION: RAIN PHENOMENOLOGY AND EMISSION BEHAVIOR

The atmospheric down-welling thermal emission is given by [3]

$$T_{DN,p}(f,\theta,H) = \sec\theta \int_0^{\tilde{}} k_{a,p}(f,z')\, T(z')\, e^{-\tau_p(f,z',H)\,\sec\theta}\, dz' \tag{1}$$

$$\tau_p(f,z',H) = \int_{z'}^{H} k_{a,p}(f,z)\, dz\,,$$

where $k_a(f,z)$ and $T(z)$ are the absorption coefficient, and the physical temperature of the atmosphere at a height z. $k_{a,p}(f,z)$ includes the attenuation due to atmospheric constituents and, eventually, clouds and rain. In the latter case, it is known that $T_{DN\,h}$ and $T_{DN\,v}$ are

Figure 1- Shape of a rain drop for equivalent radii: 0.25, 0.50 ... 3.25 mm

different due to the non-spherical shape of rain drops. This shape was computed by [4] and it is shown in figure 1 for 13 different equivalent radii. The larger the equivolumetric radii of the rain drop, the larger the difference from the spherical shape. Figure 2 shows the Laws-Parsons equivalent radii distribution vs. rain rate [5]. Note that the mean equivalent raindrop radius increases at higher rain rates, and so does their standard deviation.

This means that in the steady state of a typical rain event, when drop fusion and breaking compensate each other, there are drops of a large variety of sizes. As table 1 from [6] shows, larger drops fall faster than smaller ones. As can be appreciated, the smallest drops, say those with radii smaller than 1.75 mm, have Reynolds numbers smaller than 2000 and consequently, can be said that they fall in a shear flow, while the largest ones fall in a turbulent one. Note that the terminal velocities are comparable to usual horizontal wind velocities. This means that in the horizontal direction different regimes (shear and turbulent flows) may be found, which further complicates the mechanics of the problem.

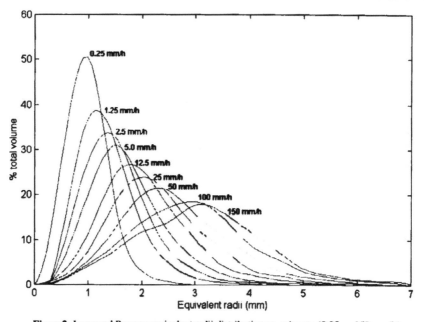

Figure2- Laws and Parsons equivalent radii distribution vs. rain rate (0.25 ... 150 mm/h)

Table 1.

Radii (a_0 [mm]), terminal velocities (V_∞ [m/s]) and Reynolds numbers ($N_R = 2a_0 V_\infty/\mu_{air}$) of different raindrops

a_0	0.25	0.50	0.75	1.00	1.25	1.50	1.75	2.00	2.25	2.50	2.75	3.00	3.25
V_∞	2.06	4.03	5.40	6.49	7.41	8.06	8.53	8.83	9.00	9.09	9.13	9.14	9.14
N_R	68.6	268	540	865	1234	1611	1989	2353	2698	3028	3345	3654	3958

When two fluids move one inside the other, two parameters characterize their relative movement: (1) $\lambda = \mu_{water}/\mu_{air}$, the ratio of the dynamic viscosities, and (2) $C = \mu_{air}Ga_0/\sigma_{water}$, the dimensionless capillary number. G [s^{-1}] is the linear shear flow, a_0 is the drop radius and σ_{water} is the water superficial tension. In the case of rain, at 20°C, $\lambda \approx 56$ and $C \approx 2.5 \times 10^{-4}aG$ [5, table1], which is very close to 0 for typical values of G and $a \approx 1$ mm. For the range of values of the above parameter, the effect of a linear shear flow in a spherical drop was first analyzed by Cox in 1969 [8]. More general analyses have been carried out recently [9], but none of them, to our knowledge, has dealt with the problem of analyzing the effects of a *turbulent* flow on a spherical or non-spherical drop. Without trying to solve the problem of the movement of a raindrop falling under shear and turbulent flows, some results found in the literature will be summarized. It is known that, for small Reynolds numbers, when a drop moves in a linear shear flow $dU/dx = -G$, being U the speed of each fluid layer, it suffers four effects:

i) <u>Deformation in the direction of the flow</u> $D = (L + B)/(L - B)$, where L and B are the longest and shortest axes in the equatorial plane [8, eqn. 6.23], and in the direction perpendicular to the flow $D_{\perp}$, which is about an order of magnitude smaller than D [9].

$$D = \frac{5(19\lambda + 16)}{4(\lambda + 1)\sqrt{(20/C)^2 + (19\lambda)^2}}$$ [2]

ii) <u>Orientation</u> of the longest axis at an angle α with respect to the normal of the shear flow [8, eqn. 6.24].

$$\alpha = \frac{\pi}{4} + \frac{1}{2}\arctg\left(\frac{19}{20}\lambda C\right)$$ [3]

iii) <u>Rotational motion</u> of the fluid inside the droplet with a certain periodicity.

iv) <u>Wobbling</u> or time-dependent deformation of the drop.

For an initially spherical drop subjected to a constant shear flow, the steady state is reached after a sinusoidal oscillation exponentially damped. For high viscosity ratios λ, the exponential damping is independent of C, while the frequency of the oscillations is proportional to C within 2% [9]. In our case, for $C \ll 1$ and $\lambda \approx 56$, for spherical drops the deformations D and $D_{\perp}$ are negligible, the orientation angle $\alpha \approx \pi/4$ loses its sense, and the steady state is reached without oscillations, as compared to the time constant of the exponential damping. In all the effects, the spherical drop acts as a rigid sphere.

In the presence of wind, raindrops migrate laterally within the air, thus in case of arbitrarily shaped raindrops (figure 1) the deformation is negligible and only a rotation takes place. For a constant shear flow in the horizontal direction, rotation angles in the range 0° to 25° have been numerically computed for the drop shapes shown in figure 1. Of course, the average angle of rotation for all raindrops is much smaller, only few degrees.

At this point it should be remembered that large drops break up into smaller ones, and small drops collapse forming larger ones, reaching a dynamic equilibrium described by the Laws-Parsons law (figure 2). In these processes, the raindrop shape oscillates. The effect of these oscillations on the polarimetric emission behavior is unknown, and therefore not considered in the analyses in the next section.

2 POLARIMETRIC EMISSION BEHAVIOR OF RAIN
2.1 Theoretical formulation

The down-welling polarimetric emission of the rain can be computed by means of the radiative transfer equation [10]:

$$\cos\theta \frac{d\overline{T}_{s}(\theta,\phi,z)}{dz} = -\overline{\overline{k}}_{e}(\theta,\phi)\,\overline{T}_{s}(\theta,\phi,z) + \overline{F}(\theta,\phi)T(z)$$

$$+ \int_{0}^{2\pi}\int_{0}^{\pi} \overline{\overline{P}}(\theta,\phi,\theta',\phi')\overline{T}_{s}(\theta',\phi',z)\,d\Omega'$$ [4]

with the boundary conditions $\overline{T}_{B}(\theta,\phi,z = h) = [0\,0\,0\,0]^{T}$, h being the height of the rain cell. This condition means that any external radiation is neglected with respect to the radiation introduced by the rain cell.

In eqn. (4), $\overline{T}_s(\theta,\phi,z) = [T_v \; T_h \; T_u \; T_v]$ is the Stokes vector containing the vertical and horizontal brightness temperatures, and T_U, T_V the so-called third and fourth Stokes parameters that measure the polarization state of the incoming wave. The extinction matrix $\overline{\overline{k}}_e(\theta,\phi)$, the phase matrix $\overline{\overline{P}}(\theta,\phi,\theta',\phi')$ and the emission vector $\overline{F}(\theta,\phi)$ depend on the forward scattering amplitudes f_{pq} [10].

The solution of eqn. (4) can be simplified by assuming that forward scattering is dominant and neglecting multiple scattering. In this case, only forward scattering amplitude factors have to be known. In order to apply this method to a realistic rain event, raindrop shapes have to be taken into account in the computation of the forward scattering amplitudes f_{pq}, and results must be averaged for all radii with a weight function given by the Laws-Parsons law.

In [6,11] a procedure is presented to compute the forward scattering amplitudes, cross-polarization factors, and wave-numbers at h/v polarization. The variation of the electric field given by the expression $\overline{E} = \overline{\overline{T}} \, \overline{E}_o$, which relates the incident electric field $\overline{E}_0$ in a layer of width z, with the outgoing electric field $\overline{E}$. The parameters of the $\overline{\overline{T}}$ matrix are given by

$$T_{11} = \cos^2\phi \; e^{\lambda_v z}(1 + G\,\mathrm{tg}^2\phi); \quad T_{12} = T_{21} = \cos^2\phi \; e^{\lambda_v z}(1-G)\mathrm{tg}\,\phi;$$
$$T_{22} = \cos^2\phi \; e^{\lambda_v z}(G + \mathrm{tg}^2\phi); \tag{5}$$

where $G = e^{(\lambda_h-\lambda_v)z}$ and ϕ is the **effective canting angle**. In the particular case of Gaussian distributions for the canting angle θ, with mean value θ_0 and a standard deviation σ_θ, and the azimuth angle γ, with mean value γ_0 and a standard deviation σ_γ, it is obtained that

$$\lambda_1 - \lambda_2 = -j\frac{1}{2}(k_x - k_y)(e^{-2\sigma_\gamma^2} \cos 2\gamma_0 + 1)e^{-2\sigma_\theta^2}, \quad \phi = \theta_0, \tag{6}$$

where k_x-k_y is the differential propagation constant. Note that the concept of effective canting angle has the physical meaning discussed in section 2: due to the large viscosity ratio, raindrops are not likely to be distorted due to a horizontal wind, but only rotated by a certain angle. Note also, that the derivation of eqn. (4), implicitly assumes that the Gaussian distribution for the canting angles is independent of the raindrop size. This is not the case since small raindrops are almost spherical, and suffer from a much smaller canting angle than the largest ones (section 2).

2.2 Simulation results

This section presents some numerical results. The scenario consists of a rain cell of 4 Km height, infinite in the horizontal plane, at a uniform temperature of 20°C, and a polarimetric radiometer measuring the down-welling atmospheric emission with 30° elevation angle. Figure 3 shows for f=10.68 GHz and a rain rate of 25 mm/h, the effect of the variation of the mean canting angle θ_0, on the four Stokes parameters, for $\gamma_0=\sigma=0°$, $\gamma_0=5°$, and $\sigma=5°$. As can be appreciated, for $\theta_0=0°$ (the longest axis of the drops is horizontal), T_h is larger than T_v due to the larger attenuation in h-polarization. When $\theta_0=\pm90°$, it is T_v that is larger. Note that for $\gamma_0\approx90°$, $T_{12}=T_{21}=0$ (eqn. (4)), which will result in zero third and fourth Stokes parameters. The third and fourth Stokes parameters show a negligible dependence on σ_θ and σ_γ, an azimuth dependence on γ_0 (averaged for all drops), and a sinusoidal dependence with the mean canting angle θ_0. For a fixed σ_θ, σ_γ and γ_0, the variation of the Stokes vector with θ_0 is given by

$$T_h = T_o + \Delta T \cos 2\theta_o; \quad T_v = T_o - \Delta T \cos 2\theta_o; \quad T_U = -T_{Uo} \sin 2\theta_o; \quad T_V = -T_{Vo} \sin 4\theta_o, \quad [5]$$

where $T_o = (T_h + T_v)/2$, and T_{U0}, T_{V0} and ΔT are the amplitudes of the modulation due to θ_0. In addition, for small rain rates $T_{U0} = \Delta T$ and the variation of T_U with $\sin 2\theta_0$ is the same as that found analytically by *Kutuza et al.* [10], for ellipsoidal raindrops.

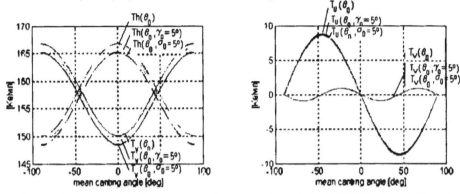

Figure 3- Dependence of the Stokes vector on the mean canting angle θ_0. f=10.68 GHz, rain rate 25 mm/h.
a) T_h and T_v,   b) T_U and T_V.

Figure 4 shows the values for T_h and T_v at $\theta_0=0°$, for T_U at $\theta_0=45°$, and for T_V at $\theta_0=22.5°$, for four different frequencies vs. the rain rate. From those parameters, any other value, at any mean canting angle, can be inferred. As can be appreciated, in the log-log plot there is a linear increase of all four parameters with increasing rain rates, but the slope of T_U and T_V is twice that of T_h and T_v. Note that, since the attenuation increases with frequency, T_h and T_v saturate at a value equal to the physical temperature (293 K) at lower rain rates for increasing frequencies. It should be pointed out that at higher rain rates the assumption of negligible multiple scattering is no longer valid and the model fails, since it predicts a saturation of T_h and T_v ($\Delta T \rightarrow 0$), but also a saturation of T_U and T_V, while their amplitude should vanish. The validity of the model is limited to the the linear region, up to 10-20 mm/h, and decreases with frequency. A very interesting feature can be seen in the evolution of T_V vs. rain rate at 34.8 GHz. T_V represents the amount of circular polarization in the incoming wave. The circular polarization comes from the different phase constant at both polarizations $\Re(k_x) \neq \Re(k_y)$. However, when the rain cell is too large, or the rain intensity is too high, the phase shift exceeds 90°, and the amount of circular polarization decreases, as can be noted in the valley at about 50 mm/h.

3   EXPERIMENTAL MEASUREMENTS

This section presents some measurements obtained with an X-band polarimetric radiometer, shown in figure 5. Table 2 lists its main performances. The elevation of the antenna beams is 30°. Figure 6 shows their orientation and that they cover three meteorological stations (white dots) of the rain gauge network of the city, as well as the one located on the same UPC Campus (ESCRA). The data available from the rain gauge network is the 5 min rain rate, and those from ESCRA the 30 s rain rate and the 5 s mean and standard deviation of the wind speed and direction. Special care has to be paid to the calibration process and antenna cross-polarization effects. A polarimetric radiometer can be viewed as a single baseline of an interferometric radiometer, in which there is a single two-polarization antenna, instead of two separate antennas. Therefore, most of the techniques used to calibrate an interferometric radiometer [11] can be applied. In our system, offset drifts are very small (e.g. figure 9e), and small phase drifts are periodically self-calibrated from the data itself.

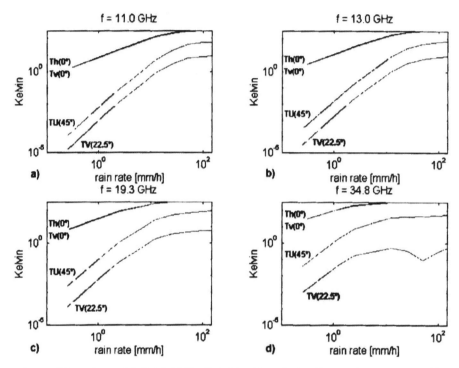

Figure 4- Dependence of $T_h(\theta_0=0°)$, $T_v(\theta_0=0°)$, $T_U(\theta_0=45°)$ and $T_V(\theta_0=22.5°)$ on the rain rate for different frequencies: a) $f = 11.0$ GHz, b) $f=13.0$ GHz, c) $f = 19.3$ GHz and d) $f = 34.8$ GHz. Predicted maximum saturation amplitudes for T_U ($\theta_0 = 45°$): a) ~ 65 K, b) ~ 72 K, c) ~ 98 K, d) ~ 65 K.

Table 2. Polarimetric radiometer parameters.

Parameter	
$\Delta\theta^{E,H}_{-3dB}$	30°, 20°
Cross-polar level	-25 dB
Side Lobe Level	-20 dB
f_0	10.68 GHz
B	40 MHz
$T_{R1,2}$	~ 90 K, 120 K
Dicke radiometer: τ (H/V switched)	0.5 s x N (N: averaging)
Polarimetric radiometer: τ . 1B/2L dig.correlators. f_s	1 ms – 65 s @ 66 MHz < 100 MHz

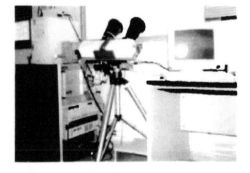

Figure 5- X-band polarimetric radiometer at the Microwave lab. (UPC-Campus Nord, Barcelona).

Presented measurements include: a) rain rate at, at least, the ESCRA meteorological station (UPC), b) the mean wind direction (N=0°, E=90°, S=180° and W=270°), c) the mean wind speed, d) T_h and T_v in Kelvin, and e) T_U and T_V in Kelvin. The Stokes vector is measured once each 15 s. Figure 7 shows the measurements of August 18, 1998, and figure 8 a zoom off the period corresponding from 7 to 8 AM. Due to the wide antenna pattern, T_h and T_v are integrated measurements of all rain cells taking place within the field of view of the antenna. At a given moment, it could be raining at one rain gauge station and not at others.

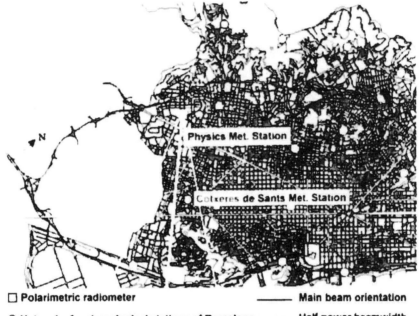

□ **Polarimetric radiometer** ———— **Main beam orientation**

○ **Network of metereological stations of Barcelona** -------- **Half power beamwidth**

Figure 6- Map of (o) Barcelona's rain gauge network and ( ) location of the polarimetric radiometer.

This effect is apparent in figures 8a and 8d, where small peaks in T_h and T_v correspond to different rain events at different locations. As expected, T_h is slightly higher than T_v. At this point it should be pointed out that in the configuration shown in figure 5, some radiation was inevitably collected from the environment through the secondary lobes, which is responsible for the high T_h and T_v values in the absence of rain.

Figure 8e shows non-zero third and fourth Stokes parameters. Due to the different rain rates measured at each station, it is difficult to infer a relationship between T_U and T_V and the wind speed. They seem to be modulated by both the rain rate (note the peaks in ESCRA rain rate data and those in T_U), as well as by the wind intensity. The T_V values, comparable to those of T_U, cannot be explained by the former theory.

Figure 9 shows the measurements corresponding to November 29, 1998. It is interesting to note that a 10 minute, light 1.5 mm/h rain produces: i) an 3 K detectable increase in T_h and T_v (not shown) over the floor set by the radiation collected through the secondary lobes, ii) a T_U value as large as in the preceding set of measurements, and iii) a very small sign-alternating T_V value.

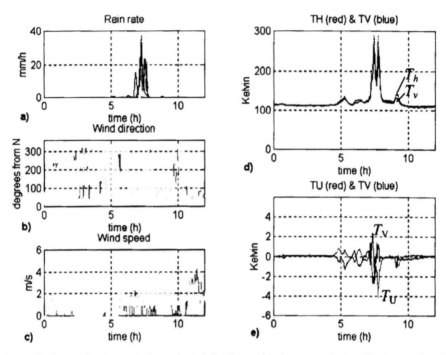

Figure 7- Measured rain rate, wind intensity and direction, and Stokes vector. Date: 0-12h, August 18, 1998

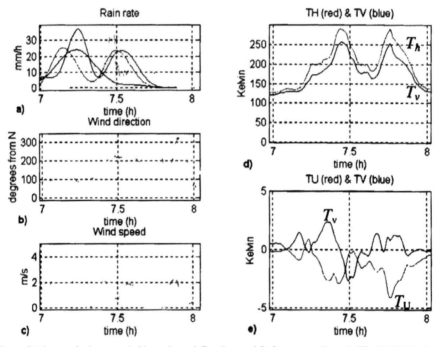

Figure 8- Measured rain rate, wind intensity and direction, and Stokes vector. Date: 0-12h, 18/8/1998, from 7 to 8 AM. Smooth rain rates curves: 5 min rain gauge network data), fine rain rate curve: 30 s ESCRA data.

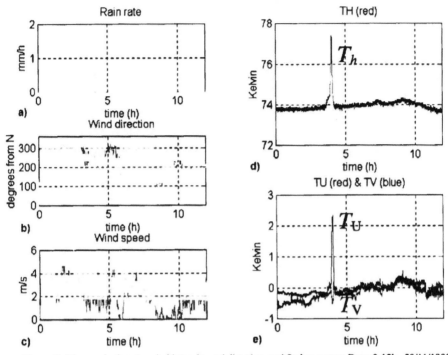

Figure 9- Measured rain rate, wind intensity and direction, and Stokes vector. Date: 0-12h, 29/11/1998.

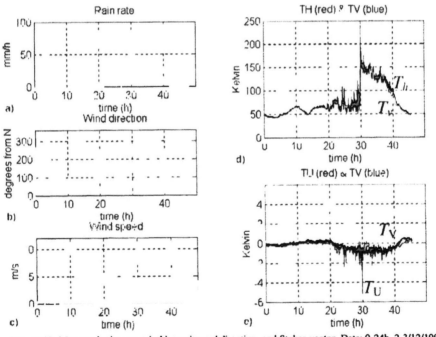

Figure 10- Measured rain rate, wind intensity and direction, and Stokes vector. Date: 0-24h, 2-3/12/1998.

These values may not be only attributed to a canting (figure 4), so other factors such as drop oscillation may be playing a role, but their effect is at present unknown.

Figure 10 shows the measurements corresponding to December 2 and 3, 1998, when an intense Mediterranean storm fell in the Barcelona area, with peak rain rates up to 90 mm/h. The T_U peaks follow the rain intensity, but their value is not as high as expected. In addition, the T_U peaks at $t = 35$-40 h, are approximately of the same amplitude as those around $t = 24$ h, and those around $t = 32$ h, even though the rain intensities were about 10, 30 and 40-60 mm/h respectively. This fact can only be interpreted through inspection of the wind speed and direction. At $t=24$ h the wind speed and direction are approximately constant and equal to 5 m/s and 50°. At $t=32$ h there is a quick decrease in the wind speed, which may be responsible for the low T_U signature (small canting angle), even though the high rain rate. Finally, in the $t=35$ to 40 h period, even though the rain rate is moderate, the wind direction changes rapidly, and so does the wind speed in the range $0 - 3$ m/s, which may be responsible for oscillations in the raindrops, translating them into the T_U signature.

4 CONCLUSIONS

This paper has presented a theoretical analysis of the down-welling polarimetric emission behavior of the rain, showing that in the absence of horizontal wind the deformation/orientation of falling raindrops is negligible, but under horizontal wind, raindrops rotate due to their non-spherical shape. For a Gaussian canting angle distribution independent of raindrop size, T_h and T_v follow a cosine dependence with the mean canting angle θ_0, while the T_U and T_V follow a sine dependence with twice and four times θ_0. The dependence of the amplitude of these variations with respect to rain rate has been computed for a 4 Km height rain cell at 20°C, at four frequencies as a function of the rain rate. Results show: (1) an increase with rain rate followed by a saturation due to the excess attenuation, (2) that the slope of the $T_{U,V}$ plots is twice that of $T_{h,v}$, (3) that T_V may vanish at intense rain rates, high frequencies and/or large rain cells (figure 4d), due to a change in the differencial phase shift.

X-band experimental measurements have shown the dependence of T_U and T_V on rain rate, even though the relationship to wind speed/direction through θ_0 and γ_0 respectively, is not yet clear and cannot by itself explain some particularly high measured values, specially of T_V. Possible causes are: the limited accuracy of the model for high rain rates, the beam-filling factor of the rain cell, transient states (wobbling) during the storm formation, and more probably, the presence of ice crystals that introduce a large phase shift among polarizations.

Acknowledgments
The authors appreciate the meterological data provided by Dr. Guillem Cortés, Assistant Professor at ESCRA, Institut de Tècniques Energètiques, UPC.

References

1. Kutuza, B.G., G.K. Zagorin, A. Hornbostel, and A. Schroth, *Physical Modeling of Passive Polarimetric Observations of the Atmosphere with respect to the Third Stokes Parameter*, Radio Science, Vol. 33, No. 3, 677-695 (1998)

2. Camps, A., F. Torres, I. Corbella, J. Bará, and X. Soler, *Calibration and experimental results of a two-dimensional interferometric radiometer laboratory prototype*, Radio Science, Vol. 32, No 5, 1821-1832 (1997)

3. Ulaby, F.T., R. K. Moore, A.K. Fung, *Microwave Remote Sensing, Active and Passive. Vol. I*, Chapter 4, ed. Artech House, Norwood MA. (1981)

4. Pruppacher, H. R., and R.L. Pitter, *A Semi-empirical Determination of the Shape of Cloud and Rain Drops*, J. Atmospheric Sciences, 28(1), 86-94 (1971)

5. Laws, J.O. and D.A. Parsons, *The Relationship of Raindrops Size to Intensity*, Trans. American Geophysical Union, 24th Annual Meeting, 452-460 (1943)

6. Oguchi, T., *Scattering Properties of Pruppacher-and-Pitter form raindrops and cross polarization due to rain: Calculations at 11, 13, 19.3 and 34.8 GHz*, Radio Science, Vol. 12, No 1, 41-51 (1977)

7. Giles, R.V., *Teoría y Problemas de Mecánica de Fluidos e Hidráulica*, ed. Mc Graw-Hill (1990).

8. Cox, R.G., *The Deformation of a Drop in a General Time-dependent Fluid Flow*, J. of Fluid Mechanics, Vol 37, part 3, 601-623 (1969)

9. Uijttewaal, W.S. and E.J. Nijhof, The Motion of a Droplet Subjected to a Linear Shear Flow Including the Presence of a Plane Wall, J. of Fluid Mechanics., Vol. 302, 45-53 (1995)

10. Tsang, L., *Polarimetric Passive Microwave Remote Sensing of Random Discrete Scatterers and Rough Surfaces*, J. of Electromagnetic Waves and Applications, Vol. 5, No. 5, 41-57 (1991)

11. Oguchi, T., *Scattering from Hydrometeors: A Survey*, Radio Science, Vol. 16, No 5, 691-730 (1981)

3. Remote sensing of clouds and precipitation

3.2 Satellite and ground based observations

Microw. Radiomet. Remote Sens. Earth's Surf. Atmosphere, pp. 339–351
P. Pampaloni and S. Paloscia (Eds)
© VSP 2000

Application of AMSU for obtaining hydrological parameters

NORMAN GRODY, FUZHONG WENG and RALPH FERRARO

NOAA/NESDIS, 5200 Auth Road, Camp Springs, MD USA 20746

Abstract- NOAA launched its most advanced sensor called the Advance Microwave Sounding Unit (AMSU) on May 13, 1998. The AMSU contains twelve channels within the 50-60 GHz oxygen band, four channels around the 183 GHz water vapor line and four window channels at 23.8, 31.4, 50.3 and 89 GHz. It was designed primarily to improve the accuracy of temperature soundings beyond that of the four-channel Microwave Sounding Unit (MSU). Besides deriving temperature profiles, the AMSU instrument is also used to measure precipitation, water vapor, cloud liquid water, snow cover, precipitation and sea ice concentration. This paper describes the AMSU algorithms used to derive these hydrological parameters.

1. INTRODUCTION

Microwave instruments are used to measure atmospheric and surface parameters under all weather conditions. As shown in Figure 1, they were first flown aboard U.S. satellites in 1972. At that time the National Aeronautic and Space Administration (NASA) began flying microwave radiometers aboard the Nimbus series of satellites. The first radiometers were flown on Nimbus-5, but had few channels with poor spatial coverage, i.e., ESMR-1 only had a 19 GHz channel and NEMS was only nadir viewing. These initial sensors were soon replaced by more advanced instruments until the Nimbus program ended in 1978 with the launch of the Scanning Multichannel Microwave Radiometer (SMMR) on Nimbus-7. Measurements from these experimental units were used to advance our knowledge of instrument design and improve the use of microwave techniques for observing the Earth from space.

Beginning in 1978, the National Oceanic and Atmospheric Administration (NOAA) launched its first microwave radiometer, called the Microwave Sounding Unit (MSU), aboard the TIROS series of polar orbiting satellites (see Fig. 1). This operational instrument was designed similar to the Scanning Microwave Spectrometer (SCAMS), which was flown on Nimbus-6 in 1973. The MSU contains four channels in the 50 to 60 GHz portion of the oxygen band to derive temperature soundings from the surface to about 50 mb. A number of identical instruments have been flown over the past twenty years, each having exceptionally high stability and precision. The high performance of the MSU has made it possible to monitor very small climatic changes in temperature, with an accuracy less than 0.1 K [1].

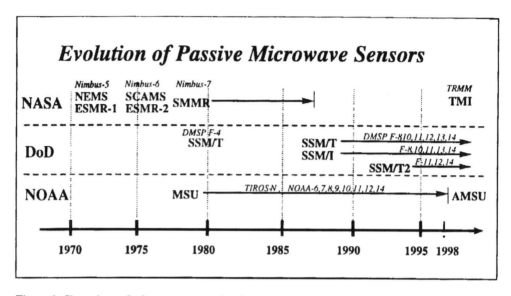

Figure 1. Chronology of microwave sensor development

During the past ten years there has been a dramatic increase in the use of microwave derived products by the worldwide community of meteorological and oceanographic organizations. This greater emphasis is primarily attributed to the launch of the Special Sensor Microwave Imager (SSM/I) in 1987 onboard the first of a series of DMSP (Defense Meteorological Satellite Program) satellites (see Fig. 1). In contrast to temperature sounders, the microwave imager contains six channels in window regions (19, 37, 85 GHz) with dual polarization, and a seventh channel centered on the 22.23 GHz water vapor line with vertical polarization. Products such as rainfall, snow cover, cloud liquid water, water vapor, sea surface winds and sea ice concentration are produced each day on a global basis using all of the SSM/I channel measurements. The SSM/I was developed by the U.S. Navy and first launched on the F-8 DMSP satellite. As a result of a shared processing agreement, all of the SSM/I products are generated by the U.S. Navy and distributed to both the U.S. Air Force and NOAA. The three agencies utilize the products in various ways to improve the analysis and forecast of weather systems.

In addition to flying the SSM/I, the DMSP satellites also carry a temperature sounder (SSM/T) and a water vapor sounder (SSM/T2), both of which were developed by the U.S. Air Force. As shown in Figure 1, the SSM/T was first launched in 1979, and contains seven channels in the 50-60 GHz oxygen band to derive temperature profiles from the surface to about 5 mb. The SSM/T2 is the latest DMSP sensor and was first flown in 1991 aboard the F-11 satellite. It contains window channels at 91 and 150 GHz, and three channels around the strong water vapor line at 183.31 GHz. These channels are used to derive moisture profiles from near the surface to around 300 mb, and to identify precipitation over land and ocean. It is noteworthy that all three instruments (SSM/I, SSM/T, SSM/T2) are currently being flown together aboard the recently launched F-14 satellite, so that radiometric measurements are now available between 19 and 183 GHz. However, the DMSP instruments all have different scan geometry (SSM/T and SSMT/2 are cross-track scanners while SSM/I is a conical scanner) so that even when they are flown together on a single satellite it is very difficult to combine the channel information. To alleviate this problem, a new sensor called the SSM/IS is being developed by the U.S. Air Force to include all of the channels (19-183 GHz) in a single instrument with the same conical scan geometry.

As shown in Figure 1, the evolution of microwave radiometers aboard satellites has increased steadily over the past twenty-seven years. Just recently, a modified version of the SSM/I (called TMI) was flown by NASA aboard the TRMM satellite in November 1997; it contains a 10 GHz channel to improve the estimates of sea surface winds [2]. Of particular importance to NOAA, is the launch in May 1998 of a twenty channel instrument called the Advanced Microwave Sounding Unit (AMSU) which consists of three modules, AMSU-A1, AMSU-A2 and AMSU-B (see Fig. 2 below).

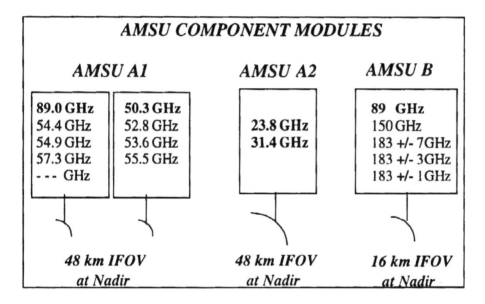

Figure 2. AMSU consists of three modules, AMSU-A1, -A2 and -B.

This long awaited instrument was designed primarily to improve the accu :y of temperature soundings beyond that of the four channel MSU. To obtain this improven: 1t, the AMSU-A1 module includes twelve channels in the 50-60 GHz portion of the oxygen band to provide temperature soundings from the surface to about 1 mb. Besides obtaining temperature soundings, the AMSU-B module contains four channels around the 183.31 GHz water vapor line for deriving moisture profiles. Furthermore, the AMSU-A1 and A2 modules include window channels at 31.4 GHz and 89 GHz to monitor surface features and precipitation, and contains a 23.8 GHz channel for obtaining the total precipitable water over oceans. Figure 2 shows the channels associated with each module, while Table 1 lists the parameters obtained from the window channel measurements along with the estimated accuracy's of each parameter. This paper describes the algorithms for obtaining the so-called Day-1 products. Following one year after launch (Day-2), we will also begin developing the remaining products listed in the Table. It is noteworthy that AMSU is the first sensor to include all of the channels (23-183 GHz) in a single instrument with the same scan geometry. Currently only the window channels are being used to develop the products listed in Table 1. However. future research will concentrate on the use of the water vapor and oxygen channels to improve the products.

Table 1. AMSU Non-Sounding Products (* Day-2 Products)

Ocean Parameters	Channels Used	Parameter Range	Accuracy (rms)
Precipitable water	23, 31 GHz	0 – 70 mm	1.0 mm
Cloud Liquid Water	23, 31, 50 GHz	0 – 3 mm	0.05 mm
Sea Ice Concentration	23, 31 GHz	0 – 100 %	15 %
Instantaneous Rain Rate	23, 31, 89 GHz	0 – 30 mm/h	2 mm/h
Cloud Ice Content *	23, 89, 150 GHz	0 - 3 mm	TBD
Sea Surface Wind *	23, 31, 50 GHz	0 - 50 m/s	TBD
Land Parameters			
Snow Cover	23, 31, 89 GHz	0 – 100 %,	5 %
Instantaneous Rain Rate	23, 89, 150 GHz	0 – 30 mm/h	3 mm/h
Snow Depth *	23, 31, 89 GHz	0 – 30 cm	TBD
Surface Temperature *	23, 31, 50 GHz	250 – 320 K	TBD
Surface Wetness *	23, 31, 50 GHz	TBD	TBD

2. COMPARISONS BETWEEN AMSU AND DMSP SENSORS

The algorithms developed for AMSU-A are primarily based on knowledge acquired using SSM/I measurements [3]. Similarly, algorithms developed for AMSU-B are based on SSMT/2 measurements [4]. For comparison purposes, Table 2 lists the channel frequencies for the SSM/I, SSM/T2 and AMSU instruments. Also listed are the different footprints (i.e., field of view at half power) viewed by each sensor. The SSM/I is a conical scanner so that the footprints are constant, independent of the azmuthal scan angle. The other sensors use a cross-track scanning mechanism so that the footprint can increase by more than a factor of two as the instrument scans from nadir to the limb position. For large-scale features such as atmospheric temperature and water vapor, the smoothing due to larger footprints is often minimal. However, when using a cross-track scanner to derive rain rates and cloud liquid water it is generally important to normalize the measurements to a common resolution. Also, when combining or comparing measurements from different sensors, the different resolutions must also be accounted for. The most commonly used approach is to simply average the product derived by various sensors to the footprint of the lowest resolution instrument.

2.1. Mixed pixels and Aliasing effects

The DMSP instruments use the full antenna aperture to collect the Earth emitted radiation. While this provides the highest resolution for a given antenna size, the footprint size varies in proportion to the wavelength, i.e., the SSM/I footprint decreases from 60 km at 19 GHz to 15 km at 85 GHz. Different resolutions can result in aliasing effects when the channels are combined to derive geophysical parameters. For example, the difference between a high and low resolution channel will be altered simply due to the different scenes being viewed by the two channels, particularly around coast lines. To eliminate aliasing effects, the antenna system for AMSU-A was designed so that all channels have the same instantaneous field of view (IFOV). Similarly the AMSU-B channels all have the same footprint (see Table 2 and Figure 2). As such, the surface emissivity and cloud measurements obtained from the window channels (23.8, 31.4, 50.3 GHz) can be used to correct the lower temperature sounding channels at 52.8 and 53.6 GHz, without having to account for differences in footprint size.

Table 2. Comparison between AMSU and DMSP Channels

| --- Channels (GHz) --- | | | --- Nadir Resolution (km) --- | | |
AMSU	SSM/I	SSMT/2	AMSU	SSM/I	SSMT/2
- - -	19.3	- - -	- - -	60	- - -
23.8	22.2	- - -	48	55	- - -
31.4	37.0	- - -	48	30	- - -
50-60	- - -	- - -	48	- - -	- - -
89.0 *	85.5	91.0	48 & 16*	15	- - -
150 *	- - -	150	16*	- - -	60
183±7 *	- - -	183±7	16*	- - -	50
183±3 *	- - -	183±3	16*	- - -	50
183±1 *	- - -	183±1	16*	- - -	50

* *From AMSU-B (note 89 GHz channels are on A & B modules)*

In the case of the SSM/I , aliasing effects result in false identification around coast lines and lake boundaries when the 19 and 85 GHz channels are subtracted to identify snow cover and precipitation. However, even if the aliasing effects are minimized by reducing the resolution of all channels to a common footprint, the false identification around coast lines still exists due to the overlap of the microwave footprint between land and ocean, i.e., mixed pixel effect. Investigators have found that the most efficient means of resolving both of these problems (aliasing and mixed pixel) is to simply eliminate the measurements within a certain distance of water boundaries. This approach has been adopted for both the SSM/I and AMSU instruments. A more physically based, and more complicated approach, would involve the retrieval of surface emissivity from the window channel measurements, and the subsequent correction of the geophysical algorithms using the retrieved emissivity.

3. AMSU ALGORITHMS

From an historical point of view it should be understood that AMSU was designed to obtain vertical profiles of temperature and water vapor. Most of the "non-sounding" products listed in Table 1 were never envisioned around 1980 when AMSU was designed. While the AMSU contains many of the important channels needed to derive the parameters listed in Table 1, it does not contain low frequency window channels (< 23 GHz) or provide polarization information. It is for this reason that one can not simply apply the SSM/I algorithms to the AMSU instrument. For example, due to the lack of polarization measurements, alternative procedures from that of the SSM/I must be sought to discriminate between desert surfaces, snow cover and precipitation. This concluding section briefly overviews the algorithms developed for the Day-1 products listed in Table 1. Each of the algorithms are described in detail, although most of them are still undergoing extensive evaluations.

3.1 Sea Ice Concentration

The sea ice concentration is obtained by assuming that the field of view contains either new ice or multiyear ice mixed with open water. As such, the fractional amount of sea ice, f, is given by

$$f = \frac{\varepsilon - \varepsilon_{water}}{\varepsilon_{ice} - \varepsilon_{water}} \tag{1}$$

where ε is the measured emissivity at the lowest AMSU frequency (23.8 GHz), ε_{water} is the emissivity of open water and ε_{ice} is the emissivity of ice. The emissivity of water around 273 K is 0.45 and the emissivity of ice varies from about 0.95 for new ice to 0.88 for multiyear ice. Multiyear ice scatters microwave radiation and is identified when $T_B(23) - T_B(31) \geq 5$ K, where $T_B(\nu)$ is the brightness temperature at frequency ν.

Based on radiative transfer simulations over land, sea ice and oceans that includes non-precipitating clouds, the emissivity at 23.8 GHz is given by

$$\varepsilon = A + B\ T_B(31) + C\ T_B(23) + D\ T_B(50) \tag{2}$$

where A =1.734 - 0.6236Cos θ, B =0.0070 + 0.0025Cos θ, C = -0.00106, D= -0.00909.

The accuracy of (1) depends on the *a-priori* emissivity of sea ice and the estimated emissivity (2). Errors increase when the AMSU field of view contains the more variable emissivity of multiyear ice. However, the largest errors are observed during the summer period when ponds of water are present on the ice. This reduces the measured emissivity, and consequently reduces the ice concentration. Infrared measurements from the AVHRR on NOAA-K will be used in the future to help identify the occurrence of summer-ice-melt. Atmospheric absorption by clouds can also result in errors, particularly during the summer when precipitation occurs at higher latitudes. The 31.4 and 50.3 GHz channels used in (2) help to remove the effects of non-precipitating clouds, but rain clouds are not accounted for since then the brightness temperatures can increase to that approaching sea ice. This results in overestimates of both the emissivity (2) and sea ice concentration (1). To filter out the effects of precipitation, a discriminate function was obtained using a partitioned data set containing clouds and sea ice. For latitudes beyond 50 degrees of the equator the following discriminate function was obtained,

$$DF1 = 2.85 + 0.020\ T_B(23) - 0.028\ T_B(50) \tag{3}$$

where values greater than 0.45 are used to filter out precipitation. Unfortunately, some actual sea ice is also removed. To improve the filter, we plan to use infrared estimates of sea surface temperature as an additional check. We also plan to make comparisons between the AMSU sea ice concentration and that derived form SSM/I as well as the Navy product, which uses visible and infrared satellite observations as well as SSM/I data.

3.2 Water vapor and Cloud liquid water

Since the sea surface emissivity can be accurately modeled, radiative transfer simulations have been used to derive algorithms for ocean products such as cloud liquid water, Q, and precipitable water, V , (e.g., see [5]). The algorithms (in units of mm) are given as follows;

$$V = \text{Cos } \theta \, [\, C_0 + C_1 \, \Psi(23.8) + C_2 \, \Psi(31.4) \,] \qquad (4)$$

$$Q = \text{Cos } \theta \, [\, D_0 + D_1 \, \Psi(23.8) + D_2 \, \Psi(31.4) \,] \qquad (5)$$

where $\Psi(v) = \ln[285 - T_B(v)]$, and $T_B(v)$ are the AMSU brightness temperature measurements at frequency v. The coefficients C_i and D_i are as follows;

$$C_0 = 247.92 \, - \, (69.235 - 44.177 \, \text{Cos } \theta \,) \, \text{Cos } \theta \, , \quad C_1 = \, - \, 116.270 \, , \quad C_2 = 73.409$$

$$D_0 = \, 8.240 \, - \, (2.622 - 1.846 \, \text{Cos } \theta \,) \, \text{Cos } \theta \quad , \quad D_1 = 0.754 \, , \quad D_2 = \, - \, 2.265$$

The scan angle (limb) corrections given in (4) and (5) are functions of Cos θ, which is a symmetrical function of θ. However, when applying (5) to the AMSU measurements, one observes consistently more liquid water on one side of nadir than the other side [6]. This asymmetry is most predominant as the antenna beam approaches the Earth's horizon. As the scan angle increases from nadir, any left-right asymmetry in the antenna side lobes results in different amounts of cold space radiation (also satellite emitted and reflected radiation) being received on opposite sides of the scan. This effect was first noted from the MSU instrument. It was accounted for using fixed adjustments as a function of beam position and may be the main source of asymmetry for AMSU. Asymmetry in the AMSU channel measurements is accounted for using empirical adjustments to the brightness temperature measurements as a function of beam position [6].

In order to extend the water vapor and cloud liquid water to high latitudes, the effects of sea ice must be identified and edited out of the data. Sea ice can increase the emissivity to that of land surfaces and result in large overestimates of both water vapor and liquid water. The algorithm for obtaining sea ice concentration has been discussed above, and uses a discriminate function (3) to identify sea ice and filter out precipitation. This function was obtained using AMSU measurements over high latitude oceans that contained varying amounts of sea ice, clouds and precipitation. Values in excess of 0.0 and 0.2 are used to filter out sea ice for the liquid water and water vapor products, respectively. As an alternate technique of identifying sea ice, the use of sea surface temperatures (less than 271 K) based on infrared satellite measurements is also being investigated.

Preliminary validation of the water vapor and liquid water parameters has been completed. The water vapor measurements were compared against radiosonde data from island stations under non-precipitating situations where Q < 0.3 mm. It is found that adjustments to (4) are needed, where the corrected water vapor algorithm is given by

$$V' = \alpha V + \beta \tag{6}$$

where α is 0.942 and β is –2.17. These α and β parameters can be associated with errors in the water vapor absorption model and instrumental biases, respectively [5]. For example, a 5 percent increase in water vapor absorption at 23.8 GHz would result in $\alpha = 0.95$. In the case of liquid water, comparisons with ground-based radiometer data [7] have not been made at this time. However, the liquid water measurements have been analyzed under clear conditions to identify any bias. By constructing a histogram of the liquid water measurements obtained under clear conditions, a small positive bias of .03 mm is observed. This bias can be removed using an equation similar to (6), i.e., $Q' = Q - 0.03$.

3.3 Precipitation over Land and Ocean

For precipitating clouds, the instantaneous rain rate can be parameterized in terms of the liquid water and average descent time of the rain drops, i.e., Rain rate = Q/τ. In actual practice, a rain rate relationship is obtained by correlating the liquid water measurements against co-located rain gauge and radar measurements. Liquid water greater than 0.3 mm is generally associated with precipitation while lower values are considered rain-free. Using radar and rain gauge data, an empirical relationship was obtained between rain rate, R, and the SSM/I measurements of liquid water [8]

$$R = 0.002(100\ Q)^{1.7} \tag{7}$$

This technique of measuring precipitation is referred to as the emission approach since it uses low frequency emission measurements over oceans. A different technique uses the high frequency scattering by millimeter-size ice particles to estimate rain rates. It was originally developed for the SSM/I and SSM/T2 and is very appealing since it is applicable over land as well as oceans. Also, since the scattering technique uses the highest frequency channels, rain rates are derived at the highest resolution.

The following scattering indices were developed to identify scattering features over land and ocean using AMSU-A channel measurements;

$$SI_{Land} = T_B(23) - T_B(89) \tag{8a}$$

$$SI_{Ocean} = -113.2 + [\ 2.41 - 0.0049\ T_B(23)]\ T_B(23) + 0.454\ T_B(31) - T_B(89) \tag{8b}$$

where $SI_{Land} \geq 3$ and $SI_{Ocean} \geq 9$ are used to identify precipitation. The scattering indices were obtained using scatter-free AMSU data over the corresponding surfaces, and regressing the high frequency channel at 89 GHz against the lower frequency channels. Scattering indices provide a measure of the scattering at high frequencies (e.g., 89 GHz) relative to low frequency channels. In order to derive rain rates the indices must be calibrated against coincident radar and rain gauge measurements in a manner similar to that mentioned previously when discussing the emission technique [8]. The resulting rain rate relationship obtained over land from AMSU measurements is given by

$$R = 0.005\ [(SI_{Land} + 18)/1.3]^{1.96} \tag{9}$$

The resulting rain rates obtained using (7) and (9) are found to be comparable to the SSM/I derived rain rates. However, these results are only preliminary and more comparisons are needed to determine the effects of angular scanning and AMSU's varying footprint size.

The AMSU-B channels at 89 and 150 GHz will be used to derive rain rates at the highest resolution (16 km at nadir). Similar channels were included on the SSM/T2, and experimental rain algorithms were developed using the following indices,

$$SI_{Land} = [42.72 + 0.85\ T_B(91)] - T_B(150) \tag{10a}$$

$$SI_{Ocean} = .013\{T_B(91) + 33.58\ ln[300 - T_B(150)] - 341.17\} . \tag{10b}$$

Over land, the scattering index, SI_{Land} , is an estimate of the scattering effect at 150 GHz relative to the 91 GHz channel. Rain rates exceeding 1 mm/h are identified when the index exceeds a 5 K threshold [4]. Over oceans the index, SI_{Ocean} , was found to approximate the liquid water measurements obtained from co-located SSM/I data. Also, values of SI_{Ocean} greater than 0.35 mm was found to correlate well with SSM/I observations of precipitation. These indices will be used to derive rain rates after calibrating them against coincident radar and rain gauge measurements.

3.4 Scattering Features over land

Over land, the index (8a) represents the first of many steps needed to identify precipitation and discriminate among other scattering features (snow cover, deserts, arid land). As shown in Table 3, precipitation generally results in the largest scattering index. The maximum values of $T_B(22v) - T_B(85v)$ are based on a variety of SSM/I observations using the vertically polarized channels at 22.2 and 85.5 GHz channels. For example, measurements over Africa were used to obtain the maximum scattering values over deserts (Sahara) and arid land (Sahel and Kalahari) while intense rain events over the United States provided the maximum value for precipitation. Incidentally, surfaces scatter largest for vertical polarization [9], so that the AMSU measurements are about 25% less for deserts and arid land, i.e., AMSU measures the average between vertical and horizontal polarization.

Table 3. Comparison of scattering properties for different materials

Materials	Scatterers	Fractional Volume	$T_B(22v) - T_B(85v)$
Precipitation	Ice Hydrometers	< 0.1	0 - 100 K
Snow Cover	Ice Crystals	0.2 – 0.4	0 - 50 K
Deserts, Arid Land	Sand & Rocks	> 0.5	0 - 20 K

Table 3 shows a consistent decrease in the scattering index as the fractional volume of scatterers increases, i.e., as one progress from diffuse media (e.g., precipitation) to denser media (snow, deserts and arid land). This is explained by first recognizing that the surface emissivity, ε, of a deep homogeneous layer of scatters decreases for increasing albedo, ω, i.e., $\varepsilon = 2\sqrt{1-\omega}/(1+\sqrt{1-\omega})$. Furthermore, and most importantly, the single scattering albedo decreases as the fractional volume of scatterers increase. To illustrate this, Figure 3 shows the calculated albedo for particles having a dielectric constant, ε_s, similar to that of ice and sand (i.e., quartz). The albedo is calculated using dense media theory [10], and is plotted as a function of fractional volume with particle radii and wavelength as a parameter (i.e., $ka = 2\pi a/\lambda$).

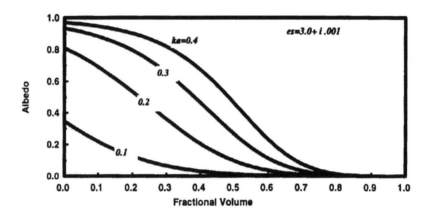

Figure 3. Single scattering albedo as a function of fractional volume

Note that the albedo increases as the particle size and frequency increase (i.e., ka increases). This is consistent with the scattering characteristic of individual particles. Of more relevance to the results in Table 3, we observe that the albedo in Figure 3 decreases sharply for increasing fractional volume. This is a consequence of dense media theory, which includes the transfer of energy from the coherent mean field to the random fluctuation field, thus increasing the absorption and decreasing the albedo for increasing fractional volume [10].

In order to identify precipitation and exclude the contributions from scattering surfaces, various filters must be used [9]. For example, snow cover can be separated from precipitation by recognizing that its physical temperature is generally less than that of precipitation. In the case of AMSU the lowest frequency (23.8 GHz) channel is used to infer the temperature beneath the snow pack. Since the emissivity of frozen soils is about 0.96, brightness temperatures less than 262 K are associated with snow cover. This same temperature thresold also filters out most cold deserts (e.g., Tibet and Mongolia) although additional conditions are needed to filter out the scattering signal due to warmer deserts and arid land such as that found over Africa. As noted in Table 3, these more-solid surfaces generally scatter less than snow cover and precipitation. Increasing the lower limit of the scattering index can remove these remaining surfaces, although light rainfall events will also be removed. It is therefore necessary to come up with a compromising set of conditions to retain most of the precipitation. The decision tree is given in the Appendix. Preliminary evaluation shows very similar distributions of rainfall occurrence when compared against the SSM/I measurements.

3.5 Snow Cover

For snow cover (also precipitation), the brightness temperature decreases as the frequency increases. As such, the index given by (8a) is used to identify snow cover under most conditions. However, unlike the low density of scatterers found in precipitation, snow cover contains densely packed ice crystals. The scattering at high frequencies becomes reduced relative to the lower frequencies due metamorphic changes in the crystalline structure as the snow ages. It is therefore important to utilize a lower frequency index to identify metamorphosed snow, i.e.,

$$SI_{31} = T_B(23) - T_B(31). \tag{11}$$

The absorption by water vapor is larger at 23 GHz than at 31 GHz, so that the index contains a residual atmospheric contribution. This unfortunately produces a positive index similar to that of snow for flooded land. In order to remove this false signature, the index is only applied when $T_B(89) < 255$ K. As indicated in the Appendix, (11) is combined with (8a) to observe the full range of snow conditions excluding melting snow. As in the case of precipitation over land, the indices represents the first of many steps [9] needed to identify and discriminate snow cover among other scattering features (e.g., precipitation, deserts). The decision tree is given in the Appendix. The product is currently being compared against the operational snow cover product generated by NESDIS which uses visible and infrared satellite data as well as available surface observations.

4. CONCLUSIONS

The paper begins by summarizing the evolution of satellite microwave radiometry in the United States. At NOAA the operational use of this technology began in 1978 with the production of temperature soundings using the MSU, and has been extended in 1998 to include surface features, clouds, precipitation and water vapor soundings using AMSU. A brief overview is given of the AMSU Day-1 product-algorithms listed in Table 1. These products were not even envisioned around 1980 when AMSU was designed. The algorithms are based on scientific knowledge gained from use of the DMSP sensors, whose data was made available to NESDIS as part of the shared processing agreement between the U.S. Navy, Air Force and NOAA. Validation of the Day-1 products is primarily based on SSM/I comparisons and water vapor obtained from radiosonde data. These comparisons demonstrated the need for improved screening of deserts for the snow cover and precipitation products. Much of the future work will also involve the use of AMSU-B higher frequency channels to improve the estimates of precipitation and cloud parameters, e.g., ice water content. Also, methods shall be developed to combine the products from different sensors (e.g., SSM/I) having different scan geometry and footprint size. This is important since it is difficult to capture the temporal changes of some of the products (e.g., precipitation, cloud liquid water) from one polar orbiting satellite system. In order to improve the spatial coverage at low latitudes and the temporal sampling it is necessary to combine the measurements from different sensors. The alternative solution of developing a geostationary microwave sensor operating at low frequencies does not appear practical at present.

APPENDIX

This Appendix contains the current version of the AMSU algorithms. The algorithms are based on the material presented in the text. In the case of precipitation and snow cover products, the algorithms also contain additional filters not presented previously, to screen out the effects of deserts on the measurements. Comments are included next to the statements to help clarify the algorithms. It should be noted that the algorithms only use the AMSU-A1 and A2 channels where T23, T31, T50, T53 and T89 represents the 23.8, 31.4, 50.3, 53.6 and 89 GHz brightness temperature measurements, respectively.

Discriminate Functions:

DF1 = 2.85 + .020 T23 - .028 T50   'Used to remove sea ice
DF2 = 5.10 + .078 T23 - .096 T50   'Used to remove warm deserts
DF3 = 10.2 + .036 T23 - .074 T50   'Used to remove cold deserts

1. Sea Ice Algorithm (ICE)
    IF Abs(Latitude) < 50   THEN RETURN
    IF DF1 < 0.45 THEN ICE =0: RETURN
        A = 1.7340 - 0.6236 CosZ : B = 0.0070 + 0.0025 CosZ
        C = -0.00106 : D = -0.00909
        E23 = A + B*T31 + C*T23 + D*T50   '23 GHz emissivity
    IF T23 - T31 >= 5 THEN EI=0.88 ELSE EI=0.95
        ICE = 100*(E23 - 0.45)/(EI - 0.45)   'Sea ice concentration

2. Water Vapor Algorithm (VAP)
    IF Abs(Latitude) > 50 AND **DF1**>0.2 THEN RETURN   'Remove sea ice
        A = 247.92 - (69.235 - 44.177 CosZ) CosZ
        B = -116.270 : C = 73.409
        D = A + B*Log(285 - T23) + C*Log(285 - T31)
        V = D CosZ   'Water vapor
        VAP = 0.942 V - 2.17   'Corrected water vapor

3. Cloud Liquid Water Algorithm (LIQ)
    IF Abs(Latitude) > 50 AND **DF1**>0.0 THEN RETURN   'Remove sea ice
        A = 8.240 - (2.622 - 1.846 CosZ) CosZ
        B = 0.754 : C = -2.265
        D = A + B*Log(285 - T23) + C*Log(285 - T31)
        Q = D CosZ   'Cloud liquid water
        LIQ = Q - 0.03   'Corrected liquid water

4. Rain Identification (0=No Rain   1=Rain)
    LAND:
        TT = 168 + 0.49 T89
        SIL = T23 - T89
    IF SIL>= 3   THEN RAIN=1   ELSE   RAIN=0

```
IF T23 =< 261  AND  T23 < TT  THEN   RAIN=0              'Remove snow cover
IF T89 > 273  OR  DF2 < 0.6  THEN   RAIN=0               'Remove warm deserts
OCEAN:
IF Latitude > 50  AND  DF1 > 0.0  THEN  RETURN                'Remove sea ice
   A =  8.240  - (2.622 - 1.846 CosZ) CosZ
   B =  0.754 : C = -2.265
   D = A + B*Log(285 - T23) + C*Log(285 - T31)
   LIQ = D CosZ                                         'Cloud liquid water
   SIW = -113.2  + (2.41 - .0049*T23)*T23 + .454*T31 - T89
IF LIQ > 0.3  OR  SIW > 9  THEN  RAIN = 1  ELSE  RAIN = 0
```

5. Snow Cover and Glacial Ice (0=No Snow   1=Snow   2=Glacial Ice)

```
      TT = 168 + 0.49 T89
      SCAT = T23 - T89 :  SC31 = T23 - T31
      SC50 = T31 - T50 :  PAR  = T50 - T53
IF T89 < 255     AND  SCAT< SC31 THEN  SCAT = SC31       'Re-Frozen Snow
IF SCAT <  3   AND     T23< 215   THEN  SNOW = 2         'Identify glacial ice
IF SCAT >= 3   THEN   SNOW=1    ELSE  SNOW = 0
IF T23 >= 262  OR  T23>= TT THEN  SNOW = 0               'Remove precipitation
IF DF3 =< 0.35  THEN SNOW=0                               'Remove deserts
IF SCAT<15 AND SC31<3 AND PAR>2 THEN SNOW=0  'High elevation deserts
IF SC31<3    AND  SC50<0 AND SCAT<9 THEN SNOW=0  'Remove frozen ground
```

REFERENCES

1. Spencer, R., J. Christy and N. Grody, 1990: Global atmospheric temperature monitoring with satellite microwave measurements: Method and results 1979-1984, *Jour. of Climate*, 3, 1111-1128.

2. Wentz, F., L. Mattox and S. Peteherych, 1986: New algorithms for microwave measurements of ocean winds applications to SEASAT and the Special Sensor Microwave Imager, *J. Geophys. Res.*, 91, 2289-2307.

3. Ferraro, R, N. Grody, D. Forsyth, R. Carey, A. Basist, J. Janowiak, F. Weng, G. Marks and R.Yanamandra, 1994: Microwave Measurements Produce Global Climatic, Hydrologic Data, *EOS Trans.*, American Geophys. Union, 75, 337-343.

4. Grody, N., F. Weng and R. Ferraro,1996: Comparison between SSM/I, SSM/T2 and radar measurements over the United States, AMS Conference on Satellite Meteorology, Atlanta Ga , 243-247.

5. Grody, N., 1980: Atmospheric Water Content over the Tropical Pacific Derived from the Nimbus-6 Scanning Microwave Spectrometer. *J. Appl. Meteor.*, 19, 986-996.

6. Weng, F., R. Ferraro and N. Grody, 1999: Effects of AMSU-A cross-track asymmetry of radiance on retrievals of atmospheric and surface parameters, *6th Specialist Meeting on Microwave Radiometery and Remote Sensing of the Environment*.

7. Weng, F., and N. Grody, 1994: Retrieval of cloud liquid water using the Special Sensor Microwave Imager (SSM/I), *J. Geophys. Res.,***99**, 25535- 25551.

8. Ferraro, R., 1997: Special sensor microwave imager derived global rainfall estimates for climatological applications, *J. Geophys. Res.*, **102**, 16715-16735.

9. Grody, N., 1991: Classification of snowcover and precipitation using the special sensor microwave imager. *J. Geophys. Res.*, **96**, 7423-7435.

10. Tsang, L., 1987: Passive remote sensing of dense nontenuous media, *J. Electromagnetic Waves and Applications*, **1**, 159-173.

Microw. Radiomet. Remote Sens. Earth's Surf. Atmosphere, pp. 353–363
P. Pampaloni and S. Paloscia (Eds)
© VSP 2000

Meteorological applications of precipitation estimation from combined SSM/I, TRMM and infrared geostationary satellite data

F. JOSEPH TURK[1], GREGORY D. ROHALY[1], JEFF HAWKINS[1], ERIC A. SMITH[2], FRANK S. MARZANO[3], ALBERTO MUGNAI[4] and VINCENZO LEVIZZANI[5]

[1]*Naval Research Laboratory, Marine Meteorology Division, Monterey, CA 93940 USA*
[2]*Department of Meteorology, Florida State University, Tallahassee, FL 32306 USA*
[3]*Dipartimento di Ingegneria Elettrica, Universita dell'Aquila, 67040 L'Aquila ITALY*
[4]*Istituto di Fisica dell'Atmosfera, Consiglio Nazionale delle Ricerche, 00133 Rome ITALY*
[5]*Istituto di Scienze dell'Atmosfera e dell'Oceano, Consiglio Nazionale delle Ricerche, 40129 Bologna ITALY*

Abstract We present a technique to statistically blend low-Earth orbiting passive microwave satellite data together with geostationary orbiting infrared satellite data in a near real-time fashion, for retrieval of instantaneous rain rate and accumulations at the geostationary update cycle. This blended geostationary-microwave technique is oriented towards rapid-update operational usage in quantitative precipitation forecasting, numerical weather prediction models, and to estimate rainfall-induced attenuation along Earth-space satellite links. Examples of rain accumulations are shown for a series of heavy rain events from Hurricane Mitch in late October 1998. To examine the model impact of assimilating a global rainfall analysis, a physical initialization of the Navy Operational Global Atmospheric Prediction System (NOGAPS) global spectral model was performed for a period of time in late September 1998, during the presence of Hurricane Georges in the tropical Atlantic Ocean. Lastly, an example where the 20 GHz path attenuation due to rainfall is derived over central Europe is presented. While the results are unique to these particular examples, it does demonstrate the capabilities and many applications of a globally complete rain rate analysis that overcomes the temporal and spatial coverage gaps characteristic of low-Earth orbiting sensors.

1. INTRODUCTION

A longstanding promise of meteorological satellite imaging systems has been improved identification and quantification of precipitation on time scales consistent with the nature and development of typical cloud rain bands. Geostationary weather satellite imaging systems provide the rapid temporal update cycle needed to capture the growth and decay of precipitating cloud systems on a scale of several kilometers. All current systems provide rapid (hourly or less) updates in the longwave infrared (IR) spectrum near 11 μm, which for optically thick clouds senses the emitted radiation from the upper cloud regions. On the other hand, microwave-based imagers are better suited to quantitative measurements of precipitation due to the well-established physical connection between the upwelling radiation and the underlying precipitation structure. The current polar-orbiting Defense Meteorological Satellite Platform (DMSP) Special Sensor Microwave Imagers (SSM/I) use swath widths of 1400 km, which leaves coverage gaps in the tropics where the majority of the Earth's precipitation occurs. In November 1997, the Tropical Rainfall Measuring Mission (TRMM) satellite was launched, which improves upon the SSM/I heritage with the addition of a polarized 10.7 GHz channel, and improved spatial resolution by a factor of about 2.5 [1].

The idea of blending low-Earth orbiting microwave and geostationary infrared data together to take advantage of the inherent advantages of each sensor is not new [2-4]. In this article, we present and demonstrate a real-time technique by which to track, at the geostationary update cycle, rain rates based on a statistical analysis of the most recent SSM/I or other microwave data in the area. We discuss three areas of application. First, to be able to identify and track accumulations from rapidly developing rain-producing storms, especially those that are heading towards coastal regions. The second application concerns development of a three-hourly global rain rate analysis that is devoid of the spatial and temporal coverage gaps characteristic of low-Earth orbiting sensor orbits and swath widths. These analyses were developed for assimilation into global spectral numerical weather prediction models via physical initialization techniques [5]. An analysis is presented during late September 1998 during the presence of Hurricane Georges. The third application is for near real-time rainfall-induced attenuation monitoring along Earth-satellite links. Given the proliferation of Ka-band and higher wireless communications, a nowcast of rain intensity anywhere in the world is needed for fade and link outage mitigation techniques. Overall results as well as shortcomings and future requirements for an improved satellite-based precipitation analysis are discussed.

2. BLENDING OF SATELLITE DATA

The constantly evolving temporal and spatial characteristics of precipitation and its relation to any satellite observations require that any statistical tuning or calibration to infrared brightness temperatures (T_B) follow the rain characteristics. Time- and space-coincident microwave and geostationary satellite data are saved each time a SSM/I or TRMM orbit pass intersects with any of the four operational geostationary satellites (GOES-East/West, GMS, and Meteosat). Every three hours, an update cycle starts and locates the most recent 24 hours of past coincident data. Separate histograms of the IR temperatures and the associated microwave-based rain rate are in built in 15-degree global boxes. The SSM/I rain rate is computed via the operational NOAA-NESDIS scheme [6], which is computed at the A-scan sampling spacing (25 km) of the SSM/I scan operation. It contains separate land- and ocean-based algorithm components, based upon a scattering-index test. For ocean backgrounds, an additional emission-based algorithm is attempted if insignificant scattering is detected via the 85 GHz channels. For TRMM data, the real-time level 2A-12 instantaneous rain rates are used directly [7]. The 2A-12 data provide rain rates from the nine TMI channels using the Goddard Profiling Algorithm (GPROF), which is based upon a Bayesian approach. The 2A-12 products are provided on the 85 GHz (high-resolution) scan grid of the TMI sampling pattern. The boxes are spaced 5-degrees apart, so that they overlap geographically and assure that the histograms transition smoothly from one box to the next. The boxes assure that the characteristics are captured for a each type of rainfall. The probability matching method (PMM) described by [8] involve the probability distribution functions (PDF) of the microwave-based rain rate and the geostationary IR brightness temperatures, respectively. For each rain rate bin, the rain PDF is matched to the IR-temperature PDF while working from the warm end in brightness temperature on upwards. The first pair to be matched is therefore the zero-rain rate IR threshold for this 15-degree region. Generally, the number of co-located points contained in each 15-degree box using the 24-hour look-back time ranges from over 10^5 within the tropics to 10^3 above and below +/- 40 degrees latitude. Lastly, a file is produced that contains all lookup tables for the 15-degree boxes, which relates the IR temperature to the microwave-based rain rate.

Figures 1a through 1c depict the histogram-matched rain rate relationship curves representing the 15-degree region centered on Central Africa, the west coast of California in the United States, and of northeastern South America, respectively. Each of the four curves shows the T_B- R relationship valid at the 0 UTC update of the global histogram-matching analysis on 21-24 February 1999. The curves depict the dynamic adjustment as new SSM/I data arrive in the previous 24 hours, especially in Figure 1b where lighter rain moved into this region after February 22 and adjusted the curve to colder IR T_B values.

Figure 1a also shows the adjustment between February 22-23, although the maximum rain rates are much higher for this equatorial tropical region. Figure 1d depicts a time series of the zero-rain rate IR temperature threshold for the three regions represented in Figures 1a through 1c. Each data point represents the value of the x-axis intercept in Figures 1a-1c at successive 3-hourly updates beginning on February 21. The thresholds follow the observation that the coldest IR values are noted over central Africa and the warmest over the California coast, and change by up to 20 degrees over this 5-day period. Typically, the zero-rain rate IR threshold can vary from as low as 210 to 220 K in the tropics to 230 to 250 K in the higher latitudes. This shows how the trends in the SSM/I-observed precipitation are tracked and dynamically adjust themselves to whatever IR T_B values are observed during the previous 24 hours. Depending upon the number of warm-end bins in the IR temperature histogram, the technique can be tuned to capture rain from warmer-based clouds.

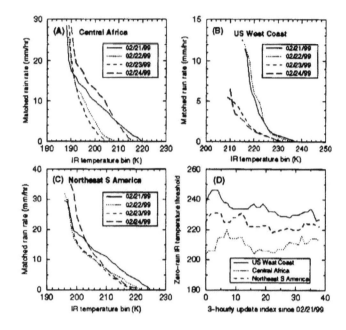

Figure 1. Histogram-adjusted microwave rain rate vs. IR brightness temperature relationships for three 15-degree regions centered on a) Central Africa, b) USA West Coast off of California, and c) Northeastern South America. Each of the four line types refer to the 0 UTC update of the histogram relationship between 21-24 February 1998. Panel (d) displays the zero-rain rate IR brightness temperature threshold for each of these three regions at three-hourly intervals beginning on 21 February 1998 at 0 UTC.

The production of instantaneous rain rates involves a relatively straightforward scan of the above-mentioned global lookup tables that relate IR brightness temperatures to the histogram-matched microwave-based rain rate. Multi-hour rain accumulations are computed by an explicit time-integration of successive instantaneous rain rate images. A well-known drawback of any infrared geostationary-type approach to rainfall estimation is the presence of high, non-raining clouds such as cirrus clouds [9]. To address this effect, two types of screens are currently used in a cascade configuration, a split window difference test (10 and 12 micron channels; the current GOES and GMS satellites possess this capability), and also a previous-time IR temperature difference. We have not yet attempted any adjustment for orographic effects, which is necessary whenever infrared data is used to track precipitating clouds across high elevation terrain.

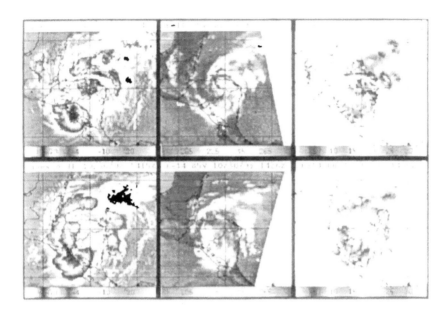

Figure 2. (a) False-color GOES-8 (GOES-East) IR images (degrees Celsius) of Hurricane Mitch corresponding to the SSM/I overpass time at 0035 UTC on 30 October 1998. (b) SSM/I 85V GHz channel image (Kelvin units) of centered on the same map projection. (c) Corresponding SSM/I-derived rain rates (mm hr-1) using the operational NOAA-NESDIS technique. Grid lines are 5-degrees apart. (d-f) Same as panels (a-c), except for SSM/I overpass time of 1426 UTC on 30 October 1998.

3. RAPID UPDATE OF RAIN ACCUMULATIONS: HURRICANE MITCH

What eventually became Hurricane Mitch initially formed in the southwest Caribbean Sea on October 21 and became the strongest October hurricane ever recorded in the Atlantic Basin. In Figure 2, two co-located GOES-8 (GOES-East) and SSM/I passes from October 30 over Central America are depicted. During this time, Mitch generated a series of strong localized rain systems over ocean and land. Especially prominent is a storm cell that parked itself over western Nicaragua on October 28 and gradually strengthened through the following days. This rain occurred near a coastline, which makes quantitative retrieval of rain problematic for the SSM/I. The NOAA-NESDIS coastal technique is a spatial test [6]. If any pixel in a surrounding 5 x 5 grid of A-scan pixels contains a coastal pixel, then the land-based algorithm is applied for the pixel. This does eliminate missing rain pixels near coasts and works well for deep convective rain in coastal regions. For example, the GOES-8 images show amazingly deep convective clouds (–80 C or colder) and 85V GHz brightness temperatures below 200 K, and the associated SSM/I passes over western Nicaragua show rain rates exceeding 25 mm hr^{-1}.

Using the blended geostationary-SSM/I technique, Figure 3 depicts the past 6, 12, 24 and 48-hour rainfall accumulations at 23 UTC on October 30. The six-hour total exceeded 150 mm, which is in accord with the > 25 mm hr^{-1} rain rates noted from Figure 2. The 48-hour rainfall totals exceed 600 mm. While we are not aware of any means to verify these accumulations at this time, the devasting flooding that resulted from these storms are evidence of the prolonged rain that accumulated over this localized region.

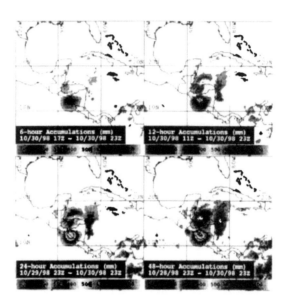

Figure 3. Previous 6, 12, 24 and 48-hour accumulated precipitation totals (in mm) over Central America valid at 2300 UTC on October 30 1998, using the blended geostationary-SSM/I technique.

4. ASSIMILATION OF RAIN RATES INTO A GLOBAL NWP MODEL

While investigations of the impact of satellite-derived rain rates upon global and regional-scale numerical forecast models have been numerous, an important consideration of all of these investigations is a means to obtain rain rates at times and locations where SSM/I data are not available. We discuss an experiment designed to investigate the impact of the blended microwave-geostationary rain rates on the forecast skill of the Navy Operational Global Atmospheric Prediction System (NOGAPS) [10] in a tropical environment. The current operational version of NOGAPS was used at a reduced T79 resolution with 24 levels. A physical initialization technique was used [5]. Physical or reverse initialization is a term used to describe the assimilation of physical variables (such as rain rates) to improve upon the analysis of the basic state variables. The physical initialization was performed as the model integrates to provide the six-hour forecast, used as the first guess, to the multivariate optimal interpolation analysis. At each grid point, the model columnar water vapor was vertically partitioned such that the convective parameterization scheme produced the same rain rate as analyzed by the satellite technique.

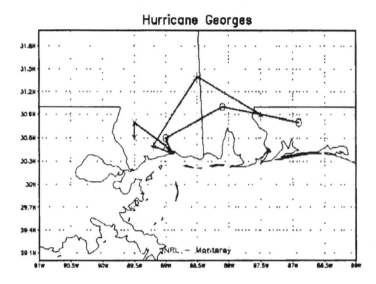

OBSERVED AND FORECASTED TRACKS

Initial Conditions: 28 September 1998 12 UTC
Position plotted at 12, 24, 36 hours

Hurricane Georges

Observed positions via NHC best track

Figure 4. Observed and forecasted tracks of Hurricane Georges beginning on 28 September 1998 at 12 UTC, at 12, 24, and 36-hr forecast times. The symbol 0 denotes the observed positions via the National Hurricane Center (NHC) best track, the × symbol denotes the control run forecast (without NOGAPS physical initialization), and the * symbol denotes the forecast track with NOGAPS physical initialization.

We analyzed the forecast model impact during an eight day period (21 –29 September 1998) from the Atlantic hurricane season. Figure 4 depicts the position and forecast track of Hurricane Georges at 12, 24 and 36-hours from 12 UTC on 28 September 1998, soon after landfall. The symbol O represents the best-track position as provided by the National Hurricane Center. The symbol X represents the forecast positions from the control run (without physical initialization). The symbol * represent these same forecast times for the model run with physical initialization. In this case, the 24- and 36-hour forecast positions show a positive impact in the storm position in relation to the best track, and most importantly, the hurricane has been correctly moved away from New Orleans and towards northern Florida. The mean overall forecast position error for Georges during the 21-29 September period in which physical initialization was applied to NOGAPS was calculated. The 24, 36 and 48-hour physically initialized forecasts show positive impacts to hurricane positioning ranging from 12 to 16 percent, although the improvement falls to nearly no-impact (3.5 percent) at 72-hours for this particular storm.

5. ESTIMATION OF ATTENUATION ALONG EARTH-SATELLITE LINKS

In recent years there has been an increasing request for large-bandwidth communication services characterized by high availability and low-fade margin requirements, which has prompted exploration of frequencies at Ka-band and above. Above 10 GHz, the atmospheric fading due to gases, clouds and rain together with scintillation can represent a strong impairment to the link budget design. In particular, rainfall due to convective storms can give up to several tens of decibels (dB) of total path attenuation thus causing severe outages. It is appealing to investigate how spaceborne remote sensing systems and derived products can be exploited to optimize the performances of a satellite communication system with low power margins, and specifically for detecting rainfall attenuation within a fade mitigation scheme. Satellite-based remote sensing measurements provide either a global or local coverage of the earth with a wide range of spatial and temporal sampling. Of special interest are the beacon frequencies at 20, 40 and 50 GHz of the Italian Italsat 1 geostationary communications satellite, first deployed in 1991.

Figure 5 depicts the 0540 UTC F-13 SSM/I overpass on 6 October 1998 over much of Europe. Small yet intense regions of convective rain are moving across central Italy, as indicating by 85V GHz brightness temperatures below 170 K. Using a cloud model-derived database of microphysical cloud structures, a power-law relationship was derived between the cloud model rain rate and the nadir path attenuation calculated by vertically integrating the specific attenuation from all hydrometeor species. This allows the rain rate from the blended technique (in mm hr^{-1}) to be converted into a total path attenuation in dB. At 20 GHz, the relationship derived was $A^{20} = 0.0928R^{1.312}$. Path attenuations exceeding 6 dB were noted over central Italy, where the land background makes rain rate estimation difficult from the SSM/I. The ability to track the attenuation with the blended satellite technique has the potential to provide a real-time analysis for fade mitigation techniques, such as up/downlink power control, or space/time diversity techniques.

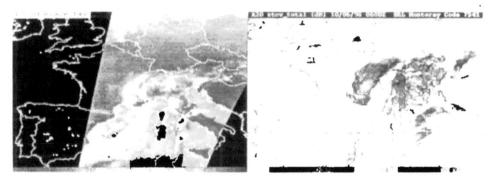

Figure 5. (a) 85V GHz image from the 0540 UTC F-13 SSM/I overpass on 6 October 1998 covering much of Europe. (b) Corresponding satellite-derived 20 GHz one-way nadir path attenuation in dB from the blended microwave-geostationary technique.

6. CONCLUSIONS

We have presented a near-real time technique for blending together geostationary infrared and low Earth-orbiting microwave-based satellite data for a rapid-update global rainfall analysis. The technique automatically adapts itself to the changing nature of the IR temperatures and microwave-based rain rates in an area, and adjusts the zero-rain rate IR temperature threshold dynamically as new SSM/I and TRMM data arrive. Analysis of rain accumulations after the landfall of Hurricane Mitch demonstrated the strengths and weaknesses of the blended geostationary-SSM/I rain rate estimation techniques near coasts. Since many of the Central American countries devastated by this hurricane are covered by a small number of SSM/I pixels, the rain rates from these flood events were not as intense in the SSM/I as the corresponding GOES-8 IR imagery showed them to be. The blended geostationary-SSM/I technique produced 2-day rain totals exceeding 600 mm (2 feet) in some places. Most importantly, the positioning of the rain was captured, which is a key advantage of using geostationary data.

The impact of the blended geostationary-SSM/I rain rates in a tropical forecast environment was examined through the use of a physical initialization of the NOGAPS global spectral forecast model during the presence of Hurricane Georges. The physically-initialized model exhibited a mean improvement in hurricane position error of nearly 16% in its 48-hour forecasts. Futhermore, the precipitation structure exhibited a much higher correlation with the satellite-derived analysis. This ability to improve the quantitative precipitation forecast near coastal areas is an important result.

The capability to monitor the attenuation due to rainfall along Earth-space satellite links was also examined. The rapid time-update of the attenuation provides the means to nowcast and compensate for possible link outages in a real-time mode. The replacement for the existing Meteosat series of geostationary satellites, the Meteosat Second

Generation (MSG) [11], is expected to yield further improvements to the screening of non-precipitating high clouds. The use of these multispectral satellite analyses together with SSM/I and mesoscale model microphysics is expected to improve the understanding of and quantification of precipitation.

ACKNOWLEDGEMENTS

The first two authors gratefully acknowledge support from the Office of Naval Research, Program Element (PE-060243N), and the Space and Naval Warfare Systems Command, PMW-185 (PE-0603207N). SSM/I data are provided via the Navy Fleet Numerical Meteorology and Oceanography Center (FNMOC) adjacent to NRL Monterey. Near real-time TRMM data arrive via the data service provided by the TRMM Science Data and Information Service (TSDIS).

REFERENCES

1. Kummerow, C., W. Barnes, T. Kozu, J. Shiue, J. Simpson, 1998: The Tropical Rainfall Measuring Mission (TRMM) sensor package. *J. Atmos. Ocean. Tech.*, 15, 809-817.

2. Vincente, G.A., R.A. Scofield, W.P. Menzel, 1998: The Operational GOES Infrared Rainfall Estimation Technique. *Bull. Amer. Meteor. Soc.*, 79, 1883-1898.

3. Levizzani, V., F. Porcu, F.S. Marzano, A. Mugnai, E.A. Smith and F. Prodi, 1996: Investigating a SSM/I microwave algorithm to calibrate Meteosat infrared instantaneous rain rate estimates. *Meteo. Appl.*, 3, 5-17.

4. Adler, R.F., G.J. Huffman, and P.R. Keehn, 1994: Global tropical rain estimates from microwave-adjusted geosynchronous infrared data. *Remote Sens. Rev.*, 11, 125-152.

5. Rohaly, G., and A.H. Van Tuyl, 1996: Impact of physical initialization on the Navy Operational Global Atmospheric Prediction System (NOGAPS). *11th AMS Conf. Numerical Weath. Pred.*, 19-23 August, Norfolk, VA, USA, 287-289.

6. Ferraro, R.R., 1997: Special sensor microwave imager derived global rainfall estimates for climatological applications. *J. Geophys. Res.*, 102, D14, 16715-16735.

7. Kummerow, C., W. S. Olson and L. Giglio. 1996: A simplified scheme for obtaining precipitation and vertical hydrometeor profiles from passive microwave sensors. *IEEE Trans. on Geosci. and Remote Sensing*, 34, 1213-1232.

8. Atlas, D., D. Rosenfeld, D.B. Wolff, 1990: Climatologically tuned reflectivity-rainrate relations and links to area-time integrals. *J. Appl. Met.*, 29, 1120-1135.

9. Arkin, P.A., R. Joyce, J.E. Janowiak, 1994: The estimation of global monthly mean rainfall using infrared satellite data: The GOES Precipitation Index (GPI). *Rem. Sens. Rev.*, 11, 107-124.

10. Hogan, T.F., and T.E. Rosmond, 1991: The description of the Naval Oceanographic Global Atmospheric Prediction System spectral forecast model. *Mon. Weath. Rev.*, 119, 1786-1815.

11. Schmetz, J., H. Woick, S. Tjemkes, M. Rattenborg, 1998: From Meteosat to Meteosat Second Generation (MSG). 9th *AMS Conf. Sat. Meteor. And Ocean.*, Paris, 25-29 May, 335-338.

Microw. Radiomet. Remote Sens. Earth's Surf. Atmosphere, pp. 365–369
P. Pampaloni and S. Paloscia (Eds)
© VSP 2000

SOM network-based retrieval algorithms for determining precipitable water and rainfall over oceans from the SSM/I measurements

HONGBIN CHEN[1], JIANCHUN BIAN[2], PEICAI YANG[1] and DAREN LU[1]

1 *LAGEO, Institute of Atmospheric Physics, CAS, Beijing 100029, China*
2 *Department of Geophysics, Peking University, Beijing 100871, China*

Abstract—A Self-Organizing feature Mapping(SOM) neural network is used to develop algorithms for retrieving the precipitable water and rain rate over oceans from SSM/I measurements. The training and test data bases are SSM/I 7-channel brightness temperatures, precipitable water derived from radiosondes, and rain rate from radar and rain gauges provided by NASDA(Japan). The SOM network was first trained using the 5/6th of the data and then tested using the rest of the data. Comparisons of retrieval results with the classical statistically-based algorithms in the literature show that the SOM neural network-based models are significantly better in retrieving both precipitable water and rain rate, especially in the low precipitable water regime.

Key words: SOM network, SSM/I, Precipitable water, Rain rate

1. INTRODUCTION

Over past 2 decades, a number of methods(mainly statistical and physical) have been used to retrieve precipitable water(PW) and rain rate (RR) over oceans from space-borne microwave radiometers[1-7]. In particular, the SSM/I onboard the DMSP satellites routinely generates the operational products of PW and RR over the oceans. However, the required retrieval accuracy (i.e., 5~10% for PW and 10~20% for RR) has not yet been globally achieved.

In this work, a Self-Organizing feature Mapping (SOM) network is used to develop the retrieval algorithms for deriving oceanic PW and rainfall from the SSM/I measurements. In the following section, the SOM network and the procedure of retrieval algorithm development are briefly presented. In Sections 3 and 4, retrieval results of oceanic precipitable water and rainfall by the SOM model are respectively examined by comparison with the operational algorithms. Finally, some conclusions are given in Section 5.

2. SOM NETWORK AND RETRIEVAL ALGORITHM DEVELOPMENT

Because of its high capability of both clustering(classification) and non-linear regression fitting, the SOM network is adopted to develop algorithms for retrieving over-ocean PW and rainfall from SSM/I brightness temperature data. Our SOM network-based retrieval algorithms can be divided into two parts: one is an SOM neural network trained for data classification and another is an ensemble of least-squares fitting relationships.

2.1 SOM Network

The Self-Organizing Map was first proposed by Kohonen in 1981[8]. The most significant difference between the SOM and other contemporary neural-model approaches is: most artificial neural networks strongly emphasize the aspect of distributed processing, and only consider spatial organization of the processing units as a secondary aspect. The map principle, on the other hand, is in some ways complementary to this idea. The intrinsic potential of this particular self-organizing process is emphasized for creating a localized, structured arrangement of representations in the basic network module. Self-Organizing Maps, or systems consisting of several map modules, have already been applied with success to many complex tasks, for example, pattern recognition, robotics, process control, telecommunications, etc.

The SOM model is a sheet-like artificial neural network(see Fig.1). Through an unsupervised learning process, the cells of the networks become specifically tuned to various input signal patterns or classes of patterns. In the basic version, at a time only one cell or one local group of cells gives the active response to the current input. The locations of the responses tend to become ordered as if some meaningful coordinate system for different input features were being created over the network. The spatial location or coordinates of a cell in the network then correspond to a particular domain of input signal patterns. Each cell or local cell group acts like a separate decoder for the same input. Thus, it is the presence or absence of an active response at that location, and not so much the exact input-output signal transformation or magnitude of the response, that provides an interpretation of the input information. So, the SOM has the special property of effectivelly creating spatially organized 'internal representations' of various features of the input signals and their abstraction.

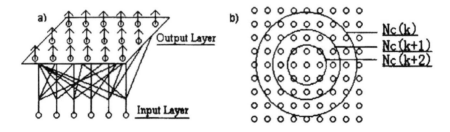

Figure 1.   Schematic illustration of a SOM network (a)
and the variation of bubble area Nc(k) with time in training process(b).

A 2-dimensional SOM network, as shown in Fig.1, is adopted to develop algorithms for retrieving precipitable water and rainfall over oceans from the SSM/I brightness temperature measurements. In the network, each input cell is connected with all cells in 2-D output layer.

In the training process, we put the learning rate $\alpha(k)=0.98/k^s$, $\beta \in [0.2,0.4]$ and the bubble area $Nc(k)=Nc(0)/k$ (see Fig. 1b). The initial weights W are given by a random number generator between 0 and 1. For an input vector X(k), a closest output cell C(i.e., the response center) is first found out and then the weights for the cells over the bubble area Nc(k) centered at C are adjusted. For next input vector X(k+1) takes place another learning cycle. After the learning process, the response center for each input data pattern is localized in the network and the weights connecting the output and input neural elements are fixed. Then, the SOM network enters the working phase.

2.2 Regression by Least-squares fitting

When a SOM network has been trained, a classification of input data is obtained, i.e. the input data can be classified into $M \times N$ groups or patterns ($M \times N$, in fact, is the number of cells in 2-D neural network). For each group, a relationship between brightness temperatures (TBs) and precipitable water PW or rainfall rate (RR) can be established by least-squares fitting. So far, the SOM network-based algorithms for retrieving PW or RR have been constructed.

3. RETRIEVALS OF PRECIPITABLE WATER BY SOM MODEL

The data for training the SOM network and for testing retrievals are the SSM/I 7-channel TBs and the values of PW derived from the radiosonde profiles. These data (about 26,400 match-ups) were preprocessed and provided by NASDA of Japan for the ADEOS II-AMSR retrieval algorithm development. In order to get better results(i.e. smaller rms errors), various combinations of SSM/I 7-channel TBs have been tried as the input data in training the network. Here, the values of PW and normalized 4-channel brightness temperatures TB_{19v}, TB_{19h}, TB_{22v}, and TB_{37v} are chosen to form the input vectors. Moreover, the nearly coincident SSM/I and radiosonde data are divided into 3 subsets: all latitudes, high latitudes($>45°$), and middle and low latitudes($\leq 45°$). The 5/6th of each data subset are used to train the SOM network ant the rest (1/6th data) are used for testing the retrieval capability of the SOM algorithm.

The oceanic precipitable water is derived from the SSM/I TBs simultaneously by the SOM model and currently operational algorithm[1-2]. The retrieved PW are then compared to the radiosonde-derived PW which are considered truth (see Figure 2). The rms errors of SSM/I retrievals in comparison with radiosonde PW are given in Table 1. It is seen that in all cases, the SOM network model provides better retrievals than the operational algorithm. Particularly, significantly less rms error in Table 1 and less bad retrievals in low regime of PW in Figure 2(b) show that the SOM network retrievals are in much better agreement with radiosonde measurements than the operational algorithm at high latitudes.

The necessity of developing the SOM network algorithms for different geographic regions is also examined. 3 SOM algorithms have been constructed respectively with all, low-middle, and high latitude data. Comparisons of retrievals show that the global algorithm(i.e. that developed based on the 5/6th of total data) can provide nearly the same results as the regional algorithms.

Table 1. Rms differences between retrieved and radiosonde PW (in g/cm²)

Region (Mean Value of PW)	All Latitudes (3.56)	Low & Middle Lat. (4.45)	High Latitudes (1.09)
Operational Algorithm	0.557	0.573	0.512
SOM Algorithm	0.492	0.530	0.286

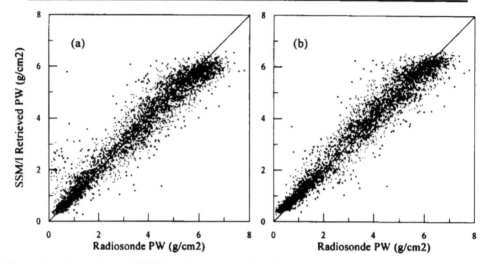

Figure 2.   Comparison of SSM/I retrieved and radiosonde-derived oceanic precipitable water: (a) retrievals by Alishouse et al. algorithm with a cubic correction of Colton and Poe[2]; (b) retrievals by the SOM model.

4. RETRIEVALS OF OCEANIC RAINFALL BY SOM MODEL

In parallel, an SOM network-based algorithm has been developed for retrieving oceanic rain rate(RR) using the SSM/I data and collocated radar-derived RR data which were also provided by NASDA of Japan.

Table 2. Comparisons of retrievals by 3 algorithms to radar rain-rate data(in mm/hr)
(RMSE: rms error between retrieved and radar-derived RR, Corr.: correlation coefficient)

Retrieval Algorithm	All data (18,185)			Without zero RR(10,702)		
	RMSE	Mean	Corr(%)	RMSE	Mean	Corr.(%)
SOM	1.25	0.97	82.8	1.62	1.62	80.0
Ferraro & Marks	1.73	0.87	71.2	2.26	1.47	67.7
Sun et al.[7]	1.64	0.95	72.5	2.13	1.58	68.7
Statistics of Radar RR data	2.23 (Var.)	0.96		2.71 (Var.)	1.64	

Based on the frequency distribution of radar RR samples, a stair-styled SOM model is constructed for different ranges of rainfall rate. Similarly, in order to get better regression and retrieval results, various combinations of SSM/I 7-channel TBs are tried as the input data. It is found that the horizontally polarized 85 GHz channel should be removed from the input data probably because this channel didn't work properly. The 5/6th of the total data is used first for training the SOM model and the rest is used for testing the retrieval. Comparisons of the retrievals against two statistically-based algorithms using a scattering index(SI) are shown in Table 2. It is seen that the SOM network-based algorithm yields the best retrievals of rainfall rate, i.e., smaller rms difference and higher correlation.

5. CONCLUSIONS

The Self-Organizing Map network has the special property of effectively creating spatially organized "internal representations" of various features of the input signals and their abstraction. Based on this neural network, the algorithms are developed respectively for retrieving oceanic precipitable water and rain rate from the SSM/I brightness temperature measurements. The 5/6th of nearly coincident and collocated SSM/I and ground-based data are used for training the SOM network and then establishing the regression relationships between PW or RR and TBs. Using the rest of the data, the test retrievals by the SOM algorithms are compared with 2 statistically-based algorithms. It is shown that the SOM network-based algorithms can provide better retrieval results than the currently operational algorithms. In addition, the consuming time by the SOM algorithm is also acceptable.

Acknowledgments This work was supported by the NASDA of Japan under an AMSR retrieval algorithm development contract and by Chinese Committee of Science for a project of developing space-borne microwave remote sensors.

REFERENCES

1. J.C. Alishouse, S.A. Snyder, J. Vongsathorn, and R.R. Ferraro. Determination of oceanic total precipitable water from the SSM/I. *IEEE Trans. Geosci. Remote Sens.*, GE-28, 811-816(1990).
2. M.C. Colton and G.A. Poe. Shared processing program, DMSP, SSM/I algorithm symposium, 8-10 June 1993. *Bull. Amer. Meteorol. Soc.*, 75, 1663-1669(1994)
3. D.L. Jackson and G.L. Stephens. A study of SSM/I-derived columnar water vapor over the global oceans. *J. Climate*, 8, 2025-2038, 1995.
4. G.W. Petty. Physical retrievals of over-ocean rain rate from multichannel microwave imagery. Part II: algorithm implementation. *Meteorol. Atmos. Phys.*, 54, 101-121(1994).
5. H.B. Chen, P.C. Wang, H.B. Sun, and D.R. Lu. Retrievals of over-ocean precipitable water from the SSM/I measurements with several regression algorithms. *Acta Meteorol. Sinica*, 12(4), 443-449(1998).
6. R.R. Ferraro and G.F. Marks. The development of SSM/I rain-rate retrieval algorithms using ground-based radar measurements. *J. Atmos. & Oceanic Tech.*, 12, 755-770(1995).
7. H.B. Sun, D.R. Lu, M.Z. Duan, J.L. Liu, H.B. Chen. Algorithm of rainfall retrieval over ocean from SSM/I data using the probability paring method. accepted by *Scientia Atmospherica Sinica*(in Chinese with English Abstract).
8. T. Cohonen. The Self-Organizing Map. *Proceedings of The IEEE*, 78, 1464-1480(1990).

Microw. Radiomet. Remote Sens. Earth's Surf. Atmosphere, pp. 371–377
P. Pampaloni and S. Paloscia (Eds)
© VSP 2000

Analysis of selected TRMM observations of heavy precipitation events

A. TASSA[1,2], S. DI MICHELE[1,2], E. D'ACUNZO[1,2], C. ACCADIA[1], S. DIETRICH[1], A. MUGNAI[1], F. MARZANO[3], G. PANEGROSSI[4] and L. ROBERTI[5]

[1] *Istituto di Fisica dell'Atmosfera - Consiglio Nazionale delle Ricerche, Roma, Italy*
[2] *Fondazione per la Meteorologia Applicata, Firenze, Italy*
[3] *Dipartimento di Ingegneria Elettrica - Universita' degli Studi dell'Aquila, L'Aquila, Italy*
[4] *Department of Atmospheric and Oceanic Sciences, University of Wisconsin-Madison, Madison, Wisconsin, USA*
[5] *Dipartimento di Elettronica - Politecnico di Torino, Torino, Italy*

Abstract - On November 27, 1997, a new meteorological satellite platform, called Tropical Rainfall Measuring Mission (TRMM), was successfully launched carrying aboard five instruments including a five-frequency (10.65, 19.35, 21.3, 37.0, and 85.5 GHz) Microwave Imager (TMI) and the first satellite-borne Precipitation Radar (PR). We have analyzed case studies concerning heavy precipitation events as observed by the above two instruments. For such cases, precipitating cloud structure retrievals, as well as surface precipitation retrievals based on TMI observations have been performed. For these computations, we have applied an inversion-type profile-based passive microwave precipitation retrieval algorithm, that we have developed in the past few years and successfully tested in various international intercomparison projects. The algorithm is based on the use of coupled cloud-radiation databases, whose cloud portions are generated by means of the time-dependent, three-dimensional, cloud/mesoscale model University of Wisconsin - Nonhydrostatic Modeling System (NMS). For each cloud model simulation, the corresponding radiation data base is generated by computing the upwelling brightness temperatures at TMI resolutions and frequencies by means of a three-dimensional polarized backward Monte Carlo radiative transfer model.

1. INTRODUCTION

The Tropical Rainfall Measuring Mission is a NASA/NASDA joint project. The satellite was launched on November 27, 1997, carrying aboard five instruments -- from the visible to the microwave range -- among which a 9-channel microwave radiometer (TRMM Microwave Imager -- TMI from now on) and a Precipitation Radar (PR).

Our retrieval approach is based on a cloud-radiation database, which is generated by applying a 3-dimensional backward Monte Carlo radiative transfer model [1] to a 3-dimensional cloud simulation. Generation of the cloud radiation database implies the choice of many options and/or values in order to provide simulated upwelling brightness temperatures (TBs) which can be compared with data, and thus can be used for surface rain rate retrieval.

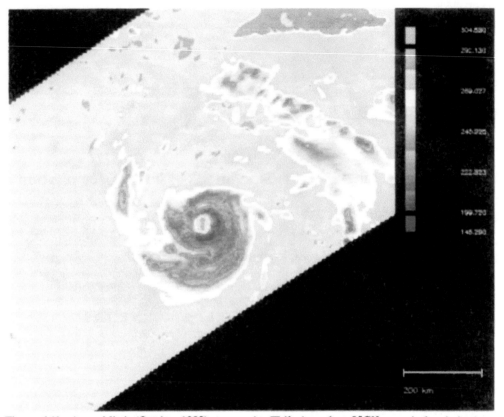

Figure 1. Hurricane Mitch (October 1998) as seen by TMI channel at 85GHz, vertical polarization. Resolution: 4.4Km X 6.6Km.

Cloud model. Our study focuses on three hurricanes which devastated the coasts of Southern U.S. and Central America between August and October 1998 -- Hurricanes Bonnie, George and Mitch. Figure 1 shows a TMI image of Hurricane Mitch (Central America, October 1998). Therefore, we have used a cloud model simulation of a tropical cyclone in its mature stage, which has been generated by the University of Wisconsin - Non-Hydrostatic Modeling System (NMS) [2] that explicitly produces six species of hydrometeors (cloud drops, rain drops, graupel particles, pristine ice crystals, ice aggregates and snow flakes) as a function of space and time.
Surface Model. The EuroTRMM official sea surface model, developed by the Université Catholique de Louvain [3], has been used.

2. DROP SIZE DISTRIBUTION PARAMETERIZATION

Drop Size Distribution (DSD) parameterization has an important effect on the simulated upwelling brightness temperatures. A sensitivity study has been made, exploring three different kinds of DSDs for rain and graupel particles, while keeping the same DSDs of the NMS simulation for the other hydrometers. Overlapping with TMI data has been tested for the databases corresponding to the three different options. First, we have used the graupel and rain

DSDs of the simulation itself (from now on, Tripoli DSDs), which are constant-slope Marshall-Palmer size distributions -- see Table 1 for details. Successively, we have tested the DSDs reported by Panegrossi et al. ([4]), which are variable Marshall-Palmer size distributions having temperature-dependent slopes over the same size ranges of Tripoli DSDs. The simulated brightness temperatures corresponding to Tripoli and Panegrossi DSD models are shown in figures 2 and 3 respectively, in terms of frequency-dependent scatterplots; as reference, TMI data for hurricane Mitch are also shown.

Figure 2 shows that even though the model manifold significantly overlaps the measurement manifold -- especially at the lower frequencies --, there are many model structures producing TBs that are too low as compared to the data; moreover, unlike the data, the simulated TBs at 85GHz show a very high correlation with those at 19 and 37 GHz. This different behavior is due to an enhanced amount of large scattering particles in the model. Panegrossi DSD modeling reduces the scattering effects by decreasing the amount of large particles in the mid-to-upper layers of the cloud, however, not even this parameterization can ensure a complete overlapping of the measured TBs by the simulated ones.

Thus, a new type of DSD has been explored (which we call truncated Marshall-Palmer distributions), that further reduces the number of large particles by linearly increasing the slope of the distribution with altitude, while linearly decreasing the maximum radius of both rain and graupel particles, as described in table 2.

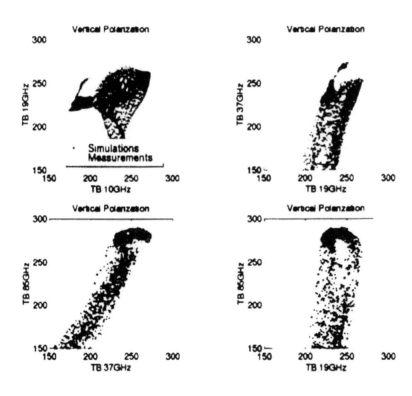

Figure 2: Scatterplots of the simulated upwelling brightness temperatures (black points) and Hurricane Mitch TMI data (grey points). The simulated database has been obtained by using Tripoli drop size distributions.

Table 1. Tripoli Drop Size Distributions.

Hydrometeor	SD function	Λ (mm^{-1})	ρ (g/cm^3)	Rmin (mm)	Rmax (mm)
Cloud	**Monodispersed**	-	1.0	0.01	0.01
Rain	**CS-MP***	1.852	1.0	0.1	3.0
Graupel	**CS-MP***	1.	0.6	0.1	5.0
Ice	**Monodispersed**	-	0.22	0.117	0.117
Snow	**CS-MP***	0.3	0.03	0.04	12.5.
Aggregates	**CS-MP***	0.3	0.03	0.04	12.5

*Constant-Slope Marshall Palmer

This new DSD model improves the overlapping of model and measurement manifolds, as shown in figure 4. However, significant differences are still apparent since the simulated temperatures show very large dynamics, as compared to the data, reaching much lower values, especially at 85GHz.
A possible explanation for this difficulty in reproducing the measured data can be that the model does not efficiently reproduce the microphysical structure of the observed storms -- especially for what concerns their ice and melting phases.

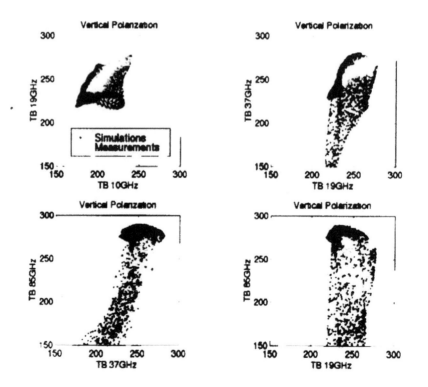

Figure 3: Same as figure 2, but for Panegrossi drop size distributions.

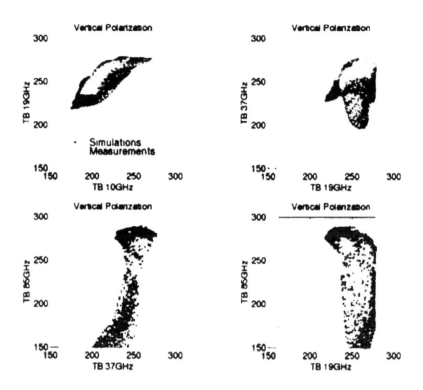

Figure 4: Same as figure 2, but for truncated Marshall-Palmer drop size distributions.

Table 2. Truncated Marshall-Palmer drop size distributions: note that the maximum radius linearly decreases with altitude for both rain and graupel particles, while the slope linearly increases.

RAIN			
Max radius at 0Km	Max Radius at 5 Km	Slope at 0Km	Slope at 5Km
3mm	0.3mm	$1.852mm^{-1}$	$5mm^{-1}$
GRAUPEL			
Max radius at 0Km	Max Radius at 10 Km	Slope at 0Km	Slope at 10Km
3mm	0.3mm	$1mm^{-1}$	$2.5mm^{-1}$

3. THE SURFACE RAINFALL RATE RETRIEVAL

For our retrieval computations, we have used the truncated Marshall-Palmer DSDs only because no retrieval can be performed with Tripoli and Panegrossi DSDs for many TMI measurements, due to the discrepancies of the two manifolds.

The inferred rainfall rate is compared, as a reference, with the product of the official TMI algorithm, the Goddard Profiling Algorithm (GPROF) [5]. Although both apply the Bayes theory of parameter estimation, the two algorithms use different assumptions. Our algorithm is a *Maximum A Posteriori* (MAP) method: given a set of measurements y_0, it searches the rainfall rate x that maximizes the conditional probability function distribution $p(x|y_0)$ [6]. By using the Bayes theorem, this quantity can be expressed as: $p(x|y_0) = p(y_0|x) \cdot p(x)/p(y_0)$, where p(x) contains the *a priori* information about the atmosphere that helps the retrieval to choose the best precipitation structure among the many that satisfy the observation. On the contrary, GPROF is a minimum mean square method: it assumes that the best estimate is the expected value of x, conditioned to the observed value y_0 of y.

Figure 5 shows the scatterplots of the rain rates retrieved using the MAP algorithm vs. those retrieved using the GPROF algorithm, for the three hurricanes considered in this study. The results for the three hurricanes have a (basically) similar behavior -- which is certainly related to the fact that the TMI measurement manifolds largely overlap each other.

The most striking feature, however, is the large spreading of the results, with the MAP algorithm generally underestimating the rain rates, as compared to GPROF algorithm -- especially for hurricane George. These differences may be due to the different estimation approaches by the two algorithms, as well as to the fact that the GPROF algorithm uses a more general database that includes the hurricane simulation as a subset. Moreover, the overall underestimation of MAP algorithm is partly due to the very large variation of the simulated brightness temperatures with respect to the measured ones; in fact, the heavy-precipitation profiles of the model simulation are never taken into account by the retrieval approach, since they are associated to TBs that are much lower than the measured ones (especially at 85GHz and 37GHz).

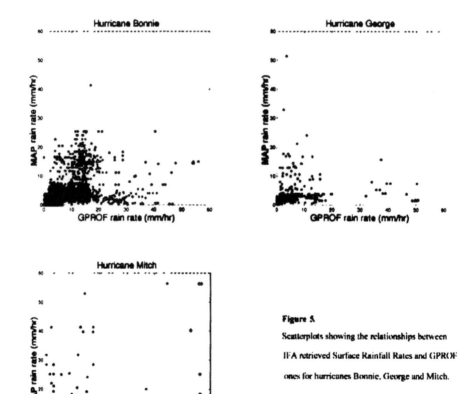

Figure 5

Scatterplots showing the relationships between IFA retrieved Surface Rainfall Rates and GPROF ones for hurricanes Bonnie, George and Mitch.

4. CONCLUSIONS

The results shown in this paper stress the very high sensitivity of the cloud-radiation database to DSD modeling. However, in spite of having explored three different DSD assumptions, discrepancies of model and measurements TB manifolds are still observed, which may be related to the fact that the cloud model produces a too large amount of ice in the upper levels, while melting is never taken into account.

ACKNOWLEDGEMENTS

TRMM is a joint NASA/NASDA mission (spacecraft launched in November 1997). We acknowledge NASA and NASDA for opening the TRMM data to EuroTRMM, a consortium of scientists from Centre d'Environnement Terrestre et Planétaire (France), German Aerospace Research Establishment (Germany), Istituto di Fisica dell'Atmosfera (Italy), Max Planck Inst. For Meteorologie (Germany), Rutherford and Appleton Lab. (U.K.), Univ. of Essex (U.K.), Univ.Catholique de Louvain (Belgium), Univ. de Munich (Germany). EuroTRMM is founded by European Commission and European Space Agency. This research has been also supported by the Italian Space Agency (ASI) and by the Italian National Group for Prevention from Hydro-geological Hazards (GNDCI).

REFERENCES

1. L. Roberti, J. Haferman, and C. Kummerow. Microwave radiative transfer through horizontally inhomogeneous precipitating clouds. *Journal Of Geophysical Research.* **99**, 16,707-16,718. (August 1994).

2. G.J. Tripoli. An explicit three-dimensional non-hydrostatic numerical simulation of a tropical cyclone. *Meteorol. Atmos. Phys.* **49**, 229-254. (1992).

3. D. Lemaire. *Non-fully developed Sea State Characteristics from Real-Aperture Radar Remote Sensing.* PhD thesis. Université Catholique de Louvain - Belgium. (December 1998)

4. G. Panegrossi, S. Dietrich, F.S. Marzano, A. Mugnai, E.A. Smith, X. Xiang, G. J. Tripoli, P.K. Wang, and J.P.V. Poiares Baptista. Use of cloud model microphysics for passive microwave-based precipitation retrieval: significance of consistancy between model and measurement manifolds. *Journal of the Atmospheric Sciences.* **55**, 1644-1673. (May 1998).

5. C. Kummerow, W.S. Olson, and L. Giglio. A Simplified Scheme for Obtaining Precipitation and Vertical Hydrometeor Profiles from Passive Microwave Sensors. *IEEE Transactions on Geosc. And Remote Sensing.* **34** (N°5). 1213-1232. (September 1996)

6. N. Pierdicca, F.S. Marzano, G. D'Auria, P. Basili, P. Ciotti, and A. Mugnai. *IEEE Transactions on Geosc. And Remote Sensing.* **34** (N°4). 831-845.(July 1996)

Microw. Radiomet. Remote Sens. Earth's Surf. Atmosphere, pp. 379–385
P. Pampaloni and S. Paloscia (Eds)
© VSP 2000

Multisensor analysis of Friuli flood event (October 5-7, 1998)

STEFANO DIETRICH[1], RENZO BECHINI[2], EMMA D'ACUNZO[1,3], SABATINO DI MICHELE[1,3], ROBERTO FABBO[2], ALBERTO MUGNAI[1], STEFANO NATALI[4], FEDERICO PORCU[4], FRANCO PRODI[4,5], LAURA ROBERTI[6] and ALESSANDRA TASSA[1,3]

[1] *Istituto di Fisica dell'Atmosfera - CNR, Roma, Italy*
[2] *ERSA-FVG/CSA - Weather Radar Operations Center, Fossalon di Grado (GO), Italy*
[3] *Fondazione per la Meteorologia Applicata, Firenze, Italy*
[4] *Dipartimento di Fisica - Università di Ferrara, Ferrara, Italy*
[5] *Gruppo Nubi e Precipitazioni - ISAO - CNR, Bologna, Italy*
[6] *Dipartimento di Elettronica - Politecnico di Torino, Torino, Italy*

Abstract - The heavy-precipitation event that occurred on October 5-7, 1998 over the Triveneto region of northern Italy, has been used here to test a combined radar-model-SSM/I retrieval technique. The method uses ground radar estimations of vertical cloud contents in conjunction with a cloud model in order to generate a cloud-radiation database for the specific event; such database is then used by a SSM/I precipitation retrieval algorithm.

1. INTRODUCTION

On the night of October 5-6, a deep through moving from the Northern Atlantic reached the Western Mediterranean Sea and affected the Triveneto region of northern Italy with an intense moist flow. The accumulated rain during the night was generally between 20 and 80 mm, and over 100 mm in the area of Udine and Gorizia. Most of this precipitation was due to local thunderstorms lasting only a few hours. On October 6, a cold drop detached from the through (cut-off), thus leading to the formation of a wide depression area over the central and western Mediterranean Sea. The depression and the associated cold front, moving from South-West to North-East, determined very moist Scirocco winds from sea level to higher altitudes, producing intense and widespread rain during the night of October 6-7 and the following day, after which the front passed by and the weather moderately improved. 260 mm of rain were registered in Udine on October 6, and the total amount of rain accumulated during the three days (October 5-7) was about 400 mm. The above event was largely observed by the GPM-500 C/F C-band Polarimetric Doppler Radar located in Fossalon di Grado (GO) [1], by the four-frequency (19.35, 22.235, 37.0, and 85.5 GHz) scanning radiometers Special Sensor Microwave/Imager (SSM/I) aboard the DMSP F-13 and F-14 polar satellites, and by the European geostationary satellite Meteosat. This event is used here to test a combined radar-model-SSM/I retrieval technique.

The strength of microwave satellite radiometry for cloud observations and precipitation measurements is universally recognized [2]. Detailed cloud/mesoscale simulations, that form the physical basis of profile-based algorithms [3], require a large amount of computational time to produce reliable microphysical outputs. Thus, cloud model simulations can be performed for a few events only, representing the thermodynamical and microphysical structures of a variety of precipitation systems. This implies that simulated cloud profiles have to be considered as an archive of possible vertical cloud profiles when the algorithm is applied to other events. On the other hand, computation of the upwelling brightness temperatures -- by means of radiative transfer (RT) schemes -- at the model grid resolution is affected by surrounding profiles inside the simulation [4], particularly when a slanted observation angle is considered -- such as for the SSM/I incident angle (53°). Thus, the resulting cloud-radiation databases depend on the spatial distribution of the microphysical profiles in the simulated events, whereas the observed event may have different spatial properties. The present work carries on the idea that radar can help to rebuild the pattern of the observed event using profiles from the archive of simulations. The cloud-radiation database, to be used by the profile-based SSM/I precipitation retrieval algorithm for the whole event under consideration, is then obtained by applying a 3-D backward Monte Carlo RT scheme [4] to the rebuilt scenario.

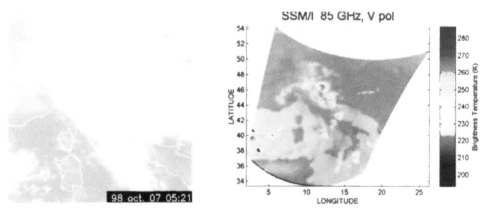

Figure 1 October 7, 1998, h 05:21 UTC: SSM/I 85 GHz measurement (right) and corresponding Meteosat-IR observation (left).

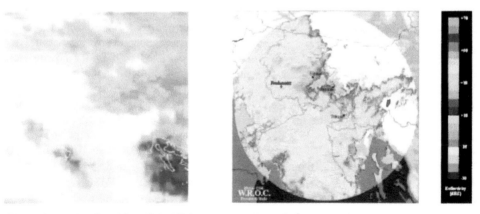

Figure 2 October 7, 1998, h 05:21 UTC: Radar Maximum Reflectivity (right) and corresponding Meteosat-SSM/I combined image (left), for the Triveneto region.

2. MULTISENSOR OBSERVATIONS

The Friuli event of October 5-7, 1998 was observed by the GPM-500 C-band polarimetric Doppler radar of Fossalon di Grado (the radar acquired data volumes every 10' starting at 19:00 of October 6), by the SSM/I, and by Meteosat. We are here considering the data for 05:21 of October 7, at which time there was an interesting SSM/I overpass (see Fig.1).

A general analysis of the structure of the cloud system, based on the multispectral observations of the two satellites, can be made by introducing a particular color scale enhancing the relative contribution of the different channels [5]. This is shown in the right panel of Fig. 2, where a color blue scale is assigned to the Meteosat-IR observations, that are sensitive to the very upper cloud layers; a green scale is assigned to the Polarization Corrected Temperature (PCT) at 85 GHz, which is related to the amount of ice particles; a red scale is assigned to the PCT at 19 GHz, that responds to the rain layer.

The analyzed system shows a convective structure embedded in a more extended cloud system. The white to light-blue central area denotes the presence of a high-top cloud with relatively high signatures at both 19 and 85 GHz, corresponding to heavy precipitation. The purple area on its southern part can be interpreted either as significant rain without a notable ice layer above it, or moist soil. In the northern part of the cloud, there is a large bluish area, indicating cirrus cloud coverage without a clear signature by either ice or rain. In general, these conclusions are confirmed by the radar observations shown in the left panel of Fig. 2.

3. SSM/I PROFILE BASED RAINFALL RATE ESTIMATION

Profile-based retrieval techniques [3] are based on the fact that precipitation retrieval is inherently a vertical structure problem, because distinct frequencies across the 10-90 GHz spectrum differentially penetrate vertically-inhomogeneous precipitating clouds, thus providing the basis for an inverse-type solution. These techniques are designed to retrieve hydrometeor profile information in keeping with the underlying physics of the problem, and incorporate cloud-radiation databases as detailed, consistent and objectively generated inputs to the numerical inversion process. The cloud portion of each database is given by the microphysical output of a cloud model simulation, and it is used as input to a Radiative Transfer (RT) scheme that computes the upwelling microwave brightness temperatures (TB's) from the simulated cloud -- since we are mainly interested in convective clouds having a large horizontal variability, a 3-D backward Monte Carlo RT model is used [4]. Then, a maximum-likelihood retrieval algorithm is used to retrieve both the vertical profile of each hydrometeor species and the surface rain rate. [6]

4. PARTICLE IDENTIFICATION AND LIQUID WATER CONTENT (ICE WATER CONTENT) ESTIMATION BY POLARIMETRIC RADAR MEASUREMENTS

The use of differential reflectivity (Z_{DR}) measurements makes it possible to distinguish between liquid and ice phase of meteorological particles. Falling raindrops are not spherical but have an oblate shape with, in still air, minor axis vertical [7], moreover axial ratio only depend on the D_e diameter of the equivalent sphere [8]. Differently, the shape of ice particles (such as ice crystals, aggregates, snow, graupel, hail) does not underlie a simple relation like that for liquid drops. A lot of bibliography exists on this argument [9], [10], [11], whose common goal is to furnish simple algorithms capable to distinguish between different meteorological particles. The basic problem is how to relate the measured quantities (in the case of the GPM-500C radar, the reflectivity Z and the differential reflectivity Z_{DR}) to the properties of hydrometeors [9]. Adding the information of the height of the melting level [10], a decision rule is sought that will partition the three-dimensional space of Z, Z_{DR}, h (height) so that each partition corresponds to a distinct hydrometeor type. Adopting this approach, we are able to distinguish among the following particle types: rain, wet snow, dry snow, ice crystal, graupel and hail. Actually, the partitions achievable with this method may partially overlay, because the triplet (Z, Z_{DR}, h) not always leads to univocally identify a single hydrometeor type. This is especially in the case of low observed reflectivity, when becomes more difficult to distinguish, for example, between dry snow, ice crystals and supercooled droplets. For reflectivity values below 20-25 dBZ the water droplets are nearly spherical, consequently the typically very small values assumed by Z_{DR} could fit equally well to water particles as well as to randomly oriented ice aggregates. In this case the particle identification gives two or more possible results. Then one or a mixture of the recognized particles may be present at the same time in the single cell observed by the radar. Estimation of the LWC (IWC) is just a matter of applying several appropriate relations, once the particle type is identified. To obtain reliable estimates of mass content from the measured radar reflectivity, the particle must be spherical or treated as equivalent spheres with diameters D that are small in comparison to the radar wavelenght. Moreover, the thermodynamic phase of the scatterers must be known in order to derive Z, because the reflectivity is a function of the dielectric factor $|K|^2$, which in turn is a function of the particle phase. The standard radar systems (like the GPM-500C) determine the water equivalent radar reflectivity (Z_e), namely the reflectivity of the target in the hypothesis that the scatterers are spherical water droplets. The conversion factor to obtain the ice equivalent radar reflectivity (Z_i) is given by $|K^2_w|/|K^2_i|$ (where $|K|^2_w=0.93$ and $|K|^2_i=0.176$ for $\xi>3$ cm), that is:

$$Z_i = 5.28\, Z_e \qquad\qquad (1)$$

where Z_e and Z_i are in mm^6m^{-3}. In decibels, the corresponding ratio is 7.2 dB. The relations used to calculate the particle LWC (IWC) are exponential functions of Z_x:

$$LWC\,(IWC) = A\, Z_x^B \qquad\qquad (2)$$

where $x=i$ for *dry snow* and *crystals*, while $x=w$ for all the other particles, Z_x is in mm^6m^{-3} and LWC in gm^{-3}.

Given in Table 1 are the coefficients A and B used for the different hydrometeor types.

Table 1. List of the coefficients A and B used in eq. (2) for the LWC (IWC) calculation.

Hydrometeor type	A	B	Reference
Rain	0.0039200	0.549	Rogers (1989) [12]
Wet snow	0.0030000	0.605	Herzegh and Hobbs (1980) [13]
Dry snow	0.0340000	0.400	Thomason (1995) [14]
Ice crystals	0.0170000	0.529	Heimsfield and Palmer (1986) [15]
Graupel	0.0010000	0.712	Kajikawa and Kiba (1978) [16]
Hail	0.0001867	0.666	Holler (1995) [17]

Fig. 3 shows LWC estimated by radar on October 7 at 05:28 UTC over 6 rain gauges included in the regional network. Each panel corresponds to a ground area of 490x490 meters, while the height and the width of the microphysical estimation are related to the 5 elevation angles (0.53°, 0.97°, 1.49°, 1.93°, and 2.90°) and to the distance from the radar site. Each vertical line represents an estimated LWC value in the hypothesis that only one hydrometeor was present.

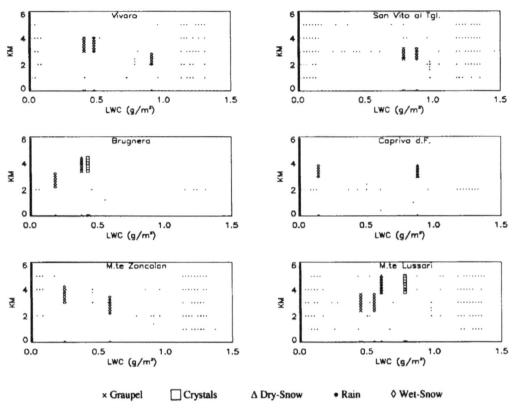

x Graupel   □ Crystals   △ Dry-Snow   • Rain   ◊ Wet-Snow

Figure 3: LWC (IWC) estimated by radar on October 7 at 05:28 UTC over 6 rain gauges

5. THE COMBINED RADAR-MODEL-SSM/I APPROACH

The procedure explained in the previous section has been used to compute the LWC's measured by the radar for all pixels of the radar grid, for the radar volume scan of October 7 at 05:28.

For each radar grid point, the radar estimates have been used to extract the most similar profile from the archive of simulated profiles, by comparing the LWC's, measured by the radar at each elevation, with the integrated LWC's of the simulated profiles over corresponding vertical ranges observed by the radar. If the so computed "gross profile" can be compared with the radar measurements (i.e., if the differences are within the range of error estimated for each hydrometeor) then the cloud profile was marked as "possible". Once the profile at the minimum distance between radar and the all "possibles" is selected in this way, a synthetic cloud model scenario can be produced by using the selected model profiles in the corresponding spatial positions of the radar grid. By applying the RT scheme to this rebuilt model scenario a cloud-radiation database for the Friuli event was then generated. The profile-based algorithm was then applied to the SSM/I overpass, in order to retrieve the rainfall rates. Fig. 4 shows the comparison among results obtained with and without the use of the radar. It is evident that the combined radar-model-SSM/I approach considerably reduces the errors of the SSM/I estimates.

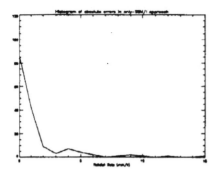 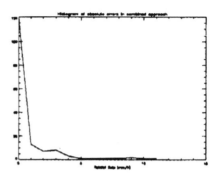

Figure 4: Histogram of absolute errors in SSM/I approach (left), histogram of absolute errors in combined radar-model-SSM/I approach (right)

6. SUMMARY AND CONCLUSIONS

Ground radar is used in conjunction with a cloud model to estimate precipitation using SSM/I.

The radar-model-SSM/I approach, illustrated in this paper, shows good potential for improving the accuracy of rainfall rate estimations by means of microphysical radar estimation in order to realize a physical calibration of the SSM/I profile-based technique.

We must notice, nevertheless, that the way to use radar measurements to estimate LWC has to be considered experimental -- the topic is worldwide under investigation; thus, the derived microphysical information is used here simply to drive the selection of simulated profiles to be used within the retrieval phase.

ACKNOWLEDGEMENTS

This research has been supported by the EU Project "Satellite and combined satellite-radar techniques in meteorological forecasting for flood events" (MEFFE), the Italian Space Agency (ASI), and the Italian National Group for Prevention from Hydro-geological Hazards (GNDCI).

REFERENCES

1 Dietrich, E.: GPM-500C/F - A Polarimetric Doppler Radar for Research, Hail Detection, General Monitoring. ERSA Friuli Venezia Giulia, Gorizia. (1994)

2 Mugnai, A., E.Smith, G.Tripoli. Foundations for statistical-physical precipitation retrieval from passive microwave satellite measurements. Part II: Emission source and generalized weighting function properties of a time dependent cloud radiation model. Journal of Applied Meteorolog.y. 32, 17-32. (January 1993).

3 Smith, E.A., C. Kummerow, and A. Mugnai: The emergence of inversion-type profile algorithms for estimation of precipitation from satellite passive microwave measurements. Remote Sensing Reviews, 11, 1-4, 211-242 (1994)

4 Roberti, L., Haferman J., and Kummerow C. Microwave radiative transfer through horizontally inhomogeneous precipitating clouds. Journal Of Geophysical Research. 99, 16,707-16,718. (1994).

5 Porcu', F., S. Dietrich, A. Mugnai, S. Natali, F. Prodi, and P. Conway: Satellite multi-frequency observations of severe convective systems in the mediterranean. Physics and Chemistry of the Earth, accepted (1999)

6 Pierdicca, N., F.S. Marzano, G. d'Auria, P.Basili, P. Ciotti, , and A. Mugnai: Precipitation retrieval from spaceborne microwave radiometers based on maximum a posteriori probability estimation. IEEE Trans. Geosci. Remote Sensing, 34, 831-846. (1996)

7 Sauvageot, H.,: Rainfall Measurement by Radar: a Review. Atmospheric Research, 35, 27-54.(1994)

8 Green, A.W.,: An Approximation for the Shape of Large Raindrops. J. Appl. Meteorol.14, 1578-1583.(1975)

9 Doviak, R.J., D.S. Zrnic,: Doppler Radar and Weather Observations. Academic Press, inc. (1993)

10 Holler, H., V.N. Bringi, J. Hubbert, and P.F. Meischner,: Particle discrimination in hailstorms. 26th Conf. on Radar Meteorology, Norman, OK, , Amer. Meteor. Soc., 594-595. .(1993)

11 Brandes, E.A., J. Vivekanandan, J.D. Tuttle, C.J. Kessinger,: Sensing thunderstorm microphysics with multiparameter radar: application for aviation. Preprints, 5th Conf. On Aviation Wea. Systems, 2-6 August, Vienna, VA, 98-102.(1993)

12 Rogers, R.R. and M.K. Yau,: A short course in cloud physics. Pergamon Press.(1989)

13 Herzegh, P.H. and P.V. Hobbs,: The mesoscale and microscale strucure and organization of clouds and precipitation in mid-latitude cyclones. II: Warm frontal clouds. J. Atmos. Sci., 37, 597-611.(1980)

14 Thomason, J.W.G. et al.,: Density and size distribution of aggregating snow particles inferred from coincident aircraft and radar observations. Preprints, 27th Conf. on Radar Meteorology, Vail, CO, Amer. Meteor. Soc., 127-129.(1995)

15 Heimsfield, A.J. and A.G. Palmer,: Relations for deriving thunderstorm anvil mass for CCOPE storm water budget estimates. J. Climate Appl. Meteor., 25, 691-702.(1986)

16 Kajikawa, M. and K. Kiba,: Observtions of the size distributions of graupel particles. Tenki, 25, 390-398.(1978)

17 Holler, H.,: Radar-derived mass-concentrations of hydrometeors for cloud model retrivals. Preprints, 27th Conf. on Radar Meteorology, Vail, CO, Amer. Meteor. Soc., 453-454.(1995)

Microw. Radiomet. Remote Sens. Earth's Surf. Atmosphere, pp. 387–396
P. Pampaloni and S. Paloscia (Eds)
© VSP 2000

SSM/I Data Analysis for Retrieving Cloud Properties: Comparisons with Ground-based Measurements

G. D'AURIA,[1] N. PIERDICCA,[1] P. BASILI,[2] S. BONAFONI,[2]
P. CIOTTI[3] and F. S. MARZANO[3]

[1] *Dipartimento di Ingegneria Elettronica, Università "La Sapienza" di Roma, via Eudossiana 18, 00184 Roma, Italy*
[2] *Dipartimento di Ingegneria Elettronica e dell'Informazione, Università di Perugia, via Duranti 93, 06125 Perugia, Italy*
[3] *Dipartimento di Ingegneria Elettrica, Università dell'Aquila, Poggio di Roio, 67040 L'Aquila, Italy*

Abstract. A retrieval methodology, based on cloud genera and radiative transfer modeling, is illustrated and applied to a large data set of Special Sensor Microwave Imager (SSM/I) passes over Southern Europe and, in particular, Central Italy. The purpose of this work is basically the verification of the retrieved cloud genera from SSM/I data, through the precipitation intensity at surface and the altitude of the cloud base. In the first case we have gathered precipitation data over more than three years from a rain gauge network along the basin of the river Tiber. The corresponding passes of the SSM/I radiometer were recorded and the data have been corrected, as much as possible, for the effect of surface and atmospheric seasonal variations. Statistical comparisons were also made by using diverse methods of rain prediction. In the second case corresponding observations of the cloud base altitude, during the SSM/I passes in cloudy conditions, were monitored by an infrared ceilometer by processing the return time of the laser beam pulse. Correspondence between the group of clouds predicted by the database automatic classifier and the range of cloud base altitude has been also verified.

1. INTRODUCTION

Microwave radiometers observing the atmosphere can determine the properties of clouds if an adequate number of frequency channels is available [1-4]. In previous works SSM/I data have been proved to be useful for the identification of cloud types and, therefore, for retrieving their hydrometeor vertical profiles [5]. The adopted inversion methodology has been based on a reliable database of cloud hydrological parameters subdivided into a chosen number of cloud types. A cloud structure database has been generated by assuming vertical profiles of four kinds of hydrometeors resulting from cloud microphysics and from dynamical models of meteorological events. This database generation has been carried out for seven cloud genera and two species according to the International Cloud Atlas of the World Meteorological Organization (WMO).

In order to identify cloud genera from multifrequency brightness temperatures measured by spaceborne microwave radiometers, the cloud database has been transformed into a radiative database making use of the radiative transfer equation together with a number of modeling assumptions [1,5]. These assumptions may limit the validity of a model-based retrieval algorithm if they are not properly ascertained; thus an extensive validation of the simulated cloud radiative database is needed. These aspects are even more pronounced for over-land case studies since satellite radiometric data are also strongly affected by the emissivity of the land beneath [6-7]. Serious problems in the retrieval accuracy can arise, especially when the chosen test site is not far from the coast and the scale of orographic and land-use variations is small.

The assessment of the database could be strictly possible only by using airborne measurements through the clouds. Due to the lack of such *in situ* measurements, a complete comparison with the physical "truth" is by no means possible. We have studied a number of partial comparisons performed with ground-based and airborne-based measurements in such a way that a number of "pieces of truth" can be ascertained.

First of all, we have compared the radiative database with available radiometric measurements. Any systematic error in the cloud model or radiative transfer model would produce a bias in the simulated T_B's with respect to the measured ones. In a previous work we have found that the T_B values measured by SSM/I in several years are fairly well represented in the database [5]. On the contrary, most of the database points corresponding to very optically thick clouds have never been observed by SSM/I in our geographical area of interest.

The cloud genera can be alternatively subdivided so that they have a single parameter that can be measured or observed from the ground, thus identifying a group of clouds. These groups are chosen so that they have supplementary features in common. For instance only some clouds produce rain (precipitation) or the range of their base altitude is between defined limits. Many other examples are possible.

In this work we have confined ourselves mainly to two aspects: the identification of clouds through their precipitation (and the relative intensity) and the range of altitude of their base. For the first experiment we have gathered precipitation data over more than three years from a rain gauge network along the basin of the river Tiber. The corresponding passes of the SSM/I radiometer were recorded. Statistical comparisons were also made by using diverse methods of rain prediction. The second experiment is concerned with the contemporaneous observation of the cloud base altitude, during the SSM/I passes in cloudy conditions. The base altitude is obtained from the return time of a laser beam pulse of an infrared ceilometer. Correspondence between the group of clouds predicted by the database automatic classifier and the range of cloud base altitude has been also verified.

2. SSM/I RETRIEVAL METHODOLOGY BASED ON CLOUD GENERA

The scarcity of *in situ* meteorological data, concerning cloud systems and precipitation, suggests tackling the forward and inverse problem by using cloud and radiative transfer models. Modeling plane-parallel clouds and simulating the related brightness temperatures (T_B's) has allowed us to generate a large cloud radiative database.

2.1 Strategy for cloud-genera radiative transfer modeling

The numerical outputs of a three-dimensional time-dependent cloud mesoscale model, named University of Wisconsin-Non-hydrostatic Modeling System (UW-NMS), have been used as starting point for generating the cloud-structure data set, explicitly describing the detailed vertical distribution of four species of hydrometeors: cloud droplets, rain drops, graupel particles, and ice particles [1]. The number of cloud layers has been reduced to at most seven homogeneous layers. The cloud structures have been classified into cloud genera and species, such as cumulonimbus (Cb), cumulonimbus with incus (Cbi), cumulus (Cu), cumulus congestus (Cuc), altocumulus (Ac), nimbostratus (Ns), stratus (St), altostratus (As), and cirrus (Ci), following the WMO definitions, the microphysical knowledge, and available information from the cloud model [5]. Note that

Ci are *high* region clouds, typically above 6 km, As, Ac, and Ns are *middle* region clouds, between 2 and 6 km, while St, Cu, and Cb are *low* region clouds, below 2 km (even though cumuliform clouds can reach altitudes up to the tropopause). Each cloud genus has been statistically characterized by a mean profile and a covariance matrix of the hydrometeor contents.

The generation of the cloud database has been based on some physically reasonable assumptions starting from the initial UW-NMS cloud simulation which refers to August conditions in Alabama. We have adopted a Monte Carlo statistical procedure to extend the number of cloud structures, based on the assumption of a Gaussian distribution of hydrometeor contents [3]. Moreover, surface temperature T, pressure p, and humidity q together with cloud layer levels were considered random parameters with a uniform distribution around the mean and bounds given by the computed variances. Any attempt to apply a modeled database to spaceborne radiometric measurements must tackle the problem of characterizing the yearly variations of atmospheric parameters. The seasonal (and geographical) adaptation of the available cloud simulation has been driven by the statistical analysis of T, q, and p profiles derived from Italian radio-sounding stations. The monthly mean and variance of surface T, p and q, together with associated height scales, zero-degree isotherm and tropopause heights have been computed at 0600 and 1800 UTC, which more or less correspond to times of SSM/I passes over Italy. The height "compression" of the summer cloud structure to produce lower winter clouds has been carried out by imposing the seasonal freezing and glaciation levels as derived by the analysis of the radio-soundings. By comparing these levels in the simulation outputs with the monthly average of the Italian radio-soundings, both the vertical levels and hydrometeor contents have been proportionally scaled separately for each month. As a result, a cloud-radiation data set, consisting of 4500x12 cloud vertical structures, and brightness temperatures at the chosen frequencies and observation angle, has been generated retaining the physical and statistical features of the input microphysical cloud model and tuning the monthly meteorological conditions as much as possible. It is worth mentioning that the plane-parallel assumption does not allow taking into account the horizontal inhomogeneity of clouds, that is the beam filling problem, which might be relevant for convective structures [8].

In order to associate T_B's to each cloud structure, a radiative transfer model, based on the fast and fairly accurate Eddington solution, has been used [5]. Within each layer, the temperature has been supposed to be linearly dependent on the height, and the gaseous absorption has been determined by means of the Liebe model. The hydrometeors have been assumed to be spherical and characterized by inverse-exponential size distributions. The land surface is treated as a Lambertian surface. The monthly mean emissivity has been computed from SSM/I images collected during clear sky conditions and corresponding radio-soundings close to the considered geographical area [7]. In order to account for the uncertainties of surface emissivity, during the statistical generation of the cloud structures we have assumed it was a Gaussian random variable with a standard deviation equal to 10% of its mean value. Thus, the uncertainty about the surface around its monthly mean is implicitly inserted in the subsequent retrieval problem as a worst case. This might reduce the contribution of the lower frequency channels in the cloud classification with respect to the ideal case of a completely known surface background. An analysis of cloud signature sensitivity to surface background is reported in [3].

A specific point we would like to stress in our approach is the generation of a cloud database that is classified according to the cloud genera defined in the Cloud Atlas. The main advantage is the possibility to retrieve the cloud genus as a further "descriptor" of the observed cloud in addition to the produced rainfall rate. This is useful as such, but it is also a further element to be compared to reference and in situ data (e.g., visual observations, base altitude and so on) for validation purposes.

2.2 Algorithms for cloud classification and parameter retrieval

It is well known that retrieving rainfall rate and cloud characteristics over land is a challenging task since for stratiform rain the emission from water particles in the cloud is hardly detected against the high emissivity of the land background. Therefore it is necessary to follow an inversion approach able to cope with non-linear relationship between predictors (the T_B values) and predictands (the cloud parameters) and the presence of errors (both instrument errors and model errors). We refer to the Bayes theory of estimation and the overall retrieval algorithm consists of a two-step approach.

The Maximum *A posteriori* Probability (MAP) criterion is used first to classify cloud genera and then to estimate the hydrometeor content profiles and related precipitation intensity. Since the precipitating clouds are supposed to be layered structures, the set of hydrometeor vertical profiles is described by a vector g, whose elements are the equivalent water contents (in g/m^3) of each hydrometeor species within each homogeneous layer. The simulated and measured multi-frequency brightness temperatures (in K) are indicated by vectors t (or t(g)) and t_m, respectively.

The cloud genus index i is obtained by minimizing the following discriminant function $d_{class}(t_m,i)$ with respect to i [5]:

$$d_{class}(t_m,i) = [t_m\text{-}<t(g)>_i]^T C_{ti}^{-1} [t_m\text{-}<t(g)>_i] + \ln[\det(C_{ti})] - \ln(P(i)) \qquad (1)$$

where $<t(g)>_i$ is the mean value of the ith class (centroid), $t_m\text{-}<t(g)>_i$ is the deviation of the measured T_B with respect to the centroid, while $\det(C_{ti})$ and C_{ti}^{-1} are the determinant and the inverse of the covariance matrix C_{ti} of the class and $P(i)$ is the a-priori probability of the ith class.

Once the individual measurements have been classified, the cloud parameters are retrieved by considering the statistical characteristics of the specific class. Since our brightness temperature values t(g) are computed by means of an electromagnetic model, the difference between t_m and t, giving the total radiometric error $\varepsilon_t=[t_m\text{-}t(g)]$, is due to both the instrumental noise and model errors. A possible choice is to assume that the error ε_t follows a multivariate Gaussian probability density function (pdf) with zero mean and so does the hydrometeor vector g with a mean m_g. Thus, by applying the Bayes theorem to the conditional pdf $p(g|t_m)$ of a cloud structure g given a measurement t_m, after some algebraic manipulation, the MAP estimation of g can be performed by minimizing the following function $d_{par}(t_m,g)$ with respect to g [3]:

$$d_{par}(t_m,g) = [t_m\text{-}t(g)]^T C_{et}^{-1}[t_m\text{-}t(g)] + (g\text{-}m_g)^T C_g^{-1}(g\text{-}m_g) \qquad (2)$$

where m_g and C_g are the mean vector and the auto-covariance matrix of g, respectively, while $\det(C_g)$ and C_g^{-1} are the determinant and the inverse of matrix C_g, specific for the considered i^{th} class. Correspondingly, C_{et} is the auto-covariance matrix (with C_{et}^{-1} its inverse) of the radiometric error vector ε_t.

Over land we classify using the mean values of vertical and horizontal polarisations of the 19, 37, and 85 GHz channels while over water, we use both the vertically and horizontally polarised channels at 37, and 85 GHz. Note that, within the first step of the algorithm aimed to cloud classification, the covariance matrix C_{ti} was far to be singular, also thanks to the choice of disregarding the 22-GHz SSM/I channel among the predictors (the 22-GHz T_B shows fairly high correlation with the 19-GHz ones). Nevertheless, some cloud classes have similar radiometric signatures so that their posterior probability is similar and the classification accuracy may be poor.

3. COMPARISON WITH GROUND-BASED MEASUREMENTS

Before illustrating the comparison, we briefly recall some characteristics of SSM/I, installed on board the DMSP platforms that fly on a near-polar sun-synchronous orbit at an altitude of about 830 km [9]. SSM/I measures upwelling T_B's at four frequency bands (19.3, 22.2, 37.0, and 85.5 GHz) and two linear polarizations (horizontal and vertical, with the exception of the 22.2 GHz channel which operates only in the vertical polarization). During each conical scanning it acquires data at an observation angle of 53.5° off-nadir and covers a swath of 1400 km. The spatial resolution of each channel is 69x43 km for the 19 GHz channel, 60x40 km for the 22.2 GHz channel, 37x29 km for the 37 GHz, 15x13 km for the 85 GHz. The radiometric resolution is better then 0.9 K in each channel. The geolocation error of the data can theoretically be better then ±10 km by using the satellite effemerides. We have used the F11 SSM/I, provided by FNMOC, Monterey, CA.

3.1 SSM/I retrievals and rain gauge network data

Even though a model-based approach, as the one we have described here, is more versatile than the empirical ones, it requires a deep investigation for the "calibration" of the cloud radiative model itself. To this aim we have used data from a rain gauge network installed in the river Tiber basin. Cumulative precipitation sampled twice an hour, with a resolution ranging in the interval between 0.1 and 0.2 mm, are routinely recorded. The area is complex, both morphologically and in terms of land use. The choice was however dictated by the availability of a long historical archive of rain rate measurements. In this work we have considered the year 1995, when there were 86 available stations. Moreover, the aim of the work was to understand the potential of microwave radiometry for detecting precipitation in typical climatological, orographic and land cover situations of Central Italy. In order to compare MAP retrieval with other techniques, we have considered the empirical regression algorithm, proposed by Ferraro and Marks, based on the use of the scattering index SI over land [2].

For the comparison with SSM/I measurements, the rain gauge stations have been geographically located and a preliminary quality test has been performed in order to identify blunder points, such as high rain rate detected by a single rain gauge during only a single sample time. For each rain gauge site the closest measurement of multifrequency brightness temperature has been identified for each pass. Such multifrequency set has been merged to the rain intensities detected by the considered rain gauge within a time interval of 5 hours, starting one hour before the satellite pass (for a total of 11 rain gauge sampling times). However, errors of ±15 minutes in the time correspondence between data exist in addition to SSM/I geolocation errors, which have been often revealed greater than nominal values. Fig. 1 represents the geographical location of the rain gauge stations within the Tiber valley.

In Figs. 2a and 2b the scatterplots between rain gauge measurements (with R>0) and SI (a) and MAP (b) estimates are presented. A relatively high dispersion of data at rain gauge level is noticeable both for MAP and SI algorithms with a correlation coefficient of 0.30 and 0.27, respectively. Figs. 2c and 2d show the same of Figs. 2(a,b), but for the average of the estimates and measurements over the Tiber basin. The correlation of the SI and MAP algorithm is about 0.42 and 0.47, respectively.

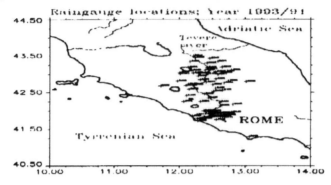

Figure 1. Rain gauge network of Tiber valley. The x- and y-axis are the East longitude and North latitude in deg.

These kinds of results have been also observed by other authors using rain gauge or meteorological radar measurements [2,6]. It emerges that the MAP algorithm tends to overestimate, while the SI tends to underestimate rain gauge measurements. The increase of the correlation at basin level is mainly due to reduction of the rain field inhomogeneity by spatially averaging over the basin. Moreover, by analysing several years we have found different and often higher correlations, even for comparisons at rain gauge level.

Many considerations can explain the low correlation between ground measurements and spaceborne MAP retrievals of rain rate.

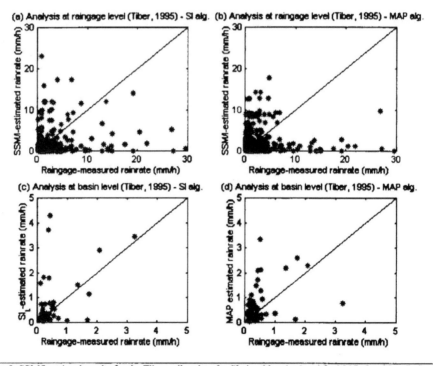

Figure 2. SSM/I retrieval results for the Tiber valley data for SI algorithm (a,c) and for MAP algorithm (b,d).

There are at least three aspects that require more investigation in the model based MAP retrieval. 1) The first one has to do with the spatial averaging of the SSM/I in the presence of non-homogeneous cloud structures and therefore a non-homogeneous precipitation field. This problem is the well-known "beam-filling" effect and the cloud genus modeling could give an interesting framework to address it. For example, *a priori* information about the co-presence of different cloud genera within SSM/I antenna footprints might be used. Averaging precipitation over the Tiber basin may at least compensate for geolocation mismatches and in fact it increases the correlation. 2) The second aspect is related to possible spatial shifts between rain cell detected by rain gauges and SSM/I. This has suggested us to make the consideration that, in the presence of widespread cloud systems, what is observed by SSM/I does not correspond to what is observed from the surface at the same time. Indeed, an improved correlation has been obtained in many cases by comparing SSM/I measurements with rain gauge data taken with a delay of 30 minutes due probably to delay in rain drop formation from the melting of ice particles and to evaporation of rain drops before reaching the ground. 3) Finally, the uncertainty on surface emissivity contributes as a source of errors in the model-based MAP algorithm.

3.2 SSM/I case study

As a case study, we have analyzed the day of 20 Feb., 1999, where stratiform clouds were present over Southern Europe, which is the region we are interested in. Fig. 3 shows the thermal infrared image of Meteosat radiometer at 1630 UTC. The corresponding SSM/I image at 1613 UTC for the 85 GHz channel (vertical and horizontal polarisations have been averaged) is shown in Fig. 4. A fairly good correspondence is noted between the high 85-GHz T_B values in the Tyrrhenian Sea and the uniform relatively warm values in the thermal infrared Meteosat channel.

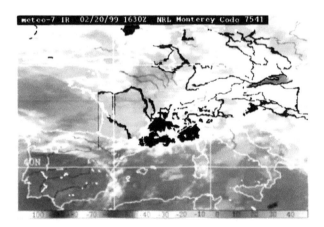

Figure 3. Meteosat thermal infrared image at 1630 of Feb. 20, 1999 (Courtesy of NRL, Monterey, CA).

The application of the Bayesian algorithm to SSM/I data is shown in Fig. 5 in terms of classification of cloud genera and in Fig. 6 in terms of retrieved surface rain rate. The area observed over Central Italy is dominated by nimbostratus and altocumulus clouds, which are not tall clouds with a cold thermal infrared equivalent temperature. This conclusion is consistent with the Meteosat observation.

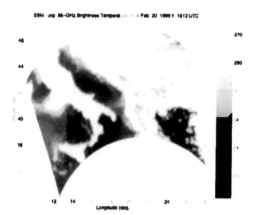

Figure 4. SSM/I unpolarised 85-GHz TB at 1613 UTC of Feb. 20, 1999.

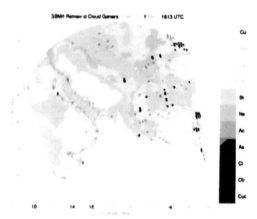

Figure 5. Cloud genera classification from SSM/I, using MAP algorithm for 20 Feb. 1999, 1613 UTC..

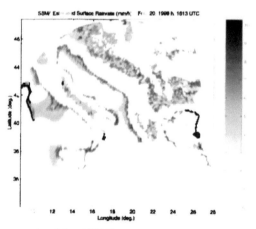

Figure 6. Surface rain rate retrieved from SSM/I by means of MAP algorithm for 20 Feb. 1999, 1613 UTC.

The retrieved rain rate map shows an effect of coast contamination, due to the large SSM/I footprint so that pixels with a mixed surface can produce an ambiguous microwave signature. Retrieved rain rates lower than 10 mm/h are typical of stratiform precipitation.

3.3 SSM/I retrievals and cloud ceilometer observations

An infrared lidar (ceilometer) has been installed at the Dept. of Electronic Engineering, University "La Sapienza" of Rome for cloud remote sensing. The CT25K instrument is manufactured by Vaisala, and has a diode laser source with center wavelength at 905 nm operating in a pulsed mode with a repetition frequency of 5.6 kHz and average power of 8.9 mW. The measurement range is of 7500 m with a range resolution of 15 m. The instrument has potentially a capability of detecting up to three cloud bases. It should be noted that our instrument is pointed off-zenith in order to observe in the direction opposite to the sea. Consequently, we consider an SSM/I pixel far enough from the coast to avoid sea contamination, also taking into consideration the geolocation error.

Figs. 7a and 7b give the time-series of the mean cloud-base height, derived from the ceilometer data averaged over 20 minutes around the SSM/I pass, as a function of the day during January and February 1999, respectively. The diamond symbols indicate detection of rain during the 20-minute interval. Figs. 7c and 7d show the estimates of the cloud class derived from the corresponding SSM/I passes acquired over Rome, using the MAP algorithm.

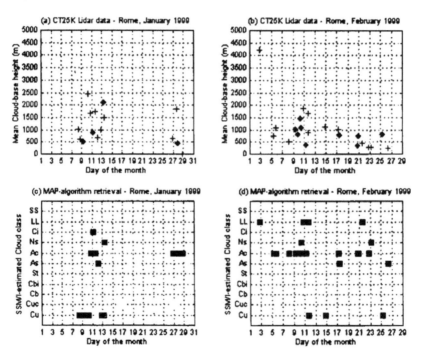

Figure 7. Lidar measurements (a;b) and MAP retrieval results (c,d) for the Rome site during Jan. and Feb. 1999.

The retrieval of low-region (such as Cu and Cb) and middle-region (such As and Ac) clouds is generally in agreement with low (between 500 and 1500 m) and middle (between 1500 and 3000 m) cloud bases detected by lidar, especially for the January cases. As expected, cirrus clouds show ambiguous results. Rain detection by lidar is in a fairly good agreement with precipitating cloud genera.

4. CONCLUSIONS

A retrieval methodology, based on a modeled cloud database and the MAP criterion, has been illustrated and applied to SSM/I data over Central Italy to classify cloud genera and estimate cloud hydrometeor profiles and associated surface rain rate. The comparison of SSM/I retrievals has been carried out both in terms of surface rain rate, by using an operational network of rain gauge stations in the Tiber valley, and in terms of cloud base height, by using a ground-based lidar.

The statistical analysis of the comparison with rain gauge data during 1995 shows correlation between estimates and rain rate measurements less than 0.5. This is similar to what has been already found in the literature but in different conditions as far as climatology, orography and land use characteristics of the investigated area are concerned.

The cloud-genera analysis can give some insight into the problem in order to improve the characterization of the cloud radiative database and to understand anomalous situations due to particular cloud systems, such as the presence of different cloud genera in the field of view of the radiometer. The comparison of cloud-base height derived from lidar can represent a tool of validation of the database itself in terms at least of the retrieved cloud genera, but it can also be a tool for acquiring cloud geometry statistics.

Further work is needed to better explain all the cases that affect the correlation in the statistical analysis we have presented here. This requires to check the assumptions made in the cloud radiative model in individual events, nevertheless the more enlarged sample of comparison events is important to lead to statistically-supported conclusions.

Acknowledgements. Rain gauge data have been provided by the Dipartimento STN, Servizio Idrografico e Mareografico Nazionale (Roma, Italy). SSM/I data have been obtained from NASA-MSFC, NOAA/NESDIS, NOAA/FNMOC, GHRC and SAA. This work has been partially supported by ASI. We thank Mr. L. Pulvirenti and Mr. A. Incalza for their help in data processing.

REFERENCES

1.   E.A. Smith, A. Mugnai, H.J. Cooper, G.J. Tripoli, and X. Xiang, "Foundations for statistical-physical precipitation retrieval from passive microwave satellite measurements. Part I: brightness temperature properties of time-dependent cloud-radiation model", *J. Appl. Meteor.*, 31, 506-531 (1992).
2.   R.R. Ferraro, and G.F. Marks, "The development of SSM/I rain-rate retrieval algorithms using ground-based radar measurements", *J. Atmos. Oceanic Technology*, 12, 755-777 (1995).
3.   N. Pierdicca, F.S. Marzano, G. d'Auria, P. Basili, P. Ciotti, and A. Mugnai "Precipitation retrieval from spaceborne microwave radiometers using maximum a posteriori probability estimation", *IEEE Trans. Geosci. Rem. Sens.*, 34, 831-846 (1996).
4.   I. Jobard and M. Desbois, "Satellite estimation of the tropical precipitation using the Meteosat and SSM/I data", *Bull. Amer. Meteor. Soc.*, 34, 285-298 (1994).
5.   G. d'Auria, F.S. Marzano, N. Pierdicca, R. Pinna Nossai, P. Basili, and P. Ciotti, "Remotely sensing cloud properties from microwave radiometric observations by using a modeled cloud database", *Radio Sci.*, 33, 369-392 (1998).
6.   M.D. Conner and G.W. Petty, "Validation and intercomparison of SSM/I rain-rate retrieval methods over the continental United States", *J. Appl. Meteor.*, 37, 679-700 (1998).
7.   C. Prigent, W.B. Rossow, and E. Matthews, "Global maps of microwave land surface emissivities: potential for land surface characterization", *Radio Sci*, 33, 745-752 (1998).
8.   C. Kummerow, "Beamfilling errors in passive microwave rainfall retrievals", *J. Appl. Meteor.*, 37, 356-370 (1998).
9. P. Hollinger, J.L. Peirce, and G.A. Poe, "SSM/I instrument evaluation", *IEEE Trans. Geosci. Remote Sens.*, 28, 781-790 (1989).

Microw. Radiomet. Remote Sens. Earth's Surf. Atmosphere, pp. 397–405
P. Pampaloni and S. Paloscia (Eds)
© VSP 2000

Rainfall retrieval from ground-based multichannel microwave radiometers

FRANK S. MARZANO,[1] ERMANNO FIONDA,[2] PIERO CIOTTI [1] and ANTONIO
MARTELLUCCI [2]

[1] *Dipartimento di Ingegneria Elettrica, Università dell'Aquila, L'Aquila (Italy)*
  Poggio di Roio, 67040 L'Aquila (Italy); Tel. 39.0862.434412; Fax. 39.0862.434414
  E-mail: marzano@ing.univaq.it; ciotti@ing.univaq.it
[2] *Fondazione Ugo Bordoni, Roma (Italy)*
  V.le Europa 190, 00144 Roma (Italy); Tel. 39.06.54802118; Fax. 39.06.54804401
  E-mail: efionda@fub.it; amartellucci@fub.it

Abstract – Inversion algorithms for ground-based retrieval of surface rainrate, integrated cloud parameters, and slant-path attenuation is proposed and tested. The retrieval methods are trained by numerical simulations of a radiative transfer model applied to microphysically-consistent precipitating cloud structures. The Eddington method is used to solve the radiative transfer equation (RTE) for plane-parallel seven-layer structures, including liquid, melted, and ice spherical hydrometeors. Polynomial regression techniques and neural network algorithms are first developed and tested on synthetic data in order to understand their potential and to select the best frequency set for rainfall and integrated cloud parameters estimation. Ground-based radiometric measurements at 13.0, 23.8, and 31.6 GHz are used for experimentally testing the retrieval algorithms. Comparison with rain gage data and rain path-attenuation measurements, derived from the three ITALSAT channels at 18.7, 39.6, and 49.5 GHz acquired at Pomezia (Rome, Italy), are performed for a selected case of light-to-moderate rainfall.

1. INTRODUCTION

In the last decades microwave radiometry has proved to be a valuable tool for retrieving atmospheric parameters both from ground-based and from satellite-borne platforms. So far, many efforts have been dedicated to spaceborne precipitation retrieval, due to the launch of the Special Sensor Microwave Imager (SSM/I) in 1987 and to the possibility of providing a rainfall retrieval on global scale using sun-synchronous platforms [1]. Ground-based microwave radiometry has been mainly investigated for estimating temperature, water vapor and cloud liquid profiles in the absence of precipitation [2].

The increasing use of multifrequency radiometers in ground-based stations, especially for communication purposes [3], raises the question of their potential for also retrieving rainfall rate and cloud parameters from ground [4-5]. This capability might also be useful when weather radars are present, since radiometric estimates might be used as a further constraint within the inversion' schemes of radar reflectivity measurements. While rain gauge cannot estimate integrated parameters, they can be used to estimate path attenuation from power law relations. However even in that case the main difference with respect to radiometer based estimates is that rain gauges can only perform point measurements while beacon as well as radiometric data refer to slant paths. Thus radiometer based estimates can offer several advantages with respect to rain gauge ones (apart the costs). The objective of this paper is to investigate the radiometric frequency sets and system configurations best suited for observing both stratiform and convective precipitation.

The estimate of atmospheric parameters by microwave radiometry may be approached both by using experimental measurements and by using simulated data. The use of experimental measurements is limited by their scarcity or even their lack, especially in rain clouds. The modeling approach is generally more versatile, even though it requires a thorough insight into the electro-magnetic interaction between the microwave radiation and the scattering medium [6]. The radiative transfer theory has been the most used approach so far to account for the multiple scattering and the vertical inhomogeneity of the atmosphere in the presence of hydrometeor scattering [7].

In this paper we develop inversion algorithms for ground-based retrieval of precipitation parameters, adopting a model-based approach. Vertical profiles of stratiform precipitation and cumuliform precipitation are modeled by means of a cloud-genera oriented microphysical-statistical generator. The Eddington method is used to solve the radiative transfer equation (RTE) for plane-parallel structures, including liquid, melted, and ice spherical hydrometeors. Polynomial regression techniques and neural network algorithms are developed and tested on synthetic data in order to understand their potential and to select the best frequency set for rainfall estimation. Finally, ground-based radiometric measurements at 13.0, 23.8, and 31.6 GHz are used for experimentally testing the retrieval algorithms. Comparison with rain gage data and rain path-attenuation measurements, derived from the three ITALSAT channels at 18.7, 39.6, and 49.5 GHz, taken at Pomezia (Rome, Italy), are performed for a selected case of light-to-moderate rainfall.

2. MODELING GROUND-BASED RAINFALL RADIOMETRIC RESPONSE

The scarcity of *in situ* meteorological data, concerning cloud and precipitation structures, suggests tackling the forward, and consequently the inverse problem, by using cloud and radiative transfer models. Modeling plane-parallel precipitating clouds and simulating the related microwave attenuations (A's) and brightness temperatures (T_B's) has allowed us to generate a large cloud radiative data set.

The numerical outputs of a three-dimensional time-dependent cloud mesoscale model have been used for generating the cloud-structure data set, explicitly describing the detailed vertical distribution of four species of hydrometeors: cloud droplets, rain drops, graupel particles, and ice particles [1]. The cloud structures have been vertically averaged to seven homogeneous layers. After classifying the cloud structures into meteorological cloud genera the cloud data set has been extended by means of a Monte Carlo statistical procedure, based on the use of the correlation matrix of the hydrometeor contents [6]. In this work we have considered both stratiform (i.e., nimbostratus, stratus and altostratus) and cumuliform (i.e., cumulonimbus and cumulus congestus) clouds producing rain [8]. As a result, a data set of 4500 cloud structures has been statistically generated retaining the physical and statistical features of the input microphysical cloud model.

In order to associate T_B's to each cloud structure, a radiative transfer model, based on the fast and fairly accurate Eddington solution, has been used [7]. Within each layer, the temperature has been assumed linearly dependent on the height, and the gaseous absorption has been computed by means of the Liebe model. The land surface emission has been modeled by a Lambertian emissivity. The hydrometeors have been assumed to be spherical and characterized by inverse-exponential size distributions. Even though ice crystals are not spherical, for ground-based observations this assumption produces negligible differences in the contribution of the iced layer to the total brightness temperature [9]. This is because of the high opacity of rain and graupel closer to the sensor. Thus, a cloud radiative data set, consisting of cloud structures, related path attenuations and brightness temperatures at the chosen frequencies and observation angles, has been simulated.

It is worth mentioning that, due to the assumptions of a plane-parallel atmosphere and single-scattering predominance for computing the atmospheric extinction, the total path attenuation at angles away from zenith can be simply obtained by applying the "cosecant" law. Moreover, this modeling framework does not allow us to take into account the horizontal inhomogeneity of precipitation, that is beam filling problem which is relevant for convective cumuliform clouds. However, in this respect it has been shown that the one-dimensional (1-D) RTE can be used to approximate three-dimensional (3-D) simulation in the frequency and rainfall ranges considered here by selecting a suitable inclined plane-parallel structure along the line-of-sight [9]. This means that, even though the internal horizontal inhomogeneity of rain clouds cannot be addressed, the above 1-D modeling framework can be easily adapted to observations of rainfall outside the rain cell.

The following figures illustrate the results of the radiative transfer simulation applied to the randomly-generated cloud data set for nimbostratus and cumulonimbus, which are the most typical cloud genera producing rain. Here we refer to the frequency bands of OLYMPUS and ITALSAT beacons, i.e. 12.5, 18.7, 29.7, 39.6, and 49.5 GHz and to most common channel frequencies of the ground-based radiometers, that is at 13.0, 20.6, 22.3, 23.8, 31.6, 36.5, 50.2, and 53.8 GHz.

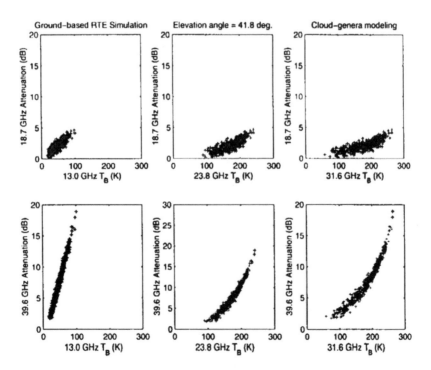

Figure 1. Simulated ground-based radiometric and attenuation signatures of stratiform rain at 41.8° elevation.

Fig. 1 shows the path attenuation for the ITALSAT channel frequencies at 18.7 and 39.6 GHz as a function of the downwelling brightness temperature at 13.0, 23.8, and 31.6 GHz, both for a 41.8° elevation angle and for the stratiform clouds only (in order to compare it with the successive Fig. 2). The choice of the observation angle refer to the experimental case discussed later on. As expected, attenuation increases with frequency and T_B tends to saturate for high values of attenuation at higher frequencies due to the large rainfall and graupel albedo. The 13.0 GHz channel has almost a linear response to path attenuation due to the lower atmospheric opacity.

3. SELECTING FREQUENCY-SET AND RAINFALL RETRIEVAL ALGORITHM

Non-linear regression and neural-network based algorithms have been developed for different sets of frequency channels, precipitation genera and seasonal meteorology. Tests on synthetic data will be shown to illustrate the potential of the proposed models.

3.1 Non-linear regression algorithm

If a linear relation is assumed between the observable vector y and the parameter vector x, then the multivariate regression estimation of x is given by [10]:

$$x - <x> = C_{xy} \, C_{yy}^{-1} \, (y - <y>) \tag{1}$$

where the angle brackets indicate average and C_{xy} and C_{yy} are the cross-covariance between x and y and the auto-covariance of y, respectively. A relevant feature of (1) is that, if the relationship between y and x is not linear, but can be expanded into a Taylor's series, then a non-linear estimation of x can still be performed. Moreover, y and x can represent any function of the desired parameter and observable vectors. Finally, techniques to constrain regression estimates have been proposed in order to ensure more robustness to test data [10].

Using the cloud radiative data set described above for training and by applying (1), rainfall rate R can be directly estimated from T_B measurements. Assuming a quadratic relationship between R and T_B, we can write as follows:

$$R = a_0 + \sum_f (a_{1f} \, T_{Bf} + a_{2f} \, T_{Bf}^2) \tag{2}$$

where the subscript f stands for the available radiometric frequencies. Similar expressions can be written for columnar hydrometeor contents C_h (with $h=c,r,g,i$ for cloud, rain, graupel, and ice species, respectively):

$$C_h = b_{0h} + \sum_f (b_{1fh} \, T_{Bf} + b_{2fh} \, T_{Bf}^2) \tag{3}$$

and the total attenuation A_c (with c the considered channel):

$$A_c = c_{0c} + \sum_f (c_{1fc} \, T_{Bf} + c_{2fc} \, T_{Bf}^2) \tag{4}$$

Alternatively, rainrate R can be also linearly estimated from the columnar rain content C_r:

$$R = d_0 + d_1 \, C_r \tag{5}$$

The use of Eq. (5) ensures the consistency among rain products.

3.2 Neural-network algorithm and frequency-set selection

As an alternative to the regression algorithms, we have also applied to the modeled database a retrieval algorithm based on a neural network [5]. We have chosen a feedforward neural network having, besides the input layer, a hidden layer of neurons with tan-sigmoid transfer functions and an output layer of neurons with linear transfer functions. Instead of the standard backpropagation, for a faster training, we used the Levenberg-Marquardt algorithm [11]. For improving the generalisation capabilities of the network, the training process was performed monitoring the error between the network outputs and the targets on a test set, independent of the training one, and stopping the process when a minimum of such error was found.

As a first test of the retrieval algorithms, we have used synthetic data, obtained from the simulated database, but independent of the data set exploited for algorithm training. The test has been performed on R and C_h parameters. Table 1 shows the results for rainrate retrieval in terms of root mean square (RMS) errors for different frequency-set configuration. Values of neural-network sensitivity to diverse initial conditions (expressed as standard deviation of the RMS's) show that, even though neural networks are generally more accurate than quadratic regressions, this performance can be affected by the dependence on initial set up of the neural network. For both algorithms the set of 6.8, 23.8, and 31.6 GHz channels is the best choice followed by the set of 13.0, 23.8, and 31.6 GHz channels.

Table 1. Root mean square (RMS) errors (in mm/h) for rainrate retrieval tests on synthetic data. Values of neural-network sensitivity (standard deviation in mm/h) to initial conditions are also given.

Frequency (GHz)	Regression RMS error	Neural Network	
		RMS error	Sensitivity
6.8	5.0	4.8	0.1
10.6	5.1	4.6	0.1
13.0	5.3	4.5	0.1
23.8, 36.5	5.7	4.9	0.5
13.0, 23.8, 31.6	4.8	4.4	0.2
23.8, 36.5, 50.2	5.5	4.6	0.2
23.8, 31.6, 53.0	5.1	4.5	0.3
6.8, 23.8, 36.5	4.6	4.4	0.1

4. TESTING THE RETRIEVAL ALGORITHM AT ITALSAT GROUND-STATION

We have chosen to consider here the quadratic regression algorithm for the experimental tests . Rainfall and path attenuation data acquired at the ITALSAT-satellite ground-station located in Pomezia (Rome, Italy) and managed by Fondazione Ugo Bordoni, have been used [12]. Since April 1994 measurements of the three ITALSAT-F1 propagation beacons (at 18.7, 39.6, and 49.5 GHz) are performed every second at an elevation angle of 41.8 degrees with a receiver-antenna of 3.5 m (that is, beamwidths from 0.2° to 0.5°). The ground station measures the amplitude and phase of copolar and crosspolar signals, at 18.7 and 39.6 GHz, and the polarization transfer matrix of the atmosphere at 49.5 GHz [13].

Concurrent measurements performed by two microwave radiometers (REC-1 and REC-2) pointed to the ITALSAT satellite, and a set of ground meteorological instruments also including a tipping bucket rain gage, are synchronously logged every 4 seconds by the ITALSAT ground station. The radiometric data are used to assess the clear-air atmospheric reference level for calibrating the ITALSAT beacon clear-air path attenuation. Radiosounding profiles are also available by the Italian Air-Force at least twice a day with the balloons launched 5 kilometers from the ITALSAT ground-station.

4.1 Description of the three-channel radiometric system

The radiometer, called REC-2, is a dual channel system manufactured by the RESCOM, Aalorg, Denmark. This radiometer is a compact self-contained configuration designed for automatic unattended operation for extended time with high measuring accuracy. The main technical characteristics are reported in Table 2.

The radiometer consists of a dual channel offset-fed reflector antenna connected to two receivers of the noise balancing type. The antenna and receiver sections are integrated in an outdoor box. The shape of the antenna reflector surface and the configuration of the wideband feed horn has been designed so that energy outside the main lobes is minimized. The reflector is of carbon fibre skin/honeycomb construction. It has a rectangular shape of 60x60 cm. Heated air blows across the reflector, preventing the formation of the dew and accumulation of light drizzle, snow, or hail on the reflector surface. Moreover, air from the heater box is directed through a tube to the feed horn window, thus avoiding water condensation and rain stagnation. The calibration is carried out by means of the tipping-curve method, usually once a month.

Table 2: Features of dual channel REC-2 radiometer, installed in Pomezia (Rome, Italy).

Specifications	Values	Specifications	Values
Operating frequencies	23.8-31.7 GHz	Azimuth axis rotation	0° to 360° degrees
Radiometric range	0 to 313 K	Rate of rotation	3°/sec
Integration time	1,2,4,8,16,32 sec	Pointing accuracy	0.5 °
Resolution	0.5 K	Dimensions	100x65x100 cm
Accuracy	3 K	Weight	200 kg
Wide dual side band	40 MHz	Power requirements	220 VAC, 50 to 60 Hz
Antenna beam width	1.9° degrees	Output	RS-232
Elevation axis rotation	-90° to +90° degrees	Control system	Personal computer

The single channel radiometer called REC-1 is an independent system designed also by the RESCOM. The operating frequency is 13 GHz. Basically, this radiometer has the same mechanic characteristics of the REC-2.

4.2 Experimental case on May 3, 1998

In order to show an example, we have selected a case of light to moderate rainfall, observed in Pomezia during May 3, 1998. A moving average with 1-minute window and 1-minute sampling period has been applied to analyze raw data.

Fig. 2 shows the measured path attenuation for the ITALSAT channel frequencies at 18.7 and 39.6 GHz as a function of the downwelling brightness temperatures at 13.0, 23.8, and 31.6 GHz, both for a 41.8° elevation angle. These results can be compared with those of Fig. 1 in order to verify the consistency between simulation and measurements, thus revealing a fairly good agreement.

The top-left panel of Fig. 3 shows the time-series of the rainfall rate estimate from the three-channel radiometer as compared to the rain gage measurements. The top-right panel shows the columnar rain and ice content estimates, while the bottom panels show the time-series of the path-attenuation estimates as compared to the corresponding ITALSAT measurements at 18.7 and 39.6 GHz.

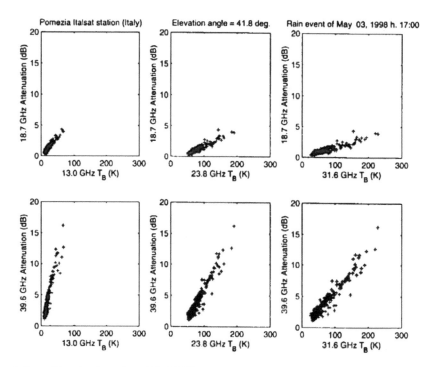

Figure 2. Ground-based radiometric and ITALSAT attenuation measurements at 41.8° elevation.

The RMS difference between estimated and measured rain rate is about 1 mm/h. For attenuation at 18.7 and 39.6 GHz the RMS is 0.4 and 0.9 dB, respectively. Note that the rain rate is measured at a given point by a rain gage and along the link path by microwave radiometers, while the comparison in terms of measured and estimated attenuation is between fairly homogeneous data. Indeed, discrepancies in the attenuation comparison can arise from the difference between the ITALSAT and radiometer antenna beamwidths (see Table 2).

5. CONCLUSIONS

Inversion algorithms for ground-based retrieval of surface rainrate and integrated cloud parameters have been proposed and tested. The retrieval methods are trained by numerical simulations of a radiative transfer model applied to microphysically-consistent precipitating cloud structures. Non-linear regression techniques and neural network algorithms have been developed and tested on synthetic data in order to understand their potential for estimating rainrate and columnar hydrometeor contents and to select the best frequency set for rainfall estimation.

Ground-based radiometric measurements at 13.0, 23.8, and 31.6 GHz have been used for an experimental test of the retrieval algorithms. Comparison with rain gage data and rain path-attenuation measurements, derived from ITALSAT channels, have been performed for a selected case of light to moderate rainfall. A fairly good agreement between estimates and measurements has been achieved.

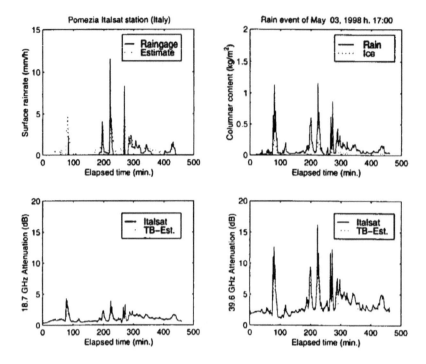

Figure 3. Temporal estimates of rainrate, columnar ice and rain, and attenuation at 18.7 and 39.6 GHz from radiometric measurements using regression algorithms. Rain gage and ITALSAT measurements are also shown.

From the analysis of ITALSAT cross-depolarization (XPD) at different frequencies and polarizations, the instantaneous raindrop size distribution and ice total content, averaged over the path, can be derived [13]. Apart from the analysis of other cases and the application of the neural network algorithm, a future goal could be to compare radiometer-derived and XPD-derived rain and ice estimates.

Acknowledgements: this work has been partially supported by the Italian Space Agency.

REFERENCES

1. N. Pierdicca, F.S. Marzano, G. d'Auria, P. Basili, P. Ciotti, and A. Mugnai, "Precipitation retrieval from spaceborne microwave radiometers based on maximum *a posteriori* probability estimation," *IEEE Trans. Geosci. Remote Sens.*, **34**, 831-846 (1996).
2. E.R. Westwater, J.B. Snider, and M.J. Falls, "Ground-based radiometric observation of atmospheric emission and attenuation at 20.6, 31.65 and 90.0 GHz: a comparison of measurements and theory", *IEEE Trans. Ant. Propagat.*, **38**, 1569-1580 (1990).
3. B.R. Arbesser-Rastburg, and A. Paraboni, "European Research on Ka-band Slant Path Propagation," *Proc. of the IEEE*, vol.85, pp. 843-852, (1997).
4. D.C. Hogg, "Rain, radiometry, and radar", *IEEE Trans. Geosci. Remote Sens.*, **27**, 576-585 (1989).

5. J.L. Vivekanandan and L. Li, "Microwave radiometric technique to retrieve vapor, liquid and ice. Part I: Development of a neural-network based inversion method", *IEEE Trans. Geosci. Remote Sensing*, **35**, 224-236 (1997).
6. G. d'Auria, F.S. Marzano, N. Pierdicca, R. Pinna Nossai, P. Basili, and P. Ciotti, "Remotely sensing cloud properties from microwave radiometric observations by using a modeled cloud data base", *Radio Sci.*, **33**, 369-392 (1998).
7. R. Wu, and J.A. Weinman, "Microwave radiances from precipitating clouds containing aspherical ice, combined phase, and liquid hydrometeors", *J. Geophys. Res.*, **89**, 7170-7178 (1984).
8. R.A. Houze, "Structures of atmospheric precipitation systems: a global survey", *Radio Sci.*, **17**, 671-689, (1982).
9. F.S. Marzano, E. Fionda, and P. Ciotti, "Simulation of radiometric and attenuation measurements along earth-satellite links in the 10 to 50 GHz band through horizontally-finite convective raincells", *Radio Sci.*, **34**, (4), (1999).
10. L.J. Crone, L.M. McMillin, and D.S. Crosby, "Constrained regression in satellite meteorology", *J. Appl. Meteor.*, **35**, 2023-2039 (1996).
11. M. T. Hagan and M. Menhaj, "Training feedforward networks with the Marquardt algorithm", IEEE Trans. On Neural Networks **5**, pp. 989-993, (1994).
12. A. Martellucci, F. Barbaliscia, and A. Aresu, "Measurements of attenuation and cross-polar discrimination performed using the ITALSAT propagation beacons at 18.7, 40 and 50 GHz", Proc. of 3rd Ka Band Utilization Conference, Sorrento (Italy), pp 113-120, Sept. 15-18, (1997).
13. A. Aresu, E. Damosso, A. Martellucci, L. Ordano, and A. Paraboni, "Depolarization of electromagnetic waves due to rain and ice: theory and experimental results", *Alta Frequenza*, **6**, 70-75, (1994).

4. New radiometric instruments and missions

4.1 Microwave sensors and calibration

Microw. Radiomet. Remote Sens. Earth's Surf. Atmosphere, pp. 409–415
P. Pampaloni and S. Paloscia (Eds)
© VSP 2000

An airborne submm radiometer for the observation of stratospheric trace gases

M. VON KÖNIG[1], H. BREMER[1], V. EYRING[1], A. GOEDE[2], H. HETZHEIM[3], Q. KLEIPOOL[2], H. KÜLLMANN[1] AND K. KÜNZI[1]

[1]*Institute of Environmental Physics (IUP), University of Bremen, P.O. Box 330440, 28334 Bremen, Germany*
[2]*Space Research Organisation of the Netherlands (SRON), Groningen, The Netherlands*
[3]*German Aerospace Research Facilities (DLR), Berlin, Germany*

Abstract–The Airborne Submm SIS Radiometer (ASUR) is a heterodyne receiver with a tunable frequency range of 604.3 GHz to 662.3 GHz, and a spectrometer bandwidth of 1.5 GHz. The receiver is flown onboard the DLR research aircraft Falcon at an altitude of 10 to 13 km to minimize absorption effects by water vapor. Looking upward at a zenith angle of 78°, ASUR detects thermal emission lines of various molecules, including some of the key species of stratospheric ozone chemistry like ClO, HCl, HNO₃, HO₂, HOCl, BrO and ozone, furthermore dynamical tracers like CH₃Cl, N₂O and water vapor. From the detected pressure broadened lines, vertical profiles of volume mixing ratio are retrieved for an altitude range of 15 to 55 km using the optimal estimation method.

In January and February 1999, the ASUR instrument participated in the European THESEO/HIMSPEC project with the Falcon based in Kiruna, northern Sweden. The participation of the ASUR instrument in the HIMSPEC-project concentrated on measuring the chlorine activation inside and across the edge of the polar vortex as well as coordinated measurements with several balloon borne instruments.

In this presentation, an overview of the ASUR instrument is given. First results of the 1999 campaign are presented, focussing on measurements of ClO, HCl and ozone inside and across the edge of the polar vortex.

1. THE ASUR INSTRUMENT

The ASUR instrument is a passive heterodyne receiver measuring rotational transitions of molecules in the submillimeter frequency range between 604.3 and 662.3 GHz. ASUR is installed onboard the research aircraft Falcon operated by the DLR. To avoid the strong absorption of tropospheric water vapor in the submm range, the measurements are taken at flight altitudes of about 12 km.

The first submillimeter receiver has been developed at the Institute of Environmental Physics (IUP), University of Bremen, in 1991[1,2,3]. The present instrument is equipped with a SIS (Superconductor-Insulator-Superconductor) liquid He-cooled Nb tunnel junction developed at the Space Research Organisation of the Netherlands (SRON) in Groningen [4]. This sensor has a frequency dependent system noise temperature ranging from 500 to 1000 K. In 1998, a new first local oscillator has been implemented. This can be tuned mechanically over a bandwidth of 33 GHz. The instrument can now be operated in the frequency range 604.3 to 662.3 GHz.

In this frequency range, thermal emission lines of various molecules of interest for the

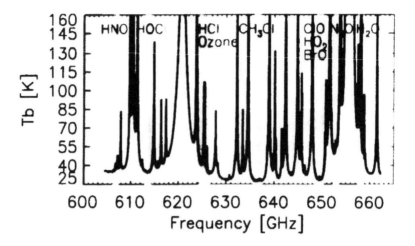

Figure 1. Radiative transfer calculation over the ASUR frequency range for an observational altitude of 12 km and a zenith angle of 78°. Grey shaded areas are frequency bands measured by ASUR in the 1999 campaign. Major species observed in these bands are indicated by molecule names.

chemistry and dynamics of the stratosphere can be detected, including some of the key species of stratospheric ozone chemistry. In figure 1, a radiative transfer calculation for the total ASUR frequency range is shown. The calculation was carried out with a radiative transfer model developed at IUP Bremen for an observational altitude of 12 km, a zenith angle of 78°, and a subarctic-winter standard atmosphere. For a description of the Bremen radiative transfer program, see [5]. Shaded grey areas are frequency bands measured by ASUR in the 1999 winter campaign. Major species observed in these bands are indicated by molecule names. The instrument allows detection of emission signals of ClO, HCl, O$_3$, N$_2$O, HO$_2$, BrO [6,7,8,9] as well as HNO$_3$, HOCl, CH$_3$Cl and water vapor.

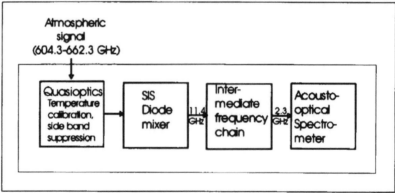

Figure 2. Schematic overview of the ASUR instrument.

From the pressure broadening of the measured lines, vertical profiles of volume mixing ratio can be obtained for an altitude range of 15 to 55 km using the optimal estimation method [10]. This method uses a priori information to stabilize the solution of the retrieval problem. The algorithm used here allows to simultaneously retrieve the vertical profiles of volume mixing ratio of several molecules as well as instrument effects like frequency shifts or baseline ripples. This method yields a typical vertical resolution of 5 to 15 km. The horizontal resolution is about 10 to 35 km depending on the necessary integration time and aircraft ground speed.

In figure 2, a schematic view of the ASUR instrument is shown. The atmospheric signal enters the instrument at a zenith angle of 78° through a special high density polyethylene window, mounted into the aircraft fuselage. The contribution of the window to the measured signal is calculated considering the known window transmission and reflection

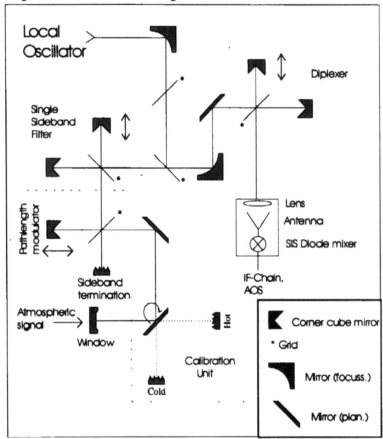

Figure 3. The ASUR quasi-optics.

as well as the measured window temperature and inside-air temperature of the aircraft. The window is slightly wedged and grooved to suppress interferences between the window surfaces. A rotating plane mirror switches between the atmospheric signal and two blackbodies at liquid nitrogen and at roomtemperature for calibration. Sideband suppression is done by using a Martin-Puplett-interferometer. A more detailed description of the quasi-optics can be found in figure 3.

Inside the dewar the signal is focussed through a polyethylene lens into the horn of the SIS tunnel junction mixer. The junction is cooled down to temperatures of 2.7 to 4.2 K using liquid helium. Within the SIS junction, the signal is converted to (11.4 ± 1) GHz. In the intermediate frequency chain the signal gets amplified and is further downconverted to the resulting frequency of (2.3 ± 0.75) GHz. This signal is then spectrally analysed with

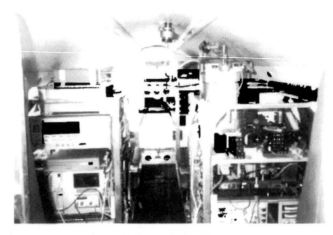

Figure 4. The ASUR instrument on board the Falcon aircraft. The rack on the right contains the quasi-optical system with the He dewar and the intermediate frequency chain. The rack on the left contains the AOS and the control software. In the background the DLR OLEX Lidar can be seen.

an acousto-optical spectrometer (AOS) developed at the Observatoire de Paris-Meudon [11]. A view of the ASUR instrument onboard the Falcon aircraft is shown in figure 4.

2. THE THESEO/HIMSPEC CAMPAIGN 1999

From January 22 to February 8, 1999, the ASUR instrument has been flown together with the DLR OLEX lidar onboard the Falcon as part of the THESEO (Third European Stratospheric Experiment on Ozone)/HIMSPEC project (see figure 5).

A major objective of HIMSPEC (High and Middle Latitude Speciation of the Nitrogen, Chlorine and Hydrogen Chemical Families by Airborne measurements) was the coordination of measurements made by various balloon and aircraft based instruments to get an almost complete set of species of the chlorine, nitrogen and hydrogen chemical families in the lower stratosphere inside the arctic polar winter vortex and at midlatitudes.

The role of the ASUR instrument as part of HIMSPEC included measurements of the chlorine activation inside and across the edge of the polar vortex as well as coordinated measurements with several balloon borne instruments. Another important feature were measurements of latitudinal cross sections of species measurable by ASUR.

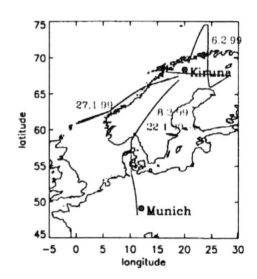

Figure 5. Flight tracks of the Falcon aircraft during the ASUR 1999 campaign.

Also, measurements of HNO₃, CH₃Cl, HOCl and water vapor were taken with the ASUR instrument for the first time.

All measurements have been taken inside or across the edge of the arctic polar vortex.

3. FIRST RESULTS OF THE 1999 CAMPAIGN

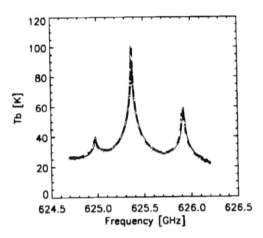

Figure 6. Black line: Measured spectrum of HCl and ozone, 27.01.1999, 65.5° North, 19.5° East. Grey dashed line: retrieved spectrum.

Figure 6 shows a spectrum taken with the ASUR instrument at a frequency of 625.447 GHz on January 27 at 65.5° North and 19.5° East, showing emission lines of both ozone (center peak) and HCl (peaks on each side). Also shown is a radiative transfer calculation of the resulting ozone and HCl retrieval, yielding a very good agreement between measurement and retrieval. Ozone and HCl are retrieved simultaneously. The resulting profiles of volume mixing ratio are shown in figures 7 (ozone) and 8 (HCl). Shaded areas indicate the measurement error due to instrument noise.

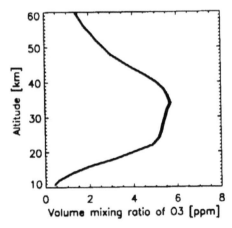

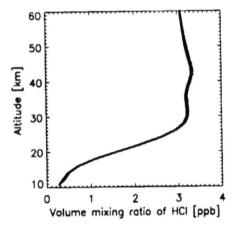

Figure 7. Bold: ozone vertical profile measured on 27.01.1999, 65.5° North, 19.5° East. Grey shaded area: Measurement error due to instrument noise.

Figure 8. Bold: to figure 7 corresponding HCl profile. Grey shaded area: measurement error due to instrument noise.

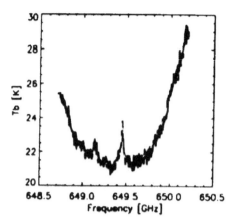

Figure 9. Bold: Spectrum of ClO, measured on 22.01.1999, 63.9°North, 15.1° East. Grey dashed line: retrieved spectrum.

In figure 9, a spectrum of ClO at 649.448 GHz is shown, taken with ASUR on January 22 at 63.9° North and 15.1° East. Small lines on the left side of the spectrum as well as the slopes on both wings of the spectrum are due to ozone. Also shown is a radiative transfer calculation of the resulting ClO and ozone retrieval. ClO and ozone are retrieved simultaneously. The resulting ClO volume mixing ratio profile is shown in figure 10. The shaded area again indicates the measurement error due to instrument noise. A value of about 0.3 ppb of ClO is observed at altitudes bewteen 20 and 26 km, significantly higher than the non-activated value of 10 to 100 ppt in this altitude

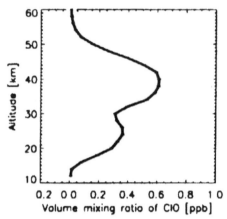

Figure 10. Bold: to figure 9 corresponding ClO vertical profile. Grey shaded area: measurement error due to instrument noise.

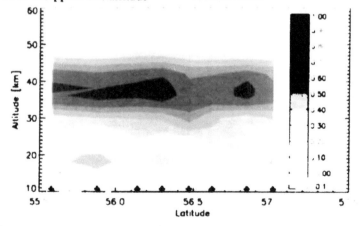

Figure 11. Latitudinal cross section of ClO measured by ASUR on 22.01.1999 on a flight from Munich to Kiruna. Crosses indicate median positions of measurements.

range.

In figure 11, a latitudinal cross section of ClO obtained on the same flight is shown. A significant chlorine activation can be observed in all measurements.

REFERENCES

1. S. Crewell, *Submillimeter-Radiometrie mit einem flugzeuggetragenen Empfänger zur Messung atmosphärischer Spurenstoffe*, PhD thesis, Bremen (1993).
2. S. Crewell, P. Hartogh, K. Künzi, H. Nett, T. Wehr, Aircraft measurements of ClO and HCl during EASOE 1991/92, *Geophys. Res. Lett.*, **21**, 1267-1279 (1994).
3. T. Wehr, S. Crewell, K. Künzi, J. Langen, J. Urban, Remote sensing of ClO and HCl over northern Scandinavia in winter 1992 with an airborne submillimeter radiometer, *Journal of Geophysical research*, **100**, 20957-20968 (1995).
4. J. Mees, S. Crewell, H. Nett, G. de Lange, H. van Stadt, J.J. Kuipers, R.A. Panhuyzen, ASUR-An Airborne SIS Receiver for Atmospheric Measurements of Trace Gases at 625 to 760 Ghz, *IEEE Transactions on Microwave Theory and Techniques*, **43** (1995).
5. S. Bühler, *Microwave Limb Sounding of the Stratosphere and Upper Troposhere*, PhD thesis, Bremen (1998).
6. J.P.J.M.M de Valk, S. Crewell, B. Franke, A. de Jonge, Q. Kleipool, H. Küllmann, J. Mees, J. Urban, J. Wohlgemuth, Four years of airborne trace gas detection in the arctic winter period with submillimeter wave radiometry, *Proceedings to the third European workshop on polar stratospheric ozone* (1995).
7. J. Urban, Measurements of the stratospheric trace gases ClO, HCl, O$_3$, N$_2$O, H$_2$O, and OH using airborne submm-wave radiometry at 650 and 2500 Ghz, *Reports on Polar Research*, **264** (1998).
8. M. von König, H. Bremer, V. Eyring, H. Küllmann, J. Urban, J. Wohlgemuth, K. Künzi, M. van den Brook, A.P.H. Goede, A.R.W. de Jonge, Q. Kleipool, N.D. Whyborn, H. Hetzheim, G. Schwaab, M.P. Chipperfield, Trace gas measurements of ClO, HCl, O$_3$ and N$_2$O in the Arctic winter, 1997, *Proceedings to the fourth European workshop on polar stratospheric ozone* (1997).
9. V. Eyring, *Modellstudien zur arktischen stratosphärischen Chemie im Vergleich mit Meßdaten*, PhD thesis, Bremen (1999).
10. C.D. Rodgers, Characterization and error analysis of profiles retrieved from remote sounding measurements, *Journal of Geophysical Research*, **95**, 5587-5595 (1990).
11. C. Rosolen, P. Dierich, D. Michet, A. Lecacheux, F. Palacin, R. Robiliard, F. Rigeaud, P. Vola, Wideband acousto optical spectrometer, *ESA workpackage 2411 Final report* (1994).

Microw. Radiomet. Remote Sens. Earth's Surf. Atmosphere, pp. 417–425
P. Pampaloni and S. Paloscia (Eds)
© VSP 2000

EMCOR: a new radiometer for the measurement of minor constituents in the frequency range of 201 to 210 GHz

D. MAIER,[1] N. KÄMPFER,[1] W. AMACHER,[1] M. WÜTHRICH,[1]
J. DE LA NOE,[2] P. RICAUD,[2] P. BARON,[2] G. BEAUDIN,[3]
C. VIGUERIE,[3] J.-R. PARDO,[4] J.-D. GALLEGO,[4] A. BARCIA,[4]
J. CERNICHARO,[4] B. ELLISON,[5] R. SIDDANS,[5] D. MATHESON,[5]
K. KÜNZI,[6] U. KLEIN,[6] B. BARRY,[6] J. LOUHI,[7] J. MALLAT,[7]
M. GUSTAFSSON,[7] A. RÄISÄNEN,[7] and A. KARPOV[8]

[1] *Institute of Applied Physics, University of Berne, Berne, Switzerland*
[2] *Observatoire de Bordeaux, Floirac, France*
[3] *Observatoire de Paris, Paris, France*
[4] *Centro Astronómico de Yebes, Guadalajara, Spain*
[5] *Rutherford Appleton Laboratory, Chilton, United Kingdom*
[6] *University of Bremen, Bremen, Germany*
[7] *Helsinki University of Technology, Espoo, Finland*
[8] *Institut de Radio Astronomie Millimétrique, Grenoble, France*

Abstract– EMCOR (European Minor COnstituent Radiometer) is a heterodyne receiver for the frequency range of 201 to 210 GHz, whose aim is to measure the emission lines of some stratospheric minor constituents involved in ozone chemistry. In order to obtain a receiver with a high sensitivity an SIS (Superconductor-Insulator-Superconductor) tunnel junction has been chosen as mixer element. Other parts of the front-end are a multi-purpose calibration system, a quasi optical system, a solid state local oscillator with electronic tuning and a HEMT pre-amplifier. The back-end consists of an acousto-optical spectrometer and the whole system is controlled by a PC. The instrument has been installed at the International Scientific Station Jungfraujoch in the Swiss Alps and tests have been carried out successfully. First measurements of O_3 and $H_2^{18}O$ are presented.

1. INTRODUCTION

The objective of the European project EMCOR (European Minor COnstituent Radiometer) was the development of a sensitive microwave heterodyne receiver for ground-based measurements of the amounts of various minor constituents of the stratosphere involved in ozone chemistry [1]. After a spectroscopic study a frequency range of about 201 to 210 GHz has been chosen. As can be seen from the calculated spectrum shown in Fig. 1 this frequency range includes the lines of many molecules which are of interest for atmospheric physics as e.g. O_3, ClO, $H_2^{18}O$, HNO_3 or N_2O.

The project started in spring 1996. After the definition and construction of the subsystems the instrument has been integrated at the University of Berne and installed at the International Scientific Station Jungfraujoch (ISSJ) at 3580 m a.s.l. in the Swiss Alps.

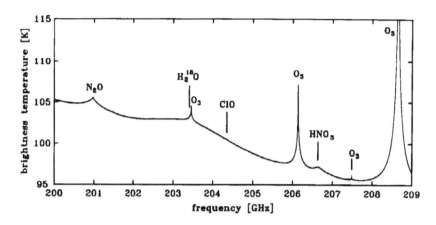

Figure 1: Spring model spectrum calculated for an observation altitude of 3580 m a.s.l. and an elevation angle of 12°.

2. INSTRUMENT

Figure 2 shows a block diagram of the system. The essential part of the instrument is an SIS mixer which is cooled together with the HEMT amplifier in a LHe cryostat. The atmospheric signal is guided through the quasi optics and then focussed together with the

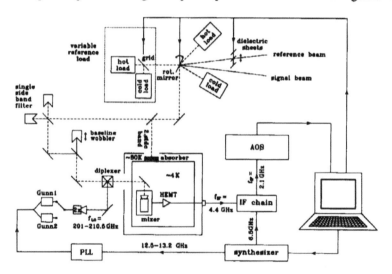

Figure 2: Block diagram of the system.

local oscillator (LO) signal onto the mixer element. At the input of the quasi optics a rotating mirror allows to switch between the atmosphere and the different calibration loads. The intermediate frequency (IF) signal is further processed in the IF chain and finally analysed using an acousto-optical spectrometer (AOS). A PC controls observation procedures and data acquisition.

2.1. Quasi optics

A schematic view of the quasi optical system is shown in **Fig. 3**. It is made up of several wire grids, plane mirrors, and ellipsoidal mirrors as focussing elements which are all together mounted on two optical plates. In order to assure a proper operation of the quasi

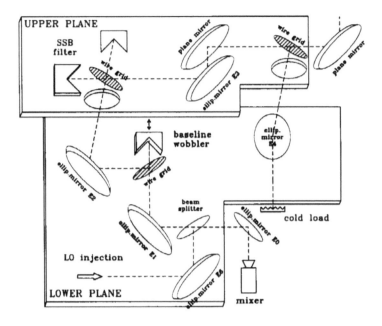

Figure 3: Schematic view of the quasi optics.

optics over the whole frequency range a Gaussian telescope technique has been employed providing beam waists and locations which are independent of frequency. The unwanted sideband is suppressed using a Martin-Puplett interferometer consisting of two roof top mirrors complemented by a wire grid. For the reduction of standing waves a baseline wobbler has been installed. Atmospheric and LO signals are coupled together using a thin Mylar foil as beam splitter.

2.2. Calibration

Since the calibration is a crucial point when measuring faint spectral lines, special care has been taken in developing the calibration unit of the system [2]. Besides the classical hot-cold calibration three different balancing methods can be employed: a beam switch technique with an atmospheric reference signal, a beam switch technique with a reference signal from a variable reference load or a frequency switch technique. The installation of three different balancing methods allows apart from choosing the most suitable technique for each line a comparison of the different methods.

2.2.1. Beam switching with an atmospheric reference signal

In a typical beam switch radiometer as described by Parrish [3] the reference signal consists of the atmospheric signal measured at a high elevation angle and attenuated by a lossy dielectric sheet. The continuum contributions of signal and reference beams are equalized by adjusting the observation angle. A schematic view of the calibration unit is shown in Fig. 4. The rotating mirror (M) switches between the signal beam at a low elevation angle and the reference beam at nearly zenith angle. Due to the conditions at the observatory this high elevation angle can only be achieved by using an external mirror (EM), which is mounted outside the building.

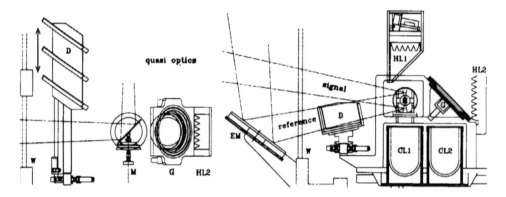

Figure 4: Schematic view of the calibration unit seen from above (left) and in a side view (right). W: window, D: dielectrics, M: rotating mirror, EM: external mirror, CL1, 2: cold loads, HL1, 2: hot loads, G: polarizing grid.

The adjustment of the observation angle might be a problem especially for the detection of faint spectral lines where the measurements have to be integrated over a long time to achieve a sufficiently good signal-to-noise ratio. In this case the final spectrum consists of a summation over measurements carried out at different observation angles which complicates the retrieval of the altitude profile. Therefore EMCOR reduces the variation of the observation angle by the use of more than one dielectric sheet.

Theoretical estimations showed that with three dielectrics the observation angle could be confined between 10 and 20 degrees still allowing balancing for tropospheric zenith opacities up to 0.4 [2]. Above this value the high amount of water vapour in the atmosphere makes measurements of weak emission lines anyway impossible. The theoretical approach is based on a one layer-troposphere and assumptions for the temperatures of the atmosphere, the window, and the dielectrics. It is therefore only suitable to estimate the optimum opacities, which then have to be adapted by testing at the observation site. As a first try three plexiglas sheets with thicknesses of 1.4, 2.15, and 5.5 mm have been chosen. Their opacities have been measured to be about 0.15, 0.2, and 0.55.

The three dielectrics are mounted on a linearly movable (D) unit which is controlled via the PC. According to the atmospheric condition the best suitable dielectric is automatically chosen and moved at Brewster's angle into the reference beam.

2.2.2. Beam switching with a variable reference load

Additionally to the beam switch technique using an atmospheric reference, a second beam switch technique using a variable reference load [4] has been installed. The advantage of this reference over the atmospheric reference is that it has no line contribution. It consists of a hot load (HL2) and a cold load (CL2) complemented by a polarizing grid (G), which is mounted at an angle of 45° with respect to the axes. Since the mixer is only sensitive to the horizontal polarization, the contributions of the two loads to the received signal T_{var} can be varied by rotating the grid and the signal seen by the mixer is

$$T_{var} = T_{cold} \cos^2 \beta + T_{hot} \sin^2 \beta, \tag{1}$$

where β is the angle between the wires as seen in the direction of signal propagation along the optical axis and the horizontal axis. This signal can also be expressed as a function of the grid angle α, which is measured in the grid plane with $\alpha = 0°$ for horizontal wires:

$$T_{var} = \frac{2\,T_{cold} + T_{hot} \cdot \tan^2 \alpha}{2 + \tan^2 \alpha}. \tag{2}$$

These equations show that T_{var} can be tuned to any value between T_{hot} and T_{cold}. In particular, T_{var} can not be tuned to a value below T_{cold}, which means for EMCOR below 77 K. But at a high mountain site the measured brightness temperatures in this frequency range can be as low 40 K. During these good weather periods one of the other two balancing methods has to be employed.

2.2.3. Frequency switching

As a third balancing method the frequency switching has been implemented. Here the reference consists of the atmospheric signal shifted in frequency. The major drawback of this technique is the limitation of the frequency shift by the requirement of a constant gain for the different frequencies, so that only very narrow lines can be observed.

2.2.4. LO system

In order to cover the whole frequency range two varactor tuned Gunn oscillators in combination with a doubler are employed [5]. The oscillator signals are frequency stabilized by

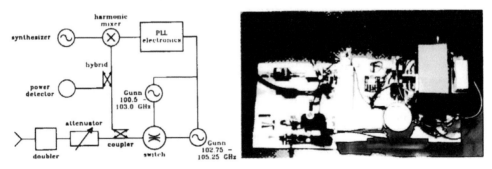

Figure 5: Block diagram (left) and photo (right) of the LO system.

a phase-locked loop using a reference frequency generated by a synthesizer. The signal is fed to the quasi optics via a Pickett horn. Figure 5 shows a block diagram (left) and a photo (right) of the LO system.

2.3. Cryogenic system with mixer and low-noise amplifier

In order to cool the superconducting tunnel junction to its operational temperature of about 4 K a cryogenic system consisting of a 10 l-LHe reservoir and a CTI Cryogenics 350 closed cycle refrigerator is employed. This system provides three cold surfaces: at 4 K, 20 K, and 80 K, and it has a hold time of about a month.

Figure 6: Photos of the cryostat (left) and the 4 K-stage with the IRAM mixer, and the HEMT amplifier mounted on the 20 K-stage (right).

Due to temporary difficulties of the Paris Observatory, which was in charge of the junction fabrication, the appropriate SIS junction is not yet available. However, first tests of the system are currently carried out with a 200 GHz mixer kindly provided by the *Institut de Radio Astronomie Millimétrique* (IRAM), Grenoble.

The IF low-noise HEMT pre-amplifier is mounted on the second cold stage at about 20 K. This amplifier is designed for a frequency range of 3.9 to 4.9 GHz [6].

The photo on the right-hand side in Fig. 6 shows the 4 K-stage. The signal comes in the window on top of the picture and is then focussed by an ellipsoidal mirror into the horn antenna attached to the mixer block. The HEMT amplifier mounted on the 20 K-stage can be seen at the bottom on the left-hand side.

3. IF chain and acousto-optical spectrometer (AOS)

In the IF chain [5] the IF signal is down-converted to a centre frequency of 2.1 GHz and further amplified to match the input requirements of the spectrometer, which is an AOS manufactured by the Observatoire de Meudon with a bandwidth of 1 GHz and 1725 channels.

4. FIRST MEASUREMENTS

After integration and tests of the system at the University of Berne, Switzerland, the system has been installed at the ISSJ.

4.1. Total power measurement

As a first test of the system the strongest line within the frequency range, the ozone line at 208.64 GHz, has been measured in a total power mode. The over 140 s integrated spectrum

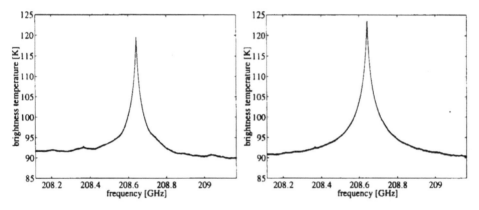

Figure 7: Measurement of the O_3-line at 208.64 GHz ($t_{int} = 140$ s, $\vartheta_{sig} = 15°$). Left: Raw spectrum. Right: Spectrum after deduction of superimposed sine waves.

measured at an elevation angle of 15° is shown in Fig. 7 on the left-hand side. The spectrum is perturbed by standing waves caused by multiple reflections in the transmission path of the receiver which then acts like a Fabry-Pérot interferometer. A preliminary retrieval of the altitude profile has been carried out using the optimal estimation method and employing the technique described in [7] which allows a simultaneous fit of superimposed sine waves. The plot on the right showing the spectrum after the deduction of the fitted sine waves demonstrates that effects from standing waves can be eliminated quite well.

4.2. Balanced measurement

So far only the beam switch technique with the variable reference load has been tested. In order to check the relation given in equation (2) the signal of the variable reference load has been measured as a function of the grid angle. The result is represented together with a fit in Fig. 8 on the left-hand side. The plot on the right shows the difference between

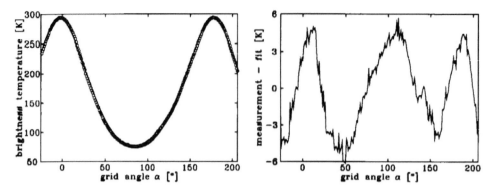

Figure 8: Left: Measured signal of the variable reference load (o) and fit (-). Right: Difference between measurement and fit.

measurement and fit. The quite large discrepancy of up to 6 K is probably due to a non-linearity effect in the read-out of the grid angle on the one hand and the non-linearity of the hot-cold calibration on the other hand. However, this does not affect the balancing because here only the derivative of the curve is needed: with its help the distance to the optimum grid position can be found from the difference between reference and continuum signal.

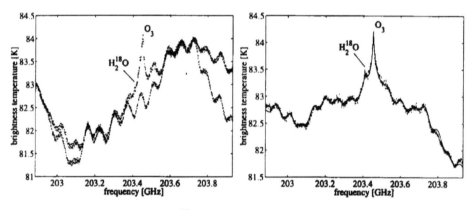

Figure 9: Measurement of the $H_2^{18}O$-line at 203.41 GHz. Left: Total power calibration of signal and reference. Right: Balanced calibration [4].

The balancing concept works well as demonstrated by the measurement of the $H_2^{18}O$-line shown in Fig. 9. In the plot on the left-hand side the hot-cold calibrated signals of the atmosphere and the variable reference (without line contribution) are shown. Balancing is done for the 300 AOS channels left from the centre. The dots in the plot on the right-hand side represent the balanced calibration using

$$T_B = \frac{sig - ref}{ref} * (\overline{T_{ref}} + T_{rec}) + \overline{T_{ref}}, \qquad (3)$$

where sig and ref are the signals of the atmosphere and the reference, respectively, T_{rec} is the receiver noise temperature, and $\overline{T_{ref}}$ is the measured mean brightness temperature of the variable reference (for details see [4]). The standing waves present in the total power measurement are considerably reduced by the balancing method. The residual standing waves can be very well fitted during the retrieval of the altitude profiles as can be seen from the calculated spectrum (solid line), which is based on the retrieved profiles plus the fitted amplitude and phase of several sine waves.

5. CONCLUSIONS

A sensitive millimetre wave heterodyne receiver for the frequency range of 201 to 210 GHz has been defined and constructed. In order to be able to measure even very faint spectral lines special efforts have been made to design a suitable calibration unit.

The radiometer has been installed at a high mountain side in the Swiss Alps. First measurements of O_3 and $H_2^{18}O$ have proven that the system is operational. Nevertheless, further improvement of the system is going on to ensure its operation on a regular basis.

Acknowledgements. The authors would like to thank Prof. H. Debrunner of the Foundation of the Alpine Scientific Stations Jungfraujoch and Gornergrat for the possibility to install the instrument at the Sphinx Observatory. Special thanks go to P. Kuster and H.R. Staub for their support at the observatory. This work was supported by BBW-contract No. 95.0433 as part of EC-contract ENV4-CT95-0137.

REFERENCES

1. J. de la Noë et al., "EMCOR," Final report, contract number ENV4-CT95-0137, March 1999

2. D. Maier and N. Kämpfer, "A multi-purpose calibration concept for the EMCOR instrument," *Proceedings of 2nd ESA Workshop on Millimetre Wave Technology and Applications: Antennas, Circuits and Systems, 27–29 May 1998, Millilab, Espoo, Finland*

3. A. Parrish, R.L. de Zafra, P.M. Solomon, and J.W. Barrett, "A ground-based technique for millimeter wave spectroscopic observations of trace constituents," *Radio Science*, **23**, 106–118 (1988)

4. R. Krupa, "Millimeterwellen-Radiometrie stratosphärischer Spurengase unter Anwendung balancierter Kalibrierung," PhD thesis, University of Karlsruhe, 1997

5. J. Louhi, M. Gustafsson, J. Mallat, and A. Räisänen, "Local oscillator and IF chain for European millimeter wave radiometer, EMCOR," *Report* **S229**, Helsinki University of Technology, October 1997

6. J.D. Gallego, R. Baeza, R. Garcia, and D. Geijo, "Measurements of EMCOR cryogenic 3.9–4.9 GHz HEMT amplifier," *Technical Report* **CAY 1997-1**, Centro Astronómico de Yebes, May 1997

7. M. Kuntz, "Retrieval of of ozone mixing ratio profiles from ground-based millimeter wave measurements disturbed by standing waves," *J. Geophys. Res.*, **102** (D18), 21,965–21,975 (1997)

Microw. Radiomet. Remote Sens. Earth's Surf. Atmosphere, pp. 427–432
P. Pampaloni and S. Paloscia (Eds)
© VSP 2000

The radiometer for atmospheric measurements (RAM)

KAI LINDNER, BERND BARRY, ULF KLEIN, JENS LANGER,
BJÖRN-MARTIN SINNHUBER, INGO WOHLTMANN and
KLAUS F. KÜNZI

Institute of Environmental Physics, University of Bremen, Bremen, Germany

Abstract   The University of Bremen operates a microwave radiometer for the detection of stratospheric ozone, chlorine monoxide and water vapor. It is located at one of the primary stations of the Network for the Detection of Stratospheric Change (NDSC) in Ny Ålesund, Spitsbergen. The instrument consists of three radiometer frontends; the backend is an acousto-optical spectrometer that is operated in a time sharing mode with all three frontends. The chlorine monoxide radiometer works at 204 GHz. It is operated in a reference beam technique. The single sideband system noise temperature at 204 GHz is about 1300 K at a physical receiver temperature of 12 K. At good weather conditions, the integration time to obtain a sufficient signal to noise ratio to detect chlorine monoxide is about 10 hours. The 22 GHz water vapor radiometer is also operated in a reference beam technique. Its system noise temperature is about 200 K at ambient receiver temperature. The ozone frontend is operated at 142 GHz in a total power mode with a single sideband system noise temperature of 3100 K at ambient temperature. The quasioptics include Martin-Puplett-Interferometers as single sideband filters, diplexers and baseline wobblers. We measure ozone and water vapor all the year. From February to April, we switch at good weather conditions from water vapor to chlorine monoxide. Chlorine monoxide mixing ratios are retrieved for the altitude range from 16 to 35 kilometers and Ozone mixing ratios from 12 to 55 kilometers. In both cases the vertical resolution is about 10 kilometers. We expect to retrieve water vapor profiles in the altitude range from 25 to 55 kilometers.

1. INTRODUCTION

The Radiometer for Atmospheric Measurements (RAM) is operated by the Institute of Environmental Physics at the University of Bremen in cooperation with the Alfred-Wegener-Institute for Polar and Marine Research. As a part of the Network for the Detection of Stratospheric Change (NDSC), we obtain stratospheric profiles of the volume mixing ratios of ozone, chlorine monoxide and water vapor over Ny-Ålesund. These three species are targeted with microwave observations in the NDSC.

The Instrument is located at the Koldewey-Station in Ny-Ålesund, Spitsbergen, Norway, at a latitude of 78°55' N and a longitude of 11°55' E. This observation site was chosen because of the special climatic conditions. The polar vortex is mostly located above Ny-Ålesund, which is essential for observing disturbed stratospheric chemistry. Figure 1 gives an impression of the RAM laboratory inside the NDSC building in Ny-Ålesund. The ozone radiometer was installed in 1993, the chlorine monoxide radiometer followed in 1994. In January 1999 we installed the new water vapor radiometer. All radiometers have individual frontends with own quasioptics and initial amplifier stages. They share one backend, consisting of an intermediate frequency chain and an acousto optical spectrometer (AOS). Our instrument is highly automated and requires only one manual calibration per day, performed by the local station engineer.

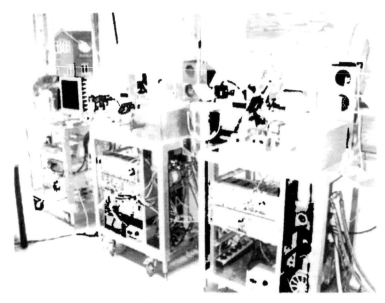

Figure 1: Inside the RAM laboratory in Ny-Ålesund. The leftmost rack is the water vapor radiometer, in the center is the ozone radiometer and on the right is the chlorine monoxide radiometer. The central rack includes the intermediate frequency chain and the acousto optical spectrometer. All these instruments observe the atmosphere through windows made of styropor.

2. MICROWAVE RADIOMETRY

The idea behind microwave radiometry is the passive measurement of thermal radiation from rotational transition lines of atmospheric trace gas molecules. The line width depends on pressure and temperature and can be used to determine the altitude of the emitter; the strength contains information on the concentration of the emitting trace gases. This information is used to generate profiles for the observed trace gases.

The emission lines are extremely weak and special techniques for amplification and detection are needed. The RAM is a superheterodyne receiver. The input signal is superimposed with the fixed frequency of a local oscillator. This leads to a down conversion of the whole frequency band to the difference frequencies under preservation of all spectral information. At this intermediate frequency, standard components are used for further signal processing. The total amplification is approximately 100 dB.

3. THE RADIOMETER FRONTENDS

The basic concept of each of our radiometers is the same – as illustrated in figure 2. The atmospheric signal from an adjustable observation angle is collected by a rotatable mirror and fed into a quasioptical system. The mirror points either to a cold load, a hot load or into the atmosphere. For reference beam measurements it can also point into the atmosphere through a plexiglass sheet as reference signal.

Common elements of our quasioptics are path length modulators for the removing of standing waves, Martin-Puplett-Interferometers as single-sideband-filters and diplexers as well as cold and hot loads for calibration purposes.

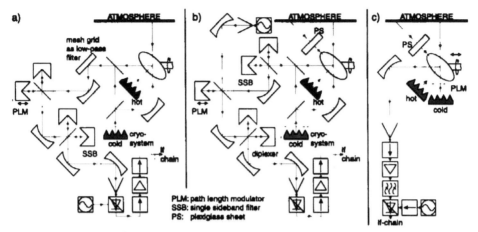

Figure 2: The schematics of the three radiometers for a) ozone, b) chlorine monoxide and c) water vapor.

The mixer converts the high frequency signal down to the intermediate frequency of 8 GHz. A phase-locked loop (PLL) is used for frequency stabilization of the local oscillators.

The Schottky diode mixer and the first HEMT amplifier stage of the ozone radiometer are presently operated at ambient temperature. The whisker contacted diode mixer and the first HEMT amplifier stage of the chlorine monoxide radiometer are mounted inside a cryo system and cooled to 13 K in order to reduce receiver noise. In contrast to the ozone radiometer, the local oscillator is optically coupled into the mixer. The water vapor radiometer is operated at ambient temperature, its low noise preamplifier is located in front of the mixer and the sideband filtering is done right behind the first amplifier.

To optimize the signal and minimize atmospheric influence, we use two different observation schemes. The total power method used for the ozone measurements compares the atmospheric signal to cold and hot loads. The reference beam method, used for chlorine monoxide and water vapor measurements compares two atmospheric signals at different observation angles [1]. A compensating plexiglass sheet in one beam accounts for the shorter path length through the atmosphere in the other beam.

4. THE RADIOMETER BACKEND

The evolution of the RAM lead to a design in which the three radiometers serially share one backend which consists of the intermediate frequency chain and an acousto optical spectrometer as illustrated in Figure 3. In the if-chain the signal is down converted from the first intermediate frequency of 8 GHz to the spectrometer input frequency of 2.1 GHz. This is done with a second mixer and local oscillator. A comb signal generator can be coupled in for frequency calibration of the spectrometer. For matching the three if-signals to the spectrometer input power level we use a computer controlled attenuator. For the ozone measurements, we use an additional second local oscillator to increase the usable spectral bandwidth.

The spectrometer, attenuator, mirrors and various high frequency switches are controlled by a computer. It also collects and preintegrates the measured spectra and stores them on the harddisk. Once per month, our data is copied on a CD-ROM and shipped to Bremen for further processing. However a preliminary ozone retrieval is constantly made in Ny-Ålesund and can be seen on our web-site in the Internet (http://www.ram.uni-bremen.de).

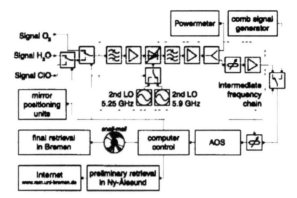

Figure 3: The structure of the RAM backend. A pair of switches selects the input signal. A portion of the signal is fed to a powermeter. A comb signal generator can be coupled in for a frequency calibration of the spectrometer. Once per month, our data is stored on CD-ROM and shipped to Bremen for processing.

One measurement cycle lasts about one hour. Only 20% of the observation time is used by the ozone radiometer, the remaining 80% are either used by the chlorine monoxide or the water vapor radiometer. Although we operate the instrument throughout the year, chlorine monoxide can only be measured in the Arctic spring, because we have to subtract day- and nighttime measurements to eliminate instrumental effects [2]. During this time, chlorine monoxide has the priority over water vapor. The latter will only be measured when the weather conditions do not allow chlorine monoxide measurements. For the remaining time of the year, we only observe ozone and water vapor. To obtain a sufficient signal to noise ratio, we have to integrate datasets of up to one hour for ozone and approximately 24 hours for chlorine monoxide, depending on tropospheric transmissivity.

5. TECHNICAL SPECIFICATIONS

Table 1 shows the technical specifications for the three RAM frontends in July 1999.

	O_3	ClO	H_2O
Receiver temperature	ambient	13 K	ambient
System noise temperature	3400 K	1150 K	200 K
Typical signal strength	15 K	50 mK	200 mK
Observation scheme	total power	reference beam	reference beam
Typical integration time	1 h	24 h	tbd
Share of a the measuring cycle	20%	80%*	80%*
Single sideband filter	Martin-Puplett	Martin-Puplett	coaxial
Observation frequency	142.2 GHz	204.4 GHz	22.2 GHz
Spectral bandwidth	1.65 GHz	0.96 GHz	0.96 GHz
Effective frequency resolution	1.6 MHz	1.6 MHz	1.6 MHz
Altitude range	12 ... 55 km	16 ... 35 km	tbd

Table 1: The technical specifications for the three radiometers. * A measuring cycle consists either of an ozone/chlorine monoxide or an ozone/water vapor measurement.

6. SAMPLE MEASUREMENTS

In Figure 4 we present a full ozone spectrum. It consists of two overlapping measurements of the center and the wing of the line to increasing the spectral bandwith up to 1.65 GHz.

Figure 5 illustrates a typical ClO spectrum. A sinusoidal baseline structure is superimposed to the emission line. To generate the spectrum, we have subtracted night-time measurements from day-time measurements, integrated over approximately 10 hours. The solid gray line represents the result of the radiative transfer calculation of the retrieved ClO distribution.

Figure 6 shows a standard water vapor spectrum with a radiative transfer calculation

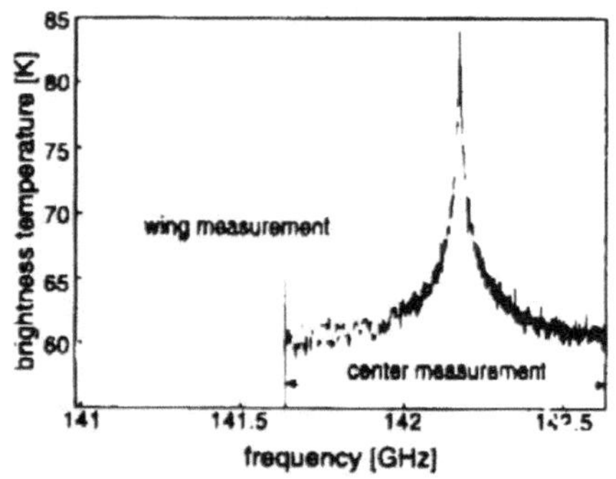

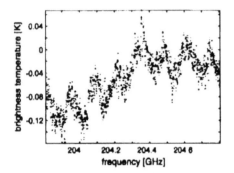

Figure 4: A typcial ozone measurement at 142.2 GHz from March 20th 1997 at 17:00. The spectrum consists of two overlapping measurements.

for comparison. The line is stronger than the ClO emission, but a baseline structure is obvious. Improvements of our hardware to reduce these effects are in progress.

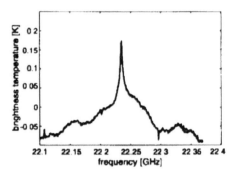

Figure 5: The dots are a typcial chlorine monoxide measurement with a superimposed baseline structure. In comparison, the gray line shows the radiative transfer calculation of the retrieved ClO distribution.

Figure 6: A typical spectral measurement of the water vapor line at 22.2 GHz. The gray line is a radiative transfer calculation for a given water vapor distribution. A baseline structure is visible.

7. RESULTS

Mixing ratio profiles are calculated from the collected spectra. This is done by using the optimal estimation method [3]. A radiative transfer calculation for a horizontally layered atmosphere is used to generate a synthetic spectrum. The variation of the trace gas concentration in these layers leads to a variation of the synthetic spectrum. This iterative process is carried out until the synthetic spectrum converges towards the measured spectrum.

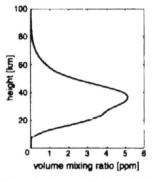

Figure 7: The retrieved ozone VMR profile from March 20th 1997 at 17:00; the gray line is an ozone sonde profile from the previous day.

Figure 7 shows a typcial ozone volume-mixing-ratio profile compared to ozone sonde data. The sonde was launched on the previous day.

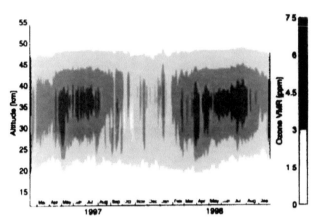

Figure 8: Ozone volume-mixing-ratio (VMR) profiles from February 1997 to March 1998 over Ny-Ålesund.

In Figure 8 we present the ozone profiles from mid February 1997 to September 1998. The plot consists of about 30,000 single Measurements and gives an impression of the seasonal ozone variations in the stratosphere.

The ozone data have been used to estimate the chemical ozone loss inside the polar vortex for the winters 1996/1997 [4] and 1997/1998 [5]. The chlorine monoxide observations of winter 1996/1997 [6] showed significantly enhanced ClO mixing ratios in the lower stratosphere from February to April.

8. REFERENCES

1. A. Parrish, R. L. de Zafra, P. M. Solomon, J. W. Barrett, *A ground-based technique for millimeter wave spectroscopic observations of stratospheric trace constituents*, Radio Science, 23 (2), 106-118, 1988.
2. R. L. de Zafra, J. M. Reeves, D. T. Shindell, *Chlorine monoxide in the Antarctic spring vortex 1. Evolution of midday vertical profiles over McMurdo Station, 1993*, Journal of Geophysical Research, 100, 13999-14007, 1995.
3. C. D. Rodgers, *Retrieval of atmospheric temperature and composition from remote measurements of thermal radiation*, Reviews of Geophysics and Space Physics, 14, 609-624, 1976.
4. B.-M. Sinnhuber, J. Langer, U. Klein, U. Raffalski, K. F. Künzi, *Ground based millimeter-wave observations of Arctic ozone depletion during winter and spring of 1996/97*, Geophysical Research Letters, 25, 3327-3330, 1998.
5. J. Langer, U. Klein, B. Barry, B.-M. Sinnhuber, I. Wohltmann, K. F. Künzi, *Chemical Ozone Depletion during Arctic Winter 1997/98 Derived from Ground-based Millimeter-wave Observations*, Geophysical Research Letters, 26, 599-602, 1999.
6. U. Raffalski, et al., *Ground based millimeter-wave observations of Arctic chlorine activation during winter and spring 1996/97*, Geophysical Research Letters, 25, 3331-3334, 1998.

Microw. Radiomet. Remote Sens. Earth's Surf. Atmosphere, pp. 433–441
P. Pampaloni and S. Paloscia (Eds)
© VSP 2000

Automatic self-calibration of ARM microwave radiometers

JAMES C. LILJEGREN

DOE Ames Laboratory, Ames, IA 50011, USA

Abstract–Microwave radiometers deployed in remote locations by the Atmospheric Radiation Measurement (ARM) Program must operate continuously and autonomously. In order to assure that their calibrations are maintained I have developed algorithms that permit these instruments to automatically self-calibrate when clear sky conditions are detected. First, basic calibration principles for these radiometers are reviewed. Algorithms that correct for misalignment of the elevation angle-scanning mirror and the finite width of the antenna beam pattern are described next. The automatic calibration scheme is then presented along with examples of its performance.

1. INTRODUCTION

The U. S. Department of Energy (DOE) Atmospheric Radiation Measurement (ARM) Program [1] has deployed dual-channel microwave radiometers in rural Oklahoma and Kansas, the north slope of Alaska, and on islands in the tropical Pacific Ocean. These radiometers are to provide continuous measurements of integrated water vapor (IWV) and integrated liquid water (ILW) amounts. Due to the remote nature of these locations, several weeks or months may elapse between maintenance visits by operations personnel. Even then, subtle problems that can adversely affect the instrument calibrations may go undetected. This necessitates expensive post-calibration and reprocessing efforts that substantially delay delivery of the data to the end users. In order to assure that the radiometer calibrations are correctly maintained, algorithms that permit them to be automatically and continuously updated have been developed and implemented.

In this paper the principles of radiometer calibration, as applied to these instruments, are first briefly reviewed. Algorithms that correct for misalignment of the elevation angle scanning mirror and for finite beam width effects are described next. Finally, the algorithms that permit the radiometer calibration to be automated are discussed and examples presented to demonstrate their performance.

2. CALIBRATION PRINCIPLES

The ARM microwave radiometers provide the equivalent blackbody brightness temperature, T_{sky} for each channel (refer to Table 1 for specifications) according to:

$$T_{sky} = T_{ref} + G\left(V_{sky} - V_{ref}\right) f_w.$$ (1)

V_{sky} is the signal recorded when the reflector is oriented toward the sky; the mirror is then pointed downward to view the internal blackbody reference target and V_{ref} is recorded. T_{ref} is the measured temperature of the reference target. The factor $f_w = 1/(1-\varepsilon)$ accounts for the polycarbonate foam window covering the mirror; ε is the window emissivity. G is the calibrated gain calculated as

$$G = T_{nd}/\left(V_{ref+nd} - V_{ref}\right).$$ (2)

V_{ref+nd} is the signal when viewing the reference target with the noise diode energized. T_{nd} is the noise injection temperature that must be determined by prior calibration.

Because the noise diode is maintained at a constant temperature (± 0.25 K), its output is constant. However, because the antenna and feedhorn are not thermally stabilized, the value of T_{nd} can exhibit a slight dependence on the temperature inside the radiometer enclosure, which is equal to the temperature of the reference target T_{ref}. This temperature dependence is determined in the calibration procedure.

Table 1. Specifications of the Radiometrics WVR-1100 microwave radiometers used by ARM.

	Vapor-sensing channel	Liquid-sensing channel
Frequency	23.8 GHz	31.4 GHz
Bandwidth	0.4 GHz	0.4 GHz
Beamwidth (FWHP)	5.9 degrees	4.5 degrees
Window emissivity	0.00164	0.00217

2.1 Tip Curves

The noise injection temperature T_{nd} is calculated by combining Eq. 1 and Eq. 2,

$$T_{nd} = \frac{T_{sky} - T_{ref}}{V_{sky} - V_{ref}}\left(V_{ref+nd} - V_{ref}\right) f_w^{-1}.$$ (3)

An independent measurement of T_{sky} is needed in order to determine T_{nd}. For horizontally homogeneous, clear sky conditions the optical thickness or opacity $\tau(\theta)$ at an elevation angle θ is proportional to the zenith opacity τ_{zen}

$$\tau(\theta) = \tau_{zen}\, m(\theta);$$ (4)

m is a mapping function that describes the ratio of the path length through the atmosphere at θ to the path length at zenith. This ratio is also known as the airmass. (For a plane-parallel atmosphere, $m = 1/\sin\theta$.) The relationship in Eq. 4 is exploited to provide an independent measurement of T_{sky}. An old or estimated value of T_{nd} is used to obtain values of T_{sky} for 10 angles corresponding to m =1, 1.5, 2, 2.5, and 3. The opacities are calculated according to

$$\tau = \ln\left(\frac{T_{mr} - T_{bg}}{T_{mr} - T_{sky}}\right). \qquad (5a)$$

T_{mr} is the atmospheric mean radiating temperature, which may be estimated from climatology (~2% accuracy) or from surface temperature and relative humidity (~1% accuracy) [2]. T_{bg} is the cosmic background radiating temperature (2.73 K).

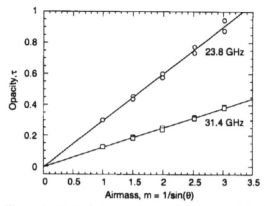

Figure 1. Typical tip curves. The divergence of the data points, especially at 23.8 GHz, is due to an offset error in the elevation angle, and is now corrected automatically.

A linear regression of τ on m (i.e., a tip curve as illustrated in Fig. 1) is then computed. The slope of the regression is an estimate of τ_{zen}. If the correlation coefficient of the regression $R \geq R_{min}$ (I use $R_{min} = 0.998$), indicating that Eq. 4 is obeyed, then the tip curve is said to be "valid" and this estimate of τ_{zen} is used to estimate T_{sky}:

$$T_{sky} = T_{bg}\, e^{-\tau} + T_{mr}\left(1 - e^{-\tau}\right). \qquad (5b)$$

This value is substituted into Eq. 3 to obtain an improved estimate of T_{nd}. This process is repeated with the new estimate of T_{nd} until the intercept of the regression converges to zero, as required by Eq. 4. (Normally, only one iteration is needed.)

3. CALIBRATION ERRORS AND CORRECTIONS

An error in the calibration δT_{nd} causes an error in the measured brightness temperature δT_{sky} given by

$$\delta T_{sky} = \delta T_{nd}\left(T_{ref} - T_{sky}\right)/T_{nd}. \qquad (6)$$

Because $|T_{ref}-T_{sky}| / T_{nd} \approx 1$, $|\delta T_{sky}| \approx \delta T_{nd}$. The ARM microwave radiometers can scan 10 elevation angles and acquire a complete tip curve in about 50 seconds. This allows about 1500 valid tip curves per day to be acquired if the sky remains clear such that $R \geq 0.998$. Although statistical errors due to an insufficient number of tip curves can be kept very small, such errors can arise from offsets in the elevation angle of the mirror which reduce R. Additionally, systematic (bias) errors can arise due to the finite width of the antenna beam pattern. Both of these issues are addressed in the automatic calibration scheme.

3.1 Mirror Alignment

Due to continuous use, the mirrors on some of the ARM radiometers have slipped as much as $1°$ on their stepper motor shafts. The resulting offset in elevation angle causes the brightness temperatures and opacities to be measured at different airmasses than specified. This problem is evident in Fig. 1. Although the regression of τ on m gives the same value of τ_{zen} as for the case of zero offset, the scatter about the regression line increases substantially. Consequently, many or most tip curves do not pass the screening criterion ($R \geq 0.998$) and the resulting calibration is based on a sharply reduced number of samples. If the screening criterion were relaxed, say to 0.995, then some cloud-contaminated tip curves could be accepted as valid which could bias the calibration.

To correct for this, the angular offset is calculated for angles $\theta \geq 150°$ and $\theta \leq 30°$:

$$\Delta\theta = \sin^{-1}\left(\tau_{zen} / \tau\right) - \theta, \quad \theta \leq 30°;$$
$$\Delta\theta = \left[180 - \sin^{-1}\left(\tau_{zen} / \tau\right)\right] - \theta, \quad \theta \geq 150°. \tag{7}$$

The median offset is then computed for each tip curve of the liquid-sensitive (31.4 GHz) channel and stored. The liquid water-sensing channel is used for this purpose rather than the vapor-sensing channel (or both) to minimize the possibility that a persistent horizontal gradient in water vapor could be mistaken for an elevation angle offset. Each hour the

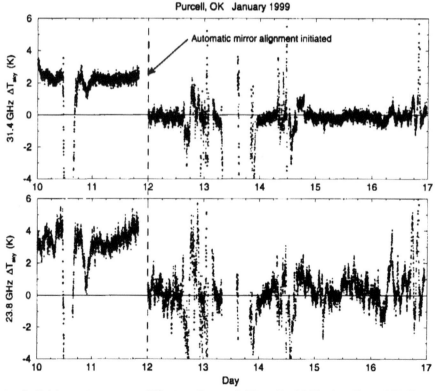

Figure 2. Brightness temperature differences for $m = 3$ (T_{sky} at $\theta = 19.5°$ minus T_{sky} at $160.5°$).

median offset angle for the most recent clear-sky tip curves (up to 1000) is computed and converted to an integer number of motor steps (0.45° per step). If the offset is a non-zero number of steps, the elevation mirror position is adjusted to account for it.

As shown in Fig. 2 (for m = 3), after this algorithm was installed, the bias in the clear-sky differences T_{sky} at $\theta = 19.5°$ minus T_{sky} at 160.5° was reduced to zero. Similarly, the correlation coefficient of the tip curve regressions increased to $R \geq 0.999$.

3.2 Beam Correction

The antenna temperature $T_{ant}(\theta_o)$ measured by a radiometer along a line-of-sight path at an elevation angle θ_o represents a convolution of $T_{sky}(\theta)$ with the antenna power pattern P,

$$T_{ant}(\theta_o) = \int T_{sky}(\theta) \, P(\theta - \theta_o) \, d\theta. \tag{8}$$

The power pattern is assumed to be radially symmetric. Azimuthal variations in T_{sky} are assumed negligible in comparison with elevational variations.

Because T_{sky} varies non-linearly with θ, T_{ant} is always greater than the value of T_{sky} at the beam center. The effective elevation angle, given by $T_{ant}(\theta) = T_{sky}(\theta_{eff})$, is always closer to the horizon than the actual angle; the effective airmass $m_{eff} = \tau_{eff} / \tau_{zen}$ is always greater than the actual value. To correct for this effect, Eq. 8 must be evaluated for a given zenith opacity τ_{zen} to yield T_{ant}, which is substituted into Eq. 5a to give τ_{eff} and finally m_{eff} is determined. Tip curve regressions can then be carried out using m_{eff}.

In order to evaluate Eq. 8, the radiometer antenna is modeled as a circular aperture with a parabolic amplitude taper because the primary beam-forming element is the Gaussian optics lens [3]:

$$g(\theta) = 8 \, J_2(\beta a \sin \theta)/(\beta a \sin \theta)^2; \quad \beta = 2\pi/\lambda. \tag{9}$$

J_2 is the second order Bessel function of the first kind, a is the radius of the aperture (7.6 cm) and λ is the wavelength. The resulting power pattern $P(\theta) = [g(\theta)]^2$ for $\lambda = 1.2605$ cm (23.8 GHz), plotted in Fig. 3a, has a half-power beam width of 6° which is very close to the value of 5.7-5.9° supplied by Radiometrics; the first sidelobe is -25 db at 10° from

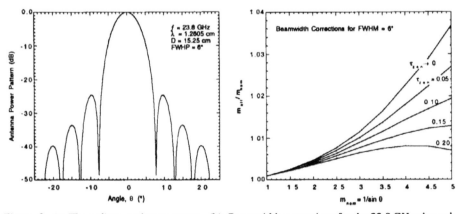

Figure 3. (a) The radiometer beam pattern. (b) Beamwidth corrections for the 23.8 GHz channel.

the main beam axis, also in agreement with the Radiometrics data. At 31.4 GHz the calculated half-power beam width is 4.5°, in close agreement with the Radiometrics measurements (4.5-4.7°).

The mirror intercepts the beam at angles ≤ 12.5° (the second null in the pattern at 23.8 GHz). An absorbing collar intercepts sidelobes beyond 21.5°. Between 12.5° and 21.5° the sidelobes "spill over" the mirror; however, the antenna gain is -33 dB or less and the contribution is negligible. Even if it were not negligible, it would not vary during the time that tip curve measurements are acquired (50 seconds). Consequently, because any "spill over" contribution is common to the blackbody and sky measurements it is eliminated by the V_{sky}-V_{ref} difference in Eq. 1. (Radiometer designs that do not permit rapid angular scanning and do not include a black-body target in their observing cycle are susceptible to "spill over" errors.)

The variation of T_{sky} with θ was modeled using Eq. 4 and Eq. 5b. Because the convolution in Eq. 8 is over all angles, the plane-parallel mapping function (m = 1/sin θ) was replaced by the Niell wet mapping function [4], which describes a spherical atmosphere with a scale height corresponding to water vapor. This is necessary because 1/sin θ → ∞ as θ → 0, so using the plane-parallel assumption would result in over-correction. The resulting corrections are presented in Fig. 3b in terms of the ratio m_{eff} / m_{nom}, where m_{nom} = 1/sin θ.

4. AUTOMATIC CALIBRATION

The first step in the automatic calibration procedure, summarized in Fig. 4, is to assess whether the sky is sufficiently clear. To do this, a 30-minute running mean and standard deviation of integrated liquid water are calculated. When the standard deviation falls below 0.008 mm (2 x RMS noise level), the radiometer begins acquiring

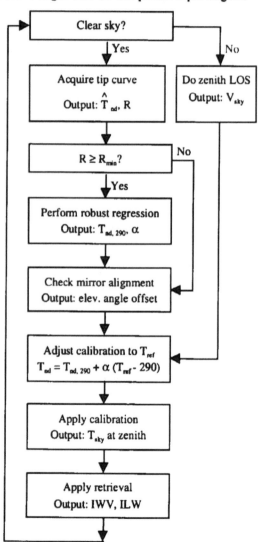

Figure 4. Flowchart of automatic calibration.

tip curves. If a tip curve is valid (R ≥ R$_{min}$), the instantaneous value of T$_{nd}$ is stored in a circular array containing the most recent (up to 3000) values of T$_{nd}$ and T$_{ref}$. Once a minimum number of valid tip curves have been acquired, a robust linear regression (least absolute deviation) of T$_{nd}$ on T$_{ref}$ is then carried out each time a new value of T$_{nd}$ is acquired. This yields a continuously updated estimate of T$_{nd, 290}$ (the value of T$_{nd}$ at T$_{ref}$ = 290 K) and a temperature coefficient α for each channel. These are used to continuously predict T$_{nd}$ from T$_{ref}$:

$$T_{nd} = T_{nd, 290} + \alpha \left(T_{ref} - 290 \right).$$ (10)

The robust regression is employed to prevent outliers from affecting the calibration. Outliers can result when the horizontal water vapor distribution is not homogeneous. When the sky is not clear the radiometer measures T$_{sky}$ along a zenith line-of-sight (LOS) only using the most recent values of T$_{nd, 290}$ and α from the automatic calibration algorithm to adjust T$_{nd}$ for the given T$_{ref}$ based on Eq. 10.

Time series of T$_{nd}$ are presented in Fig. 5 for January 10-30, 1999. A running 2-hour median of the instantaneous values is plotted to indicate the central tendency. It appears

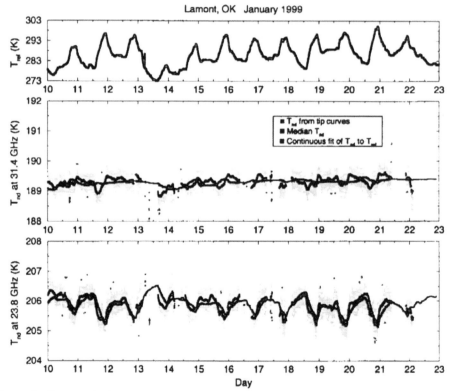

Figure 5. Top panel: time series of reference temperature T$_{ref}$. Middle and lower panels: time series of noise injection temperature T$_{nd}$ for the 31.4 and 23.8 GHz channels from individual tip curves (light grey), a 2-hour running median (black), and predicted from the continuous fit to T$_{ref}$ (dark grey). Gaps indicate cloudy sky periods when no tip curves were acquired.

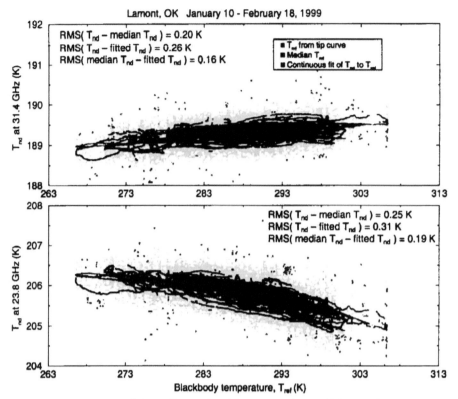

Figure 6. The dependence of T_{nd} on T_{ref} at 31.4 GHz (top panel) and 23.8 GHz (lower panel). For this instrument, the 31.4 GHz channel is nearly independent of T_{ref}.

that the value of T_{nd} predicted from T_{ref} tracks the median better at 23.8 GHz than at 31.4 GHz. However, the variations in T_{nd} at 31.4 GHz are considerably smaller than at 23.8 GHz. It is also apparent from the plots of T_{nd} vs. T_{ref} presented in Fig. 6 that the values of T_{nd} at 23.8 GHz exhibit a much greater correlation with T_{ref}. In any case, the RMS difference between the predicted values of T_{nd} and the running median is less than 0.2 K for both channels. Thus $\delta T_{nd} \approx 0.2$ K. Referring to Eq. 6, this gives $\delta T_{sky} = 0.2$-0.3 K.

5. CONCLUSIONS

A continuous, automatic self-calibration procedure has been developed and implemented for ARM microwave radiometers. Algorithms have been developed to automatically align the elevation angle-scanning mirror and correct for finite beam width effects. This procedure maintains the radiometer calibration to 0.2-0.3 K RMS.

Acknowledgement. This work was supported by the Environmental Sciences Division of the U.S. Department of Energy under the auspices of the Atmospheric Radiation Measurement (ARM) Program. The Ames Laboratory is operated for the U.S. Department of Energy by Iowa State University.

REFERENCES

1. G. M. Stokes and S. E. Schwartz, "The atmospheric radiation measurement (ARM) program: programmatic background and design of the cloud and radiation test bed," *Bull. Amer. Met. Soc.*, **75**, 1201-1221 (1994)

2. J. C. Liljegren, "Improved retrieval of cloud liquid water path," Preprints of the 10th Symposium on Meteorological Observations and Instruments, January 11-16, 1998, Phoenix, AZ.

3. W. L. Stutzman, and G. A. Thiele, *Antenna Theory and Design*, 2nd Ed., John Wiley and Sons, New York. (1998)

4. A. E. Niell, A. E., "Global mapping functions for the atmosphere delay at radio wavelengths," *J. Geophys. Res.*, **101**, 3227-3246 (1996)

Microw. Radiomet. Remote Sens. Earth's Surf. Atmosphere, pp. 443–451
P. Pampaloni and S. Paloscia (Eds)
© VSP 2000

Analog correlator for HUT polarimetric radiometer

JANNE LAHTINEN[1], OLLI KOISTINEN[2] and MARTTI HALLIKAINEN[1]

[1]*Laboratory of Space Technology, Helsinki University of Technology, P.O. Box 3000, 02015 HUT, Finland*
[2]*Metsähovi Radio Observatory, Helsinki University of Technology, Metsähovintie 114, 02540 Kylmälä, Finland*

Abstract - The analog correlator for the Helsinki University of Technology (HUT) polarimetric radiometer is presented. The fully polarimetric radiometer operates at 36.5 GHz and is part of HUTRAD (Helsinki University of Technology RADiometer), a multichannel airborne radiometer system. Analysis and measurements on correlator input matching, linearity, dynamic range, amplitude response, phase response, and stability are presented and the suitability for polarimetric radiometry is discussed.

1. INTRODUCTION

The knowledge of the polarization state of the measured brightness temperatures gives valuable information on the physics of the measured objects in passive microwave remote sensing. This knowledge can be obtained using a polarimetric radiometer the design of which can be based on coherent or incoherent approach [1]. In coherent approach the vertical and horizontal signals are cross-correlated in order to get the Stokes vector of the measured object. For the cross-correlation an analog correlator is a simple and inexpensive option since spectral information is not required for many applications in remote sensing.

In this study an analog multiplier based analog complex cross-correlator was constructed and tested for the Helsinki University of Technology (HUT) 36.5 GHz polarimetric radiometer [2]. The correlator consists of two identical correlator sub-units which are fed with in-phase and quadrature signals. The HUT polarimetric radiometer is a part of the HUTRAD (Helsinki University of Technology RADiometer) multifrequency airborne radiometer system for remote sensing [3]. The observed brightness temperatures vary from 100 K to 300 K and the polarized component is usually small. This leads to a strict requirement for the sensitivity of the correlation channels. Due to the airborne use of the radiometer and long calibration intervals also the stability of the correlator is an important factor. Further in the text the expression "correlator" refers to one sub-unit of the whole cross-correlator unless otherwise mentioned.

2. DESIGN CONSIDERATIONS

The theoretical limit for the sensitivity of the correlation channel ΔT_{min} is determined by the rms value of the correlator output signal fluctuation in the absence of correlated input signals. The sensitivity of a correlator radiometer can be obtained from the sensitivity of a single baseline in an aperture synthesis radiometer [4] by setting the baseline to zero:

$$\Delta T_{min} = \frac{K}{\sqrt{2}} \frac{\sqrt{T_{SYS,V} T_{SYS,H}}}{\sqrt{B\tau}} , \tag{1}$$

where $T_{SYS,V}$ and $T_{SYS,H}$ are respectively the system temperatures for the vertical and horizontal channels of the receiver, B is the receiver noise bandwidth [5] and τ is the integration time. The value for K is 1 for total power receivers and 2 for Dicke type receivers. In Eq. (1) it is assumed that $T_{SYS,V}, T_{SYS,H} >> T_V, T_H$. It also has to be noted that since the third (T_3) and fourth (T_4) Stokes parameters will be computed from $T_3 = 2\times\Re\{(E_V E_H^*)\}$ and $T_4 = 2\times\Im\{(E_V E_H^*)\}$, respectively, the actual sensitivity of these measurements will be twice that predicted by Eq. (1). The E_V and E_H stand for the vertically and horizontally polarized electric fields, the * denotes the complex conjugate.

The parameters influencing the correlator practical sensitivity are the dynamic range, input matching, amplitude response, phase response, and delay differences between the input signals. The non-ideal response of the correlator degrade the sensitivity with a factor D. Differences in signal delays inside the correlator are minimized in the design and can be neglected. Assuming an ideal input match the formula for the correlator sensitivity becomes:

$$\Delta T_{correlator} = \frac{T_{SYS,max}}{\dfrac{P_{in,max}}{P_{in,min}} \times D} , \tag{2}$$

where $T_{SYS,max}$ is the maximum receiver system noise temperature and $P_{in,max}/P_{in,min}$ is the correlator dynamic range. The degradation factor D [6] is:

$$D = \frac{\left| \int_0^\infty G(f) \cos[\delta(f)] df \right|}{\left[\beta \cdot \int_0^\infty G^2(f) df \right]^{\frac{1}{2}}} , \tag{3}$$

where $G(f)$ is the correlator amplitude response, $\delta(f)$ the correlator phase response, and β the passband width.

In order to achieve optimum performance the correlator practical sensitivity should be much smaller than correlator channel theoretical sensitivities for T_3 and T_4. As a design goal it was set that $\Delta T_{correlator} < \Delta T_{min}/2$. This value would ensure that the T_3 and T_4 sensitivities would degrade no more than 3%. The correlator design specifications are presented

Table 1.
Specifications for the correlator performance

Passband (MHz)	90 – 520
Sensitivity (K)	< 0.08 @ $\tau = 0.5$ s
Dynamic range (dB)	> 44.2 @ $\tau = 0.5$ s
Amplitude slope, linear (dB)	< 4.0
Amplitude ripple, p-p (dB)	< 3.2
Mean of phase variations (deg.)	< 14
Input matching, return loss (dB)	< -9.5
Degradation factor D	0.89

in Table 1. The values are derived using Equations (1) - (3). The HUT polarimetric radiometer operates in Dicke mode which sets $K = 2$. The receiver noise temperatures are 1570 K and 1210 K for vertical and horizontal channels, respectively. Antenna temperature is assumed to be 300 K.

The individual values for input matching, phase variation, amplitude flatness and ripple can not be dealt separately from each other. However, as a starting point for design those were chosen in order to keep the degradation factor over 0.97 for each parameter. This gives the worst case overall degradation factor of 0.89. The passband of the correlator was specified to cover the receiver passband, 90 – 520 MHz. The reflections in inputs cause errors in correlator amplitude and phase responses, which decrease D. Applying the worst case analysis for the influence on D a maximum return loss figure was obtained. The amplitude ripple variation figure corresponds to sine type variations. A linear slope was applied in the amplitude slope calculation.

3. CORRELATOR DESIGN

An analog multiplier based complex cross-correlator was constructed for the HUT 36.5 GHz polarimetric radiometer. The radiometer is designed for full polarimetric brightness temperature detection of remote sensing targets. Only few natural targets have a polarized component the degree of polarization being normally no more than some percent.

The complex correlator design is based on two identical correlator sub-units. The signals from the two receiver channels are fed in phase and quadrature shifted to the sub-units. As a result the sub-units are sensitive to linear and circular polarizations while the uncorrelated signals cancel out. The correlating sub-units are based on commercially available Gilbert cell analog multipliers (Analog Devices AD834) with a nominal frequency coverage of 0 – 500 MHz.

The sign of the AD834 component output signal offset follows the output signal sign. The calibration of the correlator output data requires thus a calibration equipment which is capable of generating polarized signal with both positive and negative phase shift between the correlator input channels.

4. MEASUREMENTS

4.1 Input match

Signal amplitude and phase response fluctuations are caused by input port mismatch. Resistive input matching was applied on the correlator sub-unit boards. Network analyzer measurement of the reflection coefficients for the correlator inputs is presented in Fig. 1. The input ports are quite well matched over the intended passband; the reflection coefficient varies from –41 dB to –17 dB. At higher frequencies the matching decreases rapidly. Considering the specifications the input matching would enable the use of the correlator up to about 700 MHz. The use at higher frequencies would require wider band matching.

4.2 Linearity and dynamic range

In Eq. (2) the dynamic range of a correlator is a dominant factor in determining the correlator sensitivity. The dynamic range was measured with a 300 MHz sine signal source. The in-phase signal was fed into the input ports of the correlator and the output voltage level was measured with a multimeter. An additional 6.3 dB was reduced from the obtained maximum linear input level figure. This due to the fact that the observed radiometer signal is noise but the measurement was made using a sine signal.

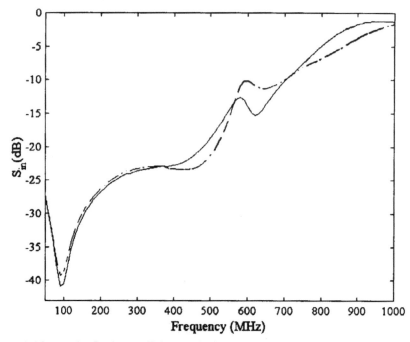

Figure 1. Measured reflection coefficients at the input ports of the correlator. The solid line and the dashed line stand for the input ports 1 and 2, respectively.

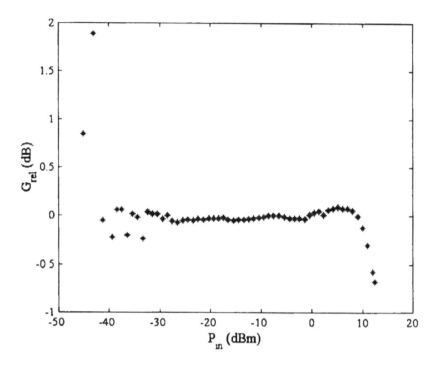

Figure 2. Measured linearity and dynamic range of the correlator.

The measured dynamic range of the constructed correlator is shown in Fig. 2. The 1 dB compression point of the device is at +13 dBm and the maximum linear output is detected with the input power P_i being +9 dBm. The low end of the device is limited to $P_i = -41$ dBm. The dynamic range is thus 44 dBm for noise signals A maximum nonlinearity of 0.18 dB p-p between −32 dBm and +9 dBm is caused probably by the measurement set-up; a set of attenuators was used to increase the dynamic range of the power meter that was used to measure P_i. The ripple at the low end is generated by quantization error of the correlator output voltage measurement. The applied integration time was 0.5 s.

4.3 Amplitude and phase response

In the correlator amplitude and phase response measurement setup the signal from a synthesized signal source was divided into two and the electrical length of one branch was tuned to find the maximum correlation. The phase difference was calculated from the off-set of the tuner, the output voltage value was used for the amplitude response calculation. The measured frequency and phase responses are presented in Fig. 3. The frequency response increases rapidly with frequency, and the response slope at the specified frequency range is 4 dB. At the specified frequency band the phase variations are 4° p-p or 1° mean. The measured phase and amplitude response unidealities cause a 3.5% degradation to the correlator sensitivity ($D = 0.965$). Due to the amplitude variations the correlator noise bandwidth decreases from 430 MHz to 403 MHz.

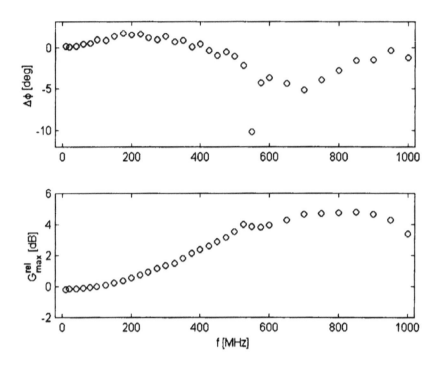

Figure 3. Measured phase response (above) and amplitude response (below) of the correlator.

4.4 Stability

The thermally induced fluctuations at the correlator output may be significant compared to the measured signal. The stability of the correlator was measured in time domain using a stable test signal, sampling the output, and calculating the Allan variance [7, 8]. The Allan variance was calculated relative to the maximum signal level and is presented in Fig. 4. The minimum variance corresponds to the longest integration time that still improves the sensitivity and is in this case obtained with 10 s. The Allan variance does not increase rapidly at very long integration times indicating that the drift of the response is minimal.

5. CONCLUSIONS

The measured characteristics of the correlator as well as the corresponding specified values are presented in Table 2. The most parameters fulfill clearly the set specifications, the most critical being the amplitude flatness and dynamic range. However, the sensitivity figure is a combination of all parameters and the sensitivity requirement is met.

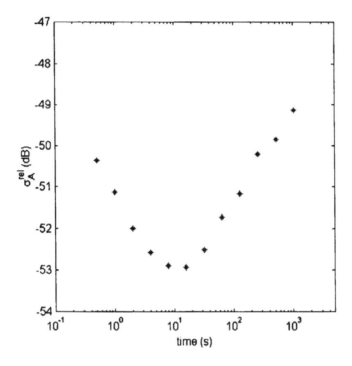

Figure 4. Measured Allan variance relative to the maximum output level.

Table 2.
Measured characteristics of the correlator at 90 – 520 MHz frequency range

	Measured value	Specified value
Noise bandwidth (MHz)	403	-
Sensitivity (K) @ τ = 0.5 s	0.07	< 0.08
Dynamic range (dB) @ τ = 0.5 s	44	> 44.2
Phase variations (mean) (deg)	1.0	< 14
Amplitude flatness (dB)	3.98	< 4.0 (linear slope)
Degradation factor D	0.965	0.89
Input matching (dB)	< −17	< −9.5
Relative stability @ τ = 0.5 s	−50 dB	-
Relative stability @ τ = 1000 s	−49 dB	-

The resulted degradation factor of 0.9650 has practically no influence in sensitivity. The input match for the correlator is significantly better than the specified value. The mismatch was thus not taken into account in calculating the degradation factor values. The stability of the correlator is very good, at the 0.5 s nominal integration time the relative stability is better than 50 dB. The best sensitivity is achieved at 10 s integration time but a very good stability is achieved also at longer integration times. The drifting of the correlator responce is thus small which is a vital characteristics when calibration period is long.

This is the case for airborne instruments without in-flight calibration, like the HUT polarimetric radiometer.

6. DISCUSSION

A continuum correlator based on a nonlinear microwave device such as a multiplier is a very simple and inexpensive way to detect wide band polarized signals. A multiplier based correlator for HUT polarimetric radiometer was constructed and tested; the measured sensitivity lie within the specified value. The correlator channel sensitivities are thus not degraded significantly due to the correlator performance.

The constructed correlator operates at best at frequencies below 500 MHz the amplitude and phase variations increasing at higher frequencies. However, the correlator can be applied for even higher frequencies: The sensitivity improvement due to the broader bandwidth compensates the influence of the relative small increase in amplitude and phase response variations at least up to 700 MHz. The correlator should be usable up to 1 GHz if a wider band matching is applied. Due to the differential structure of Gilbert cell analog multipliers the correlator is very stable: continous calibration is thus not necessary. According to its characteristics the constructed correlator is suitable for many polarimetric radiometry applications in remote sensing where wide dynamic range, good sensitivity, and high stability are important.

The analog multiplier used for the constructed correlator has the broadest bandwidth of commercially available components. Current technology level limits the use of the Gilbert cell based multipliers at frequencies below 1 GHz. However, components with noticeably broader bandwidths can be expected to become on the market in near future due to the recent advances in fast analog multiplier development [9, 10].

REFERENCES

1. C. S. Ruf, "Constrains on the polarization purity of a Stokes microwave radiometer," *Radio Science*, 33, 1617-1639 (1998).
2. J. Lahtinen, M. Hallikainen, "Polarimetric radiometer for remote sensing," *Proceedings of ESA Workshop on Millimetre Wave Technology and Applications: antennas, circuits and systems*, 361 - 365, Espoo, Finland (1998).
3. M. Hallikainen, M. Kemppinen, J. Pihlflyckt, I. Mononen, T. Auer, K. Rautiainen, J. Lahtinen, "HUTRAD: airborne multifrequency microwave radiometer," *Proceedings of ESA Workshop on Millimetre Wave Technology and Applications: antennas, circuits and systems*, 115 - 120, Espoo, Finland (1998).
4. C. S. Ruf, C. T. Swift, A. B. Tanner, D. M. Le Vine, "Interferometric synthetic aperture microwave radiometry for remote sensing of the Earth," *IEEE Transactions on Geoscience and Remote Sensing*, 26, 597-611, (1988).
5. M. E. Tiuri, "Radio astronomy receivers," *IEEE Transactions on Antennas and Propagation*, 12, 931 - 938 (1964).
6. A. R. Thompson, L. R. D'Addario, "Frequency response of a synthesis array: Performance limitations and design tolerances," *Radio Science*, 17, 357-369 (1982).

7. D. W. Allan, "Statistics of atomic frequency standards," *Proceedings of IEEE*, **54**, 221-230 (1966).
8. R. Schieder, G. Rau, B. Vowinkel, "Characterization and measurement of system stability," *Proceedings of SPIE*, **598**, 189-192 (1985).
9. K. Osafune, Y. Yamauchi, "20-GHz 5 dB-gain analog multipliers with Al-GaAs/GaAs HBT's," *IEEE Transactions on Microwave Theory and Techniques*, **42**, 518-520 (1994).
10. Y. Imai, S. Kimura, Y. Umeda, T. Enoki, "A DC to 38-GHz distributed analog multiplier using InP HEMT's," *IEEE Microwave and Guided Wave Letters*, **4**, 399-401, (1994).

Microw. Radiomet. Remote Sens. Earth's Surf. Atmosphere, pp. 453–458
P. Pampaloni and S. Paloscia (Eds)
© VSP 2000

A method for calibrating the ground-based triple-channel microwave radiometer

HONGBIN CHEN
LAGEO, Institute of Atmospheric Physics, Chinese Academy of Sciences,
Beijing 100029, China

Abstract — It has been shown from many theoretical and experimental works that for clear atmospheres, the optical depth τ in one channel measured by a ground-based triple-channel (20.6, 31.65, and 90.0 GHz) microwave radiometer can be well predicted with the two other channel τ. This implies that the brightness temperature TB in one channel may be expressed by other two channel TBs through a regression-based relation. Based on this high correlation of the 3-channel radiometric TBs, a new calibration method is proposed with which the calibration coefficients and sky TBs can be simultaneously obtained by numerically solving a non-linear equation system. The feasibility of the method has been demonstrated by numerical simulations. Further tests will be done with the real measurement data.

1. INTRODUCTION

It has been demonstrated by a number of theoretical and experimental investigations that ground-based microwave radiometer is a powerful tool for monitoring atmospheric precipitable water, cloud liquid water path, and the vertical profiles of humidity and temperature[1-7]. It will play a key role in the future atmospheric sounding system without balloon[8]. Before doing the measurements of atmospheric parameters, the microwave radiometer must be calibrated, i.e., a relation, normally linear, between the radiometric output (voltage, for example) and the sky apparent brightness temperature should be established.

In general, there now are two approaches for calibrating a microwave radiometer. The first one is called absolute calibration. Quasi-black body sources with known brightness temperatures are introduced as the radiometric input and then are related to the radiometric output. A complete absolute calibration needs at least two sources: one cold, another hot, and it is better to include the whole antenna system. A liquid nitrogen cold source or an internal noise hot source is frequently used for single point calibration, which is normal for monitoring the instrumental stability. To use two or more sources to calibrate a microwave radiometric system is very difficult because of the lack of facilities which may be very complex with a high cost. Another more widely-used calibration method is the "tipping curve" method. It is required that in the clear and steady atmosphere, a set of the radiometric outputs is obtained as a function of zenith angle by scanning the sky. In parallel, the sky brightness temperatures at different zenith angles are calculated through a radiative transfer model using the coincident radiosonde data. Plotting the calculated TBs against the radiometric outputs (voltage) will provide a calibration curve. Relatively, the tipping curve method is quite easy with low cost. However, the performance of the tipping curve calibration depends on the accuracy of radiosonde data and the radiative transfer model used and also on the horizontal homogeneity of the atmosphere.

It has been revealed by some investigations[9] that for clear atmospheres, the optical depth τ in one channel measured by a ground-based triple-channel(20.6, 31.65, and 90.0 GHz) microwave radiometer can be well predicted with the two other channel τ. This implies that the brightness temperature TB in one channel may be expressed by other two channel TBs through a regression-based relation. Based on the high correlation between the brightness temperatures (TBs) measured by a ground-based triple-channel microwave radiometer at 20.6, 31.65, and 90.0 GHz, a new calibration method is proposed and presented.

2. METHOD

It is assumed here that between the radiometric output voltage V and the sky brightness temperature TB arriving at antenna there is a linear relationship, i.e.

$$V(\theta) = A + B*TB(\theta) \qquad (1)$$

where A and B are the calibration coefficients, θ is the zenith angle of observation (antenna elevation angle = $90°-\theta$). In the case of clear and steady atmosphere, if the triple-channel radiometer scans the sky and records outputs at 6 zenith angles, we can get 18(3×6) measurements of V that can be converted to 18(3×6) equations:

$$V_i(\theta_j) = A_i + B_i*TB_i(\theta_j)$$
$$(i = 1,2,3; j = 1,2,...,6) \qquad (2)$$

where the subscripts i and j denote   respectively number of channels and viewing zenith angles. In the above set of equations, only $V_i(\theta_j)$ are known   while A_i , B_i and $TB_i(\theta_j)$ are all unknown. Totally, there are 24(=3×2+3×6) unknowns, which are more than the number of equations.

However, consider that TBs in three channels are highly correlated, the TB in one channel may be expressed by a combination of two other channel TBs.   Accordingly, the number of unknowns in eqs.(2) can be reduced to 18 (=24-6). Then, mathematically, this system of 18 non-linear equations with 18 unknowns could be resolved to give the coefficients A_i , B_i and $TB_i(\theta_j)$ as roots simultaneously.

In this calibration method, two critical points have to be clarified: 1) whether one channel TB can precisely be expressed with two other channel TBs, and 2) whether there are convergent and stable solutions for eqs.(2). The following two sections will respectively deal with these two points.

3. RELATION BETWEEN 3 CHANNEL TBs

It is shown by Westwater et al.[9] that for clear atmospheres, the measured optical depth τ in one channel can be well predicted with the two other channel τ (20.6, 31.65, and 90.0 GHz). This implies that one channel TB may be expressed by other two channel TBs through a regression-based relation, which will be confirmed by our radiative transfer simulations.

The microwave radiative transfer model of Chen[10] is used to simulate the sky TBs for three frequencies and at six zenith angles. Three frequencies are 20.6, 31.65, and 90 GHz which are commonly used in ground-based microwave radiometry. The 6 zenith angles are 0, 45, 55, 65, 70, and 75°, which are chosen in a little arbitrary way but the corresponding TBs are different enough. The model of Liebe[11] is now adopted for H_2O and O_2 absorption calculation in the modified radiative transfer model. The vertical profiles of pressure, temperature, and humidity are taken from radiosonde data of Beijing 1992. There are 122 and 132 profiles for winter and summer respectively.

The linear regression results for three channel TBs in the winter are given in Table 1 for $\theta=0°$. Note that R is the correlation coefficient and σ_e stands for the rms error(in K). It is seen that the TB in any channel is highly correlated to the other two channel TBs. Comparison of the rms errors σ_e for three channels indicates that TB at 31.65 GHz can be better derived from other two channel TBs. Thus, the relationship $TB_{31}=C_0+C_1TB_{20}+C_2TB_{90}$ will be introduced in eqs.(2) for reducing the unknown number.

Table 1. Regression results for triple-channel brightness temperatures at 0°

Regression relation	C_0	C_1	C_2	R	σ_e
$TB_{20}=C_0+C_1TB_{31}+C_2TB_{90}$	-.5599625	-.123974	.5281414	.9974	0.195
$TB_{90}=C_0+C_1TB_{20}+C_2TB_{31}$	-22.95670	.5045534	3.544956	.9994	0.191
$TB_{31}=C_0+C_1TB_{20}+C_2TB_{90}$	6.674440	.2138674	-7.145e-03	.9992	0.047

Table 2 presents the regression results for the relation $TB_{31}=C_0+C_1TB_{20}+C_2TB_{90}$ at 6 zenith angles. The values in parenthesis are for the summer season. It can be seen that until 75° the value of σ_e is very small in winter and does not exceed the instrumental error which is generally about 0.5 K. In contrast, the values of σ_e in summer are quite large.

Table 2. Regression results for the relation $TB_{31}=C_0+C_1TB_{20}+C_2TB_{90}$

θ	C_0	C_1	C_2	R	σ_e
0	6.674440	.2138674	-7.145312e-3	.99915	.047
	(3.400400)	(.2590293)	(6.891606e-3)	(.99911)	(.24)
45	8.353155	.2150180	4.758945e-3	.99907	.068
	(1.499372)	(.2960480)	(2.841852e-3)	(.99794)	(.49)
55	9.625965	.2161099	1.355537e-2	.99899	.087
	(-.9074056)	(.3153955)	(2.266381e-2)	(.99657)	(.75)
65	11.929390	.2149374	3.555796e-2	.99876	.127
	(-6.263936)	(.3150958)	(.1200420)	(.99336)	(1.34)
70	13.919080	.2110940	5.905522e-2	.99853	.167
	(-9.831649)	(.2709999)	(.2569596)	(.99049)	(1.87)
75	17.168860	.2012395	1.016761e-1	.99814	.239
	(-9.715563)	(.1411784)	(.5050906)	(.98718)	(2.62)

It is noted that there are two main   reasons to choosing 75° as maximum θ: 1) beyond
this zenith angle, the effects of atmospheric inhomogeneity and sphericity become
important, which means that the approximation of plane-parallel atmosphere is less valid,
and 2) the effects of side-lobe may occur in measurements.

4.   NUMERICAL EXPERIMENT OF CALIBRATION

Eqs.(2) will be numerically solved to see whether there are convergent solutions. Because
of the lack of measurement data, the radiometric outputs are artificially created with the
calibration coefficients given in Table 1(values in parenthesis for Summer; also note:
channel 1,2,3 now for 20.6, 90.0, and 31.65GHz). In Figure 1 is shown in solid line the
truth of TBs as a function of the zenith angle for 2 radiosonde profiles respectively in
winter(Fig.1a) and summer(Fig.1b). The instrumental noise is added to these values of TB
by a random number generator with a standard deviation of 0.5 K.
    The Newton-Raphson method [12] is applied to solving eqs.(2) which are non-linear but
with constant coefficients. This numerical method requires a first guess which may be
obtained by analyzing the radiometric outputs and the TBs computed from the (climatic)
radiosonde profiles. Table 3 presents the results of the coefficients A and B after 1,000
times of iteration, values without and with parentheses respectively for two numerical
experiments. It is found that the coefficients B and brightness temperatures TBs approach
the truth while the coefficients A do not. The sensitivity study shows that the solutions are
to some extent subject to the first guess. However, TBs are very approximate to the truth in
most of cases (see Figure 1).

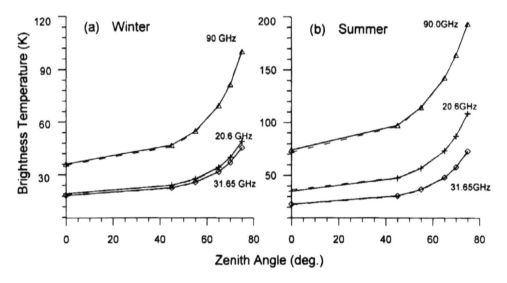

Figure 1.   Brightness temperatures as the function of zenith angle.
Solid line: truth; broken line: retrieved.

	A_1	A_2	A_3	B_1	B_2	B_3
Truth	-10.0	10.0	0.0	-2.0	-2.0	-2.0
	(5.0)	(-5.)	(5.0)	(1.5)	(1.0)	(2.0)
First Guess	-15.0	5.0	1.5	-1.5	-2.5	-2.5
	(4.0)	(-6.5)	(7.0)	(1.65)	(1.25)	(1.6)
Solutions	-7.28	7.44	-.43	-2.245	-1.95	-2.01
	(1.25)	(-2.41)	(5.94)	(1.547)	(.985)	(2.007)

Table 3. Solutions for A and B by Newton-Raphson method

5. DISCUSSIONS

On the basis of extremely high correlation of ground-based triple channel microwave radiometric brightness temperatures, i.e., one channel TB can accurately be expressed by a linear combination of two other channel TBs, a new calibration method is proposed. Its feasibility is demonstrated by numerical simulations. This method does not require any assessors and any data but only scanning the sky in a very clear and stable atmosphere. So, it can be performed very easily, economically and frequently.

It should be admitted that the method is quite subject to the first guess. That is because there is no good mathematical method to solve systems of more than one nonlinear equations. Though the coefficients A and B are not very accurately determined, the wanted sky brightness temperatures TBs are all well found, which means that A and B can be used ulteriorly since their combination results in quite accurate TBs.

In this work, the relation between sky brightness temperature and radiometric output is assumed to be linear for the purpose of simplicity. This assumption is not necessarily needed in performing the proposed calibration method. For example, if the relation between V and TB is a polynomial of 2 orders, the solutions can also be obtained. In this case, however, the measurements at other 3 angles and relevant relations between three-frequency TBs have to be added.

Acknowledgments The author wishes to thank Professor C. Wei for helpful discussions and for providing her technique report on "Calibration of Ground-Based Microwave Radiometer". Thanks also to Dr. Y. Han for valuable discussions, to Professor P.C. Yang for his help in numerically solving the systems of nonlinear equations, and to Ms. Y. Chen for typing the manuscript.

REFERENCES

1. M.T. Decker, ED R. Westwater, and F.O. Guiraud. Experimental evaluation of ground-based microwave radiometric sensing of atmospheric temperature and water vapor profiles, *J. Appl. Meteorol.*, 17, 1788-1795(1978).
2. D.C. Hogg, M.T. Decker, F.O. Guiraud, K.B. Earnshaw, D.A. Merritt, K.P. Moran, W.B. Sweezy, R.G. Strauch, E.R. Westwater, and C.G. Little. An automatic profiler for the temperature, wind, and humidity in the troposphere, *J. Appl. Meteorol.*, 22, 807-831(1983).
3. J.I.H. Askne and ED R. Westwater. A review of ground-based remote sensing of temperature and moisture by passive microwave radiometers, *IEEE Trans. Geosci.*

Remote Sensing, **GE-24**, 340-352(1986).

4. R.H. Huang and S.X. Zhou. Remote sensing of precipitable water vapor and liquid water of cloudy atmosphere by dual-wavelength microwave radiometer, *Scientia Atmospherica Sinica*, 11, 397-403(1987)(in Chinese with English abstract).

5. Y. Han, J.B. Snider, and E.R. Westwater. Observations of water vapor by ground-based microwave radiometers and Raman Lidar, *J. Geophys. Res.*, **99**, 18695-18702(1994).

6. C. Wei and D.R. Lu. A universal regression retrieval method of the ground-based microwave remote sensing of precipitable water vapor and path-integrated cloud liquid water content, *Atmospheric Research*, **34**, 309-322(1994).

7. C. Wei. Method of retrieving cloud and rain parameters in the atmosphere using the measurements of a triple-wavelength MW radiometer, *Scientia Atmospherica Sinica*, **19**, 21-30(1995)(in Chinese with English abstract).

8. Y. Han and ED. R. Westwater. Remote sensing of tropospheric water vapor and cloud liquid water by integrated ground-based sensors, *J. Atmos. Ocean. Tech.*, **12**, 1050-1059(1995).

9. ED R. Westwater, J.B. Snider, and M.J. Falls, Ground-based radiometric observations of atmospheric emission and attenuation at 20.6, 31.5, and 90.0 GHz: a comparison of measurements and theory. *IEEE Trans. Anten. Propag.*, **38**(10), 1569-1579(1990).

10. H.B. Chen. Simulation d'observations satellitaires passives en micro-onde, *Ph.D. These*, Universite de Lille I, France, No.537(1991).

11. H.J. Liebe. MPM - an atmospheric millimeter-wave propagation model, *Int. J. Infrared Millimeter waves*, **10**, 631-650(1989).

12. J.H. Mathews. Numerical Methods for Mathematics, Science, and Engineering. Prentice Hall, Inc., New Jersey(1992), p71-90.

Microw. Radiomet. Remote Sens. Earth's Surf. Atmosphere, pp. 459–464
P. Pampaloni and S. Paloscia (Eds)
© VSP 2000

Calibration methods in large interferometric radiometers devoted to Earth observation

FRANCESC TORRES, ADRIANO CAMPS, IGNASI CORBELLA, JAVIER BARÁ, NURIA DUFFO and MERCE VALL-LLOSSERA

Department of Signal Theory and Communications
Universitat Politècnica de Catalunya, Barcelona 08034, Spain

Abstract - This paper starts with a summary and classification of errors of any kind that corrupt the fundamental measurement performed by an InR (interferometric radiometer), the so-called visibilities. This summary is based on prior works from the authors concerning end-to-end modeling of the instrument. The paper follows with a trade-off analysis of the capability of different calibration approaches to remove those errors and, at the end, to recover a map of brightness temperatures. In order to perform the trade-off analysis, extensive simulations have been undertaken to analyze the approaches found in the literature: the redundant space method, used in radio astronomy, the G-matrix method, used in ESTAR, and the Noise injection method, both centralized and distributed, proposed for MIRAS. A short description is made of each one and a comparative table is presented. The trade off is presented in terms of what kind of errors can be removed by each method, hardware requirements, robustness and on-ground/on-board input data required to perform such error correction.

1. INTRODUCTION

Soil moisture is one of the key geophysical parameters for climate modeling which can be monitored from space using passive remote sensing at 1.4 GHz. Aperture synthesis using interferometric techniques is a very promising concept, first proposed by NASA in the 80's in a 1-D air-borne configuration (ESTAR) [1]. This paper is presented within the scope of a space-borne 2-D instrument, called MIRAS [2], currently under development by ESA, which aims at providing global coverage of soil moisture. An issue of main interest is concerned with error correction and calibration of the instrument, since many of the techniques inherited from radio astronomy cannot be directly applied, due to the nature of extended sources of thermal radiation, which, in the case of Earth observation applications, occupy a large field of view.

2. DEFINITION OF TERMS AND ERROR CLASSIFICATION

A baseline of an interferometer consists of two receivers and a complex correlator, as depicted in figure 1, in which G_k and G_j are the available power gains of the chains, $H_{nk}(f)$ and $H_{nj}(f)$ their normalized frequency responses, $F_{nk}(\theta,\phi)$ and $F_{nj}(\theta,\phi)$ the normalized voltage patterns of the antennas and b_k and b_j the analytic signals of the receivers' output waves. The antennas are positioned in the XY plane, at Z=0 and near the origin of coordinates. The usual polar coordinates (r,θ,ϕ) and direction cosines ($\xi=\sin\theta\,\cos\phi$, $\eta=\sin\theta\,\sin\phi$) are defined, and a thermal source is assumed to be located somewhere in the $Z>0$ hemisphere far away from the antennas. The basic measurement of the instrument is the cross correlation of the analytic signals b_k and b_j, given by:

$$\frac{1}{2}\langle b_k(t)b_j^*(t)\rangle = k_B V_{kj}\sqrt{B_k B_j}\sqrt{G_k G_j} \; ; \qquad B_k = \int_0^\infty |H_{nk}(f)|^2 df$$

here k_B is the Boltzmann constant, B_k and B_j are the noise equivalent bandwidth of the receivers and V_{kj} is the visibility function, related to the brightness temperature map of the source T_B by:

$$V_{kj} = \frac{1}{\sqrt{\Omega_k \Omega_j}} \iint_{\xi^2+\eta^2\leq 1} \frac{T_B(\xi,\eta)}{\sqrt{1-\xi^2-\eta^2}} F_{nk}(\xi,\eta)\,F_{nj}^*(\xi,\eta)\,\tilde{r}_{kj}(\tau)\,e^{-j2\pi(u_{kj}\xi+v_{kj}\eta)}d\xi d\eta$$

where Ω_k and Ω_j are the equivalent solid angles of the antennas, f_0 is an arbitrary center frequency, and the baseline u_{kj} and v_{kj} are given by the projections over the XY axes of the normalized spacing between antennas: $u_{kj}=(x_k-x_j)/\lambda_0$; $v_{kj}=(y_k-y_j)/\lambda_0$ being λ_0 the wavelength at f_0 and (x_k,y_k) (x_j,y_j) the XY coordinates of the antennas.

Now, a brief description of the different error terms is presented in order to classify and number these error terms according to their impact on the different calibration methods:

Antenna errors (1): Antenna errors affect the exploration of the scene [3-4] and require external signals (known scenes) to be corrected. Extensive simulation and

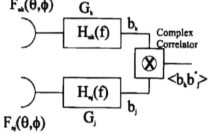

Figure 1. Basic InR receiver

laboratory measurements have identified the most significant errors as: radiation voltage phase and amplitude mismatches, mutual coupling and offset induced by receiver's noise re-radiated through mutual coupling.

Fringe-washing (2): The term $\tilde{r}_{kj}(\tau)$ is the *fringe-washing function*, which takes into account differences in the receivers' frequency responses [5]. It is given by:

$$\tilde{r}_{kj}(\tau) = \frac{e^{-j2\pi f_0\tau}}{\sqrt{B_k B_j}} \int_0^\infty H_{nk}(f)H_{nj}^*(f)e^{j2\pi f\tau}df$$

The value of the fringe-washing function at the origin ($\tau=0$) depends on the degree of non-similarity of the receivers' frequency response. It is a complex number which will be represented by its amplitude and phase:

$$\tilde{r}_{kj}(0) = g_{kj}e^{j(\theta_k-\theta_j+\theta_{kj})}$$

For identical receivers $\tilde{r}_{kj}(0)=1$. In general, however, the amplitude g_{kj} is smaller than unity. The term $\theta_k-\theta_j$ is the difference between the phase of the receivers' frequency response at the center frequency. The non-separable phase term θ_{kj} is a second order effect. However, since the image to be recovered is highly low pass in the observation domain, it has non-negligible impact in the shortest baselines, which are the most significant. The shape of the $\tilde{r}_{kj}(\tau)/\tilde{r}_{kj}(0)$ must also be measured (in amplitude and phase), since path delay within the FOV is given by $\left(u_{kj}\xi + v_{kj}\eta\right)/f_0$. It must be determined as least with the same accuracy as $F_{nk}(\theta,\phi)$ and $F_{nj}(\theta,\phi)$.

Separable Amplitude errors (3): If a digital 1bit-2level correlator is used, as is foreseen in the ESA's MIRAS instrument, the measured magnitude is actually the *normalized* cross-correlation [6]:

$$\mu_{kj} = \frac{\langle b_k b_j^* \rangle}{\sqrt{\langle |b_k|^2 \rangle}\sqrt{\langle |b_j|^2 \rangle}} = g_k g_j \frac{V_{kj}}{T_A} \quad ; \quad g_k = \sqrt{\frac{T_A}{T_A + T_{R_k}}} \quad ; \quad g_j = \sqrt{\frac{T_A}{T_A + T_{R_j}}}$$

where g_k and g_j are separable error terms that also have to be retrieved in the calibration procedure, which are directly related to the receivers equivalent noise temperature T_{Rk} and T_{Rj}. T_A is the antenna noise temperature, which is assumed to be equal for both receivers. It is measured with a well calibrated total power radiometer using an antenna located in the center of the array. Simulations proved that the separable amplitude term contributed by the fringe-washing term is negligible.

Separable phase errors (4): They are given by phase differences between receivers, such as path delay or phase differences at the downcovverters [5]. They are modeled as the separable phase terms in the fringe-washing term.

Quadrature error (5): A complex correlator is actually implemented using a phase-quadrature down converter that produces for each receiver an "in-phase" and a "quadrature". Quadrature error is produced by the error in the 90° phase shift of the Q-path local oscillator, as well as by different lengths (or group delays) between the I and Q paths of one receiver [5]. It is also produced by sampling errors in the comparators [7].

Non-separable amplitude and phase errors (6): This errors are mainly contributed by filter frequency response mismatches and correlator sampling errors. Although they are second order terms, they have a high impact in the calibration of the shortest baselines, which are the most significant.

Correlator errors (7): Threshold and skew sampling errors in the correlators give offset, quadrature and non-separable phase errors which must be corrected [7].

3. REVIEW OF CALIBRATION APPROACHES

Three approaches have been analyzed from available literature: The Redundant Space Method, used in Radio astronomy [8], the G-Matrix method (used in ESTAR) [1] and the Noise-Injection method, both centralized and distributed (proposed for MIRAS). A short description is made of each one, and a comparative table is presented.

3.1 The redundant space calibration (RSC). This method is a well-known calibration procedure in the field of radio astronomy [8]. However, in this case it must be revised since InR radiometers observing an extended source of thermal radiation, such as the Earth, cannot benefit from the properties of punctual radiosources due to the large field of view which is involved. It assumes that all errors are neglected except separable phase and amplitude errors. In this case the measured normalized complex cross-correlation between the outputs of two receivers forming a baseline μ_{kj} is related to the ideal value by:

$$\mu_{kj} = \left(g_k e^{j\theta_k}\right)\left(g_j e^{-j\theta_j}\right)\mu_{kj}^{id}$$

Now, an InR such as MIRAS is a highly redundant interferometer: the same baseline (normalized spatial separation between antennas) can be formed by different pairs of antennas. For instance, if the baseline formed by antennas numbered as "1" and "2" is redundant with the baselines formed by antennas numbered "2" and "3", "3" and "4", etc. The following set of equations can be established relating phases:

$$\phi_{12} = \theta_1 - \theta_2 + \phi_{12}^{id}; \quad \phi_{23} = \theta_2 - \theta_3 + \phi_{12}^{id}; \quad \phi_{34} = \theta_3 - \theta_4 + \phi_{12}^{id} \ldots$$

where ϕ_{12}^{id} is the ideal phase of the redundant baseline, $\phi_{12}, \phi_{23}, \phi_{34},\ldots$ are the phases of the measured visibilites, and $\theta_1, \theta_2, \theta_3, \theta_4,\ldots$ the receiver phase unknowns to be determined. The method consists of establishing all possible independent equations corresponding to the different redundant baselines and solving the linear system of equations to determine the separable phase error terms to be assigned to each receiver. An analogous procedure can be envisaged to determine the separable amplitude terms. This method is based only on the mathematical properties of the equations and on the redundancies available in the system and, thus, can be periodically performed on board while the instrument is measuring a scene. Nevertheless, if the antenna differences, the quadrature errors and the non-separable amplitude and phase terms are important, the method fails. And this is the case for an instrument such as MIRAS, where extensive simulations proved that the implications on the receiver requirements for neglecting the non-separable and quadrature errors is beyond the present state-of-the-art of the technology, and where the wide beamwidth of the antennas and the large Field of View makes antenna errors non-neglectable.

3.2 The G-Matrix calibration (GM). In an InR, the brightness temperature and the visibility samples are related by an integral equation. Former studies concerning the calibration and inversion methods in one dimensional InR (ESTAR) were based on the measurement of the actual system's impulse response (G-matrix) for a point source of thermal noise located at the position of each pixel [9]. In this approach the equation relating cross-correlation and brightness temperature is solved by direct discretization and matrix inversion. The main advantage of this method is that it corrects both receiver and antenna errors and has no hardware requirements. Nevertheless, the G-matrix must be measured on ground once, and then it must be assumed that the system impulse response does not vary in flight. When this procedure is extended to a two dimensional InR, the least squares method used during the inversion may give large radiometric errors due to small system parameter drifts. It has been found that, assuming standard parameters for the InR, the condition number of the G-matrix pseudo-inverse is very large. This condition number grows with smaller antenna pattern beamwidths, spatial decorrelation effects and array size.

3.3 Noise injection. If, for each baseline, the antennas are substituted by sources of correlated noise at equivalent temperature T_c, this is equivalent to locating a point source at equal distances from the receivers and assume omnidirectional antennas. In this case, and taking into account that now $T_A=T_c$, the measured normalized "visibility" is given simply by $\mu_{kj}^c = g_k^c g_j^c \tilde{r}_{kj}(0)$, where the superscript "c" stands for "calibration. It is, then, a method for receiver calibration, and the antenna errors are not recovered.

When a single known noise source is connected to all the receivers forming the InR (centralized noise injection -CNI-), an overdetermined set of equations can be established to recover the real parameters g_k^c, g_j^c and the complex magnitude $\tilde{r}_{kj}(0)$, without the need to consider any approximation [10]. However, for a large instrument such as MIRAS, this results in impractical mass and volume requirements for the noise distribution network.

To avoid the electrical and mechanical problems associated with the CNI method, in [11] a distributed noise injection method (DNI) was proposed. It is based on the same equations as the centralized one, but it is assumed that the separable error terms are negligible ($g_{kj} \approx 1$ and $\theta_{kj} \approx 0$). If correlated noise is injected into sets of eight adjacent antennas, a set of overdetermined equations is obtained. Solving this system allows to compute the in-phase and quadrature phase error of each one of the eight receivers. This procedure is repeated for all sets of eight antennas of the array. Once the phases have been computed, the amplitude terms (and, thus, the receivers noise temperature) are obtained with $g_{kj}=1$ and $\theta_{kj}=0$ by a similar procedure. Overlapping sets of antennas are needed to establish the reference phase and to avoid the precise knowledge of all the noise sources. Using this mechanism, the calibration is complete, except for the non-separable phase and amplitude error terms. The main advantage of this approach is that the mass and volume of the noise distribution network is drastically reduced. Also the repeatability and electrical stability of this network becomes less critical.

The main drawback of the DNI method is that it cannot correct for the non-separable error terms. To overcome this problem, an improvement of this method was proposed (i-DNI) [10]. It essentially consists of applying a centralized distribution approach to the central receivers of the Y-array, which share the same noise source, and a distributed approach to the rest of them. The equations presented allow one to calibrate all the separable and non-separable errors of the baselines formed by the central receivers, and also the separable errors of the individual receivers that are not in the central section. This is possible because the brightness temperature to be measured is smooth, and the visibility function will have large values for small baselines (signals of the central antennas) and almost zero for large ones. As a consequence, without any further hardware requirements, the central antennas can be compensated for all the errors, while the rest are only compensated for separable terms. In all the cases uncorrelated noise (matched load) is injected to each receiver to correct the offset terms.

4. TRADE-OFF BETWEEN METHODS

Table 1 compares the calibration approaches presented in the previous section with respect to the errors that are capable of being corrected. Comparison between the calibration approaches in terms of hardware requirements, robustness and possibility of on-board calibration is also shown. However, it does not appear to be an optimum calibration procedure since all of them have some drawbacks and, probably, a combination of methods will be required to account for both short term and long term parameter drifts.

Table 1. Comparison between calibration approaches

Calibration Approach	Hardware Requirements	Robustness	On Ground On Board	Error number						
				1	2	3	4	5	6	7
RSC	Low	Very Low	OG/OB			X	X		X	
GM	Low	Low/High[1]	OG	X	X	X	X	X	X	X
CNI	High	High	OG[2]/OB	X[3]	X[3]	X	X	X	X	X
DNI	Medium	Medium	OG[2]/OB	X[3]	X[3]	X	X	X		X
i-DNI	Medium	High	OG[2]/OB	X[3]	X[3]	X	X	X	X[4]	X[4]

[1] The condition number of the G-matrix pseudo-inverse grows as the array size increases
[2] The antennas measured on ground. The rest of calibration performed periodically on board.
[3] Corrected by the inversion algorithm
[4] Only shortest baselines

Noise Injection methods are very suitable to calibrate the receivers, but cannot directly deal with antenna errors. The DNI method is, in theory, the best choice but it has high hardware requirements due to the complicated noise distribution network, especially for large arrays. The main problems are related to the temperature drift: characterization, measurement of temperature along the network, etc. The distributed noise injection method solves this problem at the expense of not being capable of correcting non-separable errors. Finally, the improved noise injection method has the benefits of the distributed method in terms of hardware requirements, and corrects, to some extent, the non-separable errors. It is, then the preferred method to be used in MIRAS instrument if periodical calibration of the receivers is required to account for short term system drifts.

The G-Matrix approach does not seem as convenient as it appeared from the ability to deal with all errors. First of all it cannot be carried out on board, since it needs an external point source, which makes difficult periodical calibration of the instrument. Secondly, for a large number of antennas, the condition number of the G-matrix may become very large making the method susceptible to system drifts. However, it is the preferred method to fully test the instrument and in the case that periodical calibration of the instrument were required to account for long term system parameter drifts which were associated with the antennas.

REFERENCES

1. C. S. Ruf, C. T Swift, A. B. Tanner, D. M. LeVine, "Interferometric Synthetic Aperture Radiometry for the Remote Sensing of the Earth", IEEE Transactions on Geoscience and Remote Sensing, Vol. 26, N° 5, 597-611 (1988,.)
2. M. Martín-Neira, J. M. Goutoule. "MIRAS - A two-dimensional aperture synthesis radiometer for soil-moisture and ocean salinity observations". ESA bulletin n° 92. 95-104 (1997)
3. A. Camps, J. Bará, F. Torres, I. Corbella, J. Romeu, "Impact of Antenna Errors on the Radiometric Accuracy of Large Aperture Synthesis radiometers. Study Applied to MIRAS", Radio Science. Vol. 32, No 2, 657-668 (1997)
4. A. Camps, F. Torres, I. Corbella, J. Bará, P. de Paco, "Mutual Coupling Effects on Antenna Radiation Pattern : An Experimental Study Applied to Interferometric Radiometers", Radio Science, Vol 33, No 6, 1543-1552 (1998)
5. F. Torres, A. Camps, J. Bará, I. Corbella "Impact of receiver errors on the radiometric resolution of large two-dimensional aperture synthesis radiometers". Radio Science, volume 32, n° 2 629-641 (1997)
6. J. B. Hagen, D. T. Farley "Digital correlation techniques in Radio Science" Radio Science, volume 8, n° 8-9 775-784 (1973)
7. A. Camps, F. Torres, I. Corbella, J. Bará, J. A. Lluch. "Threshold and timing errors of 1 bit/2 level digital correlators in Earth observation synthetic aperture radiometry". Electronic Letters vol 33 n° 9 812-813 (1997)
8. NRAO "Synthesis Imaging in Radio Science" Vol 6 cap 9 NRAO Synthesis Imaging Summer school. Astronomical Society of the pacific (1989)
9. A.B.Tanner and C.T.Swift. "Calibration of a synthetic aperture radiometer", IEEE-T on Geoscience and Remote Sensing, Vol 31, N°1, pp 257-267. (1993)
10. F. Torres, A. Camps, J. Bará, I. Corbella, R. Ferrero, "On-board Phase and Module Calibration of Large Aperture Synthesis Radiometers. Study Applied to MIRAS", IEEE-T on Geoscience and Remote Sensing, Vol. 34, N°4, 1000-1009 (1996)
11. I. Corbella, F. Torres, A. Camps, J. Bará, "A new calibration technique for interferometric radiometers", The European Symposium on Aerospace Remote Sensing. Barcelona, 359-366 (1998).

4. New radiometric instruments and missions

4.2 Remote sensing missions

Microw. Radiomet. Remote Sens. Earth's Surf. Atmosphere, pp. 467–475
P. Pampaloni and S. Paloscia (Eds)
© VSP 2000

The Soil Moisture and Ocean Salinity Mission: an overview

Y.H. KERR[1], P. WALDTEUFEL[2], J.-P. WIGNERON[3], J.-M. MARTINUZZI[4], B. LAZARD[4], J.-M. GOUTOULE[5], C. TABARD[5], A. LANNES[6]

[1] *CESBIO, BPI 2801, 18 av E. Belin, 31401 Toulouse cedex 4, France*
[2] *IPSL, 10-12 av de l'Europe, 78140 Vélizy, France*
[3] *INRA Bioclimatologie, Agroparc 84914 Avignon Cedex 9, France*
[4] *CNES, 18 Avenue Edouard Belin 31401 Toulouse cedex 4, France*
[5] *MMS, 31 avenue des Cosmonautes 31402 Toulouse Cedex 4, France*
[6] *CERFACS, 42 av. Gaspard Coriolis, 31057 Toulouse cedex, France*

Abstract - The goal of this paper is to present a space mission aimed at the retrieval of soil moisture and ocean salinity from space. The mission itself is based on the use of passive microwave data at 1.4 GHz. In order to achieve a reasonable spatial resolution, the sensor is based on the interferometry concept, which allows 2-dimensional interferometry; acquisitions with a pseudo conical scan at several angles and dual polarisation. The concept is now being proposed to the Earth Explorer Opportunity Programme, under the name of SMOS. The sensor has new and very significant capabilities especially in terms of multi-angular view configuration. The main goals of this mission are: (i) to provide fields of surface soil moisture and potentially of root zone soil moisture, all over the globe and every 2 to 3 days, (ii) to provide decadal values of sea surface salinity, (iii) to be used over ice and ice caps for research purposes. The goal is to provide information globally on key parameters (moisture and salinity) for models in the fields of oceanography, meteorology, and hydrology to name the main ones.
This paper will give an overview of the SMOS Mission objectives and the main mission characteristics together with some of the technical features.

1. INTRODUCTION:MISSION RATIONALE

The main objective of the Soil Moisture and Ocean Salinity (SMOS) mission is to deliver crucial variables of the land surfaces: soil moisture, and of ocean surfaces: sea surface salinity fields. The mission should also deliver information on root zone soil moisture, vegetation, and biomass, and lead to significant research in the field of the cryosphere. The main aspects of the baseline mission are presented in this paper.

Over land, water and energy fluxes at the surface/atmosphere interface are strongly dependent upon Soil Moisture (SM). Evaporation, infiltration and runoff are driven by SM while soil moisture in the vadose zone governs the rate of water uptake by vegetation. Soil moisture is thus a key variable in the hydrologic cycle. Soil moisture, and its spatio-temporal evolution as such, is an important variable for numerical weather and climate models, and should be accounted for in hydrology and vegetation monitoring.

For the oceans, Sea Surface Salinity (SSS) plays an important role in the Northern Atlantic sub polar area, where intrusions with a low salinity influence the deep thermohaline circulation and the meridional heat transport. Variations in salinity also influence the near-surface dynamics of tropical oceans, where rainfall modifies the buoyancy of the surface layer and the tropical ocean-atmosphere heat fluxes. SSS fields and their seasonal and inter-annual variabilities are thus tracers and constraints on the water cycle and on the coupled ocean-atmosphere models.

Even though both SM and SSS are used in predictive atmospheric, oceanographic, and hydrologic models, no capability exists to date to measure directly and globally these key variables. The SMOS mission is aimed at filling this gap through the implementation of a satellite that has the potential to provide globally, frequently, and routinely this information. It is also expected that the SMOS mission will provide significant information on vegetation water content, which will be very useful for regional estimates of crop production. Finally, significant research progresses are expected over the cryosphere, through improving the assessment of the snow mantle, and of the multi-layered ice structure. These quantities are of significant importance to the global change issue. Research on sea ice will also be carried out

Significant progresses in terms of weather forecasting, climate monitoring and extreme events forecasting rely on a better quantification of both Soil Moisture (SM) and Sea Surface Salinity (SSS). Several recent groups and workshop reports conclude that further improvements now depend upon the availability of global observational information on SM and SSS.

The reason why such information are not available currently mainly stems from the fact that, while *in situ* measurements are very far from global, no dedicated, long term, space mission has been attempted so far.

The only direct way to access to SM and SSS is through the use of L band (21 cm, 1.4 GHz) microwave radiometer systems. Other means (higher frequency radiometry, optical domain, active remote sensing) suffer strong deficiencies, due to vulnerability to cloud cover and/or various perturbing factors (such as soil surface roughness or vegetation cover), as well as poor sensitivity.

Even though the concept was proved by early L band space experiments, such as the one on SKYLAB back in the 70s, no dedicated space mission followed, because achieving a suitable ground resolution ($\leq$ 50-60 km) required a prohibitive antenna size ($\geq$ 4 m). All the research work was consequently performed using either ground (PAMIR, PORTOS, etc) or airborne radiometers (e.g. PBMR, PORTOS, ESTAR).

Recent development of the so-called interferometry design, inspired from the very large baseline antenna concept (radio astronomy), makes such a venture possible. The idea consists of deploying small receivers in space (located on a deployable structure), then reconstructing a brightness temperature (T_B) field with a resolution corresponding to the spacing between the outmost receivers. The idea was put forward by D. LeVine et al., in the '80's (the ESTAR project) and validated with an airborne system. In Europe, an improved concept was next proposed to ESA. While MIRAS (Microwave Imaging Radiometer using Aperture Synthesis) capitalises on the ESTAR design, it embodies major improvements. The two-dimensional MIRAS interferometer allows measuring T_B at large incidences, for two polarisations. Moreover, the instrument records instantaneously a whole scene; as the satellite moves, a given point within the 2D field of view is observed from different view angles. One then obtains a series of independent measurements, which allows retrieving surface parameters with much improved accuracy [1].

SMOS is a mission with broad and ambitious scientific objectives. In addition, it can also be considered as a demonstrator, which should allow both to assess the potential of L-Band 2D interferometric radiometry for possible operational uses, and to pave the way for future, upgraded technical implementations. It should also be noted that such a concept is of high interest in other scientific fields (radio science to name one) and some of the results gained from the concept analysis are now put in use in radio astronomy.

2. SCIENCE OBJECTIVES

2.1 Land surfaces

The key variable to monitor over land surfaces for satisfying the above-mentioned goals is surface soil moisture. The main requirement is a sufficient accuracy to make it useful in hydrologic and meteorological models (0.04 m^3 / m^3) [1], coupled with an adequate temporal sampling. Recent studies [2], have shown that with a revisit of 2 to 4 days depending upon the region was appropriate. The spatial resolution requirements are linked to the field of application, i.e., from 100 km for climate models (Global Circulation Models) to 20 km or less for hydrology and agronomy. The coverage should be global.

To achieve these goals, it is necessary to account for vegetation contribution and correct for topographic effects and surface temperature. The SMOS concept by delivering angular information allows retrieving soil moisture in presence of vegetation [3]. The vegetation water content is retrieved together with soil moisture in the centre part of the field of view (FOV) every 6 days at most (depending on latitude) and used for soil moisture retrievals in the outer part of the FOV subsequent orbits. Surface topography is constant with time and, once known, can be accounted for. Further studies are required to assess the exact contribution of topography on the signal. Obviously variable footprint size with incidence angle will have to be accounted for. Surface temperature could be retrieved from the angular measurements but with a poor accuracy. However Wigneron et al. [1], have shown that if surface temperature is known with an accuracy of 2 K, The retrieval algorithm would be sufficiently accurate to meet the science objectives. Consequently, classical means of obtaining surface temperature (thermal infra red instruments coupled with models for cloud covered periods, or higher frequency microwave measurements) would be adequate.

Once surface moisture is known, it is possible to infer root zone soil moisture. Recent studies [4], [5] showed that such data, obtained through microwave radiometry during a continuous hydrological event (e.g. drying period), provide estimates of the corresponding surface hydraulic conductivity value. A similar approach was used in [3], where soil moisture content at field capacity was assessed from in situ measurements of w_S.

2.2 Oceans

Ocean salinity is a key parameter in determining the ocean circulation and in understanding the water cycle. It is also an important circulation tracer for water masses. Unlike other oceanographic variables, until now, it has not been possible to measure salinity from space. The only means to get consistent estimates of SSS at least at seasonal scales is the use of a satellite-based instrument.

The sensitivity of L-band (1.4 GHz) passive measurements of oceanic brightness temperature T_B to SSS is well established [6]. The dielectric constant for seawater depends on both SSS and SST [7], [8]. So, in principle, it is possible to obtain SSS information from L-band passive microwave measurements if the other factors influencing T_B can be accounted for. Then the sensitivity of T_B to SSS is 0.5K / psu for a SST of 20°C, decreasing to 0.25K / psu for a SST of 0°C [6].

Since the radiometric sensitivity achievable is of the order of 1 K, it is clear that SSS cannot be recovered to the required accuracy from a single measurement. However, if the errors contributing to the uncertainty in T_B are random, the requirement can be met by averaging the SMOS individual measurements in both space and time [6]. Similarly, monthly averages over 100 km boxes would give data comparable to the standard climatologies, such as Levitus' [9]. Lower accuracy, higher resolution measurements (typically 0.5 psu, 50 km, 3 days) provide a mean to monitor moving salinity fronts in various regions of the world: extension of the warm pool in the equatorial Pacific (1psu/200km), limit between the upwelled waters (of equatorial or coastal origin) and the subtropical waters (1 psu in 200 or 300 km), confluence of currents (1 to 2 psu Brazil-Malvinas), large river plumes (Amazone). With a SSS single retrieval to 1 psu and a spatial resolution around 30 km the data are useful for enclosed seas with high salinity contrast (e.g. the Baltic).

The averaging procedure requests excellent stability and calibration of the radiometer receiver. Factors that influence T_B in addition to SSS, and are to be corrected for, include Sea Surface Temperature (SST), surface roughness, foam, sun glint, rain, ionospheric effects and galactic/cosmic background radiation. Estimates for the uncertainties associated with some of these have been made [6], [8]. Use of L-band radiometry for the measurement of SSS from aircraft has been demonstrated (most recently by Miller [10]). SSS estimates were previously recovered from the Skylab 1.4GHz radiometer using a combination of modelling and ancillary data on SST and wind speed. Therefore the feasibility of obtaining SSS estimations with SMOS is not in doubt; the SMOS performance will undoubtedly be better than what was achieved over 20 years ago.

2.3 Cryosphere

About 10 % of the world oceans are covered by ice during some portion of the year. Sea ice influences the key large-scale processes of the Earth's climate system, involving the atmosphere, the ocean, and the radiation field [11].

Over ice caps, a key prediction of the greenhouse gas-induced climate change is that the sea ice extent will respond early to altered conditions [12]. Accurate predictions of sea level rise over the coming century will require improved knowledge of the processes controlling accumulation upon the ice sheets. The total snow accumulation over the Antarctic ice sheet is equivalent to 5-7 mm of sea level [13]. The scarcity of accumulation rate observations, both spatially and temporally, has hindered the development of this understanding. Finally, the accumulation and depletion of snow is dynamically coupled with global hydrological and climatological processes [14]. Snow cover is also a sensitive indicator of climate change: the position of the southern boundary snow cover in the Northern hemisphere is of particular significance, as it is likely to move northward as a result of a sustained climate warming [15].

3. MISSION DESCRIPTION

The baseline SMOS payload is an L band (1.4 GHz) 2D interferometric radiometer, Y shaped, with three arms 4.5 m long, for a mass (including electronics and structure) of 175 kg. The radiometer is accommodated on a generic PROTEUS platform, for a total wet mass of 475 kg (fig. 1). The folded satellite is compatible with most launchers.

It is proposed to launch SMOS on a sun synchronous (6 a.m. ascending), circular, 757 km orbit. Raw measuring performances then are: 30 to more than 90 km for ground resolution, 0.8 to 2 K for radiometric sensitivity, 1 to 3 days for temporal sampling, depending upon latitude, nature of the target and location within the instrument field of view. The mission minimum duration is 3 years (5 years expected).

SMOS aims at providing, over the open ocean, global salinity maps with an accuracy better than 0.1 PSU every few days, with a 200 km spatial resolution. Over the land surfaces, global maps of soil moisture, with an accuracy better than 0.035 m^3/m^3 every 3 days, with a space resolution better than 60 km, as well as vegetation water content with an accuracy of 0.2 kgm^{-2}.

In order to process SMOS data, corrections due to atmospheric, ionospheric and galactic effects have to be applied. SSS retrieval requires knowledge of sea surface temperature and sea roughness. Over land surfaces, knowledge of the temperature is needed. Over ice, higher frequency microwave data are useful.

Figure 1: SMOS: Artist's view

3.1 The Instrument

The instrument exhibits three deployable arms constituting a Y when deployed. In the current baseline, the antenna is composed of 3 co-planar arms 120 degrees apart. Each arm is 4.46 m long, 260 mm wide and 200 mm high. The antenna is fixed to a triangular (1.15 meter high) structure mounted on a payload module that has a square section. This payload module is used to fit electronic equipment for the instrument. Each antenna arm is composed of 23 -26 radiating elements located on the upper side of the arm. In addition there is one central element and 3 others located near the centre between the 3 arms. Each antenna element is a cylinder (diameter of 19 cm and 10 cm high). The arms' length is compatible with the PROTEUS configuration. The receiver sections, from LNA to digitisation, are integrated into a small compartment located at the back of the radiating elements. The digital receiver outputs are multiplexed by 4 by dedicated equipment located in the antenna arms and routed to the platform via optical fibres as studied in GSTP. The data are transferred to a correlator bank that generates an estimator of the visibility function. The correlator data output are CCSDS formatted by the Instrument Control Unit (ICU) which also performs the interface with the platform. A calibration unit based on a noise diode associated to a splitter and controlled by the ICU allows calibrating each visibility function.

The antenna elements are cup dipole antennas similar to those developed for the MIRAS demonstrator. The half power beam width is between 65 and 70° according to direction and polarisation. The normalised phase discrepancy between antennas is less than 5° within the instantaneous field of view. The cross polarisation level is -20dB (-30 dB inside the IFOV). The coupling between the antennas is less than 25dB.

The low noise amplification, filtering, and I and Q demodulation in MIRAS are achieved in units installed around the structure, phase locked by a clock which is distributed through an optical fibre.

The correlator unit has the same architecture as the one that has been developed for the MIRAS demonstrator.

The baseline launch vehicle (Rockot) provides ample mass-to-orbit capability as it has performance to deliver over 820 kg into a direct insertion 757 km circular, sun-synchronous orbit. With the SMOS satellite mass of 475 kg, this yields a generous margin of 72%. An option with a Vega launch vehicle may be possible, considering our current knowledge of it. Should the SMOS mission be open to USA partners, most of the US small satellite launchers could also be used.

The baseline bus is PROTEUS, developed by the CNES and Alcatel. It has simple, well-defined interfaces. The platform architecture is generic. Adaptations are limited to minor changes in software modules and launch vehicle interface. The spacecraft bus is cubic (nearly 1 meter per side) with all the equipment units accommodated on four lateral panels and the lower plate The interface with the launcher is through a specific adapter bolted to the bottom of the structure.

3.2 Operational mode

Each 300ms an image is taken. Horizontal and vertical polarisation images are interlaced. The Instrument Control Unit averages 5 images to obtain one that is formatted and sent to the platform. The equivalent integration time is 1.5s per polarisation. So two images (one per polarisation) are available every 3s. The instrument is calibrated every 10 images (15s) during 600ms: a correlated noise is injected at the receivers input for 300ms, then a non correlated noise is injected for 300ms.

Other modes are possible: the receivers can be continuously connected to only one output of the antenna (polarisation H or V), or to the calibration unit. The calibration period can be decreased or increased depending on requirements or on actual performances of the instrument. A complete acquisition sequence can be also loaded from the ground. However the integration time of the correlator unit is not programmable (300ms). The possibility to avoid correlator commutation will be studied further during phase A.

3.3 Platform and orbit

The orbit altitude was selected to minimise the accessibility delay for a global coverage assuming an altitude within the range 650-800 km in order to have a good resolution. The local time is selected to minimise the perturbation on L band signal (air, vegetation and soil temperature almost identical) and making the Faraday effect minimum, leading to a Sun-synchronous 6 am (ascending) orbit. SMOS's nominal operational orbit parameters are Sun-synchronous, 757 km altitude (+/-1 km) with time of ascending node at 6am +/- 15 min. The decrease of semi-major axis due to atmospheric drag is between 0.25 and 5 km per year depending on solar activity. SMOS requires 5 kg of hydrazine for orbit insertion, 2 kg for station keeping over 3 years (maximal solar activity). With a fuel capacity of 30 kg this yields a fully adequate margin to pursue an extended mission beyond the nominal three year mission operation. At mission end, SMOS will perform a controlled depletion/re-entry burn, thereby minimising the generation of orbital debris and eliminating the potential for ground impact.

3.4 Data acquisition and processing

The tasks for transforming the actual measurements of the instrument into the final image are summarised in the generic term "image reconstruction." The innovative design of the proposed SMOS instrument is based on the novel passive high-resolution imaging capabilities of aperture synthesis devices. In [16] the image acquisition and reconstruction basis are explained. The various experience and the wide background of the group members [17]- [21], including hardware implementations of interferometric radiometer prototypes [22], [23], system performance analysis, error correction, image reconstruction and end-to-end simulations of SMOS-type instruments, has led to the identification of three main classes of issues: error correction, image reconstruction and calibration. The different tools currently available, concerning both the theoretical and technical aspects, will be used in a complementary manner, so as to develop a novel and innovative tool for an end-to-end simulation of a SMOS-type synthesis radiometer. It will offer solutions for operational error correction, image reconstruction and calibration [19] [20], [24].

Good calibration will undoubtedly be necessary to ensure the scientific return together with accuracy. To achieve this goal, the instrument will first be calibrated with conventional means (receiver level). Then, as described above, the advantages of the instrument concept will be used to perform calibration using phase closure. Finally, through the monitoring of large areas of uniform, stable and known characteristics (such as Dome C in Antarctica), the calibration will be checked and monitored with time. Obviously the success of the calibration procedure will be strongly dependent on the instrument qualities especially in terms of similarity of elements and accurate knowledge of each element.

4. CONCLUSION

The baseline mission described in this paper is ambitious but fully achievable [25]. Actually the novel aspects of the interferometry concept has been validated by the ESTAR concept. The science background is very sound and the mission should fill existing gaps in our knowledge of surface variables (soil moisture, ocean salinity). The technological solutions have been found to be perfectly feasible thanks to several studies performed through ESA. The approach taken in SMOS is conservative with ample margins to ensure success.

It is clear nevertheless that a number of issues exist, as depicted above. The mission being very innovating, current knowledge does not allow us to always have clear-cut answers. All the known issues have been investigated and solutions proposed (see for instance http://www-sv.cict.fr/cesbio/smos/). The next step will consist in carrying out suitable field experiments to validate the solutions and check whether all has been accounted for.

The SMOS concept will be a demonstrator of L band measurements over the globe, paving the way to more ambitious concepts in terms of spatial resolution or frequency range. SMOS will undoubtedly make available long needed measurements of surface soil moisture, vegetation biomass and sea surface salinity and foster new research in these fields as well as in cryospheric studies. SMOS should be launched as a second Earth Explorer Opportunity Mission, tentatively in 2004.

REFERENCES

1. J.-P. Wigneron, P. Waldteufel, A. Chanzy, J.-C. Calvet, O. Marloie, J.-F. Hanocq and Y. Kerr, "Retrieval capabilities of L-Band 2-D interferometric radiometry over land surfaces (SMOS Mission)," this issue (1999).
2. J.-C. Calvet, J. Noilhan and P. Bessemoulin, "Retrieving the root-zone soil moisture from surface soil moisture or temperature estimates: a feasibility study based on field measurements," *J. Appl. Meteor.*, **37**, 371-386 (1998).
3. J.-P. Wigneron, A. Chanzy, J.-C. Calvet and N. Bruguier, "A simple algorithm to retrieve soil moisture and vegetation biomass using passive microwave measurements over crop fields," *Remote Sens. Environ.*, **51**, 331-341 (1995).
4. K. J. Hollenbeck,, T. J. Schmugge, G. M. Hornberger and J.R. Wang, "Identifying soil hydraulic heterogeneity by detection of relative change in passive microwave remote sensing observations," *Water Resour. Res.*, **32**, 132-148 (1996).
5. N.M. Mattikali, E.T. Engman and T.J. Jackson, "Microwave remote sensing of temporal variations of brightness temperature and near surface water content during a watershed scale field experiment and its application to the estimation of soil physical properties," *Water Resour. Res.*, **34**, 2289-2299 (1998).
6. G.S.E. Lagerloef, C.T. Swift and D.M. LeVine, "Sea surface salinity: the next remote sensing challenge," *Oceanogr.*, **8**, 44-50 (1995).
7. L.A. Klein and C.T. Swift, "An improved model for the dielectric constant of sea water at microwave frequencies," *IEEE Trans. Ant. Prop.*, **25**, 104-111 (1977).
8. C.T. Swift and R.E. McIntosh, "Considerations for microwave remote sensing of ocean-surface salinity," *IEEE Trans. Geosci. Remote Sensing*, **21**, 480-491 (1983).
9. S. Levitus, R. Burgett and T.P. Boyer, *World Ocean Atlas 1994 Volume 3: Salinity*, NESDIS 3 (1994).
10. J.L. Miller, M.A. Goodberlet and J.-B. Zaitzeff, "Airborne salinity mapper makes debut in coastal zone," *Eos, Trans. AGU*, **79**, 173&176-177 (1998).
11. F.D. Carsey (Ed.), *Microwave remote sensing of sea ice*, Geophysical monograph 68, American Geophysical Union (1992).
12. R. Stouffer, S. Manabe and K. Bryan, " Interhemispheric asymmetry in climate response to a gradual increase of atmospheric CO_2," *Nature*, **342**, 660-682 (1989).
13. *IPCC,1995*, Warrick et al., *Changes in sea level in Climate Change*, Houghton et al., Eds, International Panel on Climate Change, Cambridge University Press, 363-385 (1995).
14. A.T.C. Chang, J.L. Foster and D.K. Hall, "Nimbus-7 SMMR derived global snow cover parameters," *Annals of Glaciology*, **9**, 39-44 (1987).

15. R.G. Barry, "Possible CO_2-induced warming effects on the cryosphere," in *Climate Changes on a Yearly to Millennium Basis* (ed. N.A. Morner and W. Karlen), 571-604, Dordrecht, The Netherlands: D. Reidel (1984).

16. P. Waldteufel, E. Anterrieu, J. M. Goutoule and Y. Kerr, "Field of view characteristics of a microwave 2-D interferometric antenna, as illustrated by the MIRAS concept," this issue (1999).

17. E. Anterrieu and A. Lannes, "Image reconstruction via the G matrix," ESA report on MIRAS (rider 2) Wp 3000 (1996).

18. J. Bará, A. Camps, A. Capdevila, I. Corbella and F. Torres, "MIRAS Calibration System Definition CAS-D: Final report," ESTEC Contract No 12513/97/NL/MV (1998).

19. A. Camps, "Application of Interferometric Radiometry to Earth Observation," Ph. D. Dissertation, Universitat Politècnica de Catalunya (1996)

20. A. Lannes and E. Anterrieu, "Modelization, antenna geometry and algorithms," ESA report on MIRAS, Wps 1220 and 1810 (1994).

21. F. Torres, A. Camps, J. Bará, I. Corbella, and R. Ferrero, "On-Board Phase and Modulus Calibration of Large Aperture Synthesis Radiometers: Study Applied to MIRAS," *IEEE Trans. on Geosci. and Remote Sens.*, **34**, No 4, pp 1000-1009 (1996).

22. M. Peichl, H. Suess, M. Suess and S. Kern, "Microwave imaging of the brightness temperature distribution of extended areas in the near and far field using two-dimensional aperture synthesis with high spatial resolution," *Radio Sci.*, **33**, No 3, pp 781-801 (1998).

23. A. Camps, J. Bará, I. Corbella and F. Torres, "The Processing of Hexagonally Sampled Signals with Standard Rectangular Techniques: Application to 2D Large Aperture Synthesis Interferometric Radiometers," *IEEE Trans. on Geosci. and Remote Sens.*, **35**, No 1, 183-190 (1997).

24. A. Lannes, E. Anterrieu and K. Bouyoucef, "Fourier Interpolation and reconstruction via Shannon-type techniques, Part II: Technical developments and applications," *J. Mod. Optics*, **43**, 105-138 (1996).

25 Y.H. Kerr, et al. "*MIRAS on RAMSES: radiometry applied to soil moisture and salinity measurements*," Full proposal, A.O. Earth Explorer Opportunity Missions, ESA. (1998).

Microw. Radiomet. Remote Sens. Earth's Surf. Atmosphere, pp. 477–483
P. Pampaloni and S. Paloscia (Eds)
© VSP 2000

Field of view characteristics of a 2-D interferometric antenna, as illustrated by the MIRAS/SMOS L-band concept

P. WALDTEUFEL[1], E. ANTERRIEU[2], J. M. GOUTOULE[3], Y. KERR[4]

[1] *IPSL (CNRS), 10-12 Avenue de l'Europe, 78140 Vélizy, France*
[2] *CERFACS & OMP (CNRS), 42 av. Gaspard Coriolis, 31057 Toulouse cedex, France*
[3] *MATRA MARCONI SPACE, 31 Avenue des Cosmonautes, 31402 Toulouse, France*
[4] *CESBIO (CNES/CNRS), BPI 2801, 18 av E. Belin, 31401 Toulouse cedex, France*

Abstract - Recently developed interferometric techniques allow to design realistic space missions using L-Band radiometry, devoted to remote sensing of soil moisture (over land) and surface salinity (over the ocean) at the Earth surface. The SMOS mission, submitted to the European Space Agency, uses the 2-D interferometric MIRAS concept, developed since 1993 by ESA. The MIRAS antenna is a Y shaped structure, each arm of which consists of aligned, adjacent elementary antennas. Visibility functions are retrieved from the correlation products constructed from the signals collected by elementary antennas, then transformed into brightness temperature maps.
The field of view (FOV) is limited both by the solid angle where visibility functions can be reconstructed and by aliased Earth images. The alias free FOV roughly consists of two regions. In the "inner" swath, a large range of look angles is explored instantaneously ; the same region on Earth can thus be seen for a variety of incidence angles. This may strongly improve the accuracy and power of retrieving procedures. In the "outer" swath, the range of angles decreases : the 2-D radiometer then becomes comparable to a conical scanning radiometer. This duality may help to optimise mission scenarios in terms of revisit time and measurement performances.
Over the ocean, an averaging procedure is necessary, in order to estimate surface salinity with adequate accuracy. The total number and accuracy of instantaneous independant views then become important issues. It is shown that 2-D interferometric designs do not suffer major degradations when compared to real antennas.

1. INTERFEROMETRIC RADIOMETRY

Interferometric techniques allow, for microwave frequencies, an angular resolution comparable to that of a real antenna to be achieved ; at the same time, they need a lighter structure, thus reducing the difficulties (weight, volume, deployment) encountered when trying to put into space real, large size antennas.

For this reason, it is only the development of interferometric techniques which made it possible to design realistic L-Band radiometry space missions, primarily devoted to remote sensing of soil moisture (over land) and surface salinity (over the ocean) at the Earth surface.

The SMOS mission, submitted in 1998 to the European Space Agency, makes use of the 2-D interferometric MIRAS concept, developed since 1993 by ESA [1]. In MIRAS, the antenna is a Y shaped structure, each arm of which consists of an alignment of adjacent elementary antennas which collect H & V polarisation signals.

In remote sensing, an imaging radiometer maps the Earth brightness temperature (T_B) distribution over a given **field of view** (FOV). An ideal interferometric imaging radiometer generates this image indirectly by measuring the Fourier transform of the T_B distribution over the FOV. This measurement, referred to as the **complex visibility function V**, can be inverse Fourier transformed to produce an image.

Interferometric V data are obtained by cross correlating the signals collected by two spatially separated antennas having an overlapping field of view : see <u>Fig 1</u>.

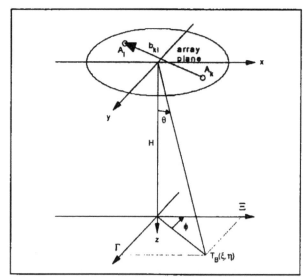

Figure 1 : geometry of interferometric measurements by an antenna array. A pair of antenna elements A_k, A_l, deployed at altitude H over Earth surface, yields the visibility function V for the spatial angular frequency

$$\vec{u}_{kl} = \vec{b}_{kl}/\lambda \; ; \quad \lambda = c/f \text{ is the}$$

central wavelength of observation.
The coordinates Ξ and Γ (across & along track) are proportional to direction cosines ξ and η :

$$\xi = \sin\theta\cos\phi \, , \, \eta = \sin\theta\sin\phi,$$

where θ and ϕ are the usual spherical co-ordinates (colatitude or elevation, and azimuth, respectively). The baseline vector b_{kl} begins at antenna element A_k and ends in A_l.

For a *n*-antenna array, the number of baselines joining antenna pairs is n (n-1)/2. Such an array gives a simultaneous access to a sampling of the V over a finite list of spatial angular frequencies, the *experimental frequency list*. This list is not necessarily non-redundant : two different **antenna** pairs with the same baseline $\vec{b}$ correspond to the same spatial angular frequency $\vec{u} = \vec{b}/\lambda$. The number $\rho(\vec{u})$ of redundant baselines for each spatial angular frequency $\vec{u}$ is the (degree of) redundancy for that frequency. As $\vec{u}_{kl} = -\vec{u}_{lk}$ (Fig 1), the experimental space frequency coverage H generated by this list is symmetric with respect to the origin.

The Y-shaped geometry and the minimal spacing d between antenna elements result in an hexagonal frequency coverage with an equal sampling step $\delta u = d/\lambda$ [2]. Then, the extension of the synthesized FOV Ω in the plane of direction cosines (<u>Fig 2</u>) is equal to

$$\Delta\xi = 2/\sqrt{3}\, \delta u \, .$$

When looking at nadir, the Earthly horizon in the plane of direction cosines is a circle with radius $\sin\theta = \sin\theta_h = R/(R+H)$, where R is the Earth radius (for H=757 Km,

$\theta_h \approx 63.4°$). For a viewing direction tilted from the nadir by an angle ε, this circle becomes an ellipse centered on $(0, \cos\theta\sin\varepsilon)$, with major and minor axes $(\sin\theta, \sin\theta\cos\varepsilon)$.

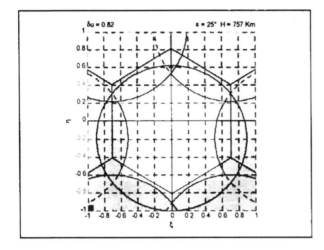

Figure 2 : Synthesized FOV and aliasing free FOV in the (ξ,η) plane. The central ellipse is the limit of the earthly horizon as seen from H=757 km altitude when the array is tilted off nadir by ε=25°. The hexagonal elementary cell of the synthesized FOV is drawn, together with its neighbours responsible for the 6 alias counterparts of the earthly horizon.

2. BRIGHTNESS TEMPERATURE RECONSTRUCTED FIELDS

2.1 The (V, T$_B$) relationship : for an ideal interferometer, visibility functions V(u,v) and scene temperatures T$_B$ (ξ,η) are related by a simple 2-D Fourier transform weighted by the antenna gain pattern $|F(\xi,\eta)|^2$, normalized in such a way that its integral over all solid angles is equal to 4π.

However, mismatches between channels affect correlation products. In a real interferometer, the relationship between the measured visibility sample and the T$_B$ distribution can be modelled by the following equation [4] :

$$V(u_{kl}, v_{kl}) = \int_{\xi^2+\eta^2 \leq 1} T_B(\xi,\eta) F_k(\xi,\eta) F_l^{\cdot}(\xi,\eta) r\left(-\frac{u_{kl}\xi+v_{kl}\eta}{f}\right) e^{-2j\pi\left(u_{kl}\xi+v_{kl}\eta\right)} \frac{d\xi d\eta}{\sqrt{1-\xi^2-\eta^2}}$$

where r is the fringe-wash function, which depends on the receiver's band-pass voltage transfer function. While offset errors, phase errors, gain errors and antenna coupling effects can be calibrated or measured on-ground or on-board, many other errors should be dealt with in the inversion algorithm. This is the case, for example, for the fringe-wash function r ; similarly, antenna patterns $F(\xi,\eta)$ include amplitude and phase ripples that may differ from one antenna to another.

In the following, examples are given of improved image reconstruction techniques, which are presently studied in view of their possible application to SMOS data.

2.2 G-matrix modelling : let G be the linear operator describing the relationship between V and T$_B$: $(GT_B)(u_{kl}, v_{kl}) = V(u_{kl}, v_{kl})$.

As **G**, restricted to the space H of H-band limited functions (i.e. functions the Fourier transform of which is contained by the experimental frequency coverage H),, is injective, it is natural to search for such a band-limited solution of the problem. The guiding idea is therefore to provide a temperature distribution T_r consistent with the experimental data $\tilde{V}$ in the G-matrix modelling frame. To this purpose, we have developed two strategies, which differ in the way to deal with redundant visibility samples.

- In the **complete** G-matrix approach, the reconstructed T_r distribution is defined as the function minimizing the least-square criterion:

$$c_1(T) = \left\| \frac{1}{\sqrt{\rho}} (\tilde{V} - GT) \right\|^2, \text{ solution to equation : } PG^* \frac{1}{\rho} GT = PG^* \frac{1}{\rho} \tilde{V}$$

Here, **P** is the low-pass projection onto the H space, $\tilde{V}$ the set of experimental visibilities, ρ the (degree of) redundancy. The quadratic optimization is carried out using the conjugate-gradients (CG) method.

- In the **averaged** G-matrix approach, redundant visibilities are averaged *via* the action of a weighted average operator **M**. One then defines the reconstructed brightness temperature distribution T_r as the solution of the equation:

$$U^{-1}MGT = U^{-1}M\tilde{V} \text{ ; where U is the Fourier transform operator.}$$

The operator $U^{-1}MG$ is not self-adjoint. In such a case, the CG cannot be used : the appropriate technique is to minimize the norm of the residue $U^{-1}(M\tilde{V} - MGT)$ in the Krylov space generated by the successive powers of $U^{-1}MG$ applied to $U^{-1}M\tilde{V}$. This is done with the aid of the generalized minimal residual (GMRes) method. The solution of this equation minimizes the least-square criterion :

$$c_2(T) = \left\| M\tilde{V} - MGT \right\|^2 \text{ ; this criterion is to be compared to } c_1.$$

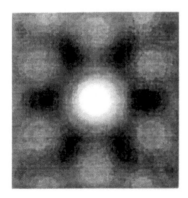 

Figure 3 : central part of uncorrected *(left)* and apodized *(right)* point spread functions.

2.3. Apodized point-spread function : in order to filter-out the effects of the instrumental point-spread function (small efficiency, high oscillations due to the frequency cut-off), a smoothed T_s field must be obtained from the original T_B values: $T_s = s * T_B$. Here, s is a centrosymmetric real-valued low-pass filter such as : (1) the fraction χ of the filter energy $\hat{s}$ present in the experimental frequency coverage H is to be maximized (i.e. $\hat{s}$ has to be small outside H in the mean-square sense) ; (2) the support D of the apodized point spread function s should be as small as possible with respect to the experimental aperture H and to the choice of χ. The aim is of course to achieve the best possible resolution.

The s function is computed on the grounds of a trade-off between resolution and efficiency, with the aid of the power method [2]. This approach leads to better results than traditional windowing functions such as Hanning, Hamming or Blackman filters ; moreover, it refers to basic physical concepts : resolution and efficiency. The obtained 3dB angular resolution θ_3^s is related to the arm length L :

$$\xi_3^* = \sin \theta_3^* = C_\theta \frac{\lambda}{L} \qquad ; \text{where } C_\theta \approx 0.8, \text{ versus } C_\theta \approx 0.6 \text{ without apodization.}$$

The cost in terms of angular resolution is to be compared to the gain on both the first side lobe levels (from −6.5dB down to −18.5dB ; see Fig 3), and the efficiency in the main lobe (from 32 % up to 90 % ; see Fig 3).

3. 2-D FIELD OF VIEW AT GROUND LEVEL

After the alias free available FOV has been characterized in the plane of the direction cosines, it can be projected on the Earth surface. Fig 4 shows the FOV, assuming a spherical Earth.

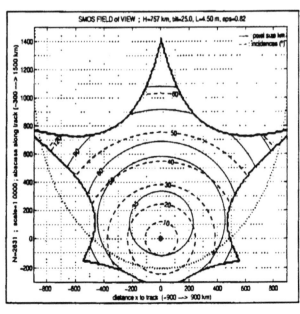

SMOS FIELD of VIEW ; H=757 km, tilt=25.0, L=4.50 m, eps=0.82

Figure 4 : Field of view on Earth surface for typical SMOS parameters (flight altitude, tilting angle of the antenna plane, arm length, spacing between antenna elements). The orbit plane is projected on the vertical axis (x - 0) ; the subsatellite point S is indicated by a square. Two families of contours are shown : incidence angles in degrees (dashed lines) and ground resolution in km (continuous lines). Since the incidence angle is directly linked to the viewing (elevation) angle, corresponding contours are circles centered on S. As for ground resolution, it varies with both elevation angle with respect to the (tilted) antenna axis and incidence angle : the contours are centered half way beween S and the intersection of the antenna axis with Earth.

The FOV appears to roughly consist of two regions. In the **central** part, observations correspond to a large range $\Delta\iota$ of incidence angles. In the FOV **edges**, $\Delta\iota$ becomes progressively narrower. When considering the total swath, the 2-D interferometer might be compared to a conical scanning instrument, able to obtain data over a 1500 km width across track, for ι values in the 50-55° bracket. The overall coverage potential is of course limited by the maximum allowed **ground resolution** : over land, it is often considered that data obtained with a resolution worse than 60 km are of reduced use [5].

Within the central FOV region, of particular interest is the zone ($\approx$ 700 km broad) where the range of incidence angles ι exceeds about 15° ; as the satellite moves, the same locations on the Earth may be observed there for various elevation (thus incidence) angles. This considerably upgrades the instrument potential, because the $T_B(\iota)$ dependance can be used to **discriminate** the influence of physical parameters [6].

4. INCREASING ACCURACY OVER HOMOGENEOUS SCENES

Over the ocean, space resolution is not a critical requirement ; measurements, in order to reach a satisfactory accuracy on sea surface salinity, will anyway have to be spatially averaged. Conversely, the number and accuracy of available independant data become major issues. For "thinned" interferometric antennas, the radiometric sensitivity ΔT includes a factor $F_A = A_{real}/A_{thin}$, where A_{thin} is the actual collecting area, and A_{real} is the area of a real antenna having the same overall size [3]. For homogeneous scenes, this may be compen-sated by the availability, within the instantaneous 2-D FOV, of many independant views.

Let us consider the **degradation factor F_N** of a 2-D interferometer with respect to a real antenna system. The ratio F_N of the number N_i of independant views present in the instantaneous FOV to F_A^2 is also the factor by which integration time should be multiplied in order to achieve a sensitivity equivalent to that of a real antenna radiometer.

For a given arm length L, when the spacing d between elementary radiating elements decreases, the FOV becomes larger (because aliased Earth images move away), and thus N_i increases. At the same time, assuming the diameter of the elements to be equal to their spacing, F_A increases (since the antenna becomes "thinner"). Since N_i, however, is ultimately limited by the radiating pattern of elementary elements as well as the Earth visibility ellipse, this results in F_N having a minimum : see Table 1.

Variation of the degradation factor F_N						
$\delta u = d/\lambda$	0.65	0.70	0.75	0.80	0.85	0.90
N_i	618	594	525	434	346	266
F_A^2	1817	1565	1361	1171	1082	996
F_N	2.94	2.63	**2.59**	2.70	3.13	3.74

Table 1 : F_N against δu ; here L = 4.5 m, ε = 25°). The views are considered as independant when separated by an angular distance 0.8 λ /L.

Technical considerations favor high δu values. Interactions between adjacent elementary antennas induce pattern distortions which become more difficult to entangle as these elements become closer ; moreover, increasing δu reduces the size (which varies as δu^{-2}) of the correlator bank on board. Table 1 shows a broad optimum around δu # 0.75.

While the sensitivity of an interferometric antenna is deteriorated, the availability of many independant data in the FOV makes up, for homogeneous targets, for most of this deterioration.

ACKNOWLEDGEMENTS : The development of the MIRAS concept was supported by ESA. This study was supported by CNES and CNRS.

REFERENCES

1. Goutoule J.M., Kraft U., Martin Neira M : Miras : preliminary design of a two-dimensional L-band aperture synthesis radiometer ; Specialist Meeting on Microwave Radiometry, Rome 1994. VSP, Utrecht, 1995

2. Lannes A. and E. Anterrieu. Image reconstruction methods for remote sensing by aperture synthesis, Proc IGARSS'94 (Pasadena, California), 2892-2903, 1994.

3. Le Vine D.M. : The sensitivity of synthetic aperture radiometers for remote sensing applications from space, Radio Sci., 25, 4, pp 441-453, 1990

4. Ruf C.S., C.T. Swift, A.B. Tanner, and D.M. Le Vine. Interferometric synthetic aperture microwave radiometry for the remote sensing of the Earth, IEEE Trans. Geosci. Remote Sens., 26(5), 597-611, 1988.

5. Kerr Y H. : the SMOS proposal to ESA, un published report, 1998.

6. Wigneron J.P., Chanzy A., Waldteufel P., Calvet J.C., Kerr Y. H. : retrieval capabilities of the microwave interferometer MIRAS over the land surface (SMOS Mission) ; Specialist Meeting on Microwave Radiometry, Firenze 1999, these proceedings.

Microw. Radiomet. Remote Sens. Earth's Surf. Atmosphere, pp. 485–492
P. Pampaloni and S. Paloscia (Eds)

Retrieval capabilities of L-Band 2-D interferometric radiometry over land surfaces (SMOS Mission)

J.-P. WIGNERON,[1] P. WALDTEUFEL,[2] A. CHANZY,[3] J.-C. CALVET,[4] O. MARLOIE, [1] J.-F. HANOCQ,[1] Y. KERR[5]

[1] *INRA Bioclimatologie , Agroparc 84914 Avignon Cedex 9, France*
[2] *IPSL , 10-12 av de l'Europe, 78140 Vélizy, France*
[3] *INRA Science du sol, Agroparc 84914 Avignon Cedex 9, France*
[4] *Météo France / CNRM, 42 av G. Coriolis, 31057 Toulouse, France*
[5] *CESBIO, CNES/CNRS/UPS, BPI 2801, 18 av E. Belin, 31401 Toulouse, France*

Abstract - This paper investigates the potential of the microwave interferometric radiometer MIRAS for monitoring surface variables over the land surface. MIRAS is the payload suggested for the space mission SMOS, recently submitted to ESA in answer to the call for Earth Explorer Opportunity Missions. The L-band MIRAS radiometer is based on an innovative bi-dimensional aperture synthesis concept. This sensor has new and significant capabilities, especially in terms of multi-angular viewing configurations. The main aspects of the retrieval capabilities of SMOS for monitoring soil moisture, vegetation biomass and surface temperature are presented in this paper. The analysis is based on model inversion. The standard error of estimate of the surface variables is computed for various configurations, as a function of both the uncertainties associated with the spatial measurements and those associated with the ancillary information used in the retrievals. The potential of SMOS, *i.e.* the possibility to retrieve one, two or three surface variables depending on the system configuration, is investigated. These questions are key issues to optimize the SMOS mission scenario, in order to meet both the scientific requirements and the technical constraints of the mission.

1. INTRODUCTION

This paper aims at analyzing the potential capabilities of the microwave interferometric radiometer MIRAS (Microwave Imaging Radiometer using Aperture Synthesis) for monitoring the surface variables over the land surface. MIRAS is the payload suggested for the space mission SMOS (Soil Moisture and Ocean Salinity), recently submitted to the European Space Agency in answer to the call for Earth Explorer Opportunity Missions. The MIRAS L-band interferometer is based on an innovative, bi-dimensional (2-D) aperture synthesis concept. The sensor has new and significant capabilities, especially in terms of multi-angular viewing configuration [1]. Also, a second frequency at C-band could possibly be added in the future to the L-band system.

The multi-angle viewing capability of a 2-D interferometric spaceborne radiometer stems from the fact that the instantaneous field of view (FOV) is bi-dimensional. As the satellite moves along its track, a series of brightness temperatures T_B emitted by the same location on Earth is thus obtained, for a range of distinct incidence angles. This multi-angle viewing capability is very interesting since it is possible to retrieve simultaneously both soil moisture w_S and vegetation optical depth τ from dual-polarization, multi-angle L-band data (such as those acquired by SMOS), as has been demonstrated by [2], over both green and senescent vegetation covers. Retrieving directly both w_S and τ is a major potential asset :

- There is no need for ancillary data to estimate vegetation attenuation τ. Usually, estimates of the optical depth τ are derived from the vegetation water content W_c (kg/m²) and from a multiplying parameter b which can be related to the vegetation characteristics, assuming $\tau \approx b\, W_c$. Computing the two parameters b and W_c at large spatial scales from ancillary remotely sensed data is not easy; therefore the multi-angular approach improves considerably the retrieval process.
- The retrieved parameter τ may be a very useful product by itself: actually this variable is a meaningful index for monitoring vegetation dynamics (development and senescence) at a global scale [3] as well as for estimating forest characteristics.

In this context, the multi-angular capabilities of SMOS appear promising. Therefore, in the perspective of the space mission, it is necessary to assess the retrieval capabilities of the radiometer, depending on the view angle system configuration. The main variables of interest to be estimated from the SMOS observations are: soil moisture w_S (m³/m³), optical depth of the vegetation layer τ (Nepers), and surface temperature T_S (K). The present study investigates the performance of the observing system when attempting to retrieve one, two or the three surface variables, depending on the system configuration.

2. MATERIAL AND METHOD

2.1 Modeling

The brightness temperature in the microwave domain T_{Bp} (p = v or h, for the vertical or horizontal polarization, respectively) is computed using a simple, first order radiative transfer model: the τ-ω model [4]. According to this model and neglecting atmospheric effects, the brightness temperature T_B of a two-layer (soil & vegetation) medium is given by:

$$T_B = [(1-\omega)(1-\gamma)\,(1+\gamma\,\Gamma_s) + (1-\Gamma_s)\,\gamma\,]\,T_s \qquad (1)$$

where T_s is the effective surface temperature; Γ_s denotes the soil reflectivity, depending mainly on soil moisture w_s; the single scattering albedo ω parameterizes scattering within the canopy layer ($\omega \approx 0$ at L-band) and γ is the canopy attenuation factor. This last quantity is a function of the optical thickness τ of the canopy and is given by:

$$\gamma = \exp(-\tau/\cos\theta) \qquad (2)$$

The microwave response is strongly dominated by soil moisture w_S (m³/m³), vegetation attenuation (τ) and effective surface temperature T_S. Other surface characteristics are needed in the retrieval process: soil surface roughness, topography, soil texture, land cover and vegetation type, etc. However, most of these parameters can be considered as stable with time, and thus be estimated or calibrated once, prior to observations, from ancillary information (soil maps, high spatial resolution data in the optical domain, digital elevation models). Also, the sensitivity of T_B to changes in soil surface roughness over agricultural areas have to be investigated at large spatial scale and it is considered in this study that these effects can be corrected from ancillary data.

The model input parameters are given in Table 1. Standard values are used for vegetation and soil parameters (the soil characteristics are those of a silty clay loam at the INRA Avignon test site). Slightly different retrieval performances would be obtained for other specific soil/vegetation conditions, with no consequences on the overall conclusions. To better evaluate the vegetation conditions corresponding to a given value of τ, the vegetation water content can be approximated from $W_C = \tau/b$, with $b \approx 0.13$ at 1.4 GHz [2], [4],[5].

Table 1.

Model input parameters

VEGETATION	VALUE
- single scattering albedo ω:	ω(1.4 GHz) = 0.0
- surface temperature T_S (K)	T_S = 290 K

SOIL (silty clay loam) :

soil texture: clay C (%) , sand S (%)	C=27%, S=11%
dry bulk density ρ_d (g/cm³)	ρ_d = 1.35
effective roughness parameters [6]:	h_{SOL} = 0.1, Q_{SOL} = 0.1

2.2 Assessment Procedure

Three steps can be distinguished in the evaluation of the standard error on the retrieved values:

(1) First, a reference vector **Tb**, which represents the actual land surface emission, is computed. For a given surface configuration (in terms of soil moisture, vegetation biomass, surface temperature), the brightness temperature is simulated from the τ-ω model for both H & V polarizations, and for a given set of N incidence angles (θ_1, θ_2, ... θ_N). The **Tb** vector is constructed from the 2N simulated variables:

$$\mathbf{Tb} = (Tb_H(\theta_1), Tb_H(\theta_2), ... Tb_H(\theta_N), Tb_V(\theta_1), Tb_V(\theta_2), ... Tb_V(\theta_N)) \qquad (3)$$

(2) Then, an "observation" vector **Tb*** is simulated, accounting for the uncertainties associated with the spaceborne measurements, in terms of random errors (ΔT) and systematic errors (bias B_S). In this study, both ΔT and B_S are set equal to 1.5 K; these values are consistent with the overall SMOS technical characteristics. Accordingly, random gaussian terms G_i (i =1,2N) (mean=B_S, standard deviation =ΔT) are added to each Tb component :

$$\mathbf{Tb^*} = (Tb_H(\theta_1)+G_1, ... Tb_H(\theta_N)+G_N, Tb_V(\theta_1)+G_{N+1},.... ,Tb_V(\theta_N)+G_{2N}) \qquad (4)$$

(3) Finally, the retrieved surface variables are computed by minimizing the root mean square error between the simulated vector **Tbm** and the "observed" vector **Tb***. The minimization routine is based on the Marquart algorithm. This routine was modified to account for uncertainties associated with input parameters estimated from ancillary information. In this study, such uncertainties may affect the effective surface temperature T_S (the assumed standard deviation is taken to be σ_{TS}=2 K in this study) and the optical depth τ (assumed standard deviation: σ_τ), whenever any of these quantities is taken to be known from external sources. Slightly modified values of the surface variables used to compute **Tb** (a multiplying factor of 0.9 is applied) are employed to initialize the inversion procedure. Using these initial values, the inversion always converges and no effects due to local minima were identified during the retrieval process. For each surface variable a global error E, which accounts for both systematic and random components is given by :

$$E^2 = E_B^2 + STD^2 \qquad (5)$$

where E_B is the error due to instrumental bias B_S, and STD is the standard deviation of estimates of the retrieved variables due to the random noise component.

2.3 Assessment issues concerning the SMOS mission

We focus on the specific SMOS case, using the SMOS baseline system configuration and assuming a circular orbit at 757 km altitude. The antenna plane is tilted by 25° with respect to the horizontal plane, in such a way that the available incidence angles θ range from 0 to about 55°-60°. The geometry of the SMOS observations is illustrated in [1]. As the satellite moves along its track, a series of brightness temperatures T_B emitted at a given location on Earth is acquired by the radiometer at varying incidence angles. Therefore, the microwave signature of each location on earth, is acquired for a range of distinct incidence angles. For each location (or "pixel"), this range is variable within the Field of View: it depends on the distance between the pixel and satellite ground track. This distance is parameterized here by the equivalent half-swath elevation angle (η_m). A large range of θ values is available in the central part of the FOV (for $\eta_m < 20\text{-}25°$), then the range becomes narrower and is about $40\text{-}55°$ for $\eta_m \approx 33°$ (Figure 1). If observations over the whole FOV or in the FOV central region are used, the associated revisit time is about 2-3 or 5-8 days, respectively.

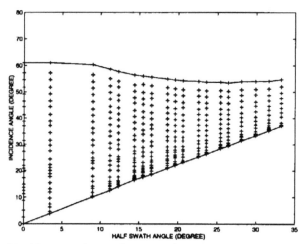

Fig. 1: Range of incidence angle as a function of the equivalent half-swath angle η_m

The evaluation of the retrieval capabilities of SMOS is carried out for four main surface conditions (SC=1, 2, 3 or 4) combining wet and dry soils with low and well-developed vegetation covers. The soil & vegetation conditions for these four cases are given in Table 2.

The well-developed vegetation cover conditions (for SC=1 & 3: τ=0.4 at L-band, i.e. $W_C \approx 3$ kg/m^2) correspond to conditions of minimal sensitivity to soil moisture w_S, due to the high attenuation effects of the vegetation layer. The value $W_C \approx 3$ kg/m^2 is that of a well-developed green crop, such as soybean or wheat.

Table 2.
Characteristics of soil and vegetation for the different surface conditions SC

Surface Confi-guration (SC)	Soil conditions w_S (m^3/m^3)	Vegetation conditions: WD= Well-Developed, LB= Low Biomass canopy	
		optical depth τ (1.4 GHz)	estimated W_C (kg/m^2)*
1	Wet Soil, w_S =0.33	WD : τ=0.4	3.0
2	Wet Soil, w_S =0.33	LB : τ=0.065	0.5
3	Dry Soil, w_S =0.04	WD : τ=0.4	3.0
4	Dry Soil, w_S =0.04	LB : τ=0.065	0.5

*W_C is estimated from τ, by dividing by the parameter b: $W_C = \tau / b$, with b=0.13

3. RESULTS AND DISCUSSION

3.1 Three-parameter retrievals

In the following, the retrieval performance of SMOS are evaluated by accounting for the actual viewing configuration system of SMOS (the range of available view angles is illustrated in Figure 1). The possibility to retrieve simultaneously the three main surface variables of interest (soil moisture w_S, vegetation optical depth τ, effective surface temperature T_S; this will be referred to as three-parameter retrievals) is investigated for the four surface conditions SC, given in Table 2. The quadratic errors E_{WS}, E_τ and E_{TS} in the retrieval of w_S, τ and T_S, respectively, are plotted in Figure 2a-c, as a function of the half-swath angle η_m. For the four SC cases, all three errors increase as η_m increases up to $33°$, i.e. as the range of available incidence angles θ becomes accordingly narrower. It can be seen that there is a rather steep increase in the errors for η_m larger than $25°$. In the following, we will therefore distinguish retrieval results for the FOV central region ($\eta_m \leq 25°$) and the FOV edges ($\eta_m > 25°$).

Concerning the FOV central region, the retrieval error for w_S is always better than $0.04\ m^3/m^3$, for all SC conditions. This is to be noted, since the $0.04\ m^3/m^3$ threshold was given as a required accuracy for studies over the land surface in the SMOS mission [7]. The retrieval error E_τ for optical depth is smaller than 0.06 for the four SC cases. The value $E_\tau = 0.06$ corresponds approximately to an amount of water $\delta W_C \approx 0.45\ kg/m^2$. It appears that best retrievals of τ can be obtained over wet soils (the error E_τ is only about 0.02 for SC=1, 2): for these conditions there is a good contrast between the low emissivity of wet soils and the high emissivity of vegetation [2]. Excluding the configuration SC=3 (over dry soils and high biomass conditions), E_τ is lower than 0.03 for $\eta_m \leq 25°$ (i.e. about $0.25\ kg/m^2$ for W_C).

The retrieval error for surface temperature E_{TS} is about 3 K for SC=1, 3, 4. However, for SC=2 (wet soil and low biomass conditions), E_{TS} is much higher (~5-6K) and then no reliable estimates of T_S can be obtained from three-parameter retrievals. It seems that the difficulty to retrieve correctly T_S for SC=2 may explain the relative higher error in the estimation of soil moisture for this case (see Figure 2a).

At the edges of the FOV ($\eta_m > 25°$) the retrieval errors are large for soil moisture (mainly for wet soils: SC=1, 2), for optical depth (mainly for dry soils: SC=3, 4) and for surface temperature. The main disadvantage of using only observations acquired in the FOV central region is that the monitoring time resolution will worsen; namely the associated revisit time will be longer (about 5-8 days instead of 2-3 days). This can be a drawback for monitoring soil moisture and surface temperature T_S. However, this is probably not a strong limitation for monitoring the vegetation parameter τ, as it is expected that τ varies slowly in time and can be correctly estimated using a revisit time of about 5-8 days.

In conclusion, good estimates of w_S and τ can be obtained from three-parameter retrievals in the central part of the FOV (associated errors are less than $0.04\ m^3/m^3$ and 0.06 for w_S and τ, respectively). The retrieval error for T_S is less than 3K for most of the surface conditions (SC=1, 3, 4), but it is about 5-6K for SC=2. The interest of such a three-parameter approach is that no ancillary information about surface temperature or vegetation characteristics, obtained from other remote sensing data, are required to retrieve soil moisture. In the following section, we investigate whether it is possible to retrieve soil moisture w_S with a good accuracy at the edges of the FOV, in order to improve the revisit time for this surface variable.

Fig. 2: Quadratic errors E_{WS} (a), E_τ (b) and E_{TS} (c) in the three-parameter retrieval of soil moisture w_S, τ and T_S, respectively, as a function of the equivalent half-swath angle η_m.

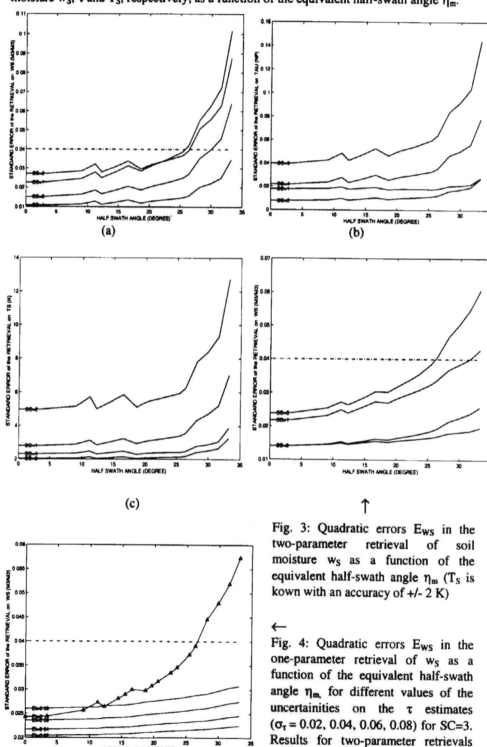

(a)

(b)

(c)

↑

Fig. 3: Quadratic errors E_{WS} in the two-parameter retrieval of soil moisture w_S as a function of the equivalent half-swath angle η_m (T_S is kown with an accuracy of +/- 2 K)

←

Fig. 4: Quadratic errors E_{WS} in the one-parameter retrieval of w_S as a function of the equivalent half-swath angle η_m for different values of the uncertainities on the τ estimates ($\sigma_\tau = 0.02, 0.04, 0.06, 0.08$) for SC=3. Results for two-parameter retrievals are also given (Δ).

3.2 Two-parameter (soil moisture w_S and vegetation optical depth τ) simultaneous retrievals

The lower accuracy in the retrieval of w_S at the edges of the FOV, especially over wet soils (for SC=1, 2), seems to be associated to the lower accuracy in the retrieval of T_S (see Figures 2a and 2c). In this section, it is assumed that T_S can be estimated from ancillary information (from remote sensing data in the thermal infrared domain, and/or meteorological outputs) with an accuracy of +/- 2 K. Simultaneous retrievals of w_S and τ are then carried out. Whereas the results for τ are close to those obtained with the three-parameter retrievals (results not shown here), results for w_S are significantly different. The quadratic errors E_{WS} in the retrieval of w_S are plotted in Figure 3 as a function of η_m. At the edges of the FOV ($\eta_m > 25°$), the error is less than 0.04 m^3/m^3 for SC=1, 2, 4. For SC=3, the error is less than 0.06 m^3/m^3. Therefore, provided estimates of T_S can be obtained with an accuracy of +/- 2 K, it is preferable to carry out two-parameter retrievals at the edge of the FOV. The higher error in the retrieval of w_S for SC=3 seems to correlated to the very low accuracy obtained in the retrieval of τ for two- and three-parameter retrievals (Figure 2b). The possibility to retrieve w_S for SC=3 is investigated from one-parameter retrievals in the following section.

3.3 One-parameter retrievals of soil moisture w_S using estimates of τ

As mentioned in section 3.1, unlike soil moisture, the vegetation optical depth which depends on the vegetation characteristics (mainly water content W_C, vegetation structure and moisture content) varies slowly in time. Therefore, accurate estimates of τ can be obtained using a lower observation frequency, *i.e.* considering the FOV central region only (a maximum error of 0.06 was obtained for τ in this area from three-parameter retrievals). These estimates of τ could then be used to retrieve soil moisture at the edges of the FOV for SC=3.

To test this, one-parameter retrievals of soil moisture w_S are carried out using estimates of τ obtained from previous three-parameter retrievals in the central part of the FOV and ancillary estimates for T_S (the associated uncertainty σ_{TS} is 2K for T_S). The quadratic errors E_{WS} are computed for different values of the uncertainties σ_τ on the estimates of τ which are used in the retrievals (σ_τ = 0.02, 0.04, 0.06, 0.08) for SC=3 (Figure 4). For an easier interpretation of the figure, the results of two-parameter retrievals are also given in the figure.

For SC=3 (for dry soils and low biomass conditions), it appears that w_S can be retrieved with a very good accuracy even if rough estimates of τ are available. Therefore, for SC=3, retrievals of w_S can be made using values of τ obtained with an error $\le$ 0.06 from previous three-parameter retrievals in the central part of the FOV.

4. CONCLUSION

The L-band MIRAS interferometer (SMOS mission) is based on an innovative bi-dimensional aperture synthesis concept. The sensor has new and very significant capabilities, especially in terms of multi-angular viewing configuration. In this study, the retrieval capabilities of SMOS, based on the actual system configuration have been assessed. From model simulations, the uncertainties on the retrieved variables is computed as a function of the uncertainties associated with both remotely-sensed data and model input parameters. This study shows how and why the 2-D interferometry concept, applied to Earth remote sensing from space, may bring a major qualitative improvement, thanks to the ability to obtain data over a significant range of viewing angles. SMOS is the first proposal which aims at demonstrating and using this feature. From the results of this study, promising retrieval capabilities can be expected from SMOS over the land surface.

REFERENCES

1. P. Waldteufel, E. Anterrieu, J. M. Goutoule and Y. Kerr, "Field of view characteristics of a microwave 2-D interferometric antenna, as illustrated by the MIRAS concept," this issue, (1999)
2. J.-P. Wigneron, A. Chanzy, J.-C. Calvet and N. Bruguier, "A simple algorithm to retrieve soil moisture and vegetation biomass using passive microwave measurements over crop fields," *Remote Sens. Environ.*, **51**, 331-341 (1995)
3. A. A. van de Griend and M. Owe, "Determination of microwave vegetation optical depth and single scattering albedo from large scale soil moisture and Nimbus/SMMR satellite observations," *Int. J. Remote Sensing*, **14**, 1875-1886 (1993)
4. F. T. Ulaby, R. K. Moore and A. K. Fung, *Microwave Remote Sensing - Active and Passive*, vol. 3., Artech House, Norwood. (1986)
5. T. J. Jackson and T. J. Schmugge, "Vegetation effects on the microwave emission of soils," *Remote Sens. Environ.*, **36**, 203-212 (1991)
6. J. R. Wang and B. Choudhury, "Remote sensing of soil moisture content over bare field at 1.4GHz frequency," *J. Geophys. Res.*, **86**, 5277-5282 (1981)
7. Y. Kerr *et al.*, *MIRAS on RAMSES: radiometry applied to soil moisture and salinity measurements*, Full proposal, A.O. Earth Explorer Opportunity Missions, ESA. (1998)

Microw. Radiomet. Remote Sens. Earth's Surf. Atmosphere, pp. 493–502
P. Pampaloni and S. Paloscia (Eds)
© VSP 2000

The role of microwave radiometry in the CLOUDS project

BIZZARRO BIZZARRI [1] and PAOLO SPERA [2] (*)

[1] *CLOUDS Project Coordinator, for Alenia Aerospazio, Rome, Italy*
[2] *Alenia Aerospazio, Rome, Italy*

(*) paper presented on behalf of the CLOUDS Study Group: Michel Desbois (LMD, France), John
Harries (ICSTM, UK), Klaus Künzi (UNIBremen, Germany), Alberto Mugnai (IFA, Italy),
Anthony Slingo (UKMO, UK), Alfonso Sutera (UNIRome, Italy), Stefano Tibaldi (ARPA/RMS,
Italy), Roberto Bordi (ALS, Italy), Peter Coppo (O.G., Italy), Norman Grant (MMS, UK), Laurent
Vial (SOFRADIR, France) and Peter Zimmermann (RPG, Germany)

Abstract - The objective of CLOUDS is to study the mission of a new satellite to provide accurate, comprehensive, consistent and frequent information on cloud structures and the associated radiative parameters, to be used by operational and research centres for the purpose of improved climate and weather forecasting. The CLOUDS concept aims at a small-medium size satellite to meet sustainability requirements as necessary to provide long-term service continuity (e.g. 15 years by a series of 3 satellites). The most unique mission feature is the observation of cloud interior (ice and liquid water) in addition to surface radiative properties. To measure cloud and radiation parameters, including the three water phases and aerosol, CLOUDS will exploit all the e.m. spectrum with several tens of channels, ranging from UV (some 330 nm) to low-medium frequency MW (some 6 GHz), passing through VIS, NIR, TIR, FIR and Sub-mm waves. Due to the requirement of frequent coverage and long-term mission duration, associated to relatively coarse resolution (5 km at most), only passive radiometers are envisaged, exploiting, however, more polarisations and dual viewing geometry (fore- and after-). The role of MW is mostly intended for cloud interior observation (ice, liquid water, precipitation and water vapour). The following channels are at present considered: (i) for precipitation and liquid water: 6.9 GHz, 10.6 GHz, 18.7 GHz, 36.5 GHz, 89.0 GHz and 150 GHz; (ii) for cloud ice: 220 GHz, 463 GHz, 683 GHz and 874 GHz; (iii) for water vapour: 23.8 GHz and three channels in the 183.3 GHz band; (iv) for inference of the vertical distribution of liquid/precipitating water: up to 4 channels in the 55 GHz band and up to 4 channels in the 119 GHz band.

1. INTRODUCTION

"CLOUDS (a Cloud and Radiation monitoring satellite)" is a new project sponsored by the EC under Framework Programme IV, Environment and Climate, Theme 3 (*Space techniques applied to environmental monitoring and research*). It is running since 1st June 1998, with 12 Partners including 7 scientific institutes and 5 industrial companies. Alenia Aerospazio provides, i.a., the project coordination. The EC support will last two years, up to completion of the Phase A (consolidation of requirements, industrial feasibility and preliminary design).

The objective of *CLOUDS* is to study the mission of a new satellite to provide accurate, comprehensive, consistent and frequent information on cloud structures and the associated radiative parameters, to be used by operational and research centres for the purpose of improved climate and weather forecasting.

Whereas a number of missions for *process study* in the cloud and radiation field are being run (*TRMM*) and continue to be planned (e.g., the ESA *Earth Radiation Mission* and the NASA *PICASSO-CENA* and *CloudSat*), the *CLOUDS* project focuses towards a <u>monitoring mission</u>. The *CLOUDS* concept aims at a *small-medium size satellite* to meet *sustainability requirements* as necessary to provide long-term service continuity (e.g. 15 years by a series of 3 satellites). The satellite would provide frequent global sampling (each 1 or 2 days, depending on whether the observation is based on emitted Earth radiation or reflected solar radiation) of a wide number of geophysical parameters on clouds and radiation, and also on land and sea surface as allowed by the instrumentation designed for the primary objective. The most unique mission feature is the observation of cloud <u>interior</u> (ice and liquid water) in addition to <u>surface</u> radiative properties. To measure cloud and radiation parameters, including the three water phases and aerosol, *CLOUDS* will exploit all the e.m. spectrum with several tens of channels, ranging from UV (some 330 nm) to low-medium frequency MW (some 6 GHz), passing through VIS, NIR, TIR, FIR and Sub-mm waves. Due to the requirement of frequent coverage and long-term mission duration, associated to relatively coarse resolution (5 km at most), <u>only passive radiometers are envisaged</u>, exploiting, however, more polarisations and dual viewing geometry (fore- and after-).

In this paper, background information on mission objectives, user requirements and mission requirements will be provided, as well as outline instrument and system requirements. The payload will include instruments operating in narrow-bands of the optical range (UV, VIS, NIR, TIR and FIR), in broad optical bands (short-wave reflected solar radiation and long-wave terrestrial emitted radiation) and in the MW/Sub-mm range. Focus will be placed on MW/Sub-mm channels, whose role is mostly intended for cloud interior observation (ice, liquid water, precipitation and water vapour).

2.   MISSION OBJECTIVES AND USER REQUIREMENTS

The short statement of CLOUDS mission objectives enables to identify the *observation fields* to be addressed, as follows:

* *clouds* and their *interior* (liquid, ice)
* the associated *radiative parameters*, including *aerosol*
* the primary sink of cloud water: *precipitation*
* the clouds precursor: *water vapour.*

These generic observation fields have to be specified in terms of *geophysical parameters* to be observed. Reference was made to the list agreed in CEOS (Committee for Earth Observation Satellites) with strong contribution from WMO, GCOS and WCRP. In *Table 1* those parameters of the CEOS list relevant to the cloud and radiation mission are reported.

Table 1 - Geophysical parameters to be observed by CLOUDS

Cloud imagery (pattern, fronts, cyclones, ash plumes)	Cloud ice - total column and gross profile
Cloud type	Cloud ice content (at cloud top)
Cloud cover	Cloud drop size (at cloud top)
Cloud top height	Cloud optical thickness
Cloud top temperature	Short-wave cloud reflectance
Precipitation rate at the ground	Long-wave cloud emissivity
Precipitation index (daily cumulative)	Short-wave outgoing radiation at Top of Atmosphere
Water vapour - total column and gross profile	Long-wave outgoing radiation at Top of Atmosphere
Cloud water (< 100 µm) - total column and gross profile	Aerosol - total column and gross profile
Cloud water (> 100 µm) - total column and gross profile	Ozone - total column (for tropopause discontinuities)

User requirements for data quality of these parameters have to be specified. According to agreement in CEOS and WMO, they are expressed in terms of: *horizontal resolution* (Δx), *vertical resolution* (Δz), *accuracy* (r.m.s. and possible limitation of bias), *observing cycle* (Δt) *and delay of availability* (δ). These quality figures can vary in a widest range, depending on the *application field*. The CLOUDS mission objectives address two equal-weight application fields: *Climate monitoring and research* and *Weather prediction and research meteorology*. In *Table 2* these applications are characterised in terms of objectives and addressed issues, typical requirements and involved institutions. In *Table 3* the reference user requirements for the CLOUDS mission are reported, inferred by forcing appropriate WMO/CEOS tables to the generic requirements shown in Table 2.

Table 2 - User application characterisation in the CLOUDS project

Application	Objectives and addressed issues	Representative requirements	Involved institutions
Climate monitoring & research	"Averaged" cloud and radiation properties for "static" purposes, to characterise the various components, to build accurate statistics, to infer balances and to monitor long-term climate changes.	$\Delta x \cong 100$ km (1) $\Delta z \cong 3$ km (2) bias $\ll$ rms (3) $\Delta t \cong 3$ h (4) $\delta \cong 3$ d	Climate research institutes R & D branches of meteo services Atmospheric research institutes Global environmental agencies
Weather prediction & research meteorology	"Instantaneous" cloud and radiation properties for "dynamic" purposes, to observe water cycle and energy transformation processes, to improve parameterization and to initialise numerical models of climate and weather.	$\Delta x \cong 30$ km (5) $\Delta z \cong 3$ km (2) rms = r.m.s.d. $\Delta t \cong 3$ h $\delta \cong 3$ h	Meteorological services Climate prediction centres Meteorological research institutes Regional environmental agencies

(1) Intended as logaritmic mean between a target of 30 km and a threshold of interest of 300 km.
(2) Reflecting the CLOUDS objective of inferring information on a gross vertical structure. i.e. 3-4 layers in the troposphere.
(3) If bias is < 20 % of r.m.s. at Δx = 30 km, the level of bias is achieved after integration over 100 km and 4.5 days.
(4) Sampled at 3 hourly intervals to capture diurnal variations: used after integration over, e.g., 1 day, 1 week, 1 month, 1 year.
(5) Intended as logaritmic mean between a target of 10 km and a threshold of interest of 100 km.

In Table 3, parameters are distinct into "basic" for the mission objectives and "support" needed to process the basic parameters. A distinction in the basic parameters is made to identify those parameters essentially measured from CLOUDS, and those also measured from METOP. Furthermore, there is a list of parameters which are relevant to the CLOUDS mission, but are assumed to be available from the overall Earth observation satellite system.

It is interesting to note that the instrumentation envisaged for CLOUDS also will provide information on parameters not part of the CLOUDS objectives. User requirements for these parameters are reported in *Table 4.*

Table 3 - Reference user requirements for the CLOUDS mission

Geophysical parameter	Horizontal resolution (1) weather	climate	Vertical resolution (2)	Accuracy r.m.s. (3)	bias (4)	Observing cycle (5)	Delay of availability (6) weather	climate	Priority
BASIC (mostly from CLOUDS)									
Cloud water (< 100 μm) total column	30 km	100 km	N/A	5 g/m²	1 g/m²	3 h	3 h	3 d	1
Cloud water (< 100 μm) gross profile	30 km	100 km	3 km	30 %	5 %	3 h	3 h	3 d	2
Cloud water (> 100 μm) total column	30 km	100 km	N/A	5 g/m²	1 g/m²	3 h	3 h	3 d	1
Cloud water (> 100 μm) gross profile	30 km	100 km	3 km	30 %	5 %	3 h	3 h	3 d	2
Cloud ice total column	30 km	100 km	N/A	0.5 g/m²	0.1 g/m²	3 h	3 h	3 d	1
Cloud ice gross profile	30 km	100 km	3 km	30 %	5 %	3 h	3 h	3 d	2
Cloud drop size (at cloud top)	30 km	100 km	N/A	5 μm	1 μm	3 h	3 h	3 d	3
Cloud ice content (at cloud top)	30 km	100 km	N/A	30 %	5 %	3 h	3 h	3 d	1
Cloud optical thickness	30 km	100 km	N/A	30 %	5 %	3 h	3 h	3 d	3
Water vapour total column	30 km	100 km	N/A	500 g/m²	100 g/m²	3 h	3 h	3 d	2
Precipitation rate at the ground	30 km	100 km	N/A	3 mm/h	0.5 mm/h	3 h	3 h	3 d	3
Precipitation index (daily cumulative)	30 km	100 km	N/A	1 mm/d	0.2 mm/d	3 h	3 h	3 d	4
Short-wave outgoing radiation at TOA (*)	30 km	100 km	N/A	3 W/m²	0.5 W/m²	3 h	3 h	3 d	1
Long-wave outgoing radiation at TOA (*)	30 km	100 km	N/A	3 W/m²	0.5 W/m²	3 h	3 h	3 d	1
Aerosol total column	30 km	100 km	N/A	30 %	5 %	3 h	3 h	3 d	2
Aerosol gross profile	30 km	100 km	3 km	30 %	5 %	3 h	3 h	3 d	3
Short-wave cloud reflectance	30 km	100 km	N/A	5 %	1 %	3 h	3 h	3 d	2
Long-wave cloud emissivity	30 km	100 km	N/A	3 %	0.5 %	3 h	3 h	3 d	2
(*) Under the same geometry as for clouds									
BASIC (also available from METOP)									
Cloud imagery	3 km	10 km	N/A	N/A	N/A	3 h	3 h	3 d	4
Cloud type	30 km	100 km	N/A	0.3 classes	0.05 classes	3 h	3 h	3 d	3
Cloud cover	30 km	100 km	N/A	5 %	1 %	3 h	3 h	3 d	1
Cloud top height	30 km	100 km	N/A	1 km	0.2 km	3 h	3 h	3 d	1
Cloud top temperature	30 km	100 km	N/A	1 K	0.2 K	3 h	3 h	3 d	2
SUPPORT TO BASIC									
Temperature gross profile	30 km	100 km	3 km	3 K	0.5 K	3 h	3 h	3 d	3
Relative humidity gross profile	30 km	100 km	3 km	30 %	5 %	3 h	3 h	3 d	2
Ozone total column	30 km	100 km	N/A	30 DU	5 DU	3 h	3 h	3 d	4
ASSUMED TO BE AVAILABLE									
(Accurate) Temperature profile	30 km	100 km	1 km	1 K	0.2 K	3 h	3 h	3 d	-
(Accurate) Relative humidity profile	30 km	100 km	1 km	10 %	2 %	3 h	3 h	3 d	-
Solar irradiance at TOA	N/A	N/A	N/A	0.5 W/m²	0.1 W/m²	3 h	3 h	3 d	-
Short-wave ougoing radiation at TOA (flux)	30 km	100 km	N/A	3 W/m²	0.5 W/m²	3 h	3 h	3 d	-
Long-wave outgoing radiation at TOA (flux)	30 km	100 km	N/A	3 W/m²	0.5 W/m²	3 h	3 h	3 d	-
Short-wave Earth surface radiation	30 km	100 km	N/A	5 W/m²	1 W/m²	3 h	3 h	3 d	-
Long-wave Earth surface radiation	30 km	100 km	N/A	5 W/m²	1 W/m²	3 h	3 h	3 d	-

(1)   Logarithmic centre of a range between a target value and a threshold of interest one order of magnitude worse.
(2)   Intended as inference of a gross vertical structure, i.e. 3-4 layers in the troposphere.
(3)   Logarithmic centre of a range between a target value and a threshold of interest half order of magnitude worse.
(4)   Required as < 20 % of rms, so that the requirement for climate use is met after integration over < 1 week.
(5)   Requirement set to account for diurnal variations. For climate use data will be integrated over longer periods.
(6)   Referred to products. Raw data are requested to be available in real time.

Table 4 - User requirements for parameters not part of the CLOUDS objectives

Geophysical parameter	Horizontal resolution (1) weather	Horizontal resolution (1) climate	Vertical resolution (2)	Accuracy r.m.s. (3)	Accuracy bias (4)	Observing cycle (5)	Delay of availability (6) weather	Delay of availability (6) climate	Priority
Wind over sea surface	30 km	100 km	N/A	3 m/s	0.5 m/s	3 h	3 h	3 d	5
Sea surface temperature	30 km	100 km	N/A	1 K	0.2 K	3 h	3 h	3 d	5
Ice/snow imagery	3 km	10 km	N/A	N/A	N/A	3 h	3 h	3 d	5
Sea-ice cover	30 km	100 km	N/A	5 %	1 %	3 h	3 h	3 d	5
Sea-ice type	30 km	100 km	N/A	0.5 classes	0.1 classes	3 h	3 h	3 d	5
Icebergs	30 km	100 km	N/A	50 %	10 %	3 h	3 h	3 d	5
Snow cover	30 km	100 km	N/A	30 %	5 %	3 h	3 h	3 d	5
Snow melting conditions	30 km	100 km	N/A	0.5 classes	0.1 classes	3 h	3 h	3 d	5
Soil moisture	30 km	100 km	N/A	30 g/kg	5 g/kg	3 h	3 h	3 d	5
Apparent Thermal Inertia (ATI)	3 km	10 km	N/A	3 K^{-1}	0.5 K^{-1}	3 h	3 h	3 d	5
Vegetation hydric stress index	30 km	100 km	N/A	30 %	5 %	3 h	3 h	3 d	5

(7) Logarithmic centre of a range between a target value and a threshold of interest one order of magnitude worse.
(8) Intended as inference of a gross vertical structure, i.e. 3-4 layers in the troposphere.
(9) Logarithmic centre of a range between a target value and a threshold of interest half order of magnitude worse.
(10) Required as < 20 % of rms, so that the requirement for climate use is met after integration over < 1 week.
(11) Requirement set to account for diurnal variations. For climate use data will be integrated over longer periods.
(12) Referred to products. Raw data are requested to be available in real time.

3. MISSION REQUIREMENTS

CLOUDS will comply with the following *generic mission requirements*:

- synergy with METOP in order to be able to frame the CLOUDS information within the basic meteorological fields, specifically accurate temperature and water vapour profiles;
- only passive instruments, to ensure a swath of at least 1400 km for a daily global coverage; also to achieve long life-time and to reduce power/weight requirements;
- exploitation of all known principles to discriminate liquid from ice water. This implies use of different polarisations and viewing angles in a number of short-wave channels as well as in MW, and use of very-high frequency MW (Sub-mm) and possibly Far IR;
- exploitation of all known principles (only using passive radiometry) to accurately measure cloud top height. This implies extensive use of TIR window channels, as well as differential radiometry with low-absorption bands (e.g., of CO_2);
- exploitation of all known principles (without having to use radar) to infer liquid or precipitating water gross profiles. This implies use of more frequencies in the MW field (both in windows and in O_2 bands) to cover different response from drops of different size located at different altitudes;
- contextual collection of basic information necessary to have a sufficiently comprehensive and self-standing data set. This implies observation of the cloud precursor (water vapour) and of the main interacting fields (radiation and aerosol which, incidentally, are not observed by METOP).

In order to provide information on the various geophysical parameters listed in Table 1, CLOUDS will have a number of radiometers operating in a number of spectral channels. In *Table 5* an indicative correspondence between user requirements and mission requirements (in terms of spectral channels, polarisations and viewing geometry) is shown.

Table 5 - User requirements and corresponding mission requirements in
respect of spectral channels, polarisations and viewing geometry

User requirements	Mission requirements for channels, polarisations and viewing
Cloud imagery and cloud cover	0.5 μm and 11 μm
Cloud top temperature and cloud emissivity	11 and 12 μm (split window), and 5.7 and 7.1 μm (water vapour)
Cloud top height	From temperature and directly by 13.4 μm (CO_2 weak absorption)
Cloud ice content at cloud top and cloud ice total column	Differential polarisation, viewing, reflectivity or emission of: 0.5 and 1.6 μm; 3.7 and 11 μm (in daylight); 90 GHz; 150 GHz; 200-1000 GHz (up to 4 channels); 15-100 μm (up to 4 channels)
Cloud optical thickness, cloud reflectance and aerosol total column	0.4 μm, 0.5 μm, 0.6 μm, 1.6 μm with more polarisations
Cloud type	Cluster analysis of more channels: 0.5 μm, 1.6 μm, 3.7 μm (fog), 8.7 μm (invisible cirrus), 11 μm and 12 μm
Total ozone	9.7 μm and the nearby 11 μm window
Precipitating core in clouds (> 100 μm)	10 GHz (monsoon), 19 GHz, 37 GHz, 90 GHz (specifically needed over land), with double polarisation
Wind speed over sea-surface, wind direction, sea-surface temperature	6 GHz, 10 GHz and 19 GHz, with double polarisation and double viewing geometry
Height of liquid water (> 100 μm) and precipitating cores	Inferred from O_2 "temperature" channels in the bands around 55 GHz (up to four channels) and 119 GHz (up to four channels)
Cloud water (< 100 μm), drop size or vertical profile, ice/liquid water	19 GHz, 37 GHz, 90 GHz, 150 GHz, 200 GHz and others TBD in the 200-1000 GHz range, with double polarisation
Water vapour profile (gross estimate)	In IR: 5.7, 7.1, 8.7 and 11+12 μm.; in MW: 23 GHz (total column), 183 GHz band (up to 3 channels)
Earth radiation budget	SW channel (0.3-4.0 μm) and LW channel (4.0-30 μm), more viewing directions and double polarisation for SW channel

Channels called for by the mission requirements have been arranged according to the following classification:

a. **"backbone channels"** of consolidated multi-purpose use, as follows:
 - optical channels (VIS/NIR/TIR) mostly for "classical" cloud observations referring to their top surface (cloud type, cover, top temperature/height, optical thickness)
 - MW channels (low-medium frequencies) mostly for cloud interior and precipitation;

b. **"special packages"** to address the following dedicated objectives:
 - improved cirrus observation in the Far IR and very-high frequency (Sub-mm) ranges
 - improved aerosol observation in the UV/VIS/NIR ranges
 - outgoing radiation at TOA by broad-band channels in the UV/VIS/NIR/TIR/FIR ranges
 - supporting temperature/humidity sounding within clouds by MW channels.

The required horizontal resolution is 5 km for all optical backbone channels and for the 90 GHz MW channel: for channels of the special packages is somewhat relaxed, to the extreme of 40 km for the broad-band radiation budget channels. The required synergy with METOP materialises in CLOUDS chasing METOP in the same orbit, de-phased by about 30 min, i.e.:

 - H = 840 km, ε = 98.7°, T = 101.7 min, LST = 10 h (constant).

4. PRELIMINARY INSTRUMENT REQUIREMENTS

At the time of this writing, parametric instrument studies and a preliminary trade-off analysis have been performed to establish instrument requirements. The following asset represents the current status (mid-1999).

Narrow-band UV, VIS and NIR "backbone" and "aerosol" channels (334.5 nm, 388 nm, 443 nm, 670 nm, 865 nm, 910 nm, 1240 nm, 1380 nm and 1610 nm) will be integrated in a single instrument, with the following features:

- conical scanning (45° from nadir), fore- and after- viewing
- triple polarisation in a number of channels
- basic sampling corresponding to 5 km resolution
- radiometric accuracy required for aerosol to be recovered by ground processing at the expenses of resolution (probably 5 km for most channels, 10-30 km for UV).

TIR/FIR channels (3.74 μm, 6.25 μm, 7.35 μm, 8.70 μm, 9.66 μm, 10.8 μm, 12.0 μm, 13.4 μm, 18.2 μm and 24.4 μm) will probably be uncooled (microbolometers). Features:

- conical scanning for preservation of same viewing geometry as for SW and MW
- 5 km resolution (relaxed to 40 km at 18 and 22 μm).

Broad-band channels (0.3-4.0 μm and 4.0-30 μm) will be implemented as follows:

- conical scanning for preservation of same viewing geometry as for SW and MW
- dual polarisation in the SW channel
- 40 km resolution
- additional conical scanner with four viewing angles (e.g., 21°, 33°, 45° and 57°), 5 km resolution, band 0.4-1.0 μm.

"Backbone" MW channels (6-90 GHz) will be implemented as in MIMR/AMSR:

- conical scanning (45° from nadir), fore- and after- viewing
- dual polarisation for all channels
- 5 km resolution at 89 GHz, degrading with decreasing frequency.

Sub-mm channels (200-900 GHz) will be implemented with the following features:

- conical scanning (45° from nadir), fore- and after- viewing
- dual polarisation for all channels
- 10 km resolution for all channels.

"Sounding" channels (in the 55, 118 and 183 GHz bands) will be added as "marginal" to either the MW or the Sub-mm package. Features:

- conical scanning for preservation of same viewing geometry as MW and Sub-mm
- a 150 GHz channel is mandatory, with dual polarisation
- 10 km resolution for all channels.

5. THE MW AND SUB-MM PACKAGES

The role of MW and Sub-mm channels, inclusive of the "sounding" channels, is clearly dominant for inferring cloud interior parameters, since channels in the optical range are nearly exclusively representative of cloud top properties. In *Table 6* a quick-look association of cloud interior parameters (inclusive of water phase) and MW/Sub-mm channels is shown. For memory, add-on observations derived from the MW package, not part of the CLOUDS objectives, also are mentioned.

Table 6 - Relevance of MW/Sub-mm channels to CLOUDS mission requirements

Precipitation and liquid water, with inference of drop size and gross profile	6.9 GHz, 10.6 GHz, 18.7 GHz, 36.5 GHz, 89.0 GHz, 150 GHz, 4 channels in the 55 GHz band, 4 channels in the 119 GHz band
Cloud ice, total and gross profile	220.5 GHz, 462.6 GHz, 682.9 GHz and 874.4 GHz
Water vapour, total and gross profile	23.8 GHz, 3 channels in the 183.3 band
"Add-on" observations	sea-surface temperature, sea-surface wind, sea-ice, soil moisture

According to the present concept, channels will be grouped in two instruments:

- one for water vapour and ice (channels 150-900 GHz) with antenna size of L = 40 cm
- one for liquid/precipitating water (channels 6-120 GHz) with antenna size of L = 160 cm
- both instruments with conical scanning, one scan / 2 s, fore- and after- views
- the channel at 89 GHz has two parallel feeds for along-track image continuity.

Table 7 provides an overview of the CLOUDS MW/Sub-mm channels specifications. It should be noted that all channels of the "sounding" class are recorded between brackets because their precise definition is still being studied. A flat requirement of 1.5 K has been adopted for the calibration accuracy of all channels

Table 7 - Overview of CLOUDS MW/Sub-mm channel specifications

Antenna diameters: L = 160 cm for channels 6-120 GHz, L = 40 cm for channels 150-900 GHz
Rate of conical scanning: one scan / 2 s; half-aperture cone: 45°; zenith angle: 53.2°
Satellite height: 840 km; LST: 10 h; ground speed: 13 km/scan; circle to be scanned: 5,700 km
Accuracy of absolute calibration: 1.5 K for all channels

Channel centre ν	Band-width $\Delta\nu$	Along-track IFOV	Along-scan IFOV	Average IFOV	Sam-ples / scan	Inte-gration time	Polari-sation	NEΔT (required)	NEΔT (estimated)
874.38 ± 6.0 GHz	3.0 GHz	13.0 km	7.8 km	10 km	800	2.5 ms	dual	1.0 K @ 240 K	2.0 K @ 240 K
682.95 ± 6.0 GHz	3.0 GHz	13.0 km	7.8 km	10 km	800	2.5 ms	dual	1.0 K @ 240 K	1.2 K @ 240 K
462.64 ± 3.0 GHz	2.0 GHz	13.0 km	7.8 km	10 km	800	2.5 ms	dual	1.0 K @ 240 K	1.0 K @ 240 K
220.50 ± 3.0 GHz	2.0 GHz	13.0 km	7.8 km	10 km	800	2.5 ms	dual	1.0 K @ 240 K	0.9 K @ 240 K
(183.31 ± 1.0 GHz)	1.0 GHz	13.0 km	7.8 km	10 km	800	2.5 ms	single	1.0 K @ 240 K	1.4 K @ 240 K
(183.31 ± 3.0 GHz)	2.0 GHz	13.0 km	7.8 km	10 km	800	2.5 ms	single	1.0 K @ 260 K	1.0 K @ 260 K
(183.31 ± 7.0 GHz)	4.0 GHz	13.0 km	7.8 km	10 km	800	2.5 ms	single	1.0 K @ 280 K	0.7 K @ 280 K
(150 GHz)	4.0 GHz	13.0 km	7.8 km	10 km	800	2.5 ms	dual	1.0 K @ 300 K	0.6 K @ 300 K
(118.75 ± 1.0 GHz)	1.0 GHz	13.0 km	7.8 km	10 km	800	2.5 ms	single	0.5 K @ 230 K	1.0 K @ 230 K
(118.75 ± 1.5 GHz)	1.0 GHz	13.0 km	7.8 km	10 km	800	2.5 ms	single	0.5 K @ 250 K	1.0 K @ 250 K
(118.75 ± 2.0 GHz)	1.0 GHz	13.0 km	7.8 km	10 km	800	2.5 ms	single	0.5 K @ 270 K	1.0 K @ 270 K
(118.75 ± 4.0 GHz)	1.0 GHz	13.0 km	7.8 km	10 km	800	2.5 ms	single	0.5 K @ 290 K	1.2 K @ 290 K
89.0 GHz	3.0 GHz	6.5 km	3.9 km	5 km	1600	1.25 ms	dual	1.0 K @ 300 K	1.0 K @ 300 K
(55 GHz)	0.5 GHz	13.0 km	7.8 km	10 km	800	2.5 ms	single	0.5 K @ 230 K	1.0 K @ 230 K
(54 GHz)	0.5 GHz	13.0 km	7.8 km	10 km	800	2.5 ms	single	0.5 K @ 250 K	1.0 K @ 250 K
(53 GHz)	0.5 GHz	13.0 km	7.8 km	10 km	800	2.5 ms	single	0.5 K @ 270 K	1.0 K @ 270 K
(50 GHz)	0.5 GHz	13.0 km	7.8 km	10 km	800	2.5 ms	single	0.5 K @ 290 K	1.0 K @ 290 K
36.5 GHz	1.0 GHz	15.8 km	9.5 km	12 km	800	2.5 ms	dual	0.7 K @ 300 K	0.7 K @ 300 K
23.8 GHz	0.4 GHz	24.3 km	14.6 km	19 km	400	5 ms	dual	0.6 K @ 250 K	0.6 K @ 250 K
18.7 GHz	0.2 GHz	30.1 km	18.6 km	24 km	400	5 ms	dual	0.5 K @ 300 K	0.8 K @ 300 K
10.6 GHz	0.1 GHz	54.6 km	32.7 km	42 km	200	10 ms	dual	0.4 K @ 300 K	0.4 K @ 300 K
6.9 GHz	0.3 GHz	83.8 km	50.3 km	65 km	200	10 ms	dual	0.3 K @ 300 K	0.2 K @ 300 K

6. PRELIMINARY SYSTEM CONCEPT

On the base of the instrument requirements, preliminary instrument concepts and system concept have been elaborated, and will be submitted to Phase A (feasibility confirmation and preliminary system design). The main system feature is that all instruments of the optical package will be optimally accommodated in a single payload rotating at one scan / 8 s, whilst the MW and Sub-mm instruments also could be optimally accommodated in a single payload rotating at one scan / 2 s. However, an alternative option is being studied, where the MW and Sub-mm instruments are distinct, to improve the probability of finding flight opportunities.

In the integrated concept, the MW/Sub-mm payload would be installed on the top surface of the platform, opposite to the Earth surface. In the separated concept, the MW radiometer would be installed on the top surface and the Sub-mm radiometer on the bottom surface.

In *Fig. 1* a possible configurations of the satellite is shown, in the hypothesis of integrated MW/Sub-mm payload. Also shown is the position of the optical payload, on the bottom face. *Fig. 2* shows the satellite in the stowed configuration, within the ogive of a Rockot launcher.

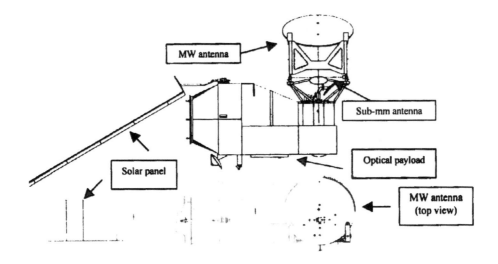

Fig. 1 - Possibile satellite configuration

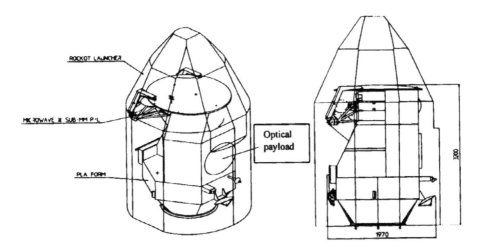

Fig. 2 - Possible accommodation of the satellite in a Rockot launcher

7.   STATUS OF THE PROJECT AND CONCLUSION

At mid-1999 the CLOUDS project is concluding the pre-Phase A study (definition of baseline instruments and system). The EC financial contribution covers up to Phase A, to last one year, for a total of 2-year duration of this mission study.

The Phase A work programme includes two interleaving activities:

- scientific studies, to document the impact of the mission objectives on the addressed applications, to assist the instruments design and to evaluate to which extent the instrumentation possible to be implemented could comply with the mission requirements. The following studies, initially oriented to identify the baseline instrumentation, will continue to run, now focusing on the evaluation of expected performances:
  - impact studies on climate models, meteorological models and theory of climate
  - clouds and aerosol (with focus on UV to NIR channels)
  - cirrus clouds (with focus on NIR to FIR channels)
  - Earth radiation budget (with focus on broad-band and multi-angle viewing channels)
  - cloud phase discrimination (with focus on Sub-mm and MW H_2O channels)
  - precipitation and liquid water (with focus on MW including O_2 channels);

- technical studies, to consolidate the instrument concepts and provide their preliminary design, including characterisation and evaluation of instrument performances, so as to enable the scientific groups to evaluate the likelihood of mission requirements to be achieved; and to provide a preliminary design of the system and of the main sub-systems, so as to assess its feasibility and evaluate the affordability elements.

If the objective of a small-medium size satellite is confirmed to be feasible and the anticipated product performances are recognised to fulfil the requirements, follow-on activity will be pursued in the framework of the joint EC/ESA/EUMETSAT concept of *Earth Watch* programmes.

REFERENCES

Due to the wide number of subjects covered by the CLOUDS mission, the number of references is impractically high, and continuously growing with the progress of the scientific studies. It is therefore preferred to relay the reader to the CLOUDS ftp sites, as follows:

- ftp://romatm9.phys.uniroma1.it/pub/clouds/basic
- ftp://nepero.ifa.rm.cnr.it/pub/clouds/basic

References relative to mission objectives can be found under the Chapters of document:

- D2B-MOD-2   (Mission Objectives Document, Issue 2).

References relative to channels selection can be found under the Appendixes of document:

- D3B-MRD-2   (Mission Requirements Document, Issue 2).

Microw. Radiomet. Remote Sens. Earth's Surf. Atmosphere, pp. 503–511
P. Pampaloni and S. Paloscia (Eds)
© VSP 2000

Non-linear inversion of Odin sub-mm observations in the lower stratosphere by neural networks

CARLOS JIMÉNEZ, PATRICK ERIKSSON and JAN ASKNE

Chalmers University of Technology, S-412 96 Göteborg, Sweden

Abstract- Odin is a small satellite employing sub-mm (480-580 GHz) radiometry of atmospheric thermal emission, in a limb sounding mode, to measure global distributions of several species important for ozone chemistry in the middle atmosphere. Although the planned standard data products are based on linear retrieval algorithms, addressing the non-linear effects in some of the Odin observation bands results in an extension of the retrieval altitude range into the lower stratosphere. This paper examines the possibility of applying a technique based on multilayer perceptrons to address the non-linear effects when inverting some of the future Odin data. The study is based on inverting simulated spectra in the 544.2-545.0 GHz band to retrieve O_3 profiles by the neural network technique and also by an iterative implementation of the optimal estimation method based on the Marquardt-Levenberg algorithm. First results show that the proposed technique performs in similar terms when compared with the iterative implementation of the optimal estimation method, but offers the advantage of performing the inversions more efficiently from a computational point of view.

1. INTRODUCTION

The Odin satellite is a Swedish initiative for a small satellite built in collaboration with Canada, France and Finland, and planned for launch at the end of 1999. Its microwave radiometer will be the first to measure thermal emission from the middle atmosphere (10-100 km) in the sub-mm range (480-580 GHz) in a limb sounding mode. Species to be measured include O_3, H_2O, H_2O_2, ClO, BrO and NO, providing both altitude determination and geographical mapping of key species involved in different atmospheric processes, with special emphasis on the ozone depletion phenomena.

Optimal estimation (OEM) [1] is the chosen method to perform the inversions. By removing channels with opacity exceeding a certain limit from a scanned spectrum, the inversion problem can be treated as linear. If higher opacities are allowed, the assumption of a linear mapping is not of general validity, and iterative inverting schemes are needed. If the limitations on a retrieval are related to non-linear effects, addressing the non-linearity usually brings an extension on the final altitude coverage [2], but the price to pay is much more demanding computations.

Neural networks (NN) have recently begun to be used for inverting remote sensing observations. Applications related to passive microwave radiometry already include e.g. retrieval of snow parameters [3], temperature profiling [4], estimation of atmospheric water vapour [5, 6], or estimation of cloud microphysics [7]. This interest is justified because NN are effective at addressing both complex statistical patterns and non-linear dependencies.

They are also interesting because they can solve non-linear atmospheric inversion problems more efficiently than classical iterative inversion schemes.

This paper examines the possibilities of applying a type of NNs, the multilayer perceptrons (MLP) to perform non-linear inversions of future Odin data. The study is conducted by retrieving O_3 profiles from simulated spectra in the 544.2-545.0 GHz band. To allow comparison, retrievals are also done by an iterative application of OEM based on the Marquardt-Levenberg algorithm. The paper is organised by presenting first the inversion methodology, followed by the description of the simulations, continued by the presentation of the results and ended by drawing some conclusions.

2. INVERSION METHODOLOGY

2.1 The inversion problem
Solving the radiative transfer differential equation for microwave radiometry yields a Fredholm integral equation of the first kind that can be expressed as

$$\int K(x, y)\, f(x)\, dx = g(y) \tag{1}$$

where $f(x)$ represents a function of the atmospheric parameters, $g(y)$ represents a function of the measured spectrum, and $K(x,y)$ describes the relation between both spaces. Solving the inversion problem consist of finding $f(x)$ from a specification of $K(x,y)$ and $g(y)$. Difficulties arise because the problem is ill-posed, so from a range of mathematically valid solutions one of them has to be selected in order to get an unique and physically acceptable solution. If Eq. (1) is discretised with a proper numerical integration formula, i.e. $Kf = g$, the ill-posedness of the problem is reflected in a poor conditioning of K.

Diverse methods have been proposed to help selecting an acceptable solution when inverting spectra from an observation. The traditional approach is using different regularization methods. One of them is OEM, where the measurement is combined with *a priori* information to select the maximum likelihood solution. The probability of obtaining a state vector x, representing the vertical profiles to be retrieved, given a measured spectrum y, can be written as [8]

$$-2\ln P(x / y) = [y - F(x)]^T S_y^{-1}[y - F(x)] + [x - x_a]^T S_x^{-1}[x - x_a] + c \tag{2}$$

where F is the forward model of the observation, x_a is the *a priori* state, S_y and S_x are the covariance matrices of the measurement uncertainties and atmospheric state respectively, and c is a constant. To select the maximum likelihood state $\hat{x}$ is equivalent to minimising the cost function given by Eq. (2). Equating the derivative of Eq. (2) to zero gives the implicit equation

$$-\hat{K}^T S_y^{-1}[y - F(\hat{x})] + S_x^{-1}(\hat{x} - x_a) = 0 \tag{3}$$

where $\hat{K} = \partial F / \partial x\big|_i = K(\hat{x})$. To numerically solve this equation different iterative schemes can be used, and Marquardt-Levenberg (see e.g. [8]) has been used here because it seems to be more robust when compared with other methods [2].

2.2 Neural Networks
A completely different approach to solve inversion problems is by using connectionist paradigms implemented by means of NNs. A NN can be seen as an interconnected

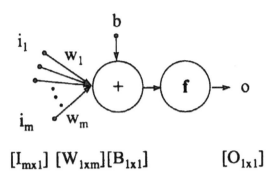

$$[I_{mx1}] \quad [W_{1xm}][B_{1x1}] \qquad\qquad [O_{1x1}]$$

Figure 1. A neural network node.

assembly of processing units called nodes. The structure of a node is sketched in Fig. 1. If W is a vector of weights, I the input vector to the node, B a bias vector and f a certain activation function, the output of a node o in matrix notation is given by

$$O = f(WI + B) \tag{5}$$

The processing ability of the NN lies in the modification of the inter-node connections when the net is subject to adaptation to a set of training patterns. Here the training patterns are simulated radiance and atmospheric states, and the NN is trained to learn the mapping between the two states.

There are different type of NNs, and MLP are selected here because they can approximate arbitrarily well any continuos functional mapping from a finite dimensional space to another [9]. A MLP consists of several layers of nodes where the input signal propagates only in the forward direction. The MLP used here have one hidden layer with sigmoidal activation functions ($f(a) = [1 + exp(a)]^{-1}$) and one output layer of linear activation functions ($f(a) = a$), where a is the input parameter of the functions. Following the notation presented in Fig. 2, the output of this MLP can be expressed as

$$O = f_O(W^O H + B^O) = f_O(W^O f_H(W^H P + B^H) + B^O) \tag{4}$$

where W^H, B^H, W^O and B^O, are the weights and bias matrices of the hidden and output layer respectively, P the input vector, H the output of the hidden layer, O the output vector, and f_H and f_O the activation functions of the hidden and output layer respectively. The number of inputs, m, and output nodes, k, are fixed by the dimension of the spectral data to be inverted and number of points of the retrieval altitude grid respectively, while the number of nodes of the hidden layer, g, is a more arbitrary number. It should be chosen not too large, to avoid overfitting problems, and not too small, to avoid compromising the mapping provided by the MLP.

Training consists of changing the weights of the NN to reduce the difference between target vectors, the true species profiles in this case, and the current output vectors, the retrieved profiles given by the NN after each training iteration. The cost function to be minimised here is the mean sum of squares of that difference for each output node, and

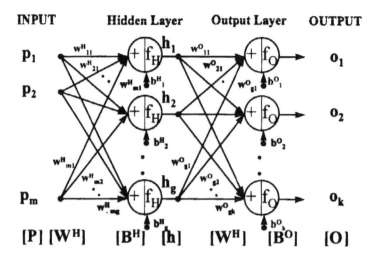

INPUT   Hidden Layer   Output Layer   OUTPUT

[P] [W^H]   [B^H] [h]   [W^H]   [B^O]   [O]

Figure 2. Multilayer perceptron with a hidden layer and an output layer.

different algorithms grouped under the name of backpropagation have been developed to calculate the gradient of this cost function. The simplest algorithm to minimise the cost is gradient descent, one iteration of this algorithm can be written:

$$W_{i+1} = W_i - \lambda_i g_i \qquad (5)$$

where W_i is a vector of current weights, λ_i is the learning rate, and g_i is the current gradient. Newton's methods substitute the constant learning rate by the Hessian matrix A (second derivatives of the cost function at the current values of the weights) and converge faster. Calculating the Hessian matrix for the MLP is very complex, so approximations of the Hessian matrix are used. If the cost function is a sum of squares, a reasonable approximation is to consider $A = JJ^T$, where J is the Jacobian matrix (first derivatives of the cost function at the current values of the weights). The training algorithm applied here uses this approximation combined with the Marquardt-Levenberg algorithm [10]. An iteration of this algorithm can be expressed as

$$W_{i+1} = W_i - \left[J_i^T J_i + \mu\, I \right]^{-1} J_i^T e \qquad (6)$$

where μ is the parameter controlling the trade off between gradient descent and Newton's method, I is the identity matrix, and the gradient is calculated as $g_i = J_i^T e$ where e is the vector of NN errors.

2.3 Reducing the dimensions of the input vector
A problem when inverting spectra is to have vectors of very large dimensions representing the spectrum. Odin spectra will typically contain more than ten thousand measured values, which cannot be the input for a NN of reasonable dimensions, so finding an appropriate technique to address this problem is a must before the MLP can be applied. The technique proposed here achieves the reduction in two steps. First a data binning, where the measured

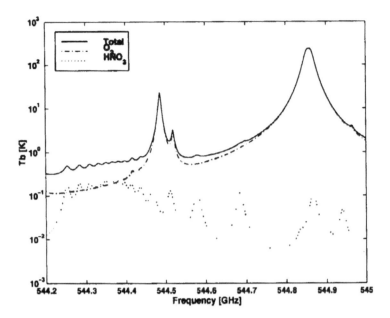

Figure 3. Typical spectra in the 544.2-545.0 GHz band.

values from neighbouring spectrometer channels are averaged to yield a single data point, is applied. This is the standard data reduction technique to be used when inverting Odin spectra by OEM, but a further reduction is still needed for the NN technique. The binning is then followed by a transformation of the resulting vector by principal component analysis (PCA). PCA uses the whole set of spectra for the different atmospheric profiles to derive a transformation matrix that maps the input space into a new space with smaller dimensions. This is done by computing the covariance matrix of the set of spectra, finding its eigenvectors and eigenvalues, and removing those eigenvectors contributing the least to the total variability of the data, i.e. only the principal components are kept. By using both techniques, binning and PCA, it is possible to reduce the dimensions of the input to the MLP by a factor of 100, with a relatively small loss in the variability of the data. A reduction technique based only in PCA for the NN technique is also possible, achieving a similar reduction, but because the simulated spectra in this study has to be inverted by both techniques, OEM and NN, the combined binning-PCA technique simplifies considerably the simulations and it has been the option used here.

3. SIMULATIONS

Inversion of synthetic spectra in the 544.2-545.0 GHz band, corresponding to one of the Odin observational standard modes, was used here for the analysis of both retrieval strategies. A typical spectra from this band can be observed in Fig. 3. This band has, besides a continuos H_2O contribution, two strong O_3 lines and a large number of relatively strong and isolated HNO_3 lines. For this study only O_3 profiles were retrieved, considering HNO_3 and H_2O as interfering species during the inversion.

To generate synthetic spectra, 1500 different atmospheric states were created by randomly disturbing an *a prori* atmospheric state representing typical mid-latitude conditions. The temperature and species mean profiles were taken from the REPROBUS chemical transport model [11], and the random temperature and species vectors calculated by applying the Choleski decomposition method (see e.g. [12]). Estimations of the expected natural variability were used to model the temperature and species uncertainties and gaussian statistics were assumed for both temperature and species distribution.

The synthetic spectra were generated by running a detailed forward model [13] on the set of atmospheric states. The geometry of the observation simulated limb-sounding scans extended upwards to an altitude of 40 km to reduce computation demands, and for the same reason, only the antenna pattern, the most relevant characteristic in this altitude range, was included in the simulations. The generated spectra were then degraded by adding calibration errors and thermal noise. A load switched receiver (see e.g. [14]) was assumed with loads set to brightness temperatures of 3 and 300 K, and the error created by assuming an uncertainty of 1 K for the larger load. The expected thermal noise for these observations is around 2.5 K, so thermal noise of this magnitude was also added to the spectra. Finally, the last step was to remove those spectral channels with optical thickness exceeding a certain limit, to fix the linearity of the problem. This was done by calculating the opacity from the tangent point to the sensor, for each tangent altitude and with the *a priori* atmosphere. The limit in maximum optical depth was set to a high value, 3.0, assuring that non-linear-effects were present and had to be addressed by the inversion method.

After generating the spectra, they were inverted by the two retrieval strategies described above. For the optimal estimation, the covariance matrices S_x and S_y had to be fixed. S_x should reflect the natural variability of the species but, in fact, it is more used as a parameter to control a trade-off between solution variance and vertical resolution. Despite the fact that the variability is perfectly known in these simulations, to represent more realistic retrievals, the standard deviation was set to an artificial 100% value and interlevel correlations were modelled by a tent function (see e.g. [15]) with correlation length of 1.5 km . The observing error matrix, S_y, was fixed by including temperature and thermal noise errors with values similar to the figures used when simulating the spectra.

The NN used for the inversion consisted of an ensemble of individuals MLP, where each MLP gave the output of a point of the retrieval altitude grid. All the MLP shared the same input vector and had a first layer of 100 inputs, a hidden layer of 20 neurones and an output layer with a single neurone. In this way, each MLP was individually trained to retrieve an altitude point. A MLP with an output vector yielding the whole altitude grid was also trained, but the first topology seemed to give slightly more accurate retrievals having an equivalent computational cost.

The reduction of the spectral data was done through two steps, as commented above. The typical spectral data from the simulated observations here had around 14000 values, a number too large even for the OEM computations. The binning technique reduced that number by approximately a factor of 10, and these were the spectra used for the OEM retrievals. For the NN method, the PCA technique made a further reduction by approximately the same factor, so a typical input vector for the NN had approximately 100 values.

Each individual MLP was trained by the backpropagation algorithm described before. Early stopping was also used to improve generalisation, i.e. to avoid overfitting situations where the NN memorised the training examples but does not generalise well to new sets. When the error on a validation set increases for a certain number of iterations, the training was stopped and weights at the minimum of the validation error are returned.

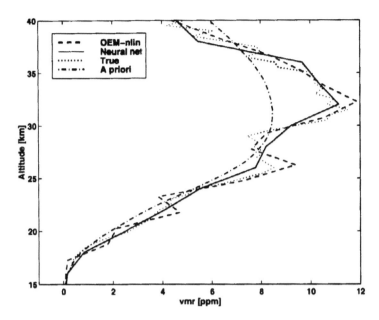

Figure 4. Typical O_3 retrieval by both OEM and NN technique. Plotted are the retrieved, true and *a priori* profiles.

4. RESULTS

A first look at individual retrievals from both methods, OEM and the NN technique, indicates that both yield reasonable retrievals. Fig. 4 shows the result of a typical inversion. The O_3 mean profile used as *a priori* state, the true O_3 distribution of the selected atmospheric state, and the corresponding retrieved profiles are plotted. It can be observed how both techniques gave an O_3 profile close to the true distribution.

The performance of the inversion methods can be evaluated by estimating the retrieval error. The difference between true and retrieved profiles was calculated for each of the retrievals and the set of errors obtained characterised by its mean and standard deviation. Fig. 5 shows these parameters. Both techniques gave comparable means and standard deviations. Looking at the standard deviation, it seems that the NN performs a bit better than the OEM at lower altitudes, and slightly worse at higher altitudes. Included in the plot is also the estimated error calculated by following Rodgers' formalism to characterise inversions [16], denoted as theoretical in the figure. This is a powerful tool for an *a priori* estimation of retrieval errors, but it can only be applied in close to linear situations, so it fails in giving an accurate indication of the total error. Also plotted are the results from a linear inversion of spectra with channels of opacity larger than 0.4 removed, showing how the non-linear inversions extend the retrieval altitude limits into the lower stratosphere by considerably decreasing the total error at those altitudes.

As discussed previously, a very high computational cost for iterative schemes, such as the OEM and Marquartd-Levenberg iterative method applied here, is one of the basic reasons to look for less demanding methods. When generating the training set and training the net, the NN technique is also very computationally demanding, but once the net is trained,

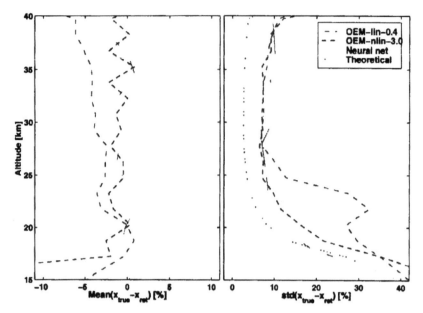

Figure 5. Mean and standard deviation of the O_3 retrieval error by both OEM and NN technique.

inversions are simple and quick matrix operations. For the number of Odin spectra to be inverted, it is clear that a properly developed retrieval technique based on NNs would be more efficient than any iterative retrieval scheme.

5. CONCLUSION

A NN technique, based on MLP and a reduction in the dimensions of the measured spectra by a combined binning-PCA technique, has shown good potential to address non-linear inversions of sub-mm spectra. This technique would contribute to the Odin retrieval capabilities by extending the profile retrievals into the lower stratosphere. Other techniques based on iterative solutions also have the potential of performing similar retrievals, but the NN technique is more efficient, from a computational point of view, when a large set of spectra has to be inverted. Further questions to be addressed are e.g. the effect of the a priori atmosphere in the training, the response of the NN to the different measurement errors, or how the errors, due to the limited capability of the forward model to describe the observation, affect the inversions.

REFERENCES

1. C.D. Rodgers, "Retrieval of atmospheric temperature and composition from remote measurements of thermal radiation," *Review of Geophys. Space Physics*, 14, 609-624 (1976)

2. P. Eriksson and C. Jimenez, "Non-linear profile retrievals for observations in the lower stratosphere with the Odin sub-mm radiometer," *Proceedings of IGARSS'98, Seattle, WA, USA*, 1420-1423 (1998)

3. T. Davids, Z. Chen, L. Tsang, J. Hwang, and A. Chang, "Retrieval of snow parameters by iterative inversion of a neural network," *IEEE Trans. Geosci. Remote Sensing*, 31, 842-852 (1993)

4. J.H. Churnside, T.A. Stermitz, and J.A. Schroeder, "Temperature profiling with neural network inversion of microwave radiometer data," *J. Atmos. Oceanic Technology*, 11, 105-109 (1994)

5. C.R. Cabrera-Mercader and D.H. Staelin, "Passive microwave relative humidity retrievals using feedforward neural networks," *IEEE Trans. Geosci. Remote Sensing*, 33, 1324-1328 (1995)

6. F. d. Frate and G.Schiavon, "Neural Networks for the retrieval of water vapour and liquid water from radiometric data," *Radio Sci.*, 33, 1373-1386 (1998)

7. L. Li, J. Vivekanandan, C. Chan, and L. Tsang, "Microwave radiometric technique to retrieve vapour, liquid and ice, Part-1 Development of a Neural Network-Based Inversion Method," *IEEE Trans. Geosci. Remote Sensing*, 35, 224-235 (1997)

8. C. Marks and C.D. Rodgers, "A retrieval method for atmospheric composition from limb emission measurements," *J. Geophys. Res.*, 98, 14939-14953 (1993)

9. M.P. Bishop, *Neural networks for pattern recognition*, Oxford University Press, New York (1995)

10. M. Hagan and M. Menhaj, "Training feedforward networks with the Marquardt algorithm," *IEEE Trans. Neural Networks*, 5, 989-993 (1994)

11. F. Lefevre, G.P. Brasseur, I. Folkins, A. Smith, and P. Simon, "Chemistry of the 1991-1992 stratospheric winter: Three dimensional model simulations," *J. Geophys. Res.*, 99, 8183-8195 (1994)

12. N. Cressie, *Statistics for spatial data*, Wiley-Interscience, New York (1993)

13. P. Eriksson, F. Merino, D. Murtagh, P. Baron, P. Ricaud and J. de la Noe, "Studies for the Odin sub-millimetre radiometer: I. Radiative transfer and instrument simulation," to appear in *Canadian J. Phys.*, Odin special issue

14. Janssen, *Atmospheric remote sensing by microwave radiometry, Wiley Series in remote Sensing*, John Wiley & Sons, Inc., New York (1993)

15. F. Merino, D. Murtagh, M. Ridal, P Eriksson, P. Baron, P. Ricaud and J. de la Noe, "Studies for the Odin sub-millimetre radiometer: III. Performance simulations," to appear in *Canadian J. Phys.*, Odin special issue

16. C.D. Rodgers, "Characterisation and error analysis of profiles retrieved from remote sensing measurements," *J. Geophysical Research*, 95, 5587-5995 (1990).

4. New radiometric instruments and missions

4.3 AMSR

Microw. Radiomet. Remote Sens. Earth's Surf. Atmosphere, pp. 515–523
P. Pampaloni and S. Paloscia (Eds)
© VSP 2000

Progress in AMSR Snow Algorithm Development

ALFRED CHANG[1] and TOSHIO KOIKE[2]

[1]*Hydrological Sciences Branch, NASA/GSFC, Greenbelt, MD 20771, USA*
[2]*Hydrosphere and Atmosphere Interaction Laboratory,
Nagaoka University of Technology, Nagaoka, Japan*

Abstract- Accurate determination of snow parameters from space is a challenging effort. Microwave radiation, which can penetrate into the snowpack, provides an opportunity to sense the snowpack parameters. Advanced Microwave Scanning Radiometer (AMSR) will be flown on-board of the Japanese Advanced Earth Observing Satellite-II (ADEOS-II) and United States Earth Observation System (EOS) PM-1 satellite. AMSR is a passive microwave radiometer with frequency ranges from 6.9 GHz to 89 GHz. With a large antenna, AMSR will provide the best spatial resolution of multi-frequency radiometer from space. This provides us an opportunity to improve the snow parameter retrieval. Currently, NASDA and NASA are developing AMSR snow retrieval algorithms. These algorithms are now being carefully tested and evaluated using the SSM/I data. Due to limited snowpack data available for comparison, this activity is progressing slowly. However, it is clear that in order to improve the snow retrieval algorithm, it is necessary to model the metamorphism history of the snowpack.

1. INTRODUCTION

The distribution of snow cover contains valuable information of the global energy and water balances. Its high albedo and thermal insulation make snow a major factor in the global energy balance. The spatial and temporal variability of snow covered area is critical for monitoring climate and global change. Snow storage strongly influences the availability of water for runoff, evaporation, and soil moisture. Moreover, to forecast the snowmelt runoff and the potential of flooding, it is necessary to monitor the snow water equivalent of the snowpacks.

All snow surfaces display a unique "scattering" signature in the microwave region. Surface emissivity decreases with increasing frequency. This snow-scattering signature can be used to identify and separate snow from all other surfaces [1]. One observes a progressive decrease in the high frequency (37 GHz) measurement as the snow water equivalent value increases to a maximum of 200 mm. Wetness in the snowpack, however, will mask out most of this signature. As the snow gets deeper and begins to metamorphosis due to temperature gradients, melting, and re-freezing, the emissivity displays wide variations depending on the ice conditions within the snowpack. The most dramatic change occurs when an ice-crust forms near the top or the bottom the snowpack. It is evident that the variation of emissivity with frequency can offer discrimination between various surface types. Polarization measurements also could provide information on snow conditions by identifying the presence of sub-surface ice layers [2].

Accurate determination of snow parameters from space is a challenging effort. Over the years, many different techniques have been used to account for the complicated snow parameters such as the density, stratigraphy, snow grain size, temperature variation of the snowpack. Forest type, fractional forest cover and land use type also need to be considered in developing an improved retrieval algorithm. However, snow is such a dynamic variable, snowpack parameter keeps changing once the snow is deposited on the earth surface. Snow retrieval algorithm has to take into account the time history of the evolution of the snowpack.

Advanced Microwave Scanning Radiometer (AMSR) will be flown on-board of the Japanese Advanced Earth Observing Satellite-II (ADEOS-II) and the United States Earth Observation System (EOS) PM-1 satellite. With a large antenna, AMSR will provide the best spatial resolution of multi-frequency radiometer from space. This provides us an opportunity to study the surface features that were not be resolved before.

Currently, NASDA and NASA are developing AMSR snow retrieval algorithms. These algorithms are now being carefully tested and evaluated using the SSM/I data. Due to limited snowpack data available for comparison and validation, this activity is progressing slowly.

2. AMSR INSTRUMENT

AMSR is a passive microwave radiometer with frequency ranges from 6.9 GHz to 89 GHz. It will be carried on board the ADEOS-II and EOS PM-1 satellites to be launched in later part of year 2000. There are eight frequencies for ADEOS-II AMSR, while there are six frequencies for EOS PM-1 AMSR-E. The spatial resolution varies from approximately 50 km at 6.9 GHz to 5 km at 89 GHz. The antenna aperture is 2 m for AMSR and 1.6 m for AMSR-E. The antenna beams scan by continuous rotation along a conical surface, which intersects the earth's surface at an angle of 55°. The swath width is about 1600 km. Calibration data are taken during the remainder of the rotation. It measures the radiation in both horizontal and vertical polarizations. AMSR on ADEOS-II has two additional channels in the oxygen band of microwave spectra (50.3 and 52.8 GHz). These two channels only measure the vertically polarized component. Frequency, spatial resolution and polarization of the AMSR and AMSR-E sensors are tabulated in Table 1.

Table 1. Frequency (GHz), resolution (km) and polarization characteristics of AMSR and AMSR-E

ADEOS II AMSR							
Frequency	6.9	10.65	18.7	23.8	36.5	89.0	50.3	52.8
Resolution	50	50	25	25	15	5	10	10
Polarization	V&H	V&H	V&H	V&H	V&H	V&H	V	V

EOS PM1 AMSR-E					
Frequency	6.9	10.65	18.7	23.8	36.5	89.0
Resolution	58	38	24	24	13	5
Polarization	V&H	V&H	V&H	V&H	V&H	V&H

3. EXISTING ALGORITHMS

Microwave techniques are the most promising for measuring the SWE because microwave can penetrate into snowpacks. Snow crystals are effective scatters of microwave radiation. The deeper the snowpack, the more crystals are available to scatter microwave energy away from the sensor, thus lower the brightness temperature. Based on this physical phenomenon many algorithms have been developed using either Nimbus-7 SMMR data or the more recent SSM/I data [3-6]. SSM/I data have been used semi-operationally to infer snow water equivalent values over Canadian Prairie with some success [6]. A typical algorithm used by many researchers is employed to demonstrate algorithm development.

Microwave radiation emanates from features on or near the surface of the Earth that is proportional to the product of the physical temperature and the emissivity of the surface. The measured value referred to as the brightness temperature can simply be expressed as:

$$T_B = (R \, T_{sky} + (1 - R) \, T_{surf}) \, e^{-\tau} + T_{atm} \qquad (1)$$

where $e^{-\tau}$ is the atmospheric transmissivity and R is the surface reflectivity, T_{sky} is the sky radiation, T_{surf} is the surface emission, and T_{atm} is the emission from the intervening atmosphere. In the microwave region both T_{sky} and T_{atm} are small in a fair weather condition and can be neglected. Thus, the observed T_B is directly related to surface features.

Based on radiative transfer calculation [4], a relationship between brightness temperature and the number of snow crystals was developed for SWE retrieval. The differences between the 19 and 37 GHz brightness temperature is linearly related to the snow water equivalent values when SWE is less than 200 mm [4]. The scattering information comes largely from the 37 GHz signal. The 18 GHz signal is used to minimize the varying surface temperature. The SWE - brightness temperature relationship of a homogeneous snow layer can generally be expressed as follows;

$$SWE = A \times (T_{19} - T_{37}) + B \qquad mm, \qquad (2)$$

where SWE is the snow water equivalent in mm of equivalent water, T_{19} and T_{37} the brightness temperature for the 19 and 37 GHz. A and B are constants depending on the snow conditions and sensors [3-6]. Both vertical and horizontal polarization will give generally similar results in Eq (2). Due to differences in the surface snow characteristics, researchers have used either vertical or horizontal polarization [7-8] in retrieving the SWE.

The brightness temperature difference for forest covered areas will cancel out if the emissivities of forest for both the high scattering and the low scattering channels are about the same. This is based on the findings that the emissivities for forest in Finland at 37 and 19 GHz are very similar and have the values of 0.9 to 0.92 [9]. Thus, only the snow-covered fractions will contribute to the brightness temperature difference. For a footprint with f fraction of forest cover and (1 - f) fraction snow cover, Eq (2) will become

$$SWE = A * \Delta T_B / (1 - f) + B \qquad mm. \qquad (3)$$

Eqs (2) and (3) can be used to infer snow water equivalent values for relatively homogeneous snowpack with some success [10]. More recently, Kruopis et al. [11] found that forest stem volume could also affect the brightness temperature, hence it requires more study. Due to the large footprints of microwave sensors, mixed pixel will further complicate the retrieval. Forest cover and other types of surface affect the brightness temperature differently. To improve the accuracy of estimated SWE values, it is necessary to understand the changes within the snowpack due to mixed signatures.

4. ALGORITHM INTER-COMPARISON STUDY

In order to assess the quality of the pre-launch snow retrieval algorithms, Japanese Earth Observation Research Center (EORC) compiled a four years (1992-1995) data set based on in-situ measurements and SSM/I data. Snow parameters and climatology data from one hundred climatology stations spread over the Northern Hemisphere were compiled. Figure 1 depicts the station location. Snow parameter data includes snow depth, general surface condition, and snow temperature. Climatology data includes air temperature, dew point temperature, wind speed and direction and pressure. Five closest SSM/I footprints were extracted from the swath data to form the coincident data set.

Selected 100 Stations

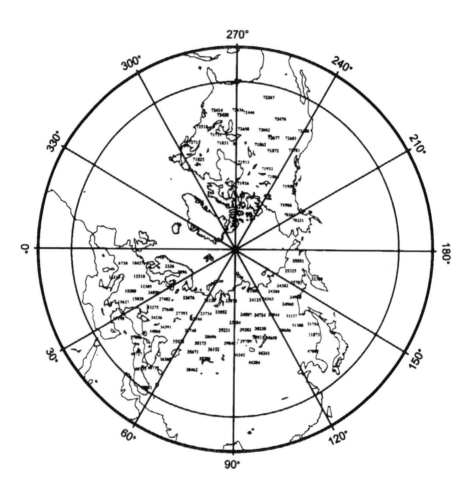

Figure 1. Station number and location of 100 selected stations.

There are three algorithms selected by EORC for snow algorithm test. At the time of the test, all algorithms are still under development and subject to changes. Table 2 is a summary of these three algorithms. Chang's algorithm is built on the model developed for the SMMR data [4]. SMMR algorithm was modified to take into account different snow conditions in different snow regions and with forest cover corrections. The snow classification system of Sturm et al. [12] is used to classify snow and Robinson and Kukla [13] maximum snow albedo map is used as a proxy for percent forest cover. For a rough estimates of surface temperature, Basist et al. [14] technique is used. Kokie's algorithm is derived from simulated brightness temperature calculated from a radiative transfer model for porous media. Vegetation effect is corrected by applying the monthly NDVI values from ISLSCP data sets. Tsang's algorithm is based on a dense media radiative transfer model to simulate the 19V, 19H, 37V, and 37H brightness temperatures [15]. A neural network is used to extract the snow depth, density, grain size and snow water equivalent information. Initial estimates of snow parameters were input to a hydrologic snowmelt model to account for the time evaluation of the snowpack. From time to time, climatology data are needed to update the hydrologic model. Due to the requirement of this iterative loop, this algorithm was considered difficult to implement for global application and not tested by NASDA.

Table 2. Summary of inter-comparison of snow algorithms

	Chang	Kokie	Tsang
Physical basis	Radiative transfer for dense media	Radiative transfer model for porous media plus vegetation effects	Radiative transfer for dense media
Inversion	Classification and statistical regression	Lookup table generated by interpolation of the result of model	Iterative inversion by using neural network
Algorithm	1) retrieve snow conditions 2) estimate surface temperature 3) snow classifications 4) select eq. And apply it	Snow depth and temperature calculated from a lookup table for each NDVI	1) NN training 2) Input initial values into NN 3) Comparison with observed TB 4) Iterative inversion
Input data	19V/H, 22V, 37V/H, 85V of SSM.I snow and forest condition	10H, 37V/H of SSM/I monthly mean NDVI and air temperature from ISLSCP CD-ROM	19V/H, 37V/H of SSM/I initial guesses derived from SHM by inputting weather data
Outputs	SWE, SD	SWE, SD	SD, density, grain size
Operational consideration	Global snow mapping	Global snow mapping	Difficult to implement for global product

A statistical analysis of Chang's and Kokie's snow retrievals was performed for the period of January 20 to 25, 1993 (Figure 2). Basically, both algorithms relied on radiative transfer calculations to derive the SWE brightness temperature relationships. Thus, they gave very similar results. Due to the different data filtering techniques, Chang's algorithm was able to retrieve snow depth from 93 stations and the mean absolute difference between the observations and the estimates is 25.2 cm. Kokie's algorithm retrieved snow depth from 69 stations and the mean absolute difference between the observations and estimations is 24.5 cm. If retrievals from all stations and for the entire period were considered, Chang's algorithm gave a slightly better mean absolute difference than Kokie's algorithm. Both algorithms were not able to infer accurately the snow information from deep snowpacks. This is probably due to the limited penetration of 37 GHz radiation. For snow depth larger than 60 cm, these algorithms retrieved snow water equivalent values less than half of the true values.

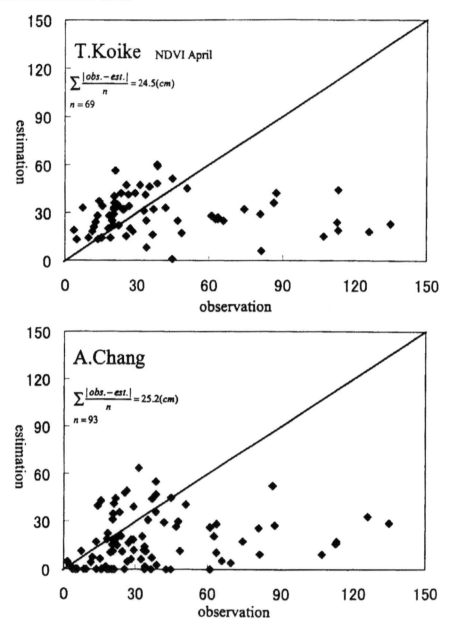

Figure 2. Scatter plot of estimated snow depth (cm) and observed snow depth (cm).

5. FUTURE ALGORITHM

As a first step in improving the snow algorithm, a decision tree approach will be used to identify the various snow types based on the frequency and polarization measurements, similar to those reported in [1]. Taking the advantage of large penetration depth into the snowpack, combinations of 10 and 19 GHz radiation will be used to detect the deep snowpack. This step will serve as the preliminary identification of snow and its water equivalent. Particularly it is necessary to distinguish the shallow snow and wet snow cover.

It is learned that snow grain size changes as the metamorphism advances. Snowpacks on the earth's surface memorize the changes of these physical parameters. Recently Josberger and Mognard [16] developed a technique using measured air temperature as a proxy for snowpack grain size. A temperature gradient index, accumulated daily, is found to correlate very well with the brightness temperature differences. This result can be applied to the cold snowpacks found in Alaska, Siberia, and Canada, where strong snowpack temperature gradients dominate the metamorphic processes.

Due to the rapid change of the snowpack parameters, it is necessary to constrain the retrieval in order to obtain accurate snow estimates. A priori statistics of snowpack information will be one way to stabilize the retrieval processes. An ensemble of brightness temperature with varying snow depth, density, grain size distribution, temperature, fractional forest cover, frequency and polarization will be generated using a dense media radiative transfer model. Based on the classification in the previous step, cases with brightness temperature closest to measured values will be selected. A final decision tree approach will be used to select the snow parameters that best match the measured multi-frequency brightness temperatures.

Research is still required to better delineate between light precipitation and shallow snow cover, which can have similar signatures. Microwave snow detection techniques can be affected by wet and melting snow. Some progress has been made using polarization measurement [17] to detect wet snow. It is well known that wet snow emits like a black body. Microwave measurements may be able to delineate the patchy snowmelt, which increase the brightness according to the percentage of melting areas. A detailed test effort is now being formulated to quantify the effect of snow wetness.

6. CONCLUSION

Microwave sensors onboard of spacecraft obtained useful data for snow parameter retrieval. Based on the preliminary comparison results, it is possible to conclude that (1) it is necessary to account for the forest coverage to improve snow water equivalent estimates, and (2) the physical temperature and the wetness of the snowpack are needed to accurately determine the snow water equivalent.

Forest is the first one that needs our attentions. Snowpack melting does cause difficulty in retrieving snow water equivalent. Also, snow grain size may contribute to large errors. These problems will be addressed in the new algorithm currently under development.

An interim algorithm is now being tested and refined in hope they will be ready for the ADEOS-II and EOS PM-1 launch date. It is a challenging but necessary task to develop robust snow retrieval algorithms that will give a reliable snow water equivalent estimates and allow for a quality, long term snow data set.

REFERENCES

1. Grody, N.C., Classification of snowcover and precipitation using the special sensor microwave imager. *J. Geophys. Res., 96*, 7423-7435, (1991).
2. C. Matzler, "Passive microwave signatures of landscapes in winter," *Meteorol. Atmos. Phys., 54*, 241-260, (1994).
3. K.F. Kunzi, S. Patel and H. Rott, "Snow cover parameters retrieved from the Nimbus-7 scanning multichannel microwave radiometer (SSMR) data. *Rev. Geophys. , 22*, 452-467, (1984).
4. Chang, A.T.C., J.L. Foster and D.K. Hall, Nimbus-7 derived global snow cover parameters, *Annuals of Glaciology, 9*, 39-44, (1987).
5. Rott, H. and J. Aschbacher, On the use of satellite microwave radiometers for large-scale hydrology, *Proc. IASH 3rd Int. Assembly on Remote Sensing and large Scale Global Processes*, Baltimore, 21-30, (1989).
6. Goodison, B.E., and A.E. Walker, Use of snow cover derived from satellite passive microwave data as an indicator of climate change. *Annals of Glaciology, 17*, 137-142, (1993).
7. Hallikainen, M.T., and P.A. Jolma, Comparison of algorithms for retrieval of snow water equivalent from Nimbus-7 SMMR data in Finland. *IEEE Trans. on Geoscience and Remote Sensing, 30*, 124-131, (1992).
8. Goodison, B.E., and A.E. Walker, Canadian development and use of snow cover information from passive microwave satellite data, *Passive Microwave Remote Sensing of Land-Atmosphere Interactions,(Eds. Choudhury, Kerr, Njoku and Pampaloni), VSP*, 245-262, (1994).
9. Hallikainen, M.T., P.A. Jolma and J.M. Hyyppa, Satellite microwave radiometry of forest and surface types in Finland. *IEEE Trans. on Geoscience and Remote Sensing, 26*, 622-628, (1988).
10. Chang, A., J. Foster and A. Rango, Utilization of surface cover composition to improve the microwave determination of snow water equivalent in a mountainous basin. *International Journal of Remote Sensing, 11*, 2311-2319, (1991).
11. Kruopis, N., J. Koskinen, J. Praks, A.N. Arslan, H. Alasalmi, and M. Hallikainen,, Investigation of passive microwave signatures over snow-covered forest areas. Proc. IEEE IGARSS'98, 1544-1546, (1998).
12. Sturm, M., J. Holmgren, and G.E. Liston, A seasonal snow cover classification system for local to global applications. *J. Climate, 8*, 1261-1283, (1995).
13. Robinson, D.A., and G. Kukla, Maximum surface albedo of seasonally snow-covered lands in the northern hemisphere. *Journal of. Climate and Applied Meteorology, 24*, 402-411, (1985).
14. Basist, A., N.C. Grody, T.C. Peterson and C.N. Williams, Using the Special Sensor Microwave/Imager to monitor land surface temperature, wetness, and snow cover. *Journal of Applied Meteorology, 37*, 888-911, (1998).
15. Wilson, L.L., L. Tsang, J.N. Hwang, and C.T. Chen, Mapping snow water equivalent by combining a spatially distributed snow hydrology model with passive microwave remote-sensing data. *IEEE Trans. GRS, 37*, 690-704, (1999).
16. Josberger, E.G., and N.M Mognard, A passive microwave snow depth algorithm with a proxy for snow metamorphism, submitted to *Journal of Hydrologic Processes*, (1998).

17. Walker, A.E. and B.E. Goodison, Discrimination of a wet snow cover using passive microwave satellite data. *Annals of Glaciology, 17*, 307-311, (1993).

Microw. Radiomet. Remote Sens. Earth's Surf. Atmosphere, pp. 525–533
P. Pampaloni and S. Paloscia (Eds)
© VSP 2000

Retrieval of soil moisture from AMSR data

ENI NJOKU,[1] TOSHIO KOIKE,[2] THOMAS JACKSON,[3] and SIMONETTA
PALOSCIA[4]

[1]*Jet Propulsion Laboratory, Pasadena, CA, USA*
[2]*Nagaoka University of Technology, Nagaoka, JAPAN*
[3]*USDA/ARS/Hydrology Laboratory, Beltsville, MD, USA*
[4]*CNR/Istituto di Ricerca sulle Onde Elettromagnetiche, Firenze, ITALY*

Abstract—In this paper we discuss the potential and problems of soil moisture sensing using AMSR data that will become available in late 2000 or early 2001. The Advanced Microwave Scanning Radiometer (AMSR) will be the first spaceborne radiometer since the Nimbus-7 SMMR to include a frequency at C-band (6.9 GHz). The ability to penetrate vegetation, and to sense deeper in the soil, increases with wavelength. The AMSR is thus expected to have a better soil moisture sensing capability than the DMSP SSM/I or TRMM Microwave Imager (TMI) instruments which have lowest frequencies of 19.35 and 10.7 GHz, respectively. The spatial resolution of the AMSR at 6.9 GHz is approximately 60 km, a factor of two better than the SMMR. While not optimal for soil moisture sensing, the AMSR should provide useful information over low-vegetated areas, and will serve as a valuable precursor to future proposed L-band soil moisture sensors. Vegetation and snow cover, frozen ground, topography, open water, and footprint heterogeneity are surface characteristics that must be considered in estimating soil moisture. Ancillary data that can provide information on these characteristics will be useful in improving the retrievals. Descriptions and preliminary tests of different soil moisture retrieval approaches are discussed here. Experiments using ground-based, airborne, and satellite measurements are planned for continued development of the retrieval algorithms, and for validation of the derived soil moisture products after launch.

1. INTRODUCTION

The Advanced Microwave Scanning Radiometer (AMSR) will provide new data of interest for land surface studies, particularly for soil moisture sensing. The AMSR is currently planned for launch on the ADEOS-II and EOS PM-1 satellites in late 2000. The AMSR will operate with vertical and horizontal polarizations at frequencies of 6.9, 10.7, 18.7, 23.8, 36.5, and 89 GHz. It will be the first spaceborne radiometer since the SMMR in 1987 to include a C-band (~7 GHz) frequency. The TRMM Microwave Imager (TMI), launched in 1998, includes an X-band (10.7 GHz) frequency, but coverage is limited to the tropics (between ±~35° latitude). Penetration of vegetation and soil is greater at lower frequencies, thus the AMSR is expected to have better soil moisture sensing capability than either the SSM/I (with a lowest frequency of 19.35 GHz) or the TMI. The spatial resolution of the AMSR (~60 km) will be about a factor of two better than the SMMR, and is comparable to the grid scales of current climate models. While not optimal for soil moisture, the AMSR should provide useful information over low-vegetated areas, and will serve as a valuable precursor to future proposed L-band soil moisture sensors. The ADEOS-II and EOS PM-1 satellites will have equator crossing times of 10:30 am and 1:30 pm, respectively. By combining data from the AMSR instruments on both satellites an evaluation can be done of the improvement in soil moisture estimation obtainable by sampling at different points in the diurnal surface temperature cycle.

There are significant challenges to be overcome in developing algorithms for AMSR soil moisture sensing. Validation of the derived soil moisture products will also pose some problems. Vegetation and snow cover, frozen ground, topography, open water, and sub-footprint heterogeneity need to be considered in the retrievals. Extrapolation of the retrieved surface moisture to deeper soil layers is desired for some hydrologic applications. Registration of the AMSR data to an earth-fixed grid will be advantageous in using ancillary data sets, such as soil texture, topography, vegetation index, and climate data, to address some of these problems.

Approaches currently being developed for AMSR soil moisture retrieval are described in this paper. Since soil moisture is not currently measured from space, and is considered a challenging measurement, a number of different algorithms are being considered. Evaluation of these approaches is expected to extend through the post-launch validation phase. Experiments using in-situ surface measurements, airborne sensors, and satellite data are planned for further development of the retrieval algorithms, and for validation of the derived products after launch. The ADEOS-II and EOS AMSR validation activities will be conducted in conjunction with ongoing agency investigations in hydrometeorology and climate, and will include field experiments and data acquisitions planned by international programs such as the GEWEX Coordinated Enhanced Observing Period (CEOP) experiments.

2. MODELING

The research framework for soil moisture sensing using passive microwaves has been well established [1], [2], and includes an extensive history of field experiments and modeling studies. Most of these investigations have been performed at 1.4 GHz (L-band), with fewer studies done at the higher frequencies at which AMSR operates (C- and X-band). Algorithms for retrieving soil moisture from AMSR data require microwave models valid at the frequencies, viewing angle, and spatial resolution of the AMSR measurements. The models need to be relatively simple for satellite application yet physically meaningful. Some model parameters need to be estimated using ancillary data or other a-priori information.

The emission model most appropriate for AMSR soil moisture retrieval represents the land surface as a homogeneous, single-scattering vegetation layer above the soil. This model has been discussed by many authors [3], [4]. The soil is characterized by a surface reflectivity (r_p) and effective temperature (T_s). The reflectivity is related to the soil moisture (m_v) by the Fresnel expressions, modified for surface roughness. The vegetation acts as an attenuating and emitting layer, and is characterized by the opacity (τ_c), temperature (T_c), and single-scattering albedo (ω). The observed brightness temperature (T_{b_p}) is then given by

$$T_{b_p} = T_s [1 - r_p] \exp(-\tau_c) + T_c (1 - \omega) [1 - \exp(-\tau_c)] [1 + r_p \exp(-\tau_c)] \quad (1)$$

where p refers to either vertical or horizontal polarization. The vegetation opacity is related to the vegetation water content, and less directly to the leaf area index. The single-scattering albedo is small, and is usually neglected, but can be estimated if necessary as a calibration parameter. The effects of atmospheric absorption and emission are not usually significant at X-band frequencies and below.

Equation (1) represents the dominant physical mechanisms for surface emission and scattering and keeps to a minimum the number of free parameters that must be known or estimated. The following factors must be considered in applying the model since they affect the retrieval accuracy. First, the single scattering albedo and vegetation opacity can exhibit polarization dependence related to the orientations of leaves, stalks, and branches within the vegetation volume. Second, the effect of surface roughness on soil reflectivity is difficult to parameterize since few studies have been made of surface roughness at the AMSR footprint scale. Finally, since land surfaces are heterogeneous, the modeled and retrieved parameters must be interpreted as averages over the horizontal footprint and the vertical sensing depth [5]. At ~7–10 GHz the AMSR footprint is ~60 km, and the soil moisture sensing depth is on the order of less than a centimeter.

The most important relationship for soil moisture sensing is that between the soil dielectric constant (hence reflectivity and emissivity) and the soil volumetric moisture [6], [7]. This relationship is influenced by soil texture (represented by the fractional contents of sand, silt, and clay), and surface roughness. Ancillary data on soil texture and topography are thus useful in improving the accuracy of the retrievals.

The dependence of vegetation opacity on columnar water content (w_c) and leaf area index (*LAI*) has been studied both theoretically and experimentally. Two expressions developed independently to represent these relationships are [8], [9]

$$\tau_c = b \, w_c / \cos\theta \quad (2)$$

$$\tau_c = [(k/\sqrt{\lambda}) \ln(1 + w_c)] / \cos\theta = [(k/y\sqrt{\lambda}) \, LAI] / \cos\theta \quad (3)$$

where the $\cos\theta$ factor accounts for the slant observation path through the vegetation.

In the first expression, b is a parameter that depends on vegetation type and is approximately proportional to frequency. As frequency increases, the frequency dependence of b decreases, and its dependence on canopy structure increases [10], [11]. Thus, ancillary data on vegetation type or classification are useful to calibrate b for global applications. In the second expression, a logarithmic function is used to express the dependence of vegetation opacity on water content, and a linear dependence is used to relate the opacity to leaf area index. The wavelength dependence is expressed explicitly. The coefficients k and y depend on vegetation characteristics and must be determined experimentally. Again, ancillary data are useful.

Variability in land surface temperature causes uncertainty in retrieving soil moisture. One method to correct for this is to use ancillary information from meteorological surface air temperature data, or forecast model land-surface temperatures. Another option is to use brightness temperature-derived indices which are, to first order, independent of surface temperature, instead of brightness temperatures in the retrieval algorithm. Indices of interest include a spectral gradient index (I_{sw}), and a polarization index (PI), which can be defined as

$$I_{sw} = \frac{(T_{b_i} - T_{b_j})}{\frac{1}{2}(T_{b_i} + T_{b_j})} \tag{4}$$

$$PI = \frac{(T_{b_v} - T_{b_h})}{\frac{1}{2}(T_{b_v} + T_{b_h})} \tag{5}$$

where I_{sw} is defined for a specific polarization, and for frequency i greater than j, and PI is defined for a specific frequency. The different sensitivities of these indices to soil moisture and vegetation can be exploited in the retrievals. For instance, lower frequencies penetrate deeper through vegetation and soil, are less sensitive to surface roughness, and are more sensitive to soil moisture. Hence, I_{sw} typically increases with soil moisture. Similarly, PI typically increases with soil moisture since the brightness temperature at horizontal polarization is more sensitive to soil moisture than at vertical polarization. PI also decreases with increasing vegetation and roughness, and is less sensitive to soil moisture at higher frequencies [9], [12].

3. ANCILLARY DATA

Ancillary data are useful for improving the accuracy of soil moisture retrievals, and are critical for some approaches. Much of the required data are readily available, but from diverse sources and in varying formats and resolutions. These data sets must be quality-controlled and registered to the satellite data. Some of the required data are static, such as soil texture and topography, while others vary with different time scales, such as vegetation cover and surface temperature. The data must be used carefully, since in most cases the data parameters do not correspond directly to parameters of the microwave models. For instance, vegetation index (NDVI) is related indirectly to vegetation water content and leaf area index, while in-situ surface air temperature, and forecast model land-surface temperature, are related indirectly to the microwave-observed land surface temperature. Table 1 lists some of the candidate ancillary data sets and their sources.

4. RETRIEVAL ALGORITHMS

Modeling and experimental data suggest that the AMSR C-band (6.9 GHz) horizontally polarized channel will be the most sensitive to soil moisture. An algorithm using this channel alone can be used to retrieve soil moisture if ancillary data are available to correct for surface temperature, vegetation, surface roughness, and soil texture. An algorithm of this form has been applied to soil moisture estimation using L-band (1.4 GHz) aircraft data [13].

By using normalized indices as defined in Equations (4) and (5), the reliance on ancillary surface temperature data can be avoided. One method is to use the indices PI and I_{sw}, at appropriate frequencies to estimate the vegetation correction and soil moisture. The two indices are chosen using frequencies that provide different sensitivities to soil moisture and vegetation, so that they can be used to retrieve these parameters independently. Alternatively, two *PI* indices at different frequencies, e.g. 6.9 and 10.7 GHz, can be used to estimate the vegetation correction and soil moisture. To implement

Table 1: Ancillary data sets for use in retrieval of soil moisture from AMSR data.

Data Set	Original Resolution	Source	Comments
Surface Topography	30 arc-sec	USGS/EDC	GTOPO30 Global DEM
Soil Texture	0.08° x 0.08°	Global GRASS1	Based on Zobler [17]
Land Cover	1 km	USGS/EDC	Loveland et al. [18]
NDVI	1 km x 1 km	AVHRR, MODIS	
Surface Temperature	2° x 2°	GSFC/DAO, NCEP, ECMWF	

the polarization index methods, table look-up or iterative schemes can be used. These may be based on a simple forward model such as Equation (1) or on empirical data. Alternatively, a nonlinear regression equation can be developed using the indices.

As an alternative to using ancillary data for the surface temperature corrections, and to using brightness temperature indices to normalize the surface temperature effects, one can use the higher frequency AMSR channels to provide information and correct for the surface temperature variability. At higher frequencies, and vertical polarization, the brightness temperature is less sensitive to soil moisture and vegetation, and more sensitive to surface temperature [14]–[16]. Thus, the 6.9, 10.7, and 18 GHz channels at both polarizations (six channels of information) can be used to estimate surface temperature, vegetation moisture content, and soil moisture independently. An iterative retrieval scheme based on the model of Equation (1) can be used to implement this approach. In practice, ancillary data on surface temperature (as discussed above) can be incorporated into the retrieval scheme as a-priori information or constraints on the retrieval.

The AMSR data will require screening for land surface type prior to implementing the soil moisture retrieval algorithms. Land cover databases and classification algorithms will be used to identify locations of dense vegetation, open water, snow cover, variable topography, and other adverse retrieval conditions. For these purposes, it is advantageous to pre-process the ancillary data sets and brightness temperature data to the same grid. The data sets, classification algorithms, and soil moisture retrieval algorithms, can then be managed similar to a conventional geographic information system.

5. ALGORITHM TESTS

The soil moisture retrieval approaches discussed above have been tested to varying degrees using theoretical simulations, experimental data from ground-based and airborne campaigns, and satellite data. Descriptions of this research can be found in the literature [4], [9], [13]. In [4], model simulations were performed to estimate the expected soil moisture retrieval accuracy. The simulations included a wide range of soil moisture, vegetation water content, and surface temperature conditions, and assumed expected sensor noise (ΔT) of 0.3K. The results showed that an accuracy of 0.06 g cm^{-3} should be achievable for vegetation cover less than ~1.5 kg m^{-2}. This estimate is consistent with sensitivity studies of attenuation by vegetation canopies [19].

Realistic tests of algorithm performance using satellite data are difficult. The best satellite data for simulating AMSR C- and X-band measurements are the Nimbus-7 SMMR data covering the time period 1978–1987. Unfortunately, there are no suitable in-situ measurements available during this period for validating SMMR-derived soil moisture retrievals. However, an intercomparison test of the different approaches was carried out recently using a data set provided by the Japanese National Space Development Agency (NASDA) of in-situ soil moisture measurements from sites in the former Soviet Union. Soil moisture data from 79 stations, sampled every 10 days, were compared with corresponding soil moisture values retrieved from SMMR data. The in-situ data were measured at point locations, and represent averages over the top 10 cm of soil. No site data on vegetation cover, topography, or surface temperature were available. The SMMR measurements were footprint averages over an approximately 120-km horizontal scale and the top centimeter of the surface. The discrepancies between the in-situ and SMMR spatial sampling, and the lack of ancillary information on surface characteristics, adversely affect the quality of the intercomparisons. Figure 1 shows one example of the results obtained.

The results indicate some agreement between the satellite and in-situ observations, but there is significant scatter (rms of ~8% volumetric soil moisture, or 0.08 g cm^{-3}). All the algorithms tested performed more or less in this range, although tuning to specific sites yielded better results. As mentioned above, the soil moisture accuracy to be expected from AMSR is better than 0.06 g cm^{-3} (in areas where the vegetation cover is less than ~1.5 kg m^{-2}), with better accuracy expected for lower vegetation cover and where the surface characteristics are well known. In view of these sampling and data quality issues the comparison results of Figure 1 are consistent with expectation. A major effort will be required to develop suitable validation data at different global locations to evaluate the quality of the AMSR soil moisture estimates.

6. FIELD EXPERIMENTS AND VALIDATION

Validation plans for the AMSR soil moisture products are currently under development. An initial version of the EOS-PM1 AMSR soil moisture validation plan is available at the web site http://eospso.gsfc.nasa.gov/validation/valpage.html. The validation plan proposes a combination of enhanced in-situ ground measurements at dedicated test sites, ground-based and airborne radiometer measurements, model output data, and intercomparisons with other spaceborne sensors. The dedicated test sites will require enhanced in-situ instrumentation to provide dense sampling over the spatial extent of a few AMSR footprints (~200 x 200 km). These sites need to be automated to provide continuous sampling at 6-hourly intervals, permitting interpretation of diurnal effects on the surface soil moisture. A primary validation site is the U.S. Southern Great Plains (SGP) region centered in Oklahoma. This site was the location of previous L-band soil moisture field experiments, and is the location of the L- and C-band field experiments recently conducted in July 1999 (SGP'99). Information on the SGP'99 experiment is available at the web site http://hydrolab.arsusda.gov/sgp99/.

Validation of soil moisture over areas larger than the dedicated sites will be difficult since the only independent estimates of soil moisture at these scales are products derived from hydrometeorological models, computed from knowledge of surface land cover, soil and topographic characteristics, and surface-air budgets of energy and water. These models have inherent errors that are not yet well understood. Studies using these model products will be pursued in collaboration with data assimilation groups and the operational forecast centers (e.g. NCEP, ECMWF). In addition, the in-situ data collection and modeling activities of international programs such as the GEWEX Coordinated Enhanced Observing Period (CEOP) experiments will be utilized to the extent possible.

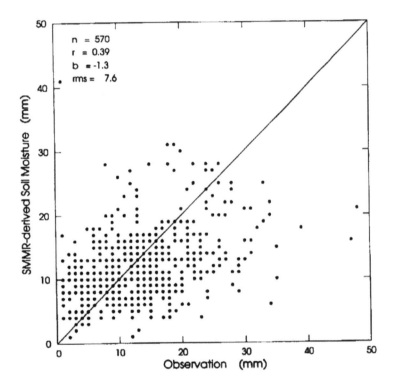

Figure 1. Comparison between soil moisture estimates from SMMR and in-situ observations. In-situ data are from sites over the former Soviet Union.

7. CONCLUSIONS

In this paper, some approaches currently under study for retrieving soil moisture from AMSR data have been described. The theory and experimental background for these approaches is well established. However, the complexity of natural terrain makes the retrieval of soil moisture a difficult problem. Continued experiments involving enhanced in-situ surface observations, airborne sensors, and satellite data are necessary for further improvement of the AMSR retrieval algorithms and for validation of the derived soil moisture products after launch.

ACKNOWLEDGMENTS

This research was supported by the NASA EOS-PM1 project and the NASDA ADEOS-II project. The research was carried out in part at the Jet Propulsion Laboratory, California Institute of Technology, under contract with the National Aeronautics and Space Administration.

REFERENCES

1. J. R. Wang and B. J. Choudhury, "Passive microwave radiation from soil: Examples of emission models and observations," In: *Passive Microwave Remote Sensing of Land-Atmosphere Interactions (B. Choudhury, Y. Kerr, E. Njoku, and P. Pampaloni, Eds.)*, 423–460, VSP, Utrecht, The Netherlands (1995).

2. Y. H. Kerr and J. P. Wigneron, "Vegetation models and observations - A review," In: *Passive Microwave Remote Sensing of Land-Atmosphere Interactions (B. Choudhury, Y. Kerr, E. Njoku, and P. Pampaloni, Eds.)*, 317–344, VSP, Utrecht, The Netherlands (1995).

3. T. Mo, B. J. Choudhury, T. J. Schmugge, J. R. Wang, and T. J. Jackson, "A model for microwave emission from vegetation covered fields," *J. Geophys. Res.*, **87**, 11229–11237 (1982).

4. E. G. Njoku and L. Li, "Retrieval of land surface parameters using passive microwave measurements at 6–18 GHz," *IEEE Trans. Geosci. Rem. Sens.*, **37**, 79–93 (1999).

5. E. G. Njoku and D. Entekhabi, "Passive microwave remote sensing of soil moisture," *J. Hydrology*, **184**, 101–129 (1996).

6. J. R. Wang and T. J. Schmugge, "An empirical model for the complex dielectric permittivity of soil as a function of water content," *IEEE Trans. Geosci. Rem. Sens.*, **18**, 288-295 (1980).

7. M. C. Dobson, F. T. Ulaby, M. T. Hallikainen, and M. A. El-Rayes, "Microwave dielectric behaviour of wet soil – Part II: Dielectric mixing models," *IEEE Trans. Geosci. Rem. Sens.*, **23**, 35-46 (1985).

8. T. J. Jackson and T. J. Schmugge, "Vegetation effects on the microwave emission from soils," *Rem. Sens. Environ.*, **36**, 203-212 (1991).

9. S. Paloscia and P. Pampaloni, "Microwave polarization index for monitoring vegetation growth," *IEEE Trans. Geosci. Rem. Sens.*, **26**, 617-621 (1988).

10. U. Wegmuller, C. Matzler, and E. Njoku, "Canopy opacity models," In: *Passive Microwave Remote Sensing of Land-Atmosphere Interactions (B. Choudhury, Y. Kerr, E. Njoku, and P. Pampaloni, Eds.)*, 375–387, VSP, Utrecht, The Netherlands (1995).

11. D. LeVine and M. Karam, "Dependence of attenuation in a vegetation canopy on frequency and plant water content," *IEEE Trans. Geosci. Rem. Sens.*, **34**, 1090–1096 (1996).

12. Y. H. Kerr and E. G. Njoku, "A semiempirical model for interpreting microwave emission from semiarid land surfaces as seen from space," *IEEE Trans. Geosci. Rem. Sens.*, **28**, 384-393 (1990).

13. T. J. Jackson and D. E. LeVine, "Mapping surface soil moisture using an aircraft-based passive microwave instrument: algorithm and example," *J. Hydrology*, **184**, 85–99 (1996).

14. J.-C. Calvet, J.-P. Wigneron, E. Mougin, Y. H. Kerr, and J. S. Brito, "Plant water content and temperature of the Amazon forest from satellite microwave radiometry," *IEEE Trans. Geosci. Rem. Sens.*, **32**, 397–408 (1994).

15. E. G. Njoku, "Surface temperature estimation over land using satellite microwave radiometry," In: *Passive Microwave Remote Sensing of Land-Atmosphere Interactions (B. Choudhury, Y. Kerr, E. Njoku, and P. Pampaloni, Eds.)*, 509–530, VSP, Utrecht, The Netherlands (1995).

16. J. Pulliainen, J. Grendell, and M. Hallikainen, "Retrieval of surface temperature in Boreal forest zone from SSM/I data," *IEEE Trans. Geosci. Rem. Sens.*, **35**, 1188–1200 (1994).

17. L. Zobler, "A world soil file for global climate modeling," *NASA Technical Memorandum #87802*, National Aeronautics and Space Administration, Washington, DC, 1986.

18. T. R. Loveland and A. S. Belward, "The IGBP-DIS global 1-km land cover data set, DISCover: first results," *Int. J. Rem. Sens.*, **18**, 3291–3295, (1997).

19. T. J. Jackson, "Measuring surface soil moisture using passive microwave remote sensing," *Hydrological Processes*, **7**, 139–152 (1993).

Microw. Radiomet. Remote Sens. Earth's Surf. Atmosphere, pp. 535–540
P. Pampaloni and S. Paloscia (Eds)

AMSR/AMSR-E algorithm developments in NASDA - water vapor, cloud liquid water, sea surface wind speed, and SST

AKIRA SHIBATA [1] and TADAHORO HAYASAKA [2]

[1]*Earth Observation Research Center, NASDA, Roppóngi 1-9-9, 106-0032 Japan*
[2]*Center for Atmospheric and Oceanic Studies, Tohoku University, Sendai, 980-0845 Japan*

Abstract — The National Space Development Agency of Japan (NASDA) has developed two satellite-borne microwave radiometers, the Advanced Microwave Scanning Radiometer (AMSR) and AMSR-E. Various geophysical products, mainly related to water on the Earth, will be retrieved from the two AMSR instruments, and NASDA will produce eight geophysical products (total water vapor, total cloud liquid water, precipitation, sea surface wind speed, sea surface temperature, sea ice concentration, snow depth, and soil moisture). Algorithms used after the launch were solicited under the first Research Announcement (RA). Proposed algorithms were compared by using common data sets between satellite (aircraft) data and truth data. The day-1 algorithms were selected from those candidates. The Day-1 algorithms will be tuned for six months after the launch, and used for 18 months. Re-selection of standard algorithms is planned 18 months after the launch.

1. INTRODUCTION

The National Space Development Agency of Japan (NASDA) has developed two satellite-borne microwave radiometer. One is the Advanced Microwave Scanning Radiometer (AMSR). AMSR will fly aboard NASDA's ADvanced Earth Observation Satellite -II (ADEOS-II) which will be launched in Nov. 2000, and will have a local observation time of 10:30 a.m. The other radiometer is AMSR-E. AMSR-E will fly aboard NASA's Earth Observing System PM-1 (EOS-PM1) which will be launched in Dec. 2000, and will have a local observation time of 1:30 p.m. Hereafter "AMSR" will refer to both sensors.

Compared with other sensors such as the Special Sensor Microwave / Imager (SSM/I) aboard the US Defense Meteorological Satellite (DMSP) [1], AMSR have lower frequencies of 6 and 10GHz, and a larger antenna (2 m diameter for ADEOS-II and 1.6 m diameter for EOS-PM1).

NASDA plans to produce eight geophysical products from AMSR data. Those products, total water vapor, total cloud liquid water, precipitation, sea surface wind speed, sea surface temperature (SST), sea ice concentration, snow depth, and soil moisture, would be used in both research and operations. In research work, they would primarily be used to study the water and energy cycle as a part of global climate system. They would also be used to detect possible signals of long-term climate change, through making continuous observations by AMSR and a proposed AMSR follow-on aboard the proposed ADEOS-III.

In operations such as weather forecasting, data on water vapor, cloud liquid water, precipitation, and wind speed would contribute to improving short range (within a week) weather forecasting. In addition, data on SST, sea ice concentration, snow depth, and soil moisture would contribute improving medium-range (several weeks) weather forecasting.

Though many types of algorithms have been developed during the last few decades [2-4], additional algorithms must be developed to produce accurate geophysical products. For some of the eight products, some established techniques might be useful for developing algorithms. In those cases, algorithm can be developed easily. Even so, accuracy of retrieved products could be improved by using AMSR's new frequencies of 6 and 10 GHz. Also, sea surface wind speed might be estimated accurately even under rain, and precipitation might be estimated accurately even in heavy rain. Furthermore, several types of algorithms will be necessary for new products such as SST and soil moisture, and many trials must be conducted.

Within several decades, satellite data will be assimilated for numerical models. Of the several kinds of satellite data, passive microwave radiometer data are thought to be most suitable for weather and climate numerical models. For AMSR, parameters analyzed by weather forecasting models will be used to retrieve geophysical products. Vertical profile of air temperature and humidity would be particularly useful. However, NASDA only use elementary retrieval techniques for AMSR data; meteorological agencies would be expected to perform more complete data assimilation.

Section 2 of this paper will describe algorithms previously developed by NASDA. Section 3 will present the day-1 algorithms which will be used for 18 months after the launch, particularly those for water vapor, cloud liquid water, wind speed, and SST products. Section 4 will discuss future activities and present a summary.

2. ALGORITHM DEVELOPMENTS

NASDA issued the first Research Announcement (RA) in October 1995 to solicit standard algorithms. Standard algorithms will be installed at NASDA's Earth Observation Center to produce level 2 geophysical products. Several requirements are for standard algorithms. They must provide accurate products, run on appropriate computer facilities, be as simple as if possible, and run as independently as possible. Principal Investigators (PIs) must also support algorithm installation before launch and algorithm tuning after launch. Standard algorithms may use air temperature, humidity, and other data obtained from global analysis is by the Japan Meteorological Agency (JMA) weather forecasting model.

A total of 28 PIs was selected by NASDA from the first RA. Table 1 relates PIs and geophysical products. Several PIs proposed algorithms for each product. It was necessary to select one or two algorithms as candidates of standard ones. The selection was a hard work for NASDA, and also might make disappointed results for PIs who were not selected.

Table 1.
PIs proposing algorithms for geophysical products

	Vapor	Cloud	Precip.	Wind	SST	Sea ice	Snow	Soil
Ferraro	*	*	*					*
Wentz	*	*	*	*	*			
Vonder Haar	*	*						
Takeuchi	*	*	*					
Lu	*	*	*					
Petty	*	*	*	*	*			
Hayasaka	*	*	*					
Klapisz	*	*	*					
Takayama	*	*						
Liu		*	*					
Spencer			*					
Adler			*					
Wilheit			*					
Aonashi			*					
Mitnik	*	*		*	*			
Swift					*			
Shibata				*	*			
Arai					*			
Cavalieri						*		
Comiso						*		
Nishio						*		
Gudmandsen						*		
Chang							*	
Tsang							*	
Koike							*	*
Jackson								*
Paloscia								*
Njoku								*

NASDA has produced several match-up data sets from satellite (or airborne) data and truth data (see Table 2). Satellite data were obtained from the SSM/I aboard DMSP, the Scanning Multichannel Microwave Radiometer (SMMR) aboard Nimbus-7, the TRMM Microwave Imager (TMI) aboard Tropical Rainfall Measuring Mission (TRMM) [5]. Airborne data were obtained by NASDA's Airborne Microwave Radiometer (AMR). Truth data differed by product as shown in Table 2. The match-up data sets in Table 2 are available from NASDA.

Table 2.
Match up data sets used for algorithm intercomparison

Product	Satellite	Truth data
Water vapor	SSM/I	GTS sonde data
	SSM/I	Keifu maru
Cloud water	SSM/I	GMS images
Precipitation	SSM/I	Radar around Japan
Wind speed	SSM/I	Buoy of NBDC, TAO, and JMA
SST	AMR	Infrared data on airplane
	TMI	Reynolds SST [6]
Sea ice	SSM/I	AVHRR in Okhotsk and the Antarctic
Snow	SSM/I	GTS snow depth
Soil moisture	SSM/I	Russian soil moisture data
	SMMR	Russian soil moisture data

GTS; Global Telecommunication Systems
GMS; Geostationary Meteorological Satellite
NBDC; National Buoy Data Center
TAO; Tropical Atmospheric Ocean

PIs were given match-up data sets containing both satellite and truth data, and calculated parameters from satellite data using their algorithms. Results reported by PIs were compared with truth data, and algorithm performance was compared. In the first step, two or three candidates for standard algorithms were selected, and the performance of selected algorithms was checked for computational time, memory, and so on. In the second step, one or two candidates were selected. The selection schedule differed slightly among products, but the main milestones are listed below.

Making match up data sets	June 1996 - Dec. 1996
Data set distribution	Jan. 1997
PIs reports	June 1997
Intercomparison of PIs' reports	July 1997 - Oct. 1997
Three candidates selected	Jan. 1998
Testing candidates algorithms	Feb. 1998 - Sept. 1998
One or two candidates selected	Dec. 1998

These activities were led by five Japanese scientists and supported by personnel at NASDA's Earth Observation Research Center (EORC).

3. DAY-1 ALGORITHMS

The algorithm comparison showed little difference in the performance of algorithms. PIs knew the truth data and could tune their own algorithm by using that data. In spite of the small difference in performance, the day-1 algorithms were determined as shown in Table 3. The algorithms selected demonstrated slightly better performance than others. PI's supports are also considered, such as whether he or she would keep report and program deadlines.

Table 3.

Day-1 algorithms

Water vapor	Takeuchi
Cloud liquid water	Wentz
Precipitation	Petty, Liu
Wind speed	Shibata
SST	Wentz, Shibata
Sea ice	Comiso
Snow	Chang, Koike
Soil moisture	Njoku, Jackson, Paloscia, Koike

First, selected algorithms will be installed at EORC. They will be tuned using real AMSR data for the first 6 months after launch. For products with several algorithm candidates, one algorithm will be selected by using real AMSR and truth data. At EORC, personnel will then make experimental calculations. Second, these algorithms will be installed at EOC and will be used to generate geophysical products operationally.

Characteristics of algorithms for four of the eight products will be described. Takeuchi's algorithm for water vapor uses air temperature at a lower level. Since emission by water vapor is expressed as a multiple of absorption coefficient by air temperature, it depends on air temperature. Even if the total water vapor is the same, the emission depends on the air temperature. He adjusts its effect by using the air temperature obtained from a weather forecasting model. Wentz's algorithm[7] for cloud liquid water demonstrated best performance in the algorithm comparison. In comparison with GMS images, the variance of cloud liquid water was the minimum. For sea surface wind speed algorithm, Shibata adopted a graphical method[8], and achieved a slightly better result. Coefficients used in the graphical method change with changes in SST. Wentz's and Shibata's algorithms for SST led almost the same performance, although the two methods are quite different.

4. FUTURE WORKS and SUMMARY

This paper has described algorithms developed by NASDA. Candidates for standard algorithms were compared by using common match-up data sets. Although performance differences among algorithms were small in the algorithm comparison, one or two algorithms were selected as the day-1 algorithms for each product. These day-1 algorithms will be used for 18 months after launch, after which algorithms will be reselected.

NASA's AMSR-E science team (its name has been changed several times) began developing algorithms in the late 1980s, and has already own algorithms. Several of the NASA team's algorithms were selected as NASDA's day-1 algorithms (Wentz, Comiso, Chang, and Njoku). In algorithm comparison, algorithms proposed by NASA scientists demonstrated better performance.

At present, NASA scientists and NASDA Japanese scientists meet once a year to exchange information about project status including algorithm development. Algorithm development in the US and Japan will continue in parallel until 18 months after the launch. At that time, two algorithms may be joined if appropriate.

This is the first time that such a large antenna will rotate so rapidly (40 rpm) in space. NASDA engineers and its contractor (Mitsubishi Electric Corporation) have therefore been endeavoring to ensure the successful rotation of the AMSR. If the antenna rotates normally, AMSR can be expected to produce new data such as at 6 and 10 GHz. AMSR's design life is 5 years for mechanical parts.

NASDA issued the second RA in March 1999 to solicit research algorithms, science applications, and so on. NASDA will provide all AMSR data to PIs selected in the second RA. Furthermore, NASDA plans to distribute Level 3 AMSR data through the Internet.

REFERENCES

1. J. Hollinger, et al., "DMSP Special Sensor Microwave/ Imager Calibration/Validation," *Naval Research Laboratory*, Vol.1 (1989).
2. J. Hollinger, et al., "DMSP Special Sensor Microwave/ Imager Calibration/Validation," *Naval Research Laboratory*, Vol.2 (1991).
3. E. C. Barrett, "The first WetNet Precipitation Intercomparison Project (PIP-1)," *Remote Sensing Reviews*, 11 (1994).
4. E. A. Smith, et al., "Results of WetNet PIP-2 Project," *J. Atoms. Scien.*, **55**, No.9, 1483-1536 (1998).
5. J. Simpson, et al., "On the Tropical Rainfall Measuring Mission (TRMM)," *Meteorol. Atmos. Phys.*, **60**, 19-36 (1996).
6. R. W. Reynolds and T. M. Smith, "Improved global sea surface temperature analyses," *J. Climate*, 7, 929-948 (1994).
7. F. J. Wentz, "Algorithm theoretical basis document," *RSS Report* 120296, *Remote Sensing System* (1996).
8. A. Shibata, "Passive microwave remote sensing on ocean surface," *Meteorological Research Note*, 187, 53-63 (1996).

Microw. Radiomet. Remote Sens. Earth's Surf. Atmosphere, pp. 541–549
P. Pampaloni and S. Paloscia (Eds)
© VSP 2000

Resampling of AMSR observations for spatial resolution matching

PETER ASHCROFT AND FRANK J. WENTZ

Remote Sensing Systems, 438 First St. Suite 200, Santa Rosa, California 95401, USA

Abstract - Multiple frequency space or aircraft-based radiometers using a common antenna produce footprints of differing sizes due to diffraction. This effect hampers physical retrieval algorithms that use observations of more than one frequency. In order to alleviate this problem for the AMSR instruments (PM-1 and ADEOS-II platforms), actual observations will be linearly combined to produce a data set of *effective* observations constructed at a series of common spatial resolutions. This process will effectively decouple spatial resolution effects from other frequency-dependent radiative transfer effects. In preparation for operational application when AMSR data becomes available, the weighting coefficients will be computed prior to launch using the Backus-Gilbert method of optimal interpolation and measured instrument antenna patterns. As a precursor to use with the AMSR instrument, the technique is currently being applied to TMI observations. The tradeoff of antenna pattern matching versus noise amplification is discussed, and the statistical difference between the resampled data and the original data is described. We also briefly consider the impact of uncertainty in the antenna pattern shape or footprint location errors on the choice of weighting coefficients.

1. INTRODUCTION

Due to diffraction or differences in antenna pointing directions, microwave radiometer antenna gain patterns of differing frequency may differ in size or location on the Earth surface while physical inversion algorithms may assume that observations describe consistent locations. In such circumstances, it is desirable to spatially adjust observations so that they describe the same locations.   This process is termed "resampling", "antenna pattern matching," or "antenna pattern correction".

Bringing observations of differing frequencies to a common spatial resolution can partially alleviate the effect of heterogeneity within the antenna footprint. Although some heterogeneity may persist across the common footprint, the effect will be less than if the footprints were not coincident. The importance of spatially consistent observations is well recognized. (For example, see North and Polyak [1] for a discussion in the context of rain rate retrieval.)

AMSR instruments

The first AMSR instrument will be launched aboard the Japanese ADEOS-II platform, currently scheduled for the year 2000. A second instrument (AMSR-E), is also currently scheduled for launch in the year 2000 on the NASA PM-1 platform. AMSR-E observations will be collected in scans of 196 observations over a width of 1445 km. The spacing between observations is approximately 10 km for all but the highest frequency, and approximately 5 km for the highest frequency. Due to the conical scan geometry, the density of observations

P. Ashcroft and F. J. Wentz

increases at the edges of the scan. Table 1 summarizes the range of instantaneous footprint sizes for AMSR-E.

Table 1: 3 dB AMSR-E footprint dimensions

Frequency	IFOV (km x km)		
6.9 GHz	75	X	43
10.7 GHz	48	X	27
18.7 GHz	27	X	16
23.8 GHz	31	X	18
36.5 GHz	14	X	8
89.0 GHz	6	X	4

2. BACKUS-GILBERT METHOD

The resampling technique consists of linearly combining actual observations. The weighting coefficients are calculated using the Backus-Gilbert method. This technique, originally developed in the context of inversion of seismic signals propagated through the Earth, approximates a given function as a linear superposition of other functions. One of the most significant obstacles to implementation of the Backus-Gilbert method is that calculation of weighting coefficients requires calculation and inversion of large matrices, and is consequently time consuming. This problem is avoided in computing the AMSR data set because the antenna patterns and the relative geometry of the observations are known *a priori*. The weighting coefficients can therefore be calculated before observations are actually collected, leaving only the relatively simple task of applying the weighting coefficients when the data is collected.

The Backus-Gilbert method was first suggested as a means of modifying the spatial characteristics of radiometer observations by Stogryn [2]. Poe subsequently used the method to demonstrate that observations could effectively be translated from one location to another [3]. Poe did not attempt to modify the effective resolution of the antenna pattern, nor did he trade off noise amplification against size of the constructed pattern. Since then, other investigators have also employed the technique, (see, for example, [4] and [5]).

Description of method

Following preliminary processing, measured brightness temperatures can be expressed as

$$\overline{T}_{Bi} = \int T_B(\rho) G_i(\rho) dA,$$

where $T_B(\rho)$ is the brightness temperature of the Earth at location ρ, and $G_i(\rho)$ is the antenna gain pattern corresponding to observation *i*. (Antenna gain patterns used to generate weighting coefficients are calculated to a radius of approximately 2.5 times the IFOV 3 dB beam width.) The constructed brightness temperature $\hat{T}_B$ is defined as a weighted sum of actual observations

$$\hat{T}_B = \sum_{i=1}^{N} a_i \overline{T}_{Bi}$$

$$= \int T_B(\rho) \sum_{i=1}^{N} a_i G_i(\rho) dA$$

The challenge is to choose weighting coefficients such that the constructed brightness temperature and the corresponding effective antenna pattern have desirable characteristics. The ideal effective antenna pattern should closely match some specified target pattern, and minimally amplify noise. In general, these two objectives are incompatible, requiring some compromise. The Backus-Gilbert method accomplishes this compromise by providing a solution vector of weights that minimize a sum of the integrated squared error in the fit, and variance due to noise amplification. (A "noise factor" is defined as the square root of the sum of the squares of the weighing coefficients. Assuming that the errors of the original measurements are independent, this means that a noise factor of 0.2 will produce a resampled observation with measurement error only 20% of the original measurements.) A smoothing factor specifies the relative importance of fit versus noise factor. See [2] for further discussion.

Figure 1 shows the tradeoff of fit and noise amplification for one combination of AMSR channels at a range of observation positions in the scan. For this figure, fit is defined as the integrated magnitude of the mismatch between the actual antenna pattern and the reconstruction. Because scan geometry changes across the scan, the precise relationship between smoothing, noise, and fit will also evolve. Note that the fit error is large at the extreme edge of the scan regardless of choice of smoothing factor, but that the fit rapidly improves away from the edges of the scan.

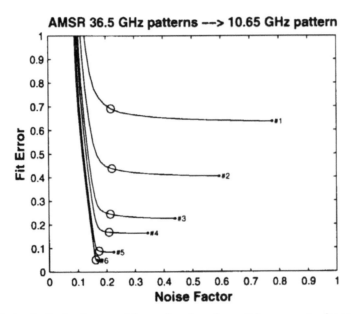

Figure 1: Tradeoff of noise factor and fit as a function of smoothing parameter for one combination of channels from the edge of the scan (position #1) and moving towards the center. The solid end point of each line represents the effect of no smoothing (optimal fit), and the open circle indicates the compromise of noise factor and fit used for the resampled data set. Nearly overlapping curves corresponding to positions 7 through 98 are omitted for visual clarity, but are very similar to that of position #6.

The relationship of Figure 1 was used to choose a smoothing factor at each scan position for each combination of channels. In general, we were guided by the principle that it was undesirable to reduce fit error beyond the uncertainty in the actual antenna patterns, especially at the cost of increased noise.

Another example is shown in Figure 2, this time with 6.9 GHz observations resampled onto the same resolution. For this particular case, the 10 km observation intervals oversample the relatively large 6.9 GHz antenna patterns. As a result, it is possible to attain significant reductions in noise by combining nearly overlapping observations with very little compromise in fit.

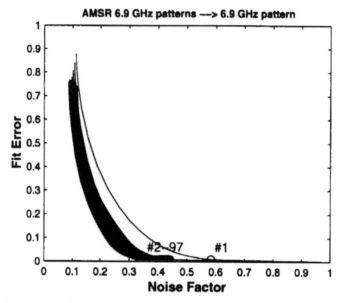

Figure 2: Tradeoff of noise factor and fit for another combination of channels.

Orbital simulations have shown that the distance between adjacent scans measured along the satellite track is nearly constant for a given position in the scan despite the eccentricity of the satellite orbit and the oblateness of the Earth. Thus, analysis confirms that a single set of weighting coefficients that are appropriate at one point of the orbit will suffice throughout all orbits.

After the computationally intensive task of calculating the weighting coefficients is performed, negligible weights are then dropped and the remainder are renormalized. This step significantly reduces the computational burden when the weights are applied. For a typical case, (resampling of the AMSR 36.5 GHz observations onto the 6.9 GHz resolution), the number of neighboring observations actually used to construct a single effective observation is on the order of 100.

3.   RESULTS

Description of AMSR resampled data set

In order to avoid noise amplification, the resampled AMSR-E data set (Level2A) will not include any enhanced resolution observations. The decision not to include any resolution enhancement in the standard Level2A data set is driven by the belief that users will be reluctant to accept any increase in noise. Instead, each set of actual observations will be resampled to one of several lower resolutions corresponding to that of the 6.9 GHz, 10.7 GHz, 18.7 GHz, 36.5 GHz, and 89.0 GHz channels. The Level2A data set will include all of the original observations, as well as observations resampled to match one of the five standard resolutions. While we anticipate that the resampled observations will generally be used in place of the original observations, the original observations will be included in the Level2A data set so that users will only need the one data set.

Prior to resampling, quality checks will be applied to calibration measurements, (hot counts, cold counts, and hot load thermistor measurements), and to the Earth observations themselves. In each case, a range of acceptable values will be defined, and measurements falling outside the range, (whether single Earth observations or an entire scan of calibration data), will not be included. Resampling will occur if the sum of the weighting coefficients corresponding to the missing data is below a predetermined threshold.

Resampling of TMI observations

Although the AMSR instruments will not be launched until late in the year 2000, the resampling technique has already been applied to TMI (TRMM Microwave Imager) data. Figure 3 through Figure 5 illustrate the effect of resampling for actual TMI data at the mouth of the Amazon River over the course of three days in December of 1997. Figure 3 and Figure 4 illustrate two different observation frequencies, highlighting the effect of differing spatial resolutions of the channels as well as the differing sensitivities of the two channels to meteorology. As expected, the 37 GHz observations better capture the high spatial frequency components of coastlines, and rivers, but they also capture the atmospheric moisture to which the 10.7 GHz channel is relatively insensitive.

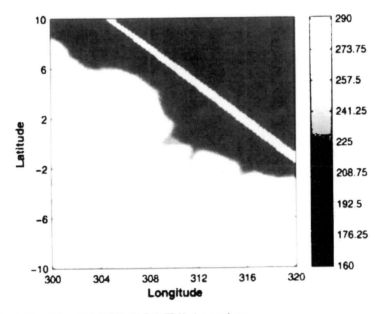

Figure 3: Overlaid swaths of 10.7 GHz V Pol. TMI observations.

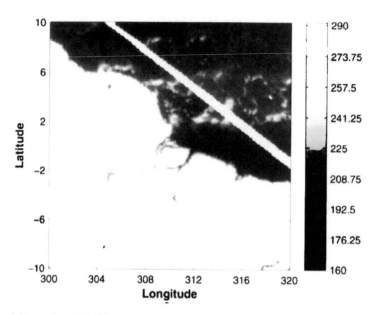

Figure 4: Overlaid swaths of 37 GHz H Pol. TMI observations.

Figure 5 is version of Figure 4 that has been resampled to the 10.7 GHz resolution. Aside from such relatively minor effects as measurement noise on the original observations and errors in the assumed antenna patterns, all the remaining differences between Figure 5 and Figure 3 are due to the differing radiative transfer characteristics of the two channels rather than some artifact of spatial resolution. The spatial frequency differences have been effectively removed.

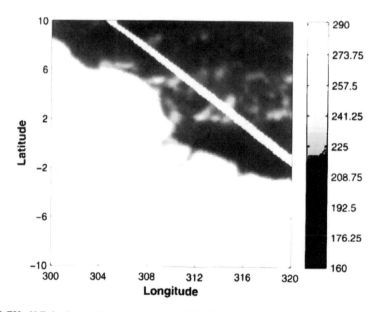

Figure 5: 37 GHz H Pol. observations resampled to 10.7 GHz resolution.

The effect of resampling is highlighted in Figure 6. As illustrated by the figure, areas of spatial variation on the scale of the footprints, (areas highlighted by the spatial filtering process of resampling) include coastlines and rivers as well as the meteorological features over the ocean.

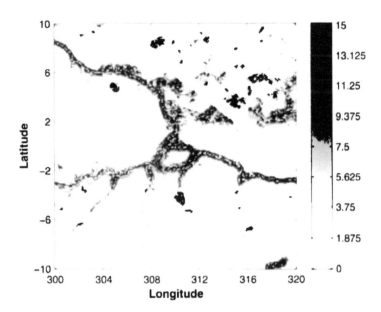

Figure 6: Magnitude of brightness temperature difference between actual and resampled 37 GHz observations (Figure 4 and Figure 5).

In order to better quantify the statistical effect of resampling, actual TMI observations at 37 GHz were compared to versions that had been resampled to a resolution matching that of the 10.7 GHz antenna (as in Figure 5). These two channels have footprint sizes on the order of 13 km and 52 km respectively. Thus, comparison of these actual and resampled data sets highlights geophysical features of a spatial scale in that range. After excluding observations over land or near coastlines so that the remainder represented only ocean observations, a histogram of the differences was plotted as shown in Figure 7. The histogram suggests that the statistical distribution of these differences is not Normal, but instead has 'heavy tails'. This is confirmed more clearly in Figure 8. Although the standard deviation was less than four degrees, the original and resampled sets differed by as much as 20 degrees 1% of the time, and sometimes by more than 40 degrees. Examination of several sets of orbits showed that these results are fairly typical.

548 P. Ashcroft and F. J. Wentz

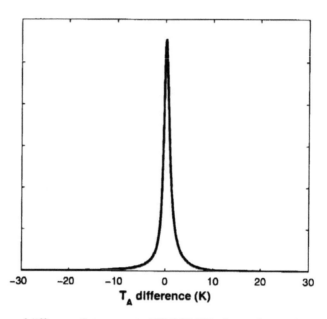

Figure 7: Histogram of differences between actual TMI 37 GHz observations and versions resampled to 10.7 GHz resolution.

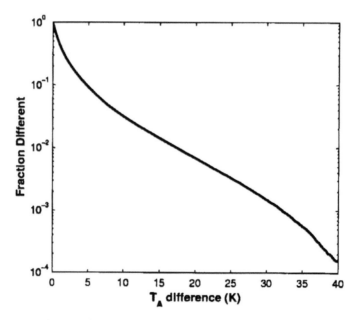

Figure 8: Fraction of observations in which the magnitude of the difference due to resampling is greater than a specified threshold.

It is important to remember that while the fraction of ocean observations for which spatial resampling makes a significant difference is small, these observations represent precisely those events that may be the most meteorologically important. For example, they correspond to meteorological fronts and storm systems. The effect of resampling (whether expressed as a standard deviation, or in terms of occasional large differences) serves as a reminder that a single forward radiative transfer model will not serve for all spatial scales.

4. CONCLUSION

A resampled set of AMSR observations will be produced and distributed as a standard data product. This resampled (Level2A) data set will completely contain the original Level1B data set. For most observations, the difference between actual and resampled observations will be small, but in a small fraction of cases the difference will be significant. As this small fraction of significantly different observations might correspond precisely to compact and intense meteorological events of great interest, the resampled data set should prove useful. The next step of this research will be to quantify the effect of resampling on estimates of such physical phenomena as rain rates, water vapor, sea surface temperature, and wind speed.

In addition to operational application of this technique for the AMSR instrument, this resampling method might also be applied operationally to other instruments such as TMI.

REFERENCES

1. G. R. North and I. Polyak, "Spatial Correlation of Beam-Filling Error in Microwave Rain-Rate Retrievals," *Journal of Atmospheric and Oceanic Technology*, vol. 13, pp. 1101-1106, 1996.
2. A. Stogryn, "Estimates of Brightness Temperatures from Scanning Radiometer Data," *IEEE Transactions on Antennas and Propagation*, vol. AP-26, pp. 720-726, 1978.
3. G. A. Poe, "Optimum Interpolation of Imaging Microwave Radiometer Data," *IEEE Transactions on Geoscience and Remote Sensing*, vol. 28, pp. 800-810, 1990.
4. W. D. Robinson, C. Kummerow, and W. S. Olson, "A Technique for Enhancing and Matching the Resolution of Microwave Measurements from the SSM/I Instrument," *IEEE Transactions on Geoscience and Remote Sensing*, vol. 30, pp. 419--429, 1992.
5. M. R. Farrar and E. A. Smith, "Spatial Resolution Enhancement of Terrestrial Features Using Deconvolved SSM/I Microwave Brightness Temperatures," *IEEE Transactions on Geoscience and Remote Sensing*, vol. 30, pp. 349--355, 1992.

Author Index

Abraham, S.	89	Ellison, B.	417
Accadia, C.	371	English, S.	263
Amacher, W.	471	Eriksson, P.	503
Anterrieu, E.	477	Evtushenko, A.	313
Ashcroft, P.	541	Eymard, L.	47, 235
Askne, J.	503	Eyring, V.	409
Avelin, J.	81	Fabbo, R.	379
Barà, J.	325, 459	Fedor, L. S.	13
Barbaliscia, F.	203	Ferraro, R.	255, 339
Barcia, A.	417	Fionda, E.	203, 397
Baron, P.	417	Fischer, J.	71
Barry, B.	417, 427	Gallego, J. D.	417
Basili, P.	387	Goede, A.	409
Beaudin, G.	417	Goutoule, J. M.	467, 477
Bechini, R.	379	Gradinarsky, L.	183
Bennartz, R.	71	Grody, N.	255, 339
Bian, J.	365	Guissard, A.	213
Bizzarri, B.	493	Gurney, R. J.	97
Bonafoni, S.	387	Güstafsson, M.	417
Bosisio, A. V.	271	Hallikainen, M.	443
Bourras, D.	47	Han, Y.	129, 145
Bremer, H.	409	Hanocq, J.-F.	485
Burke, E. J.	97, 107	Hawkins, J.	353
Calvet, J.-C.	485	Hayasaka, T.	535
Camps, A.	325, 459	Hetzheim, H.	409
Capacci, D.	247	Heygster, G.	39
Capdevila, J.	325	Hornbostel, A.	313
Capsoni, C.	271	Irisov, V. G.	13
Cernicharo, J.	417	Jackson, T.	525
Chang, A.	515	Jiménez, C.	503
Chang, P. S.	21	Johnsen, K.-P.	39
Chanzy, A.	485	Kadygrov, E. N.	193
Chen, H.	365, 453	Kämpfer, N.	213, 417
Ciotti, P.	387, 397	Karpov, A.	417
Connor, L. N.	21	Kaufmann, P.	213
Corbella, I.	325, 459	Kerr, Y.	467, 477, 485
Costa, J. E. R.	213	Klapisz, C.	283, 291
D'Acunzo, E.	371, 379	Klein, U.	417, 427
D'Auria, G.	387	Kleipool, Q.	409
De Castro, G. C. G.	213	Koike, T.	515, 525
De La Noë, J.	417	Koistinen, O.	443
Dibben, P.	263	Koldaev, A.	193
Dietrich, S.	371, 379	Küllmann, H.	409
Di Michele, S.	371, 379	Künzi, K.	409, 417, 427
Duffo, N.	325, 459	Kutuza, B.	313
Elgered, G.	183	Kuzmin, A.	3

Lahtinen, J.	443	Renshaw, R.	263
Langer, J.	427	Ricaud, P.	417
Lannes, A.	467	Roberti, L.	371, 379
Laursen, B.	29	Rohaly, G. D.	353
Lazard, B.	467	Rovira, M.	213
Lei, H.	165	Rubinstein, I. G.	299
Leuski, V. E.	13	Ruf, C. S.	137
Levato, H.	213	Ruisi, R.	59, 119
Le Vine, D. M.	89	Schroth, A.	313
Levizzani, V.	353	Sharma, R.	81
Li, Y.	165	Shen, Z.	165
Liang, G.	165	Shibata, A.	535
Liljegren, J. C.	155, 433	Siddans, R.	417
Lindner, K.	427	Sihvola, A.	81
Louhi, J.	417	Simmonds, L. P.	97, 107
Lu, D.	365	Sinnhuber, B.-M.	427
Lüdi, A.	213	Skou, N.	29
Macelloni, G.	59, 119	Smith, E. A.	263, 353
Magun, A.	213	Sobachkin, A.	313
Maier, D.	417	Søbjærg, S.	29
Mallat, J.	417	Solheim, F.	129
Mallet, C.	283, 291	Spera, P.	493
Marloie, O.	485	Susini, C.	59
Martellucci, A.	397	Tassa, A.	371, 379
Martin, L.	213	Thomas, C.	47
Martinuzzi, J.-M.	467	Torres, F.	325, 459
Marzano, F. S.	353, 371, 387, 397	Trokhimovski, Y.	3, 13
Masullo, P. G.	203	Turk, F. J.	353
Matheson D.	417	Vall-llossera, M.	325, 459
Mätzler, C.	213	Viguerie, C.	417
Mauri, M.	271	Viltard, N.	291
Mironov, A.	193	Von König, M.	409
Moreau, E.	283	Waldteufel, P.	467, 477, 485
Mugnai, A.	353, 371, 379	Wang, P.	173
Natali, S.	379	Wei, C.	165, 173
Njoku, E.	525	Weng, F.	255, 339
Paape, K.	71	Wentz, F. J.	541
Paloscia, S.	59, 119, 525	Westwater, E. R.	13, 129, 145
Pampaloni, P.	59, 119	Wigneron, J.-P.	59, 467, 485
Pardo, J.-R.	417	Wohltmann, I.	427
Pearson, D.	97	Wu, Y.	173
Pierdicca, N.	387	Wüthrich, M.	417
Porcú, F.	247, 379	Xang, Y.-Y.	173
Pospelov, M.	3	Xue, Y.	183
Poulsen, C.	263	Yang, P.	365
Prodi, F.	247, 379	Zagorin, G.	313
Räisänen, A.	417		
Rayer, P.	263		

T - #0463 - 071024 - C568 - 246/174/25 - PB - 9780367447441 - Gloss Lamination